Water Science and Technology Library

Volume 96

The aim of the *Water Science and Technology Library* is to provide a forum for dissemination of the state-of-the-art of topics of current interest in the area of water science and technology. This is accomplished through publication of reference books and monographs, authored or edited. Occasionally also proceedings volumes are accepted for publication in the series. *Water Science and Technology Library* encompasses a wide range of topics dealing with science as well as socio-economic aspects of water, environment, and ecology. Both the water quantity and quality issues are relevant and are embraced by *Water Science and Technology Library*. The emphasis may be on either the scientific content, or techniques of solution, or both. There is increasing emphasis these days on processes and *Water Science and Technology Library* is committed to promoting this emphasis by publishing books emphasizing scientific discussions of physical, chemical, and/or biological aspects of water resources. Likewise, current or emerging solution techniques receive high priority. Interdisciplinary coverage is encouraged. Case studies contributing to our knowledge of water science and technology are also embraced by the series. Innovative ideas and novel techniques are of particular interest.

Comments or suggestions for future volumes are welcomed.

Vijay P. Singh, Department of Biological and Agricultural Engineering & Zachry Department of Civil and Environment Engineering, Texas A&M University, USA
Email: vsingh@tamu.edu

More information about this series at http://www.springer.com/series/6689

Ashish Pandey · S. K. Mishra · M. L. Kansal ·
R. D. Singh · V. P. Singh
Editors

Water Management and Water Governance

Hydrological Modeling

Springer

Editors
Ashish Pandey
Indian Institute of Technology Roorkee
Roorkee, India

M. L. Kansal
Department of Water Resources
Development and Management
Indian Institute of Technology Roorkee
Roorkee, India

V. P. Singh
Texas A&M University
College Station, TX, USA

S. K. Mishra
Department of Water Resources
Development and Management
Indian Institute of Technology Roorkee
Roorkee, India

R. D. Singh
Indian Institute of Technology Roorkee
Roorkee, India

ISSN 0921-092X ISSN 1872-4663 (electronic)
Water Science and Technology Library
ISBN 978-3-030-58050-6 ISBN 978-3-030-58051-3 (eBook)
https://doi.org/10.1007/978-3-030-58051-3

This Springer imprint is published by the registered company Springer Nature Switzerland AG
The registered company address is: Gewerbestrasse 11, 6330 Cham, Switzerland

Contents

About the Editors

Prof. Ashish Pandey did his B.Tech. in Agricultural Engineering from JNKVV, Jabalpur, and obtained M.Tech. in Soil and Water Engineering from JNKVV, Jabalpur. He received Ph.D. in Agricultural Engineering from IIT Kharagpur. He joined IIT Roorkee in 2007 and is currently working as Professor at the Department of WRD&M, IIT Roorkee. His research interests include irrigation water management, soil and water conservation engineering, hydrological modeling of watershed, remote sensing and GIS applications in water resources. He guided ten Ph.D. and 78 M.Tech. students at IIT Roorkee. He published 155 research papers in peer-reviewed and high impact international/national journals/seminars/conferences/ symposia. He has also co-authored a textbook on *Introductory Soil and Water Conservation Engineering*.

He also served as Guest Editor for two issues of *Journal of Hydrologic Engineering* (ASCE). He is Editor, *Indian Journal of Soil Conservation*. His laurels include Eminent Engineers Award-2015 given by the Institute of Engineers (India), Uttarakhand State Centre, DAAD scholarship and ASPEE scholarship. He was offered prestigious, JSPS Postdoctoral Fellowship for Foreign Researchers, Japan, and BOYSCAST fellowship of DST, GOI. He is a fellow member of (1). Institution of Engineers (India); (2). Indian Association of Hydrologists (IAH) and (3). Indian Water Resources Society (IWRS).

Dr. S. K. Mishra is a 1984 Civil Engineering graduate of the then Moti Lal Nehru Regional Engineering College (presently MNNIT), Allahabad. He obtained his M.Tech. degree in Hydraulics and Water Resources from IIT Kanpur in 1986 and doctoral degree from the then University of Roorkee (presently IIT Roorkee) in 1999. He served the National Institute of Hydrology Roorkee at various scientific positions during 1987–2004. During the period, he also visited Louisiana State University, USA, as a postdoctoral fellow during 2000–2001 and Department of Civil Engineering, IIT Bombay, as Visiting Faculty during 2002–2003.

He joined IIT Roorkee in 2004 and is presently working as Professor. Besides having been Head of the Department of Water Resources Development and Management, he is also presently holding Bharat Singh Chair of the Ministry of Jal Shakti, Government of India. He is specialized in the fields of hydraulics and water resources, environmental engineering, design of irrigation and drainage works, dam break analysis, surface water hydrology. He has published more 250 technical articles in various international/national journals/seminars/conferences/ symposia. His reference book on *Soil Conservation Service Curve Number Methodology* published by Kluver Academic Publishers, the Netherlands, has received a significant attention of the hydrologists/agriculturists/soil water conservationists around the globe. Of late, he has been associated with the *Journal of Hydrologic Engineering* as Guest Editor of two special issues and Executive Vice President and Editor of Indian Water Resources Society. He has visited several countries during the period and is on the role of several national/international professional bodies. Among several others, the Eminent Engineers Award and Dr. Rajendra Prasad Award are worth citing.

Dr. M. L. Kansal currently working as NEEPCO Chair Professor and Head in the department of Water Resources Development and Management at Indian Institute of Technology (IIT) Roorkee (India). He is a Civil Engineering graduate with post-graduation in Water Resources Engineering. He obtained his Ph.D. from Delhi University (India) and holds the post-graduate diploma in Operations Management. Previously, he worked as Associate Professor in the department WRD&M, IIT Roorkee, and served at Delhi Technical University, Delhi (erstwhile, Delhi College of Engineering, Delhi), NIT Kurukshetra, IIT Delhi and NIH, Rookree (India) at various levels. He has published more than 150 research papers and two books. He has got best paper awards from Indian Building Congress, Indian Water Works Association, and received star performer award from IIT Roorkee. He acted as a reviewer for several international journals and research agencies. He has visited various countries as International Expert and Visiting Professor. He acted as International Expert for RCUWM of UNESCO and IUCN, etc. He is working as Executive Vice President of Indian Water Resources Society and served as expert panel member for All India Council for Technical Education (AICTE), India. He has contributed substantially by providing consultancy services to various national and international agencies of repute. He has worked in various administrative capacities such as Associate Dean of Students Welfare and as Chairman, Co-ordinating Committee of Bhawans at Indian Institute of Technology Roorkee and National Expert for several bodies.

R. D. Singh did B.E. Civil Engineering and M.E. Civil Engineering with Specialization in Hydraulics and Irrigation Engineering from University of Roorkee (Now IIT Roorkee). He did M.Sc. in Hydrology from University College Galway, Ireland. Presently, he is working as Visiting Professor, Department of Water Resources Development and Management, IIT Roorkee. He worked as Director, National Institute of Hydrology (NIH), Roorkee, for more than 9 years. Before taking over as Director NIH, he was holding the charge of Nodal Officer Hydrology Project-II, a World Bank Funded Project for Peninsular region of India completed by NIH during the year 2016. During his service at NIH, he had worked on more than eighty sponsored/consultancy projects for solving the real-life problems in water sector. He had also worked as well as guided eleven international collaborative projects at NIH.

He has research and development experience of more than 40 years in different areas of hydrology and water resources. He has an extensive experience in flood estimation, flood management, drought management, hydrological modeling, environmental impact assessment and climate change and its impact on water resources, etc. He has published more than 316 research papers in the reputed international and national journals, international and national seminar/symposia, workshops, etc. He has received C.B.I.P. Medal, Institution of Engineers Certificate of merit, Union Ministry of Irrigation Award and best Scientist Award form from NIH, Roorkee. He guided three Ph.D. and fourteen M.E. and M.Tech. and one M.Phil. dissertations. He has widely traveled abroad for different assignments.

Prof. V. P. Singh is University Distinguished Professor, Regents Professor and Caroline and William N. Lehrer Distinguished Chair in Water Engineering at Texas A&M University. He received his B.S., M.S., Ph.D. and D.Sc. degrees in engineering. He is a registered professional engineer, a registered professional hydrologist and an honorary diplomate of ASCE-AAWRE. He is a distinguished member of ASCE, a distinguished fellow of AGGS and an honorary member of AWRA. He has published extensively in the areas of hydrology, irrigation engineering, hydraulics, groundwater, water quality and water resources (more than 1290 journal articles; 30 textbooks; 70 edited reference books; 105 book chapters; and 315 conference papers). He has received more than 95 national and international awards, including three honorary doctorates. He is a member of 11 international science/engineering academies. He has served as President of the American Institute of Hydrology (AIH), Chair of Watershed Council of American Society of Civil Engineers, and is currently President of American Academy of Water Resources Engineers. He has served/serves as Editor-in-Chief of three journals and two book series and serves on editorial boards of more than 25 journals and three book series.

Chapter 1
Integrated Watershed Management and GIS: A Case Study

Sabita Madhvi Singh, Pabitra Ranjan Maiti, and Neetu Singh

Abstract To optimize the use of available land area and to meet the multiple demands of food and forest cover, planning and process use a series of cooperative, iterative steps to characterize existing conditions, identify and prioritize problems of land and water resources, define management objectives, and develop and implement protection or remediation strategies as necessary through watershed development-based programs. A watershed plan is a strategy and a work plan for achieving water resource goals that provide assessment and management information for a geographically defined watershed. Remote sensing and GIS, which plays an important role in detecting and monitoring the physical characteristics of watershed area by reflected and emitted radiation, is also efficiently used for watershed management. In the present study an attempt has been made to study the extensive use of GIS in watershed management. For this purpose, Khajuri Watershed is selected which is situated at the border of Marihan block. The study area is almost fully covered by metamorphic rocks, which include different limestone, gneiss and sandstone. The area's topography is generally mountainous and rugged dissected by many distributaries of Khajuri river. The catchment covers an area of about 126 km^2 (30,000 acres or 12,600 ha). Three commercially available GIS software tools are used in this project, which are ARCMAP v 8.1, ArcView and ERDAS IMAGINE v 8.5. ARCMAP is used as a tool to create and manage geographic information, and ArcView is used as a tool to visualize, explore, query, interpolate, update and analyze geographic information. As

S. M. Singh (✉)
Ministry of Jal Shakti, GOI, New Delhi, India
e-mail: sabitamadhvi@gmail.com

P. R. Maiti
Department of Civil Engineering, Indian Institute of Technology, Banaras Hindu University, Varanasi, India
e-mail: prmaiti.civ@itbhu.ac.in

N. Singh
Centre for Technology Alternatives for Rural Areas, IITB-Monash Research Academy, Mumbai, India
e-mail: neetusingh2701@gmail.com

A. Pandey et al. (eds.), *Water Management and Water Governance*, Water Science and Technology Library 96,
https://doi.org/10.1007/978-3-030-58051-3_1

the total area is very large, it is sub-divided into sub-watersheds and micro watersheds to determine the available water resources of the study area, current water demand and future projection and technological options for water resource management and, in turn, to improve water availability and management of crop yield of a watershed, largely for optimum utilization of water resources to derive maximum benefits.

Keywords Watershed · GIS · Micro-watershed · Remote sensing

1.1 Introduction

Remote Sensing is defined as the technology by which the characteristics of the objects of interest can be identified, measured or analyzed without direct contact. Electromagnetic radiation which is emitted or reflected from an object is the usual source for remote sensing data (Lillesand et al. 2015). A device that detects the electromagnetic radiation emitted or reflected from an object is called a "remote sensor" or "sensor". Cameras, scanners are a few examples of remote sensors. A vehicle thatcarries the sensor is called a "platform". Aircraft or satellites are used as platforms (Campbell and Wynne 2011).

To generate a host of inter-related data and studied in relation to each other requires micro-watershed level planning. Remotely sensed data provide useful, helpful and up-to-date spatial and temporal information on physical terrain parameters and natural resources (Chowdary et al. 2009). Geographical Information System (GIS) with its capability of integration and analysis of spatial location and organizes layers of information into visualizations using maps have proved to be an effective tool in planning for micro-watershed development (Makhamreh2011).

Geographical Information System (GIS), which is typically required for hydrological studies has evolved as a highly sophisticated data management and analyzing system to collect and store large data. Thus, remote sensing and GIS reveal deeper insights into data, such as patterns, relationships and situations helping users for efficient management of water resources make smarter decisions (Strager et al. 2010). The synoptic view provided by remote sensing and the analysis capability provided by GIS offers the most appropriate method for studying these resources. (Jain et al. 2000; Khan et al. 2001) have said that land resource development programs are applied generally on a watershed basis. Delineation of watersheds of a large drainage basin and their assessment is required for proper planning and management of natural resources for sustainable crop production (Willis et al. 1989). Rao et al. (2004) described that watershed management as the rational utilization of the land and water resources for optimum production with minimum loss of natural resources. The basic principle of watershed management is to use the land according to its capability and treat the land according to the needs for the sustainable development of the people living in that area (Thiruvengadachari et al. 1994). The land that was used beyond its capability generates adverse effects on the environment like soil degradation in the form of erosion, ground water depletion, etc. In the present study, optimization and watershed management are carried out for Khajuri Watershed and few results are presented to give an idea about the application of GIS in watershed management.

1.2 Study Area

The study area is located in the southeastern part of Mirzapur District of Uttar Pradesh, bounded by Varanasi and Sant Ravidasnagar Bhadohi, district in north and Chandauli district in the east, by Sonbhadra, Siddhi and Rewa (Madhya Pradesh) districts in south and southeast, respectively, and by Allahabad in the west. Mirzapur has five development blocks with 694 villages.

1.2.1 Location of the Khajuri Watershed

Khajuri Watershed is situated at the border of City Block and Marihan block as shown in Fig. 1.1. The catchment is located between 25°0′ 0″ and 25°5′ 0″ N latitude and 82°30′ 0″ to 82°40′ 0″ E longitude. It covers an area about 126 km^2 (30,000 acres or 12,600 ha). As the total area is very large for the study as a watershed, so here on the basis of subwatershed and micro-watershed study is done and the result is also shown accordingly. The study area also Barkachhakhurd and Barkachhakalan where the South Campus of Banaras Hindu University Barkachha KVK centre is located. In this subwatershed Khajuri, Tand, Bhukwa, Belahra, Hardi Kurd, Phullari and Bahmandeva villages are also studied. Thus, the total area of around 9000 ha for the present study is considered in which the main emphasis is given on Barkaccha KVK centre of around 1104 ha.

1.2.2 Physiography and Drainage System

It is located between the forest areas of Barkachha Reserve Forest on the East and the Danti Forest Reserved Forest on the west and has a seasonally dry climate dominated by dry deciduous forest and dry savannah grasslands. It is situated approximately 160 m height from Khajuri River flowing along the western end of subwatershed. The lowest elevation is about 90 m from the mean sea level (MSL). The study area has a complex alignment in distribution and orientation, structure, relief, slope, morphology, climate and vegetation. This is an impact of lineament on the drainage routes. The main river system in the study area is the Ganga river system with 2760 km^2 covering 55.74% of the district with the coverage of Khajuri tributary for a length of 50 km and of basin area 171 km^2. The drainage pattern is dendritic and some of the tributaries of Khajuri drain the eastern terrain of the whole area.

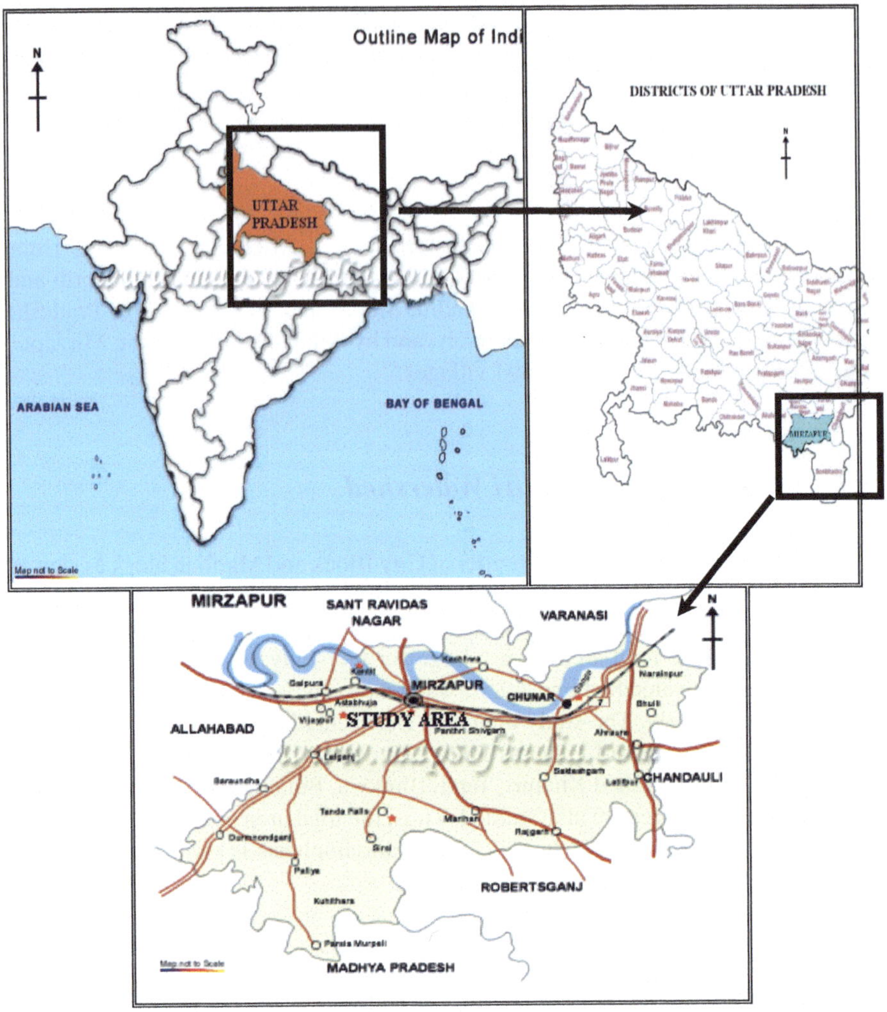

Fig. 1.1 Location of study area

1.2.3 Geology

The study area is almost fully covered by metamorphic rocks, which include different limestone, gneiss and sandstone. The area's topography is generally mountainous and rugged dissected by many distributaries of Khajuri river. It is dominated by the Vindhyan plateau system. The study area comes under the Kaimur series of variegated sandstone, limestone, inter-bedded shale with or without alluvium.

1.2.4 Geomorphology

Geomorphology affects availability of soil moisture, depth, the structure of soil and landholding patterns and yield output. The geomorphic unit of the area is the Vindhyan Upland. The landform features of the area include:

(a) Buried pediment: This area is very good for cultivation and the agricultural quality of the land is regarded as very good. The groundwater potential is also considered very good. These are flat surfaces of the plateau area with thin
(b) Cover of unconsolidated materials mainly gravel, soil (alluvium/colluvium) or weathered rocks. Hence they are considered as good for rice cultivation.
(c) Pediment: These are open rock surfaces and running water is the main agent of its formation. These are transitional zones between hills and plains. These are gently sloping at 1–7°. These are divided into two main classes:
    1. Pediment with vegetation: These have land use of open forest, groves, etc. The agricultural land quality is considered poor. Also, the ground water potential is low.
    2. Pediment with stony surface: These are lands considered as stony waste. The agricultural quality and ground water potential are also very poor.
(d) Dissected plateau: A plateau with various streams cutting across its terrain creates a dissected plateau. Features like gorges, valleys, and scarps lands create an undulated topography. These are forest areas and the agricultural land quality is considered to be moderate but the ground water potential is poor.
(e) Valley fill: Lateral erosion landform with and underlying bed rock with an erosional surface. This land is considered a good quality agricultural land with poor ground water quality.

1.3 Methodology

(i) Data Analysis and Thematic Layer Generation using GIS
(ii) Satellite Image classification of the area.

ArcView v 3.1a: This software has been developed by ESRI Inc. It is one of the leading software for GIS and mapping. ArcView gives the power to visualize, query, explore and analyze data geographically. In this project ArcView is use for display of raster map. Digitizing different features, visualizing the different features, linking attributes of the digitized data with the features and for querying the data geographically and for finding the attribute of any feature of the map.

ArcGIS v 8.1a: This software has been developed ESRI Inc. In this project all the thematic layers have been prepared using this software. Here Arc Hydro Tools are used for delineation of watershed. Arc Hydro is an ArcGIS-based system prepared to support water resources applications. It consists of two key components:

1. Arc Hydro Data Model
2. Arc Hydro Tools.

The tools are accessed through the Arc Hydro toolbar, where they are grouped by functional key into six menus.

Terrain Preprocessing: Digital Elevation Model (DEM) tools deal processing. Prepare spatial information that is mostly used for analysis.

Terrain Morphology: Perform initial analysis of a non-dendritic terrain these tools are used.

Watershed Processing: Dealing with watershed and subwatershed delineation and basin characteristic determination this tool in menu work. They operate on top of the spatial data prepared in the terrain preprocessing stage of delineation.

Attribute Tools: These tools provide many functionalities for generation of some of the key attributes (fields) in the Arc Hydro data model. Some of the tools require existence of a geometric network.

ERDAS IMAGINE v 8.5: This software has been developed by ESRI Inc. USA. This system incorporates the function of both image processing and GIS. This function includes viewing, importing, altering, and analyzing raster and vector data set. In the present study, ERDAS Imagine is used for making Digital Elevation Model of the area, analyzing the IRS data and delineating the sub-watersheds and for classification of satellite data for landuse and other features (ERDAS 2000).

1.3.1 Data Analysis in GIS

The methodology which has been developed is dependent on spatial topographic data, which are partitioned into number of layers for the area under consideration. The data for each tile is divided into many layers of information with common themes or structures. Three commercially available GIS software tools are used in this project: ARCMAP v 8.1, ArcView and ERDAS IMAGINE v 8.5. ARCMAP v 8.1 is used as a tool to create and manage geographic information, and ArcView is used as a tool to query, visualize, explore and analyze geographic information. With the aid of ARCMAP and Arc View, land use and soil type data for the study area are merged to obtain a different number of thematic layers. All the thematic maps are converted into digital form, using a scanner (Mas et al. 2015). The dataset is converted from vector to raster form. The data are imported into ARCMAP GIS v 8.1 on IBM RS6000 workstation and different thematic layers are edited to create an error-free digital database. Several thematic maps are needing to be derived from the basic soil survey data. The attributes chosen are going to be those that influence plant growth through soil–water-air-plant relationships (Gao 2008). These maps help land users to develop an action plan for various areas like conserving soils, ameliorating degraded soils and for using soils in line with their capability (Jain et al. 2000).

(a) **Digitization Steps**:

Scanning of SOI toposheet: The toposheets are vector maps, which are converted to raster form after scanning. This does not contain any information and are merely pixels of colour composites of red, green and blue. The scanned map area does not represent the location on the surface of the earth and is inadequate to perform analysis.

Geo-referencing: The map is then geo-referenced, a process of defining the raster data in map co-ordinates it allows overlaying and analyzing of geographic data. A co-ordinate system or map projection system to define the point, line and area features. Geographic projection systems of the world are selected with Asia, GCS Kalianpur 1975 as the co-ordinate map projection system to align raster to existing spatial data such as streams, contours, roads, etc. Ground control points (GCPs): four ground control points were identified and a ground control shapefile are created in ARCVIEW software. The ARCMAP software wraps the raster by rubber sheeting by taking the reference of ground control points. The transformation is done, followed by interpreting the root mean square error. Resampling is performed by nearest neighbour first to give a final geo-referenced map (Scott and Janikas 2010).

Digitization: The thematic maps are digitized using ARCMAP. Digitization can be said as the conversion of images into digital format (Hamar et al. 1996). The objects stored in a GIS may be of three types, referred to as the geometric base elements:

1. Points: representing objects without any aerial extent such as drilling sites, wells, rainfall stations, etc.
2. Lines: representing linear objects such as roads, railways, rivers, telephone lines, etc.
3. Polygons: representing areas such as soil classification, land cover classes, etc.

Creation of thematic layer: Each layer file is saved as a separate shapefile based on the theme and properties as summarized in Table 1.1.

Thematic layer generation: The following thematic layers are prepared in ARCMAP version 8.1 software developed by ESRI, U.S.A.

Table 1.1 Digitization of various land features in GIS

Feature class	Mode of digitization
Contour	Line
Drainage	Line
Water bodies	Polygon
Road	Line
Soil	Polygon
LULC	Polygon

(a) **Location of Khajuri watershed**

The main map of the study area is prepared from the Survey of India toposheet at a scale of 1:50,000 as shown in Fig. 1.2. Water bodies, drainage networks, roads, etc. are digitized.

(b) **Contour map**

To display the topography of the area an elevation map is digitized from the toposheet. The contour intervals of 10 m interval were mapped from the toposheet. A variation from 90 to 160 m is observed, giving information on the undulating relief features in the area.

(c) **Drainage and water bodies**

A layer file depicting the water resources within the area are created in ARCMAP as shown in Fig. 1.3. The area is drained by the tributaries of the Khajuri river flowing along eastern region. The waterbodies include wells, water harvesting tanks and ponds within the area.

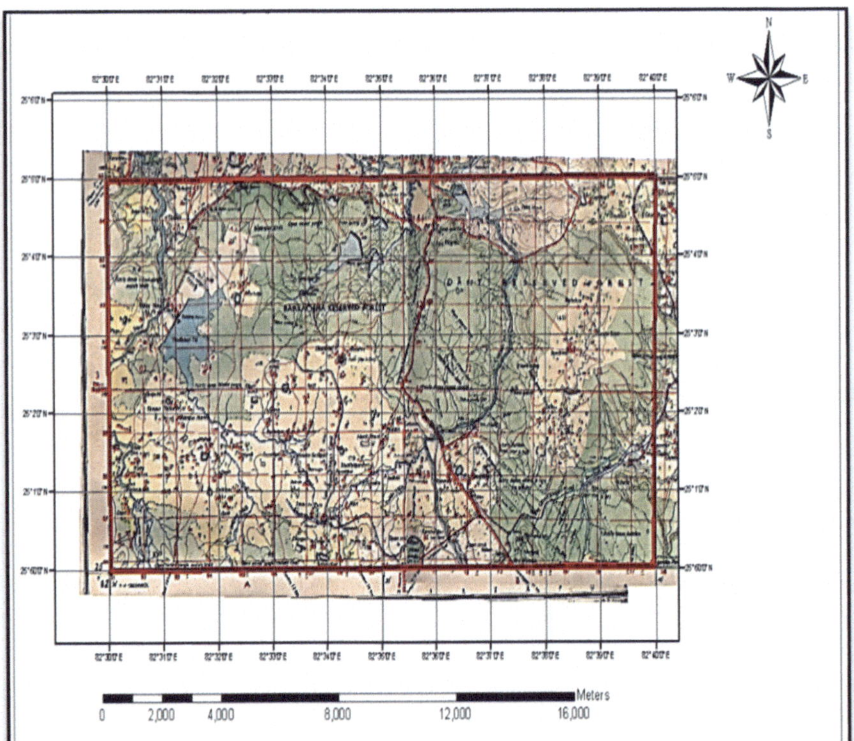

Fig. 1.2 SOI toposheet 63K/12

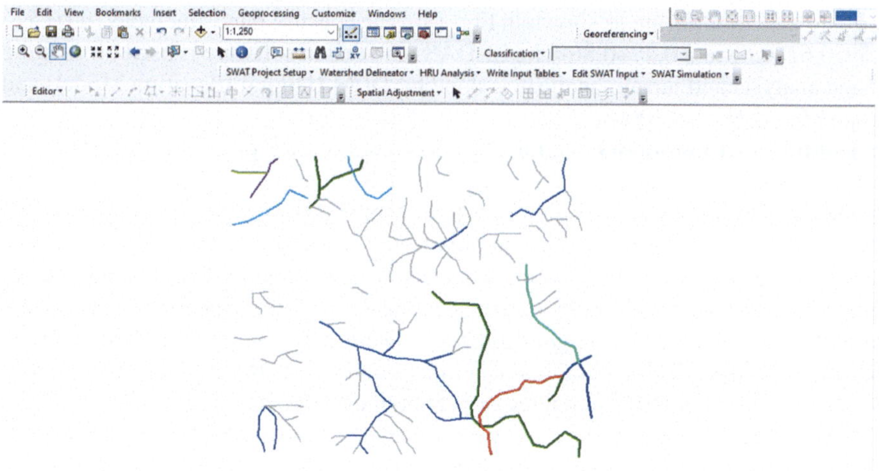

Fig. 1.3 Drainage pattern

(d) **Slope map**

The area has an undulating relief features with slopes varying from very gently sloping land of 1–3% to steeping 15–25% slopes. The slope map is developed from the digitized soil map provided by the Department of Agriculture, B.H.U.

(e) **Soil texture**

The layer file giving information about the various soil textures found in the area using the polygon features and snapping environment. There different types of soil classes in the map.

(f) **Soil depth**

The soil depth is an important attribute of a land area for ecological planning. Hence a soil depth map is created using the data provided in the soil map of the B.H.U south campus at Barkachha.

(g) **Soil erosion**

Soil erosion details are also mapped from the soil map of Barkachha campus area. It identifies the areas prone to slight, moderate and severe erosion.

1.3.2 Data Analysis Using Remote Sensing

IRS—P6 RESOURCESAT—LISS—IV image dated 27th September 2005 is used for the classification of the area. 3 band multi-spectral LISS IV camera with a spatial resolution better than 5.86 m and a swath of around 25 km across track steerability for selected area monitoring as shown in Fig. 1.4. The spectral bands includes (μ) 0.52–0.59, 0.62–0.68 and 0.77–0.86. The ERDAS 8.5 version software is used for satellite image analysis and terrain analysis is done by DEM (Digital Elevation Model) and Slope Map (ERDAS 2000).

Digital elevation model: Digital elevation models are typically used to represent terrain relief, slope, drainage pattern, etc. It is also referred to as digital terrain model (DTM). For the creation of Digital Elevation Model (DEM), the contour map of the area is prepared by digitizing the contours from the Survey of India toposheet. The data preparation is done by creating a surface in 3D surfacing features that creates three dimensional image files by rubber sheeting the feature shape file imported from ArcGIS. For this purpose, height differences need to be computed in both *X* and *Y* directions, as overall slope gradient is a function of height differences. A slope map can be created from DEM in degree or percentage (Arun 2013).

Slope map: Change in elevation over a distance expressed as slope. Distance is the size of the pixel in this case. Slope is most often expressed as a percentage, but can also be calculated in degrees.

Visual interpretation: Visual Interpretation makes use of following basic characteristics:

(i) Tone (colour)
(ii) Texture
(iii) Shape
(iv) Size
(v) Shadow
(vi) Pattern
(vii) Height
(viii) Association

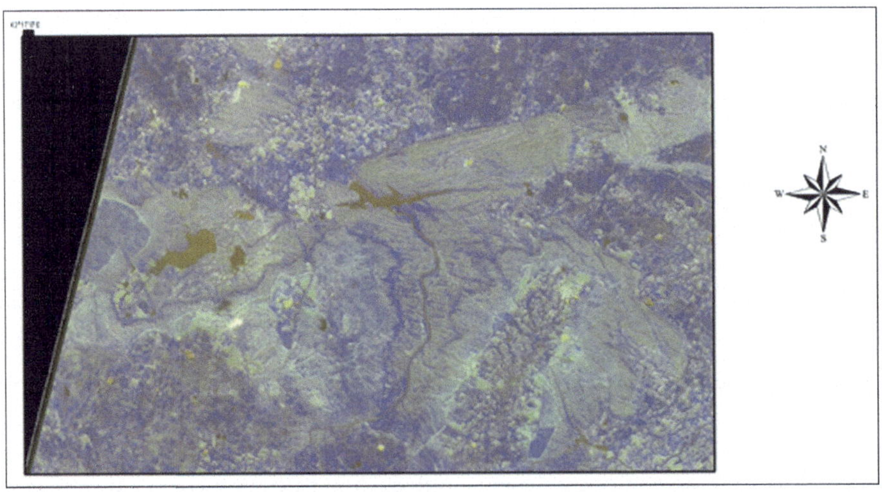

Fig. 1.4 IRS P6(LISS IV) image of study area

Table 1.2 Interpretation key for land cover mapping using remote sensing data

Classes	Tone	Texture	Shape	Association
Forest	Red	Medium-smooth	Irregular	High altitudes
Water body	Blue-black and light blue	Smooth	Linear Irregular polygon	Valley bottom and low altitudes
Barren areas	Brown-creamy white	Medium-smooth	Irregular	None
Shrub and grassland Vegetation	Shades of red	Smooth	Irregular patches	Rocky slopes and water bodies
Hard Rocky slopes	Black-brown	Medium-coarse	Irregular	High hills
Rocky slopes	Creamy white	Coarse	Irregular	High hills
Agriculture	Pink red	Smooth	Geometrical polygons	Habitation

Although the study area falls amidst the Danti and Barkachha reserve area as represented in the SOI toposheet, but the recent satellite image of the area shows rampant degradation and scanty vegetation in the study area. The only forest left is along the ridge line at higher elevations as summarized in Table 1.2. Most of the area is under scanty vegetation with patched of barren land.

Natural colour composite: The landuse map is prepared by digital image processing of Indian remote sensing satellite, IRS-P6, LISS IV data of 27th September 2005. ERDAS 8.5 classified the image into the following different landuse classes by 0.94 convergence factor and six iterations as forest and vegetation, degraded forest, agriculture, barren areas, water bodies in the signature editor (Schrader and Duniway 2011). Natural colour composite is produced by combining bands 3, 2 and 1 as red, green and blue, respectively, as shown in Fig. 1.5.

Landuse classification: ARCMAP classification of raster data is the process of sorting pixels into a finite number based on their data file values, individual classes, or categories of data. If a pixel satisfies a certain set of criteria, then the pixel is assigned to the class that corresponds to that criteria (Rozenstein and Karnieli 2011). There are two ways to classify pixels into different categories:

- Supervised
- Unsupervised.

In unsupervised classification any individual pixel is compared to every discrete cluster to see which one it is closest to. A map of all pixels in the image, classified as to which cluster every pixel is most likely to belong, is created. In supervised classification, spectral signatures are created from specified locations in the image. These specified locations are provided with the generic name 'training sites' and are defined by the user.

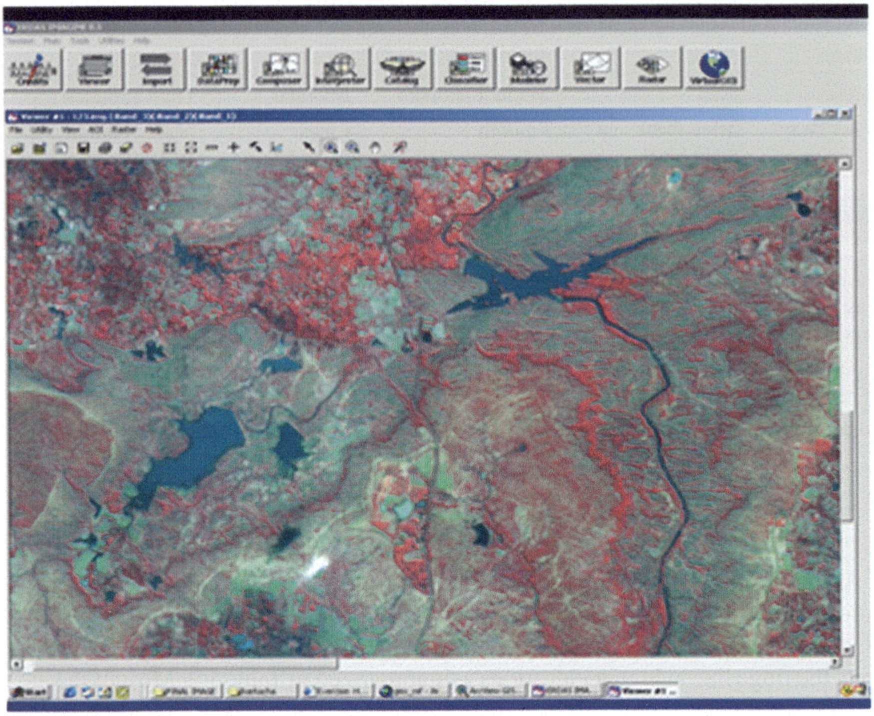

Fig. 1.5 ERDAS image of Khajuri watershed

Supervised classification, the interpreter knows beforehand what classes, are there and whether each is in one or more locations within the scene. These are located on the image and the areas containing examples of the class are circumscribed (making them training sites) along with this, the statistical analysis is performed on the multi-band data for each such class. The parameters choose to differentiate for separation will completely depend on analysis. An area within an image generally, multiple pixels in the same cluster correspond to visually correlating map patterns to their ground counterparts or an already known ground feature or class (Scott and Janikas 2010).

In the present study supervised classification is used. The supervised classification is summarized in Table 1.3 with 40% of the area under rocky terrain. The occurrence of wasteland in approximately 1000 ha of land. The forest cover is left in only about 14% of the area with patches of barren land. The midland with wasteland and grazing lands in the slope show signs of erosion.

Agricultural land: The agricultural area is observed or located by the shapes of fields which used to be regular, with red characteristic tone, and its association with few water bodies.

Table 1.3 Supervised classification of land use classes

S. No	Landuse class	No: of pixels	Apprx. area (ha)	Apprx. area (%)
1	Natural water body	6222	21	1
2	Barren areas	14,912	55	3
3	Forest	63,360	231	14
4	Vegetation	70,306	236	15
5	Rocky terrain	192,601	648	41
6	Wasteland	126,475	425	27

Forest: Forest is observed and located by its dense coverage because of the thick canopy, which may remain green throughout the year according to its type as summarized in Table 1.4. This type of land cover is identified by its tone varying from red to dark red. It mostly shows irregular shape and smooth texture (Thiruvengalacha and Sakthivadivel 1997). Based on the different characteristics the forest cover of the study area is categorized in different types which are:

a. Dense forest cover
b. Open forest cover
c. Degraded forest cover
d. Waste land.

a. **Dense forest**:

Dense forests are characterized by their crown cover which lies up to 40 per cent and over. In the study area, dense forests are located mostly in the eastern part of study area and are found to be confined between the higher and medium altitude areas.

b. **Open forest**:

Open forests are found to have the concentration of trees comparably less. In the study area, the open forest cover is found mainly in northwestern and western and along with this in medium and low altitude areas.

Table 1.4 Forest cover categorization

Land cover	Crown cover	Tone	Association	Texture	Shape
Dense forest	>=40%	Dark red	–	Smooth	Irregular
Open forest cover	(10–40)%	Dark red to Medium red	Dense forest cover	Medium	Irregular
Degraded forest cover	<=10%	Yellow	Other forest cover	–	–
Waste land	0	Brownish and creamy white to yellow	–	–	–

c. **Degraded forest**:

Degraded forests are characterized by limited crown cover of about 10 per cent. It is mainly composed of shrubs, bushes and smaller trees.

d. **Waste lands**:

Waste lands are those which possess zero or limited ability to support vegetation. The different types of waste lands are found in study area. The common feature was the increase in the occurrence of barren and rocky exposures often they are devoid of soil cover and vegetation. They are irregular in shape. They are found close to hill forests in the form of openings, sometimes as scattered like isolated exposures. In the study area they are found as patches of degraded forest land along the ridge line at lower elevations.

e. **Water body**:

It is composed of both natural and artificial water holding structures like.

- Rivers or streams
- Lakes
- Tanks
- Reservoirs.

The water features are found to have bluish-black in tone in the satellite imagery. The low water surfaced especially the newly built water harvesting ponds appear light blue and deep-water features and deep-water bodies are observed by their tone of dark blue. In the study area, many tanks were observed in the western region to support the KVK activities.

1.3.3 Analysis of Landuse in GIS

The various soil and landuse characteristics digitized in ARCMAP are used to analyze the current soil and landuse relationships and to identify the areas of concern.

Slope: Maximum slope encountered in the area is 15–25% with occurrence of rocky surfaces. The area has an undulating and rugged terrain with maximum area under flat land. Table 1.5 shows that more than 60% of the area has flatland or mild slope (upto 5%). Therefore if suitable soil depth is available, a large part of the landscape can harbour vegetation provided protection and conservation measures are taken.

Soil depth: About 40% of the area is under shallow to very shallow soils and are highly susceptible to erosion. Table 1.6 indicates availability of more than 400 ha of land of deep to very deep soils, thus supporting future prospects of agriculture and cultivation in the area.

Soil texture: The different classes of soil texture influences soil water infiltration, retention and drainage conditions of the area. Table 1.7 gives the soil distribution

Table 1.5 Slope distribution of present area

S. No	Slope (%)	Frequency of soil units	Area (ha)	Area%
1	1–3	83	537.27	53
2	3–5	28	70.5	14
3	5–10	7	227.15	22
4	10–15	3	62.77	6
5	15–25	2	32.82	3

Table 1.6 Soil depth profile

S.No	Soil depth	Frequency of soil units	Area (ha)	Area%
1	Very deep (>90 cm)	26	260.18	26
2	Deep (45–90 cm)	36	147.04	14
3	Moderately deep (22.5–45 cm)	36	201.76	20
4	Shallow to very shallow	24	396.88	39

Table 1.7 Distribution of soil texture

S.No	Soil texture	Frequency of soil units	Area (ha)	Area%
1	Clay loam to silty clay	12	65.95	6.50
2	Silty clay loam to clay	42	317.25	31.39
3	Loam	35	195.61	19.32
4	Loam to clay loam	10	29.97	2.95
5	Sandy loam with 70% rock exposure	24	396.88	39.27

pattern in the area with maximum area under loam soil conditions; therefore soil texture is not a limiting factor in the area. Sandy loam soil along the slopes covers about 396 ha with 70% rock exposure.

Soil erosion: The major part of the area is under erosion degradation. Table 1.8 indicates a strong threat of soil erosion with moderate to severe erosion affecting about 50% of the area. Therefore if serious conservation measures are not undertaken the erosion may become a serious limiting factor for vegetation development in the area. The added benefit of erosion control will be prolonged life of water harvesting structures. This problem is particularly severe in the steep areas with rocky terrain and scant vegetation cover, thus needing special attention (Grigg 1996).

Land use and land capability (LULC): The current land use capability for cultivation is for more than 450 ha as indicated by Table 1.9. This accounts for more than 45% of the land area. Hence the land is suitable for productive use if proper conservation measures are taken for sustainable use of the land. (Klingebiel and Montgomery 1961).

Table 1.8 Soil erosion pattern

S. No	Soil erosion	Frequency of soil units	Area (ha)	Area%
1	Slight	64	413.17	40.84
2	Moderate	36	195.61	19.32
3	Moderate-severe	24	396.88	39.27

Table 1.9 Land capability classification

LU/LC Class	Limitations	Possible land use	Frequency of soil units	Area (ha)	Area%
II	Moderate limitations of the cropland	Intense cultivation of the field crops	9	388.37	38.34
III	Severe limitation of the cropland	Moderate cultivation of the field crops	9	28.29	2.81
IV	Very severe limitations of the cropland	Limited cultivation of the field crops	1	51.13	5.06
V	Slight to moderate limitations of the grassland	Intense grazing found	14	19.35	1.90
VI	Severe limitations of the grassland	Moderate grazing found		78.93	7.81
VII	Very severe limitations of the grasslands	Limited grazing found	24	42.79	4.24
VIII	Non-agricultural land	Wildlife and Forest	7	396.88	39.27

(Land capability classification, 1961 USDA Agricultural Handbook).

1.4 Conclusion

The present study is an effort to assess and analyze the available water resources, prevailing current demand and future projection and along with these technological options for water resource management to improve water availability and management of the efficiency of crop yield of a watershed. For optimum utilization of water resources and getting maximum benefit, extensive study has been done. The following conclusions may be summarized from this project work:

(i) The various sources of water are identified and quantified. The Khajuri stream is located in this area and our focus area is Khajuri Watershed and its five sub-watersheds.
(ii) Remote sensing and GIS have been found useful tools in identification and categorization of watersheds on the basis of natural resources and their limitations.

Thematic information can be derived from spatial analysis of remote sensing helpful in the assessment of developmental plans before implemented in the field. An approach for an effective tool for selection of the best management plan to be implemented.

(iii) Remote sensing data compute runoff estimation to provide a quick result for decision-makers, before any experimentation of quantification is taken up. This gives more reliable results and also less time-consuming method. This method can be used effectively in the design of stormwater drains and small water control projects. In this study, GIS has been used as a useful tool to give real authentic topology information.

Acknowledgements The authors are thankful to the Department of Civil Engineering and remote sensing facilities at Civil Engineering lab, IIT BHU for providing the required facilities.

References

Arun PV (2013) A comparative analysis of different DEM interpolation methods. The Egypt J Remote Sens Space Sci 16(2):133–139

Campbell JB, Wynne RH (2011) Introduction to remote sensing. Guilford Press

Chowdary VM, Ramakrishnan D, Srivastava YK, Chandran V, Jeyaram A (2009) Integrated water resource development plan for sustainable management of Mayurakshi watershed, India using remote sensing and GIS. Water Resour Manage 23(8):1581–1602

ERDAS (2000) "Macro-Language Reference Manual" ERDAS IMAGINE V8.5, ERDAS Inc, Atlanta, GA, U.S.A.

Gao J (2008) Digital analysis of remotely sensed imagery. McGraw-Hill Professional

Grigg N (1996) Water resources management. Principles, regulations, and cases. McGraw-Hill, New York

Hamar D, Ferencz C, Lichtenberger J, Tarcsai G, Ferencz-Arkos I (1996) Yield estimation for corn and wheat in the hungarian great plain using landsat MSS data. Int J Remote Sens 17(9):1689–1699

Jain SK, Seth SM, Jain MK (2000) "Remote sensing and GIS Application studies" National Institute of Hydrology. www.gisdevelopment.net

Khan MA, Gupta VP, Moharana PC (2001) Watershed prioritization using remote sensing and geographical information system: a case study from Guhiya India. J Arid Environ 49(3):465–475

Klingebiel AA, Montgomery PH (1961) Land-capability classification (No. 210). Soil Conservation Service, US Department of Agriculture.

Lillesand T, Kiefer RW, Chipman J (2015) Remote sensing and image interpretation. Wiley

Makhamreh Z (2011) Using remote sensing approach and surface landscape conditions for optimization of watershed management in Mediterranean regions. Phys Chem Earth, Parts A/B/C 36(5–6):213–220

Mas E, Bricker J, Kure S, Adriano B, Yi C, Suppasri A, Koshimura S (2015) Field survey report and satellite image interpretation of the 2013 Super Typhoon Haiyan in the Philippines. Nat Hazards Earth Syst Sci 15(4)

Rao SVN, Bhallamudi SM, Thandaveswara BS, Mishra GC (2004) Conjunctive use of surface and groundwater for coastal and deltaic systems. J Water Res Plan Manage 130(3):255–267

Rozenstein O, Karnieli A (2011) Comparison of methods for land-use classification incorporating remote sensing and GIS inputs. Appl Geogr 31(2):533–544

Schrader TS, Duniway MC (2011) Image interpreter tool: an ArcGIS tool for estimating vegetation cover from high-resolution imagery. Rangelands 33(4):35–40

Scott LM, Janikas MV (2010) Spatial statistics in ArcGIS. In: Handbook of applied spatial analysis (pp 27–41). Springer, Berlin, Heidelberg.

Strager MP, Fletcher JJ, Strager JM, Yuill CB, Eli RN, Petty JT, Lamont SJ (2010) Watershed analysis with GIS: the watershed characterization and modeling system software application. Comput Geosci 36(7):970–976

Thiruvengalachari S, Sakthivadivel R (1997) Satellite remote sensing techniques to irrigation system performance assessment: a case study in India" Research Report 9. Int Irriga Manag Insitute, Colombo

Thiruvengadachari S, Jonna S, Murthy CS, Raju PV, Hakeem A (1994) Evaluation of system performance of bhadra project command area through satellite remote sensing techniques during 1993–94 Rabi season. Project report National Remote Sensing Agency, Hyderabad, India

Willis R, Finney BA, Zhang D (1989) Water resources management in north China plain. J Water Res Plan Manag 115(5):598–615

Chapter 2
Prioritization of Sub-watersheds Based on Morphometric Parameter Analysis Using Geospatial Technology

Kanak Moharir, Chaitanya Pande, R. S. Patode, M. B. Nagdeve, and Abhay M. Varade

Abstract Watershed prioritization, particularly in the context of watershed plans and management programmes, is part of the water resources development. Actually, morphometric analysis aided by geospatial technology is accomplished in prioritizing sub-watersheds according to their natural features of availability of resources. The information relating to the study area's geomorphology and erosion factors are used in the area and to prepare the local models of the ungauged seven sub-watersheds, which otherwise lack adequate hydrological database. The region under investigation has observed the rising increases in the groundwater table. As a result of the failure to look at the interrelated criteria that influence water-show planning, the watershed development plans so far implemented in the region fail. For this respect, for the pre-oral measurement of seven sub-sheds, the integrated methodology involving morphometric aspects derived from geo-spatial technology is used. The basic mathematical equations used in a GIS setting have been used to measure a series of the aspects of hydrology in the rotation, stream order, stream length, stream frequency, drainage densities, texture ratio, shape factor, circulatory ratio, elongation ratio, bifurcation and compactness ratioArcGIS10.3 was used to perform the geospatial research. The results divided the entire watershed into four priority areas, namely Big, medium, small, and very weak. The results are relevant in order to establish soil and water conservation plans in the Man river basin in Akola and Buldhana districts in Central Indian, as well as inadequate production and management of groundwater.

K. Moharir (✉)
Department of Geology, Sant Gadge Baba Amravati University, Amravati, Maharashtra 446002, India
e-mail: kanak.moharir11@gmail.com

C. Pande · R. S. Patode · M. B. Nagdeve
All India Coordinated Research Center for Dryland Agriculture, Dr. PDKV, Akola, Maharashtra, India

A. M. Varade
Department of Geology, RTM Nagpur University, Nagpur 440001, India

A. Pandey et al. (eds.), *Water Management and Water Governance*, Water Science and Technology Library 96,
https://doi.org/10.1007/978-3-030-58051-3_2

Keywords Morphometric analysis · Geospatial · Watershed prioritization · Sustainable water · Akola and Buldhana districts · Maharashtra · Central India

2.1 Introduction

This area of study planned conservation of ecological resources for a long time, i.e., land, soil, and water were reduced day after day as a consequence of climate change and increasing population pressure on the earth's surface. The right scientific planning and management of these resources need for huge data for the organization of natural resources (Moharir et al. 2017; Pande et al. 2019; Pande 2020a–d). Therefore, watershed geomorphological characters are used to develop regional hydrological modeling in the region of basaltic hard rock for the solution of various hydrological problems in ungauged water-sheds and inappropriate results. Geoinformatics approaches merged different technologies, including remote sensing, the GIS, and the global positioning system (Bera and Bandyopadhyay 2013). Geographic Information System (GIS) technology applications are very effective, time-saving, and spatial planning suitable. The GIS techniques can handle huge datasets, but they also solve several complex questions in addition to facilitating spatial data retrieval and query. For effective management, it was necessary to properly understand the hydrological behavior of water-shelter growth. Thus, a detailed analysis of each watershed is required to create a management plan that requires comprehensive data. Most watersheds in India on the present drainage line have to be specially designed to meet farming needs. Inadequate data collection can be rendered by morphometric watershed analyzes. the morphometric characteristic of a watershed is its characteristics, and its hydrological activity can be synthesized helpfully (Moharir et al. 2014; Pande and Moharir 2015; Moharir et al. 2017). Due to certain geo-environmental or economic factors, it is very difficult to grow a wide area on one level. The region must therefore be given priority when the software system is being used. During the research, the watershed was characterized and prioritized by the use of remote sensing and GIS techniques (Biswas et al. 1999; Gajbhije et al. 2014; Pande et al. 2017a). The region has been divided into seven sub-watersheds, with geospatial technology as a priority for the Man Watershed. For water management and prioritization of watersheds are of significant significance (Javeed et al. 2009; Khan et al. 2001). This involves defining and assessing the water basin that helps to employ faster and indirect methods and creates linkages to excessive erosion losses. This can be helpful, as is the case if it is located remotely and no other direct monitoring system is available. If the funds to execute a conservation plan are small, priority should be provided to erosion-prone areas in the catchment. The area is more likely and more important in the management of the sediment to contribute towards the high volume of sediment and erosion.

2.2 Study Area

In the districts of Akola and Buldhana in Maharashtra in India (Fig. 2.1) the analysis is situated between 59″ N latitude and 76° 41′ 23″ E longitude. In the hard rock basaltic zone, the area is underneath the surface. Under the drylands, full farmland has been cultivated. The study shows an area of 750–850 annual precipitation during the rainfall seasons. There is very little effort to comprehend hydrogeological factors that control water quality, and therefore attempts were created to undermine the groundwater system of the Man River Basin. The management of the river basin uses detailed research in the fields of stratigraphy and geochemistry. In the Man river basin, the drainage pattern was observed to sub-dendritic (Moharir et al. 2017; Pande et al. 2017a, b; Khadri et al. 2016).

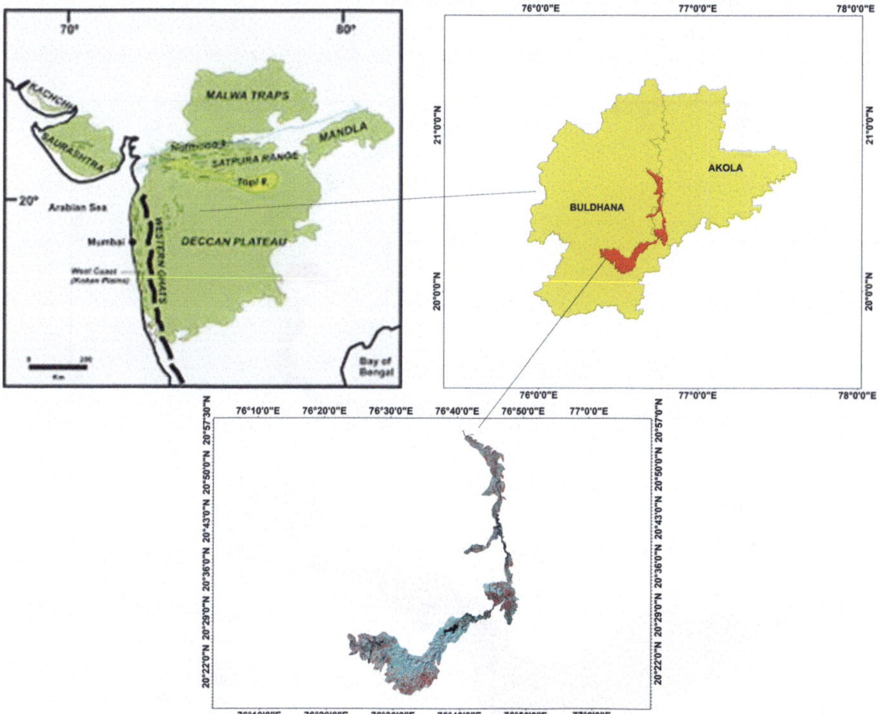

Fig. 2.1 Location map

2.3 Methodology

The watershed limit was determined with ARC GIS 10.1 (Fig. 2.3) by the digital elevation model. The Man River Basin, prepared by Arc hydro-tools was subdivided such as seven sub-watersheds (Fig. 2.2). The morphometric study was conducted using seven water conservation priorities of sub-watersheds. GIS data and mathematical equations were helpful to measure morphometrical parameters during research (Pande et al. 2018). The following parameters have been included: location, perimeter, stream order, stream length, stream number, and elevation, obtained from the digital drainage map coverage. Finally, Arc GIS 10.3, and standard formula were developed for the measurement of bifurcation, drainage distance, frequency of streams, texture ratio, shaping factor, circulation ratio, elongation, and compactness ratio. During the analysis of drainage parameters, the methodology was used for the estimation of the priority methods was shown in the figure. The second of the fourth show in Fig. 2.3 and Fig. 2.4.

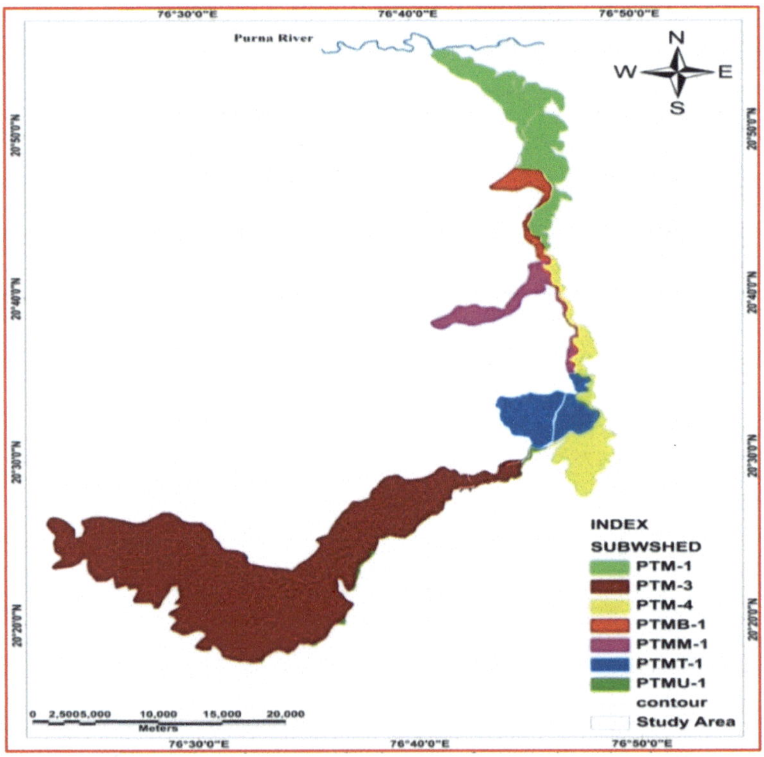

Fig. 2.2 Sub-watershed map of Man river basin

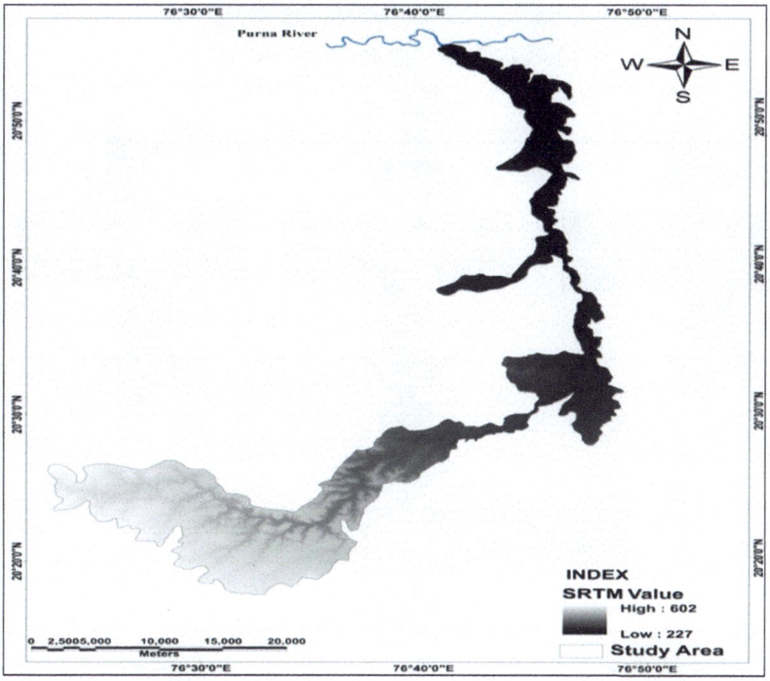

Fig. 2.3 SRTM map of Man river basin

2.4 Results and Discussions

A morphometrical analysis is required and helps in the identification of various hydraulic characteristics of stream basins, i.e., patterns, form, streaming phases, bedrock permeability, stream safety, and helps to correlate these with lithology (Nookaratnam et al. 2005; Vittala et al. 2008). This analysis was alienated into three river basin sections. The first section deals with stream numbers, stream order, and the stream longitudes, and the description of watershed area, perimetric and longitudinal GIS environment by means of mathematical equations and GIS data (see Table 2.1). The second part discusses the different linear and shapes morphometric parameters that define sub-watersheds, contributing to the understanding of the hydrological comportments of sub-watersheds and soil erosion in the particular sub-watersheds. The third section discusses the importance of water changes based on these linear and morphometric type parameters.

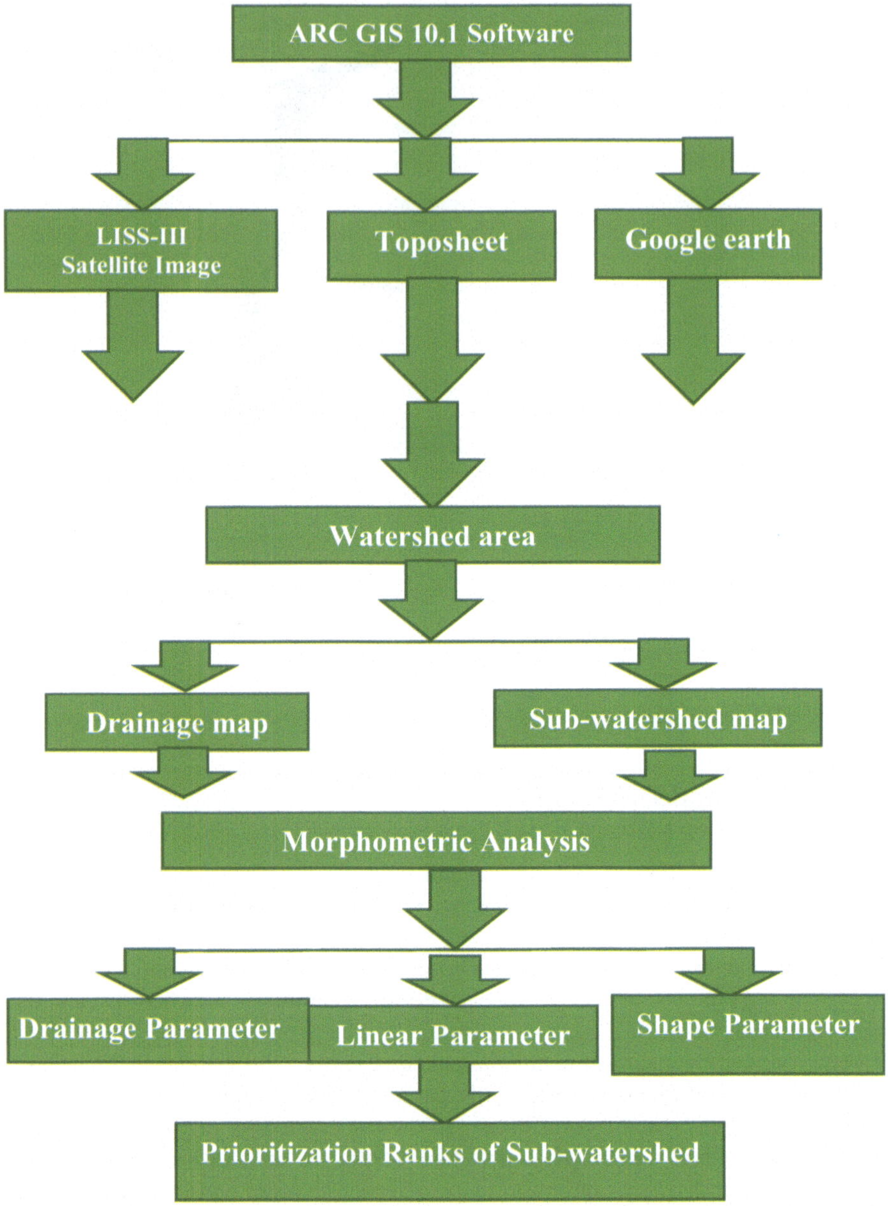

Fig. 2.4 Flow chart of prioritization ranks

Table 2.1 Morphometric parameter analysis of Man River basin comparative characteristics

S. No.	Morphometric parameter	Formula	Result
1	Area (km^2)	GIS Software Analysis	447.80
2	Perimeter (km)	GIS Software Analysis	315.58
3	Elevation	GIS Software Analysis	272.16–560.77
4	Length of basin (km)	GIS Software Analysis	115.43
5	No. of streams	Nu = $N_1 + N_2 + \ldots + N_n$	1454
6	Total stream length (km)	Lu = $L_1 + L_2 \ldots + L_n$	1134.36
7	Bifurcation ratio (Rb)	–	4.80–23.81
8	Drainage density (Dd)	–	2.53
9	Stream frequency (Fs)	$F_s = N_u/A$	3.25
10	Circulatory ratio (Rc)	$R_c = 12.57 * (A/P^2)$	0.06
11	Form factor (Rf)	$F_f = A/L_b{}^2$	0.024
12	Elongation ratio (Re)	$R_e = 2/L_b * (A/\pi)^{0.5}$	0.21

2.5 Morphometric Analysis

A study of the morphometric quantity analysis should be estimation to given that the derived basin variables are ratios or dimensional numbers, thereby providing an effective, non-scale comparison. This area was performed using geospatial technology with wise sub-watershed morphometric parameters of drainage and remote data (Table 2.2).

Table 2.2 Sub-watershed-wise morphometric parameters

Sub watershed	Area (km^2)	Perimeter (km)	Elevation		Length of basin (km)	No. of streams	Total stream length (km)
			Min	Max			
PTM-1	57.02	57.02	227	289	26.75	253	133.13
PTM-3	280.39	281.50	289	602	45.36	900	664.02
PTM-4	35.20	35.20	227	358	27.38	93	73.15
PTMB-1	12.23	12.23	227	289	14.87	57	29.03
PTMM-1	20.62	20.62	227	358	14.41	47	44.68
PTMT-1	35.45	35.45	289	358	10.11	96	76.96
PTMU-1	2.55	2.55	289	558	31.95	1	115.43

Table 2.3 Sub-watershed-wise morphometric parameters

Sub watershed	Bifurcation ratio (Rb)	Drainage density (Dd)	Stream frequency (Fs)	Circulatory ratio (Rc)	Form factor (Rf)	Elongation ratio (Re)
PTM-1	18.75	4.12	5.13	0.22	0.07	0.31
PTM-3	20.83	1.65	3.22	4.48	0.13	0.41
PTM-4	15.75	4.77	2.72	0.15	0.04	0.24
PTMB-1	10.56	10.10	4.34	1.02	0.05	0.26
PTMM-1	10.50	4.70	2.18	0.60	0.009	0.35
PTMT-1	14.17	3.89	2.73	0.35	0.34	0.66
PTMU-1	2	2.30	1.17	4.93	0.002	0.05

2.5.1 Stream Number (Nu) and Stream Orders

A dendritic drainage type, which indicates a uniform underground layer of the study is included in the Man river basin. The current study assessed stream ordering by using digital streams from top sheets and satellite pictures (Strahler 1964). Table 2.3 shows the sequential range of stream numbers and their linear features. The drainage pattern study of the Man River Basin has shown that the region has a tectonic control dam. The total stream number was located in the first order and with a reduction in stream number the stream order increases. On the drainage line with Arc GIS software 10.1, the stream order map was developed (Fig. 2.2). The basin's current order varies from the first to the fifth level (Fig. 2.5). ARC GIS software 10.1 applicable to the new law was used during the stream orders in the Man River Basin (Horton 1932). The average stream length is found to be maximum in streams of the first order and decreases with increased stream size. This change in streams may indicate the variability of high altitude streams as well as lithological changes. There is approximately 1134.36 km gross stream length in the Man River Basin (Table 2.1).In GIS Setting the average stream length (Lsm) and the ratio is determined. The definition of streams in the water-age system is the water pattern, which in turn mainly replicates the structural lithological checks of the rocks underneath. The research area has dendritic drainage patterns and remote sensing information has been recognized with stream lengths and other hydrological characteristics.

2.5.2 Stream Length (Lu), Mean Stream Length (Lsm), and Stream Length Ratio (RL)

The different parameter types are the same. For the prioritizing of watersheds by GIS software, stream length, a mean stream length, and stream length ratio was calculated. Commonly, the entire stream segment length decreases with increasing

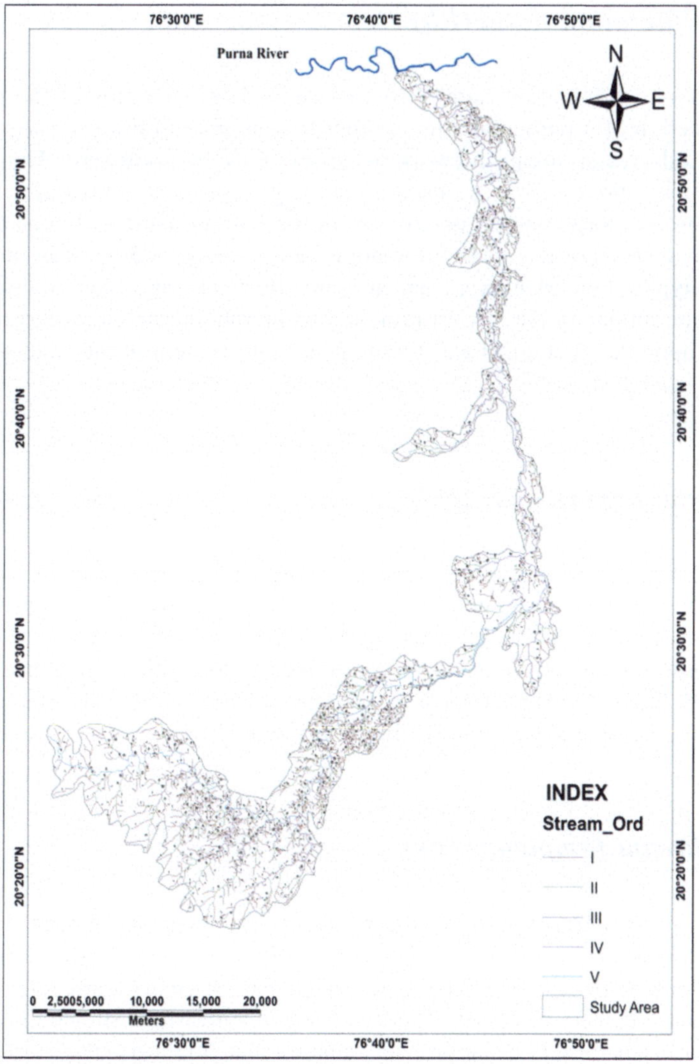

Fig. 2.5 Drainage map with stream order of the Man river basin

stream order. The longitude and ratio of the stream of this study are the most important considerations to understand the hydrologic properties of the river system since the rock formations in a lake are so permeable. It indicates also whether the hydrological characteristics of the rock surfaces underneath the bay are modified considerably (Singh and Singh 1997; Javed et al. 2009; Khadri and Pande 2015). Hydrogeological, physiographical, and geological characteristics determine the relationship between bifurcation and stream length ratio.

2.5.3 Bifurcation Ratio (Rb)

The analysis of bifurcation ratio (Rb) was defined as per ratios of the drainage network section of a particular stream order to a number of drainage segments of the afterward greater stream network using Arc GIS 10.3 software. The analysis of bifurcation ratio values was observed ranging in between 4.80 and 8 and it is measured to be morphometric parameters in the basaltic hard rock area (Strahler 1957). The analysis of the mean bifurcation ratio is found at 5.95. It is shown that the stream types of the Man river basin area are effected using structural instabilities of Rb is not similar in the stream orders. The morphometric characteristics have depended upon the geological and lithological improvement of the watershed area (Tables 2.1 and 2.3).

2.5.4 Drainage Density (Dd)

Horton (1932) was presented for drainage density as an appearance to show the understanding of the arrangement of drainage channels network in the basaltic hard rock region. Total stream length was observed from the stream segments for total drainage orders as per unit area and the controlled for slope values gradient and to the comparative relief of the hard rock area. The study of drainage density was computed of 2.53 using geospatial technology (Tables 2.1 and 2.3).

2.5.5 Stream Frequency (Fs)

For each unit area Horton (1932), the whole stream segments of total orders are Stream Frequency (Fs), or Channel Frequency. The f_s values show a helpful correlation with drainage density which indicates an intensification in the stream population as the drainage density increases. The frequency of a measured stream in a basin of 3.25 indicates a helpful association with value for area drainage, suggesting that the drainage population is increasing (Table 2.2).

2.5.6 Elongation Ratio (Re)

It is the average of the circular width of basin with the actual drainage basin longitude (Schumm 1956). The Re values usually range from 0.6 to 1.0 in various climatic and geological circumstances. Ranges near to 1.0 are typical for most downside relief regions, with values ranging from 0.6 to 0.8 typically connected with upper side relief and steep ground slopes (Strahler 1964). Such proportions can be classified into the three types of circular (>0.9), oval (0.9–0.8), elongated (<0.7). (c) circular (>0.9).This means that the system belongs to the expanded form reservoir and low relief (Tables 2.2 and 2.3).

2.5.7 Circularity Ratio (Rc)

The RC ratio of Miller (1953) has defined to be the ratio of area for circle area is being the same diameter within the basin. The length and frequency of rivers, geographical formations, land use, atmosphere, elevation, relief and slope an impact Rc. Rc is the most popular. The basin's circularity is 0.06. This suggests lengthy and more permeable geological structures that are homogenous. This observation shows that the basin is extended in form, low runoff, and highly permeable sub-solar conditions (Table 2.3). The circularity ratio is observed indicates elongated in shape.

2.6 Prioritization of Sub-watersheds

Therefore, the prioritization of the watershed is the order in which various sub-watersheds ought to be occupied to manage and to protect the land. A proper mechanism for prioritizing sub-watersheds must therefore be developed. The prioritization of morpho-metric parameters was done using Arc GIS software 10.3 (Fig. 2.6) from the digital elevation model. All sub-sheds were given priority in four categories based on a percentage for each sub-watershed cultivated area and drainage density to facilitate the phase-by-phase implementation of the watershed management programme (Durbude et al. 2001; Pandey et al. 2004). In addition, Pandey et al. (2007) defined these groups based on the average slope. This priority setting also requires a number of data types (Javed et al. 2009). Greater values of linear parameters improve the potential for runoffs and thus eroding, while lower values of form parameters provide higher unit sediment rates. Thus, the whole sub-watershed has been rated according to the ranges of various geomorphological factors (Table 2.4). Thus, by computing the composite parameter valves, the priority rating for all sub-watersheds of basin was carried out. Highest priority has been given to the subwatershed show with the lowest compound parameters. The morphometric analysis is one of the best methods to measure current water dispersion and erosion trend problems over the catchment. Geospatial technology also provides useful natural resource knowledge and regional physical field parameters. Drainage patterns, drainage orders, catchment lines, and other stores for the study area were used in this analysis with the remote sensing and geographic information system (GIS). In a variety of plans, both geospatial and conventional sources, are implemented with GIS' combining ability and its analysis of spatial and multiple-layered information, which are basic parameters for planning sustainable water resources.

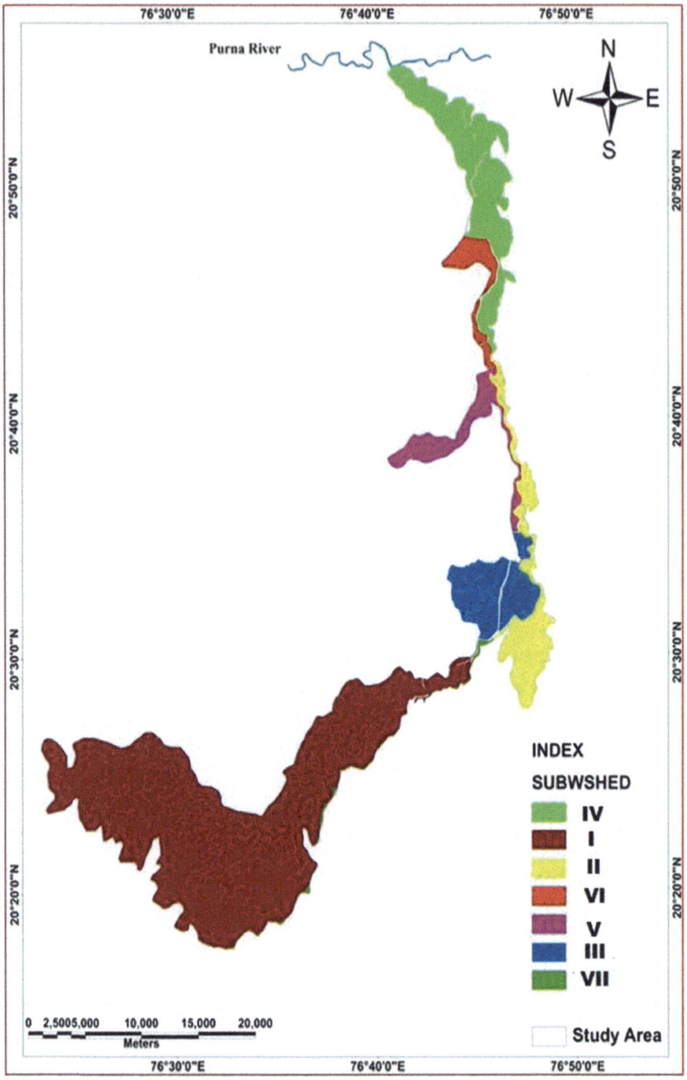

Fig.2.6 Prioritized sub-watershed and priority rank using morphometric analysis

2.7 Conclusion

In the current research, topographical maps, SRTM (30 m resolution), and IRS-LISS III data were utilized for the quantitative morphometric analysis. The remote sensing and GIS technologies were therefore more successful at understanding the morphological and parameter characteristics of the single sub-water change. This examination of the variations in parameters was carried out in spatial, local, and

Table 2.4 Prioritization sub-watersheds using morphological parameters

Sub watershed	Bifurcation ratio (Rb)	Drainage density (Dd)	Stream frequency (Fs)	Circulatory ratio (Rc)	Form factor (Rf)	Elongation ratio (Re)	Prioritization ranks
PTM-1	2	4	1	6	3	4	IV
PTM-3	1	7	3	2	2	2	I
PTM-4	3	2	5	7	5	6	II
PTMB-1	5	1	2	3	4	5	VI
PTMM-1	6	3	6	4	6	3	V
PTMT-1	4	5	4	5	1	1	III
PTMU-1	7	6	7	1	7	7	VII

relief ways of the watershed. The characteristics showed that every sub-watershed is a priority in the Man river basin. For the study of watershed planning and development using remote sensing and GIS technologies, studies of seven sub-watersheds were considered. Five stream orders from the drainage network with Arc GIS software were identified in this basin. The analysis of relief and slope aspects of morphometric parameters affected the drainage fragments of the basaltic terrain rocks in dryland conditions. A total of 1454 drainage segments were observed from remote sensing data in the GIS region, and drainage orders were divided into the first order, the second-order, etc. The most possible in-situ dryland conservative measures of soil and water erosion were also carried out innovative and computable methods for prioritizing sub shows and soil erosion and results for sustainable development and increasing production of crop yields under dryland conditions and basaltic hard rocks were presented in the study. Active preservation of land and water protection and rainwater harvesting systems in order to effectively store, management, and preparing watersheds has to be undertaken for the sake of observed sub-watersheds (Patode et al. 2017). As a water management planner, a public body, and a private enterprise, it was helpful to research the target water change to take soil and water conservation steps or to complete the activities of water collection in the Akola river basin and the Buldhana districts of Maharashtra, India.

References

Bera K, Bandyopadhyay J (2013) Prioritization of watershed using morphometric analysis through geoinformatics technology: a case study of Dungra sub-watershed, West Bengal, India. Int J Adv Remote Sens GIS 1(3):1–8

Biswas S, Sudhakar S, Desai VR (1999) Prioritisation of subwatersheds based on morphometric analysis of drainage basin-a remote sensing and GIS approach. J Indian Soc Remot Sens 27:155–166

Durbede DG, Purandara BK, Sharma A (2001) Estimation of surface runoffpotential of a watershed in semi and environment-a case study. J Indian Soc Remote Sens 29(1–2):47–58

Gajbhiye S, Sharma SK, Meshram C (2014) Prioritization of watershed through sediment yield index using RS and GIS approach. Int J U E Serv Sci Technol 7(6):47–60

Horton RE (1932) Drainage basin characteristics. Trans Am Geophys Union 13:350–361

Javeed A, Khanday MY, Ahmed R (2009) Prioritization of sub-watersheds based on morphometric and land use analysis using remote sensing and GIS techniques. J Indian Soc Remote Sens 37:261–274

Khadri SFR, Pande C (2015) Remote sensing based hydro- geomorphological mapping of Mahesh River Basin, Akola and Buldhana Districts, Maharashtra, India, Int J Geol Earth Environ Sci 5(2):178–187. ISSN: 2277-2081 (online)

Khadri SFR, Pande C (2016) Ground water flow modeling for calibrating steady state using MODFLOW software: a case study of Mahesh River basin, India, Model Earth Syst Environ 2:39. https://doi.org/10.1007/s40808-015-0049-7

Khan MA, Gupta VP, Moharana PC (2001) Watershed prioritization using remote sensing and geographical information system: a case study from Guhiya, India. J Arid Environ 49:465–475

Miller VC (1953) A quantitative geomorphic study of drainage basin characteristics in the Clinch Mountain area, Virginia and Tennessee. Project NR 389042, Tech. Rept. 3. Columbia University, Department of Geology, ONR, Geography Branch, NY

Moharir KN, Pande CB (2014) Analysis of morphometric parameters using remote-sensing and GIS techniques in the lonar nala in Akola district Maharashtra India. Int J Tech Res Eng 1(10)

Moharir K, Pande C, Patil S (2017) Inverse modelling of aquifer parameters in basaltic rock with the help of pumping test method using MODFLOW software. Geosci Front 8(6):1385–1395

Moharir K, Pande C, Varade AM, Pande R (2017) Morphometric analysis in Koldari watershed of Buldhana district (MS), India using geoinformatics techniques. J Geomat 11(1)

Nookaratnam K, Srivastava YK, Venkateswarao V, Amminedu E, Murthy KSR (2005) Check dam positioning by prioritization of micro watersheds using SYI model and morphometric analysis-Remote sensing and GIS perspective. J Indian Soc Remote Sens 33(1):25–28

Pande CB, Moharir K (2015) GIS based quantitative morphmetric analysis and its consequences: a case study from Shanur River Basin, Maharashtra, India. Appl Water Sci 7(2):861–871

Pandey A, Chowdary VM, Mal BC (2004) Morphological study of watershed using geographical information system. In: Proceeding of international conference and emerging technologies in agricultural and food engineering (etae. 2004), IIT Kharagpur, pp 34–40

Pandey VK, Pandey A, Panda SN (2007) Application of remote sensing and GIS for watershed characterization—a case study of Banikdin watershed. (Eastern India). Asian J Geoinf 3(7):3–15

Pande CB, Patode RS, Moharir KN (2017a) Morphometric analysis usingremote sensing and GIS techniques a case study of devdari watershed, PaturTq., Akola District, Maharashtra. Trends Bioscie 10(1)

Pande CB, Khadri SFR, Moharir NK, Patode RS (2017b) Assessment of groundwater potential zonation of Mahesh River basin Akola and Buldhana districts, Maharashtra, India using remote sensing and GIS techniques. Sustain Water Resour Manag 1–15. https://doi.org/10.1007/s40899-017-0193-5

Pande CB, Moharir KN, Pande R (2018) Assessment of morphometric and hypsometric study for watershed development using spatial technology-a case study of Wardha river basin in the Maharashtra, India. Int J River Basin Manag 4(4):1–36

Pande CB, Moharir KN, Singh SK, Varade AM (2019) An integrated approach to delineate the groundwater potential zones in Devdari watershed area of Akola district, Maharashtra, Central India in Environment, Development, and Sustainability Springer Journal. https://doi.org/10.1007/s10668-019-00409-1

Pande CB (2020a) Introduction. In: Sustainable watershed development. SpringerBriefs in water science and technology. Springer, Cham. https://doi.org/10.1007/978-3-030-47244-3_1

Pande CB (2020b) Watershed management and development. In: Sustainable watershed development. SpringerBriefs in water science and technology. Springer, Cham. https://doi.org/10.1007/978-3-030-47244-3_2

Pande CB (2020c) Thematic mapping for watershed development. In: Sustainable watershed development. SpringerBriefs in water science and technology. Springer, Cham. https://doi.org/10.1007/978-3-030-47244-3_3

Pande CB (2020d) Sustainable watershed development planning. In: Sustainable watershed development. SpringerBriefs in water science and technology. Springer, Cham. https://doi.org/10.1007/978-3-030-47244-3_4

Patode PCB, Nagdeve BM, Moharir KN, Wankhade RM (2017) Planning of conservation measures for watershed management and development by using geospatial technology–a case study of Patur Watershed in Akola District of Maharashtra. Current World Environ 12(3):708–716

Schumn SA (1956) Evolution of drainage systems and slopes in badlands at Perth Amboy, New Jersey. Geol Soc Am Bull 67(5):597–646

Singh S, Singh MC (1997) Morphometric analysis of Kanhar River basin. Nat Geo J India 43(1):31–43

Strahler AN (1957) Quantitative analysis of watershed geomorphology. Trans Am Geophys Union 38(6):913–920. https://doi.org/10.1029/TR038i006p00913

Strahler AN (1964) Quantitave geomorphology of drainage basins and channel networks. In: Handbook of applied hydrology, section 4-II. McGraw Hill Book Company, New York

Vittala SS, Govindaiah S, Gowda HH (2008) Prioritization of subwatersheds for sustainable development and management of natural resources: an integrated approach using remote sensing. GIS Socio-Econon Data Curr Sci 95(3):345–354

Chapter 3
Study of Morphological Changes in Deltaic River of Odisha Using GIS

Anil Kumar Kar, Rashmirekha Hembram, and Harihar Mohanty

Abstract Delta of Odisha is highly populated and commercially, agriculturally developed. The vast agricultural land and habitation near the meandering part of the river are subjected to the wrath of the changing course of the river. Flood is a regular phenomenon and saucer-shaped morphology makes the area vulnerable to flood due to overtopping and breaching as well as waterlogging. Many times, such incidents had occurred in such areas engulfing vast agricultural areas and villages. Therefore, structural flood protection measures are needed to prevent such morphological catastrophes. Few strategic locations with respect to morphological changes were identified on the deltaic river of Odisha by using Remote Sensing and Geographic Information System. This paper focuses on the change detection of a deltaic river and understanding the morphological response of channel variation in discharge characteristics over a period of time. Using ArcGIS software and toposheet, spatiotemporal variation of land use and land cover (LULC) scenario is studied. River instability is analyzed using Mueller's Sinuosity index. The results predict the areas to be at risk of erosion and progress of accretion in the future.

Keywords Change detection · Meandering · GIS · LULC · Mueller's sinuosity index

---

A. K. Kar (✉) · R. Hembram
Department of Civil Engineering, VSSUT, Burla, Sambalpur 768018, Odisha, India
e-mail: anilkarhy@gmail.com

R. Hembram
e-mail: rekharashmi80@gmail.com

H. Mohanty
Water Resources Department, Government of Odisha, Bhubaneswar 751001, India
e-mail: 2011mohanty@gmail.com

A. Pandey et al. (eds.), *Water Management and Water Governance*, Water Science and Technology Library 96,
https://doi.org/10.1007/978-3-030-58051-3_3

3.1 Introduction

Riverbank areas at delta are most vulnerable due to continuous exposure to the gradual and sudden physical process of anthropogenic and natural processes, such as tidal wave action, sea-level rise, flooding, erosion-sedimentation and embankment subsidence. Each and every deltaic river characterized by several morphogenetic features evolved under the impact of natural and human activity factors. Many researchers put their attention to river bank changes, land use changes, economic growth, climatic changes and natural phenomena. The river course change relative to land is seasonal changes of river morphology considerably significant but anthropogenic causes had a greater influence in the changes of river morphology (Nandi and Ghosh 2014). In order to identify the rate of morphological changes along the river bank and river bank shifting over time to time from spatial–temporal study. The river instability can be detected from "channel pattern" the term is used to describe the plain view of a river channel from an airplane.

The term 'Sinuous' or 'Sinuosity' is generally used for the purpose, which is the form between the straight and meandering character of the river. Sinuosity is the ratio of thalweg length to the airline distance which is defined as the degree of changes of the river (Ezizshi 1999) and Muller's sinuosity indexes are derived by dividing the length of a reach as measured along a channel by the length of reach as measured along the valley (Muller 1968). Morphological changing channel pattern of Damodar River was analyzed for its spatial–temporal changes (Ghosh and Mistri 2012). Meandering and Braiding character were compared on Ganga River (Pal 2017). Land use land changes (LULC) along with human activity, economical activity/development are measured for river shifting (Fan et al. 2008). Shoreline changes detection of the Mississippi delta also done by Blum and Roberts (2009) and Wetland loss in world deltas by Coleman et al. (2008). Variation of sinuosity index and radius of curvature calculated by the remote sensing and GIS software at Khowai river by Debanath et al. (2017).

In this research work, we have considered the deltaic river of Odisha, such as the Budhabalang River and Baitarani River which have undergone the process of channel morphological changes at certain critical points. For the change detection measurement, GIS software and Google Earth images are used for spatial analysis and Muller sinuosity index. This study speaks about important features like (1) understanding the morphology of the deltaic river of Odisha. (2) Historical changes effects on locality and consequences. (3) Types of the river as a term of sinuosity using of Muller sinuosity index. (4) Identify the vulnerability area of the river. (5) Remedial measures and immediate solution.

3.2 Study Area

The area of this study is the deltaic river of a part of Budhabalanga River and the Baitarani River in Odisha state of India. Budhabalanga River lies between latitude 21°55′ to 21°53′and longitude 86°23′30″ to 86°23′ and it flows through the district Mayurbhanj and Balasore in Odisha, Bank of this river is steep and it falls into the Bay of Bengal. Populations of this basin area are 20, 08,156 and total catchment is 4840 km^2 and annual rainfall is 3295 mm maximum and 544 mm minimum. Baitarani River lies between latitude 21°22′ to 21°15′ and longitude 86°21′ to 86°7′ and one of the six major rivers and it is a source of agricultural irrigation. The total catchment area of the Baitarani River is 14,218 km^2 with a population 43, 63,092 and annual rainfall is 3094 mm maximum and 642 mm minimum. Flooding is a regular phenomenon in the basin area and huge loss occurs to life and property. Riverbank also changes due to the physical processes, so it is realized that the part of the river Budhabalanga (Fig. 3.1) at Barehipani fall (Jurunda to Similipal reserved forest) and the River Baitarani (Fig. 3.2) at Kamando to Palaspur area are vulnerable to the morphological changes. So that necessary protection measures can be taken regarding this in order to save valuable lives and property.

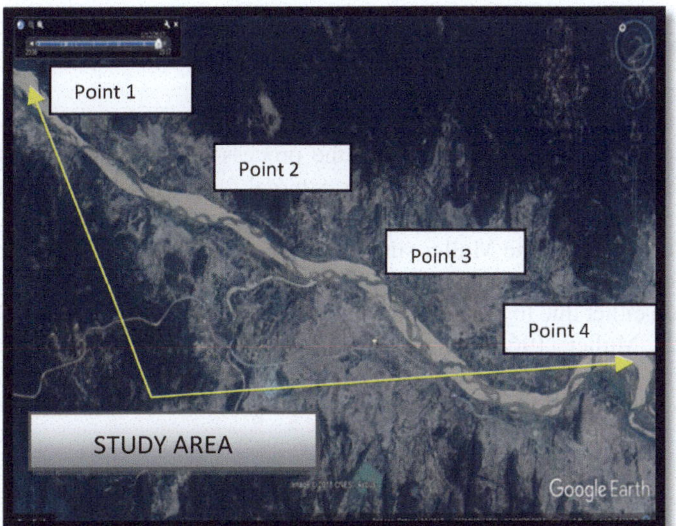

Fig. 3.1 Google Earth images of Baitarani River

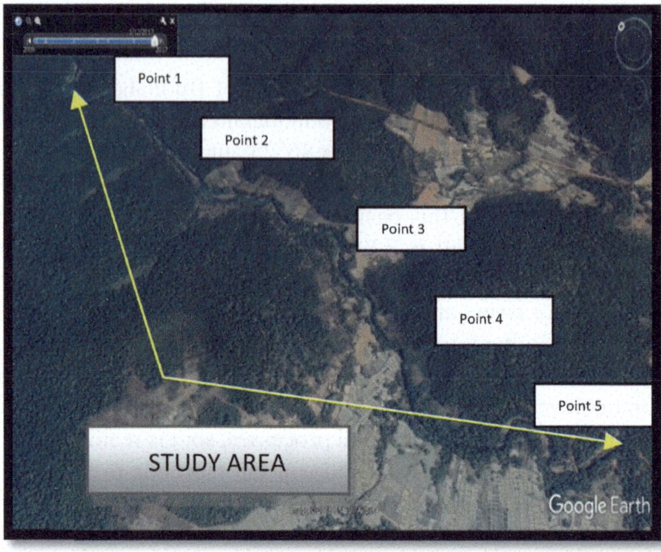

Fig. 3.2 Google Earth images of Budhabalanga River

3.3 Methodology

Due to the physical activity and hydraulic properties the river changes its morphology from the previous path of the river and the process continues. It directly affects the hydrogeomorphic dynamics and other external factors which is well measured by the term 'Sinuosity index'. Other indexes can be measured indirectly. Another mostly applied index is the Muller sinuosity index, which has universal acceptance because of its account for what percentage of a channel departure from a straight-line course is safe either due to hydraulic factor or topographic interferences (Ghosh and Mistri 2012). Muller (1968) has redefined the index as hydraulic sinuosity index (i.e., that freely developed by the channel influenced by valley-wall alignment) and topographic sinuosity (i.e., that imparted by the geometry of the valley) shown in Fig. 3.3.

In Fig. 3.3,

$$\text{CL(Channel Length)} = \text{CD}$$

$$\text{Air Length} = \text{XY}$$

$$\text{VL(Valley Length)} = \frac{\text{AA}' + \text{BB}'}{2}$$

$$\text{CI(Channel Index)} = \frac{\text{CL}}{\text{Air Length}}$$

Fig. 3.3 Parameters taken for Muller's Sinuosity Index

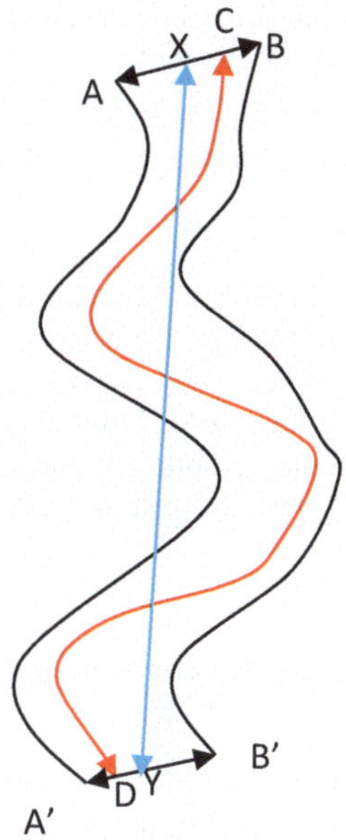

$$\text{VI(Valley Index)} = \frac{\text{VL}}{\text{Air Length}}$$

$$\text{HSI(Hydraulic Sinuosity Index)} = \text{\% equivalent of} \frac{\text{CI} - \text{VI}}{\text{CI} - 1}$$

$$\text{TSI(Topographic Sinuosity Index)} = \text{\% equivalent of} \frac{\text{VI} - 1}{\text{CI} - 1}$$

$$\text{SSI(Standard Sinuosiry Index)} = \frac{\text{CI}}{\text{VI}}$$

where

CL	The length of channel (thalweg) in the stream understudy.
VL	The valley length along with steam, the length of a line which is everywhere midway between the base of the valley walls.
Air Length	The shortest air distance between the source and mouth of the stream.

The standard sinuosity values are shown in Table 3.1.

In this paper, the study area of the base topography has sub-divided into a number of parts. The river Budhabalanga divided into the four reaches from near Barehipani fall (Jurunda, to near Simlipal reserved forest). The Baitarani River divided into three reaches from Kamando, to Palaspur. GIS software and Google Earth images at different time intervals have been used for the spatiotemporal analysis. Taking the Survey of India toposheet of the year 1974–76 for river Budhabalanga and toposheet of the year 1981–82 for River Baitarani the changes are determined. The toposheets are geo-referenced in WGS-1984 projection system and then the river bank outlines

Table 3.1 Channel pattern in term of Sinuosity

Type	Sinuosity
Straight	<1.05
Sinuous	>1.05
Meandering	>1.5
Braided	>1.3
Anastomosing	>2.0

are digitized. Then the channel length, valley length, and air length of the channel are measured and for further calculation. The topographic images and temporal Google Earth's images are compared to analyze the change detection.

3.4 Results and Discussion

Here, the change detection of riverbank, the topographic images of 1976 and 1982 is taken as initial data and the 2017 Google Earth images as the final one. River Budhabalanga divided into four reaches, from near Barehipani fall Juruda to Nikhirda, Nikhirda to Hill block, Hill block to Noana, and Noana to near Simlipal reserved forest for a total length of 2.95 km. River Baitarani divide into three reaches, from Kamando to Bandha, Bandha to Biragobindapur, Biragobindapur to Palaspur for a total length of 16 km.

Different indices of river Budhabalanga are first calculated as a base from toposheet of 1974–76 and recorded in Table 3.2a and then from Google Earth image of 2009 (Table 3.2b) and again from the latest available Google Earth image of 2017 (Table 3.2c). These four reaches are considered as one and the same indices are calculated and recorded in Table 3.3.

The significant deviations in the Standard Sinuosity Index (SSI) over these reaches are plotted in Fig. 3.4a and the temporal variations of SSI are recorded in Fig. 3.4b.

Similarly, different indices of river Baitarani are calculated as a base from toposheet of 1974–76 and recorded in Table 3.4a and then from Google Earth image of 2009 (Table 3.4b) and again from the latest available Google Earth image of 2017 (Table 3.4c). These four reaches are considered as one and the same indices are calculated and recorded in Table 3.5.

The significant deviations in the Standard Sinuosity Index (SSI) over these reaches are plotted in Fig. 3.5a and the temporal variations of SSI are recorded in Fig. 3.5b.

From Tables 3.2 and 3.3, the idea about four reaches of Budhabalanga River in 1974–76 have been studied with increasing Channel Index (CI) from1.18 to 2.54. Similarly, Valley Index (VI) is increased from 1.13 to 2.5. But Standard Sinuosity Index (SSI) of four reaches is decreasing from 1.04 to 1.02. In the year 1974–76, the river is partly sinuous from Hillblock to Noana area but after crossing the Noana it has moved in narrow path towards the Similipal reserved forest area. Standard Sinuosity Index (SSI) of four reaches is constantly decreased from 27 to 3% and Topographic

Table 3.2 Sinuosity values of four reaches of River Budhabalanga (Length in km)

Sl. No	Name of the reaches	CL	VL	AIR length	CI	VI	SSI	HIS (%)	TSI (%)
(a) Toposheet (1974–76)									
1	Bend 1 (Near Barehipani fall to Nikhira)	0.72	0.69	0.61	1.18	1.13	1.04	27	73
2	Bend 2 Nikhirda to Hillblock	0.74	0.73	0.7	1.06	1.04	1.01	25	75
3	Bend 3 Hillblock to Noana	1.1	1.05	0.95	1.16	1.11	1.05	33	67
4	Bend 4 Noana to Near Simlipal reserved forest	1.22	1.2	0.48	2.54	2.5	1.02	3	97
(b) Google Earth images 2009									
1	Bend 1 (Near Barehipani fall to Nikhira)	0.94	0.86	0.74	1.27	1.16	1.09	40	60
2	Bend 2 Nikhirda to Hillblock	0.65	0.61	0.57	1.14	1.07	1.06	50	50
3	Bend 3 Hillblock to Noana	1.1	1.05	0.91	1.21	1.15	1.05	26	74
4	Bend 4 Noana to Near Simlipal reserved forest	1.41	1.38	0.98	1.44	1.41	1.02	7	93
(c) Google Earth images 2017									
1	Bend 1 (Near Barehipani fall to Nikhira)	0.69	0.64	0.58	1.19	1.10	1.08	45	55
2	Bend 2 (Nikhirda to Hillblock)	1	0.96	0.91	1.09	1.05	1.04	44	56
3	Bend 3 (Hillblock to Noana)	0.80	0.71	0.68	1.17	1.04	1.13	75	25
4	Bend 4 (Noana to Near Simlipal reserved forest)	1.2	0.97	0.84	1.43	1.15	1.24	64	36

Table 3.3 Total SSI values of Budhabalanga River (Barehipani fall to Simlipal reserved forest)

Sl. No	Year	CL	VL	AIR length	CI	VI	SSI	HIS (%)	TSI (%)
1	1974–76	3.78	3.67	2.74	1.38	1.34	1.03	11	89
2	2009	3.69	3.28	3.01	1.23	1.09	1.13	61	39
3	2017	4.1	3.9	3.2	1.28	1.22	1.05	21	79

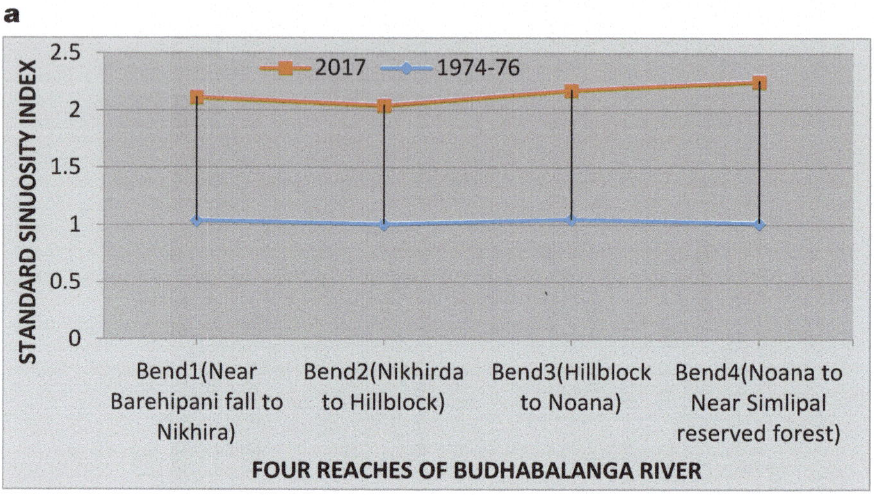

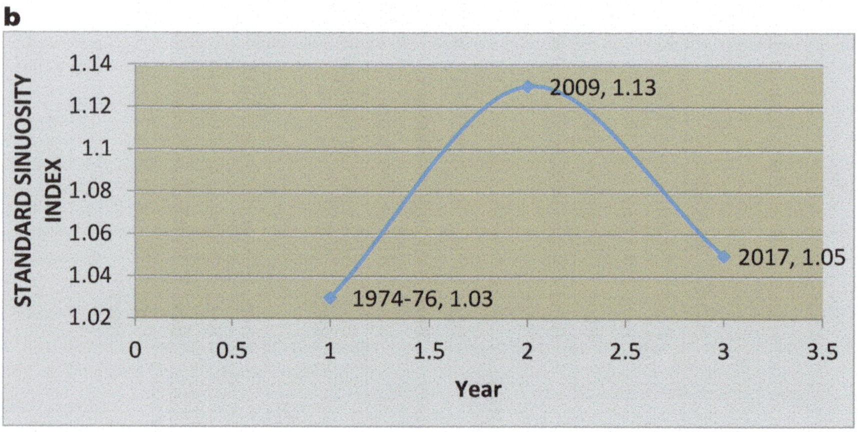

Fig. 3.4 **a** Significant deviation of SSI of Budhabalanga River in between 1976–2017. **b** Temporal patterns of changing SSI for Budhabalanga River

Table 3.4 Sinuosity values of three reaches of River Baitarani (Length in km)

SL NO	Name of the reaches	CL	VL	AIR length	CI	VI	SSI	HIS (%)	TSI (%)
(a) Toposheet (1981–82)									
1	Bend 1 (Kamando to Bandha)	5.84	5.52	5.48	1.07	1.01	1.05	89	11
2	Bend 2 (Bandha to Birgobindapur)	6.8	6.57	6.38	1.07	1.03	1.03	55	45
3	Bend 3 (Birgobindapur to Palaspur)	4.81	4.73	4.39	1.09	1.07	1.02	19	81
(b) Google Earth images 2009									
1	Bend 1 (Kamando to Bandha)	6.34	5.74	5.73	1.11	1.00	1.1	98	2
2	Bend 2 (Bandha to Birgobindapur)	7.12	6.68	6.48	1.09	1.03	1.06	69	31
3	Bend 3 (Birgobindapur to Palaspur)	4.86	4.59	4.46	1.09	1.03	1.06	67	33
(c) Google Earth images 2017									
1	Bend 1 (Kamando to Bandha)	6.1	5.68	5.68	1.07	1.00	1.07	98	2
2	Bend 2 (Bandha to Birgobindapur)	7.19	6.72	6.51	1.10	1.03	1.07	69	31
3	Bend 3 (Birgobindapur to Palaspur)	4.95	4.42	3.88	1.26	1.12	1.12	50	50

Table 3.5 Total SSI values of Baitarani River (Kamando to Palaspur)

Sl. No	Year	CL	VL	AIR length	CI	VI	SSI	HIS (%)	TSI (%)
1	1981–82	17.45	16.82	16.25	1.07	1.03	1.04	53	47
2	2009	18.32	17.01	16.67	1.09	1.02	1.07	79	21
3	2017	18.24	16.82	16.07	1.15	1.05	1.08	65	35

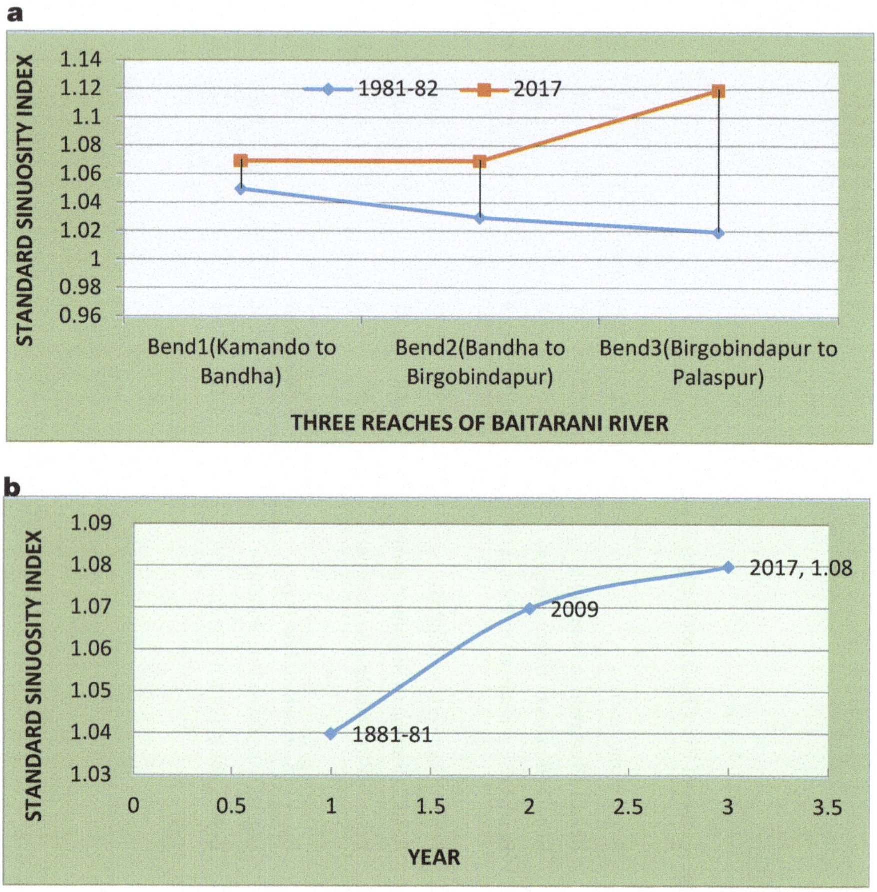

Fig. 3.5 **a** Significant deviation of SSI of Baitarani River in between 1981–82 to 2017. **b** Temporal patterns of changing SSO for Baitarani River

Sinuosity Index (TSI) is increased from 73 to 97%. Which indicates that river is an almost natural condition and sinuosity of the river is controlled by the topographic factor, tight valley alignment, cohesive bank material, coarse bed configuration, and other hydraulic factors. The reaches four HIS value is 3 and TSI value is 97 which indicates that topographic influence is more than the hydraulic influence.

After 33 years of time span 2009, there are rapidly changes in the channel pattern and the river is getting sinuous pattern due to the changing channel index and valley index. From Barehipani fall to Noana sinuosity value is more than and equal to 1.05, but approximately the same as the previous condition because the river HIS is increased. But at the reaches, four topographic influences are more.

In 2017 surprisingly SSI is increased (<1.05) due to the CI and VI is changed. Also, HIS is increased gradually and TSI decreases. Sinuosity index is varied 1.24 (meandering), it means the river widens its valley and developing within the river with high influence of hydraulic factor. After taking the entire channel segment (Fig. 3.4b), it obtained that SSI of 2009 is 1.13 which was only 1.03 in 1974–76 years.

In between 1976 to 2017 the channel length of the river Budhabalanga is increased 3.78 to 4.1 km and the valley length also increased 3.67 to 3.9 km. from Table 3.2 temporal channel pattern of four reaches have found that period of 1976 river is very much straight in pane and more influenced of TSI. But in 2007, river is sinuous value is more, which is the sign of braiding and TSI is dominated by HIS. It implies the increasing flow velocity in monsoon, thalweg shifting lateral migration, and probability of changing course or bank are occurred.

Similarly, Tables 3.4 and 3.5 provide the base morphology about river Baitarani basing on toposheet of 1981–82. The channel index and valley index are increased and sinuosity value is decreased. In this case, TSI is dominated HIS. Therefore, it is indicating that the river is almost straight. But in the year 2009–2017, TSI dominated by HIS and SSI value is obtained more than 1.05 which indicates that the river is meandering. At the first reach, it shows that HIS is 98 and TSI is 2, which indicates that the hydraulic influence is more than topographic influence. Taking that of the entire channel segment (Fig. 3.5b), it obtained that SSI of 2017 is 1.08 which was only 1.04 in 1982. It is the sign of meandering which may due to the high velocity, excess water discharge and other factors. Therefore, after some year river may change its course again and again and it needs protection works at salient points.

Channel sinuosity appears to increase with the water and sediment discharge from straight to sinuous single channel but it decreases towards the transition to braiding parts of channel. During monsoon the excess water discharge and high velocity are resulting in initial degradation (thalweg shifting), leading to channel widening (bank erosion) and aggradations (formation of island).

3.5 Conclusions

Generally, the natural alluvial river has high degree of meandering and SSI. From the above analysis, we found that in the year 1976 river Budhabalanga was straight due to the dominance of TSI. Similarly, in 1982 the river Baitarani shows the same nature. But in 2017 both the river's sinuosity is affected by hydraulic factors due to the dominance of HSI. Above study as per images of 2017 indicate that the river changes their course pattern with the time of span, so most venerable part of the river Budhabalanga is Hillblock area to Similipal reserved forest (SSI is 1.13 to 1.24) and that of river Baitarani is Biragobindapur to Palaspur (SSI is 1.13). As we know the upper part of the catchment of the river is having a steep slope, sediment generated due to high velocity, and gets deposited at the down steam thus reducing the river's capacity at downstream. Proper catchment treatment is highly necessary for sediment reduction.

The above study of the river shows the valleys are straight but thalweg show meandering pattern, which indicates channel is shifting frequently. Therefore, it is necessary to think about acceptable potential changes in channel shifting and the probability of overtopping, breaching of river in the future, and saving the vulnerable areas from engulfing within the river. Afforesting, providing different types of river training works, river straightening to diverting the flowing water and the preparedness strategies are some of the possible solutions for saving the valuable life and property at the vulnerable area.

References

Blum MD, Roberts HH (2009) Drowning of the mississippi delta due to in-sufficient sediment supply and global sea-level rise. Nat Geosci 2(7):488–491

Coleman JM, Huh OK, Braud D Jr (2008) Wetland loss in world deltas. J Coastal Res 24(sp1):1–14

Debnath J, Pan ND, Ahmed I, Bhowmik M (2017) Channel migration and its impact on land use/land cover using RS and GIS: a study on Khowai River of Tripura, North-East India. Egypt J Remote Sens Space Sci 20(2):197–210

Ezizshi AC (1999) An appraisals of the existing descriptive measures of river channel patterns. J Environ Sc 3(2):253–257

Fan FL, Wang YP, Wang ZS (2008) Temporal and spatial change detecting (1998–2003) and predicting of land use and land cover in core corridor of Pearl River Delta (China) by using TM and ETM+images. Environ Monit Assess 137:127–147

Ghosh S, Mistri B (2012) Hydrogeomorphic significance of sinuosity index in relation to river instability: a case study of Damodar River, West Bengal, India. Int J Adv Earth Sci 1(2):49–57

Muller JR (1968) An introduction to the Hydraulic and Topographic Sinuosity Index. J Annal Assoc Am Geogr 58:371–385

Nandi D, Ghosh C (2014) Morphodynamic changes around subarnarekha coast: Study using multi change satellite data. Asian J Sci Technol 5(20):620–624

Pal R (2017) Meandering-braiding aspects of the middle-lower part of the Gan-ga River, India. J Ind Geophys Union (May 2017), 21(3):191–197

Chapter 4
Land Use Classification Using Remotely Sensed Images: A Case Study of Eastern Sone Canal—Bihar

Zeenat Ara

Abstract Land cover/land use classification has evolved as a rapid, reliable, and competent tool to observe conditions and alterations in land cover/land use at a global scale. Land use classification maps have been widely used in water resources engineering as it influences the available water resources in an area. This paper presents the land use/land cover pattern of the eastern Sone River Basin of Bihar, India. Landsat ETM + satellite data was utilized for analyzing and classifying the land cover in the eastern Sone Canal command area. ERDAS imagine 9.0 software was used for processing and analysis of the imagery. False color composite (FCC) maps were developed to identify the various land features, and supervised classification was performed using maximum likelihood technique. Land use/land cover map prepared for the study region showed seven classes, namely agricultural land, forest cover, fallow land, settlements, river bed, waterbody, and barren land. The overall accuracy obtained for the classification technique used was assessed to be 80% with overall kappa statistics as 0.749.

Keywords Land use · Land cover · GIS · Remote sensing · Eastern Sone Canal River basin

4.1 Introduction

The term land cover reveals the surface cover of certain pieces of land. Land use classification is an important component of water resource engineering as the runoff response from the catchment area depends on its land use characteristic (Muzzammil et al. 2015, 2018; Zakwan et al. 2016; Zakwan 2016). Land use of an area also exercises a significant influence on the volume of groundwater available in an area. As such, land use of an area influences the entire ecosystem of an area. Authentic updated knowledge of land use and the environmental conditions of an area play a vital

Z. Ara (✉)
Civil Engineering Deparment, MANUU, Hyderabad 500032, India
e-mail: zeenatiitd@rediffmail.com

A. Pandey et al. (eds.), *Water Management and Water Governance*, Water Science and Technology Library 96,
https://doi.org/10.1007/978-3-030-58051-3_4

role in environmental planning and management (Zakwan and Muzzammil 2016; Pandey et al. 2018; Ara and Zakwan 2018). Realizing the importance of land use characteristics of a catchment, several attempts have been made for the classification of land using the remote sensing technique in the recent past.

Haralick and Shanmugam (1973) suggested that images may be grouped on the basis of textural data. Contrast, angular second moment, and inverse difference moment are the various measures that are generally utilized to analyze the images. Jensen et al. (1979) applied the measures suggested by Haralick and Shanmugam (1973) for analyzing the Landsat imagery of land cover which resulted in nearly 80% efficiency. Campbell (1987) discussed that the most commonly used classification techniques are unsupervised and supervised classification. Unsupervised classification has been defined as the classification of natural classes, structures from the multispectral image while the supervised classification is defined as processes of utilizing known patterns for assigning pixels to classes. Fung and Chan (1994) reported that each land use class is associated with specific intensity, based on which spectral groups are distinguished for image classification. Reddy and Reddy (1996) did a performance analysis of IRS bands for land use/land cover classification systems using the maximum likelihood classifier. Bastiaanssen and Bos (1999) suggested the application of remote sensing technique for evaluating irrigation performance indicators, and this approach defines spatial scales more clearly in comparison with the classical approach based on collected discharge measurements. Bastiaanssen and Bos (1999) and Sakthivadivel et al. (1999) analyzed the efficiency of the Bhakra Irrigation System, India, based on RS data, reported spatial variation of crop productivity, and aligned this variability more toward hydrological setting instead of water delivery efficiency. Su (2000) presented information about land use classification (LUC) system. In this system, different land use, land cover areas are classified according to their implications on the hydrologic process. Chang et al. (2017) used Landsat image for estimating evapotranspiration. The specific objective of the present study is to carry out land classification through a methodological framework to obtain better land use classification and highlight its application to a real-life study area (part of Sone Canal command area).

4.2 Description of Study Area

The study area selected for this paper is Eastern Sone Canal (Fig. 4.1), part of the Sone Canal Irrigation Project. Dating back to 1874, Sone Canal system is among the oldest canal in India. This canal originates from Sone River. River Sone has its origin in the plateau area of Amarkantak in the District of Sahdol in Madhya Pradesh very near to the River Narmada and Mahanadi and mainly passes through hilly areas until it enters the Gangetic valley opposite Akbarpur in Rohtas District in Bihar (Zakwan

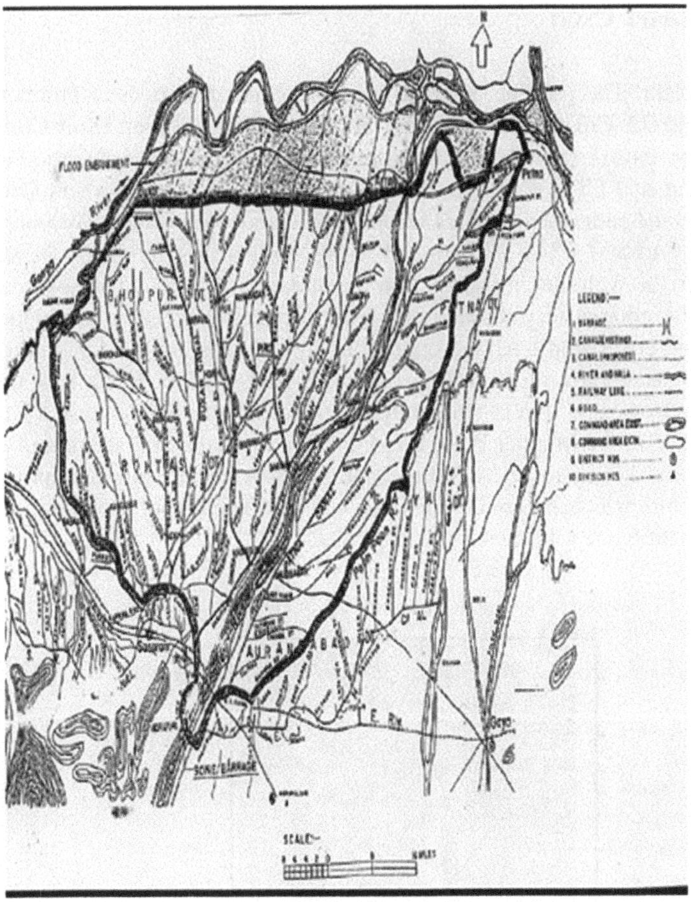

Fig. 4.1 Sone Canal command area

et al. 2018). It runs along the districts of Bhojpur, Aurangabad, Gaya Jehanabad, and Arwal until it merges in the River Ganges near Maner in Patna district. The total catchment area of Sone River is about 70,196 sq km of which 15,965 sq km lies in Bihar. The study area is bounded by latitude $2412'0''$ to $25°14'0''$ ‘and longitude $84°30'0''$ to $85°0'0''$. The climatic condition is generally hot except during the months of December to February. The area experiences heavy rainfall during the months of June to September. The average rainfall in the region is approximately 300 mm.

4.3 Dataset Used

Ancillary data: The present study utilized topological map sheet number 72-D/5, 72–D/9, and 72–C/8 with a scale of 1:50,000 (SOI), soil map (Soil Conservation Department, Patna), rainfall data (Statistic and Monitoring Department, Patna).

The Landsat 7 ETM + images dated January 25, 2001, were downloaded for the full scene of the path and row 141/43 from global land cover facility Web site. Satellite image of Landsat 7 ETM + .Path-141, row-043, and resolution–30 m was downloaded from the Web site (https://glcf.umd.edu/data/) (Fig. 4.2). The raw images were processed for geometric, radiometric, and atmospheric corrections as per the requirement of land cover study. All the operations from importing the data to the analysis of the data were carried out in RS software's ERDAS 9.0 (Earth Resources Data Analysis System) and ILWIS (Integrated Land and Water Information System). The Enhanced Thematic Mapper Plus (ETM +) is a multispectral scanning radiometer carried onboard of Landsat 7 satellite. It provides better measurement precision and spatial resolution as compared to the other sensors. The sensor provides seven bands with two different resolutions.

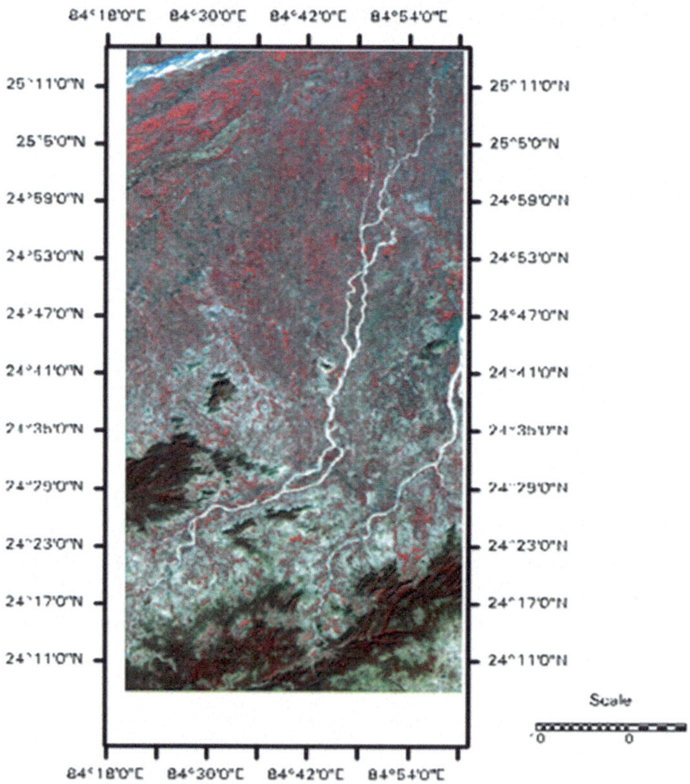

Fig. 4.2 Satellite image of study area

4.4 Methodology

The image can be interpreted at various levels to start with FCC image, that is the image produced with no optional processing, may be interpreted. Further, the image may be classified into groups based on similarity/dissimilarity indices. Similarity indices are decided based on one or other set of distant parameters. Alternatively, after initial grouping, one may choose one or more specific groups (clusters) of interest and obtain more information on these thereby refining the possibilities for interpretation.

4.4.1 Normalized Difference Vegetation Index (NDVI)

For determining the intensity of green cover over a piece of land, observations were made to identify various colors based on the wavelength of visible and near-infrared sunlight reflected from the vegetation. NDVI was estimated based on reflections of visible and near-infrared (NIR) of vegetal cover. Thick vegetal cover absorbs a major proportion of visible light and reflects almost all the NIR radiations. On the other hand, scanty vegetal cover reflects a major proportion of visible light and absorbs NIR radiations. NDVI values for a given pixel varies from (−1) to (+1), but none of the land cover areas gave the value of zero. Zero value of NDVI stans for the complete absence of vegetal cover while values adjacent to + 1 (0.1–0.6) represent the maximum density of green leaves. NDVI is the ratio of the difference and aggregate of the reflections of NIR emissions and red emissions as presented in Eq. 4.1. For densely vegetated regions, the NDVI spans over a range of 0.1–0.6, depending on the thickness and greenness of vegetal cover. Clouds, water, and snow are associated with greater visible reflections as compared to NIR reflections and as such represent negative NDVI. Identical reflectance for NIR and visible bands are observed for rocks, boulders, and bare soil regions resulting in NDVI close to zero.

$$\mathrm{NDVI} = (\mathrm{NIR} - \mathrm{RED})/(\mathrm{NIR} + \mathrm{RED}) \tag{4.1}$$

Equation (4.1) is more appropriate for Landsat TM images, and for IRS LISS III, we can use

$$\mathrm{NDVI} = (\mathrm{SWIR} - \mathrm{NIR})/(\mathrm{SWIR} + \mathrm{NIR}) \tag{4.2}$$

NDVI values significantly decline as the water stress increases in the vegetal cover. However, when available moisture declines spatially (resulting from geographic variability of climate) or temporally (because of drought stress), AET rates are replaced by PET rates leading to the cutback of photosynthesis activity which boosts the reflectance leading to an overall decline in NDVI values. With an increase in water stress, NDVI declines steeply from 0.9 to 0.

4.4.2 Application to Sone Canal Command Area

The satellite data was used to prepare the landcover map for the Sone Canal command area. First four bands of the ETM + sensor were required for this study; hence, these bands were downloaded and imported to ERDAS 9.0. A sub-map of Landsat image of the Sone Canal command area was extracted and geo-referenced utilizing the toposheet of Sone Canal command area or reprojects the image using ERDAS 9.0 software.

4.5 Results and Discussion

The methodology described earlier was applied to Landsat image of Eastern Sone Canal area. The satellite image is shown in Figs. 4.1 and 4.2. An FCC was prepared with R: G: B as band 4:3:2. The graph for number of pixels with pixel values is shown in Figs. 4.3, 4.4, and 4.5 for red, green, and blue bands, respectively.

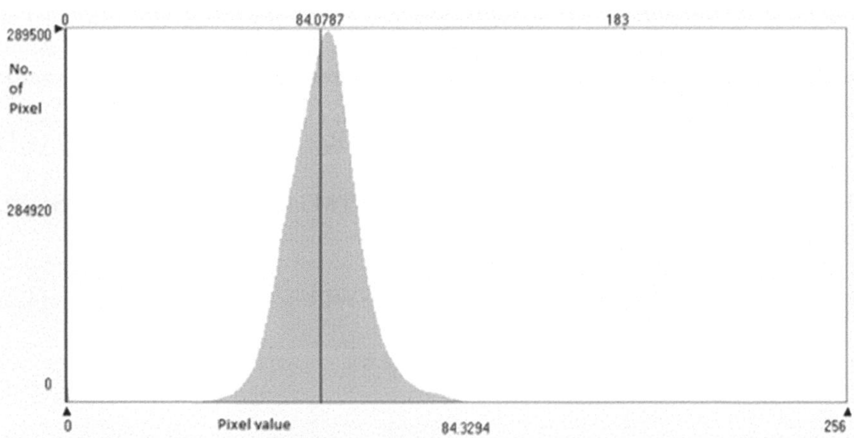

Fig. 4.3 Pixel value versus no. of pixels for red (Band 4)

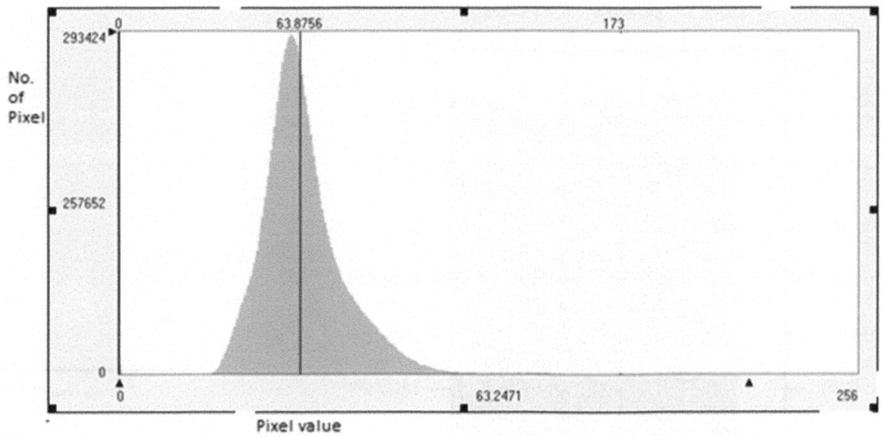

Fig. 4.4 Pixel value versus no. of pixels for green (Band 3)

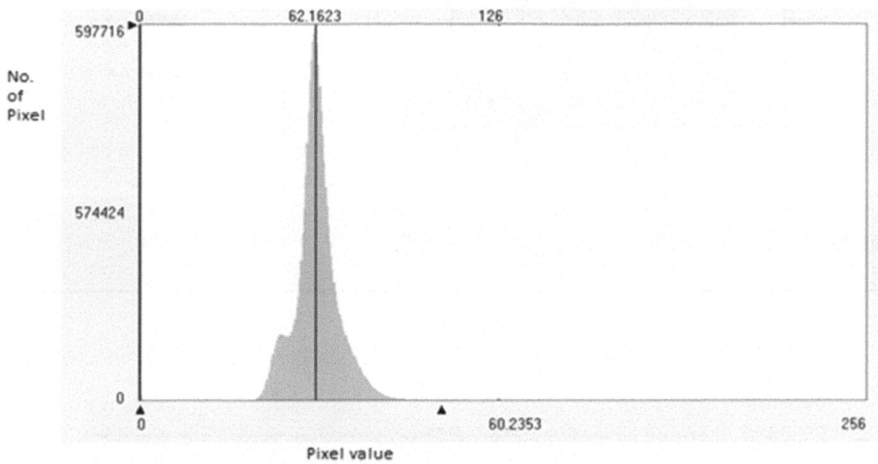

Fig. 4.5 Pixel value versus no. of pixels for blue (Band 2)

4.5.1 Land Use Classification

The image classification is therefore an important tool for examination of the digital images. Then, supervised classification of the image was performed based on maximum likelihood technique available in ERDAS 9.0. After performing image processing operation in ERDAS software, land use/land cover maps were created as shown in Fig. 4.6 while the percentage occupation is shown in Fig. 4.7. Of the classified images, the maximum likelihood method was found to be more applicable and reliable for satellite image classification purposes. Based on the results of land use, classification the Sone Canal command area was classified into seven different land

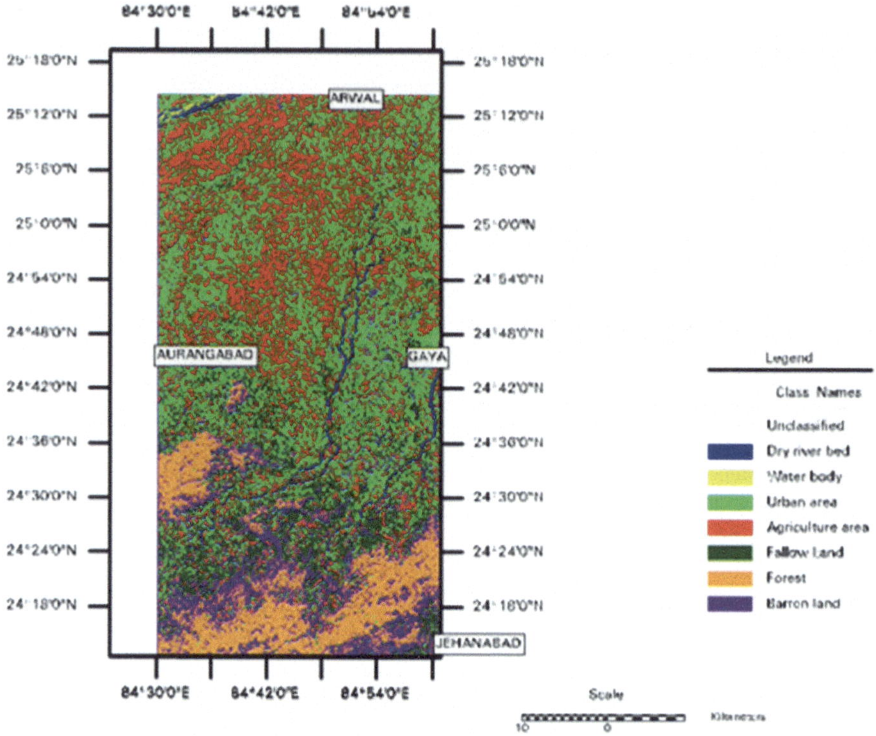

Fig. 4.6 Land use/land cover map of the study area

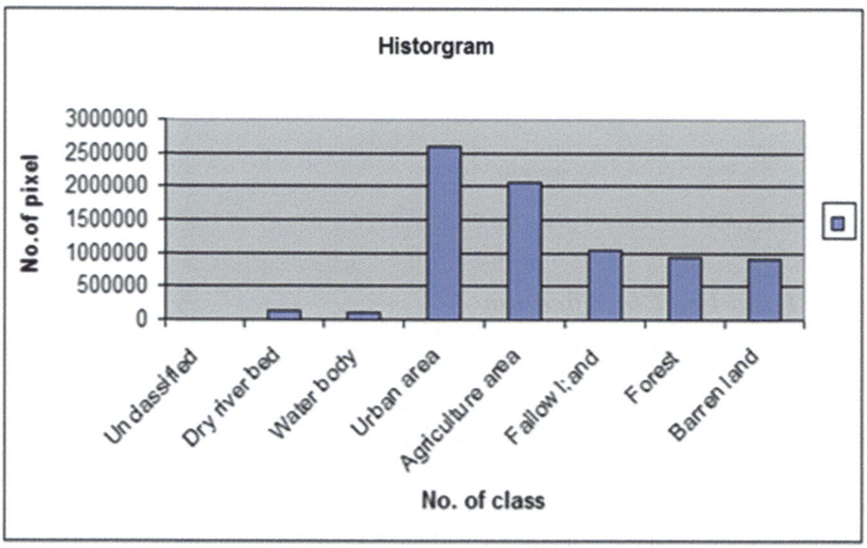

Fig. 4.7 Different types of land use with their area occupancy

Table 4.1 Comparison of actual and computed values for various class area

Sr. No	Class_Name	Actual area (km^2)	Computed area (km^2)	Error (%)
1	Agriculture area	1491.91	1440.08	3.4
2	Urban area		1813.62	
3	Forest	712.37	654.33	8.1
4	Barren land	676.72	626.554	7.4
5	Dry river bed		87.540	
6	Water body		79.39	
7	Fallow land	758.743	723.57	4.6

use classes. The total calculated area of these classes was calculated as 5425.08 km^2. The largest classified area of about 1800 km^2 (33%) was identified as urban area. Table 4.1 presents a comparison of land use data to computed values. NDVI values for the Eastern Sone Canal command are ranged from −0.16 to 0.67. Based on the results of the present study, three crops namely paddy, oilseed, and pulses could be segregated in the agriculture class in the Eastern Sone Canal command area.

4.5.2 Accuracy Assessment and Error Matrix

In order to determine the reliability of the classified image, "Accuracy Assessment" was performed in ERDAS IMAGINE 9.0. Hundred random points automatically selected from the referenced topographic map were used for the assessment and presented in Table 4.2 under the head "Accuracy Assessment". For accuracy assessment, the specific class values were matched to the classified class values in the accuracy matrix. The accuracy and kappa efficiency were estimated based on the user's accuracy and producer accuracy for every identified class. From the 100 arbitrary points checked, 80 points were found to be correctly classified, resulting in classification accuracy of about 80%. The kappa statistics of about 75% was obtained for the study area. Accuracy assessment reinforces our assumption that enhanced data assured better land use classification.

Fairly satisfying results were obtained in case of supervised classification for discriminating agriculture, forest, urban area, waterbody, barren land; but poor efficiency was obtained for classifying agricultural and urban resulting from significant variations in spectral signatures. The highest producers and user accuracy were achieved in the classification of fallow land (94.74%) and barren land (90%), respectively. The lowest producers and user accuracy were for agriculture area (83.33%) and urban area (72.22%), respectively). Mean pixel value of the seven classes generated by supervised classification is shown in Fig. 4.8. The roads of the Sone Canal command area were not identified because of close reflectance values with adjacent raster cells.

Table 4.2 Classification accuracy assessment report error matrix

Reference tables

Classified data	Unclassified	Dry river bed	Water body	Urban area
Unclassified	1	1	0	0
Dry river bed	0	0	0	0
Water body	0	0	0	0
Urban area	0	0	8	26
Agriculture area	0	0	2	2
Fallow land	0	0	0	0
Forest	0	0	1	0
Barren land	0	0	0	1
Column Total	1	1	11	1
Unclassified	0	0	0	0
Dry river bed	0	0	0	0
Water body	0	0	0	0
Urban area	0	1	0	1
Agriculture area	15	0	1	0
Fallow land	2	18	0	0
Forest	1	0	10	0
Barren land	0	0	0	9
Column Total	18	19	11	10

Accuracy totals

Class	Reference totals	Classified totals	Number correct	Producers accuracy	Users accuracy	Kappa
Unclassified	1	1	1	–	–	1.00
Dry river bed	1	1	1	100.00	100.00	1.00
Water body	11	0	0	–	–	0.00
Urban area	29	36	26	89.66	72.22	0.61
Agriculture area	18	20	15	83.33	85.00	0.70
Fallow land	19	20	18	94.74	90.00	0.88
Forest	11	12	10	90.91	83.33	0.81
Barren land	10	10	9	90.00	90.00	0.89

4.5.3 NDVI of Study Area

The Normalized Difference Vegetation Index (NDVI) is a measure of the amount of greenness in the vegetation cover of a watershed. Fig. 4.9a represents NDVI of the study area calculated as agricultural and non-agricultural part. Agriculture area is 26.54%, and non-agricultural area is 73.46%. NDVI ranges of non-agricultural

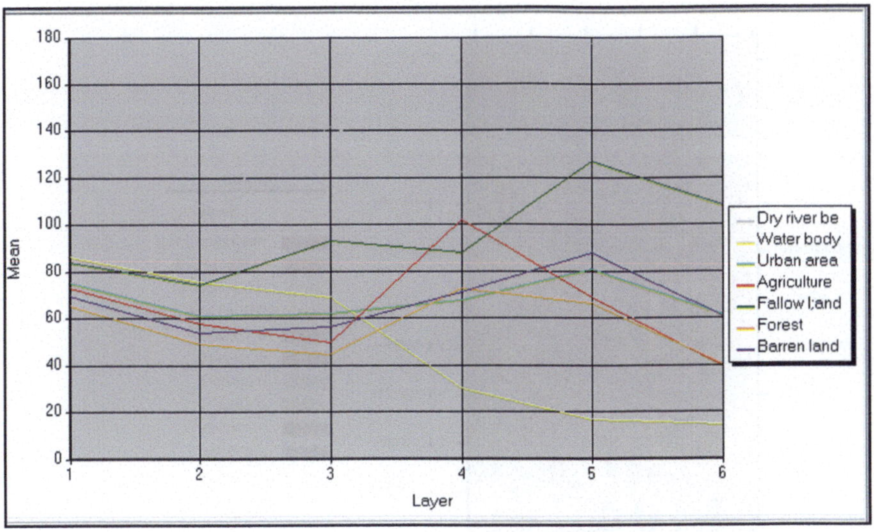

Fig. 4.8 Mean pixel value of the seven classes generated by supervised classification

area are (−0.634921 to 0.0981443) and agriculture area is (0.103271 to 0.39547) which is shown in Fig. 4.9b. The graph for NDVI value vs number pixel is shown in Fig. 4.10. Agriculture area contains paddy, oil, and pulse crops. In this, 62.022% area is paddy, 10.15% area is oil, and 27.82% area is a pulse as shown in Fig. 4.11.

4.6 Conclusions

Over the years, the application of remote sensing has increased in various aspects of hydrology and water resources. Land use classification maps have been widely used in water resource engineering as it influences the available water resources in an area. In the present study, Landsat imagery was used for land use/land cover map which shows agricultural area, urban area, forest cover, fallow land, dry river bed, and water body. Supervised classification of maximum likelihood technique is used in this study, seven land use/land cover classes were produced. NDVI value ranged from −0.16 to 0.67 from the study region during the image acquisition time. The overall accuracy for the image is 80% and overall kappa statistics 0.7494 which shows that the supervised classification using the enhanced data provides reliable results. Based on the results of the present study, three crops, namely paddy, oilseed, and pulses, could be segregated in the agriculture class in the Eastern Sone Canal command area.

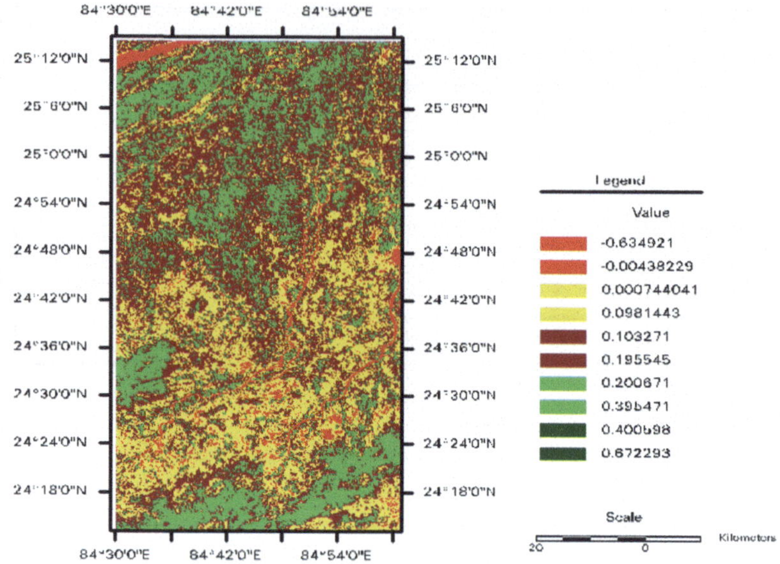

(a)

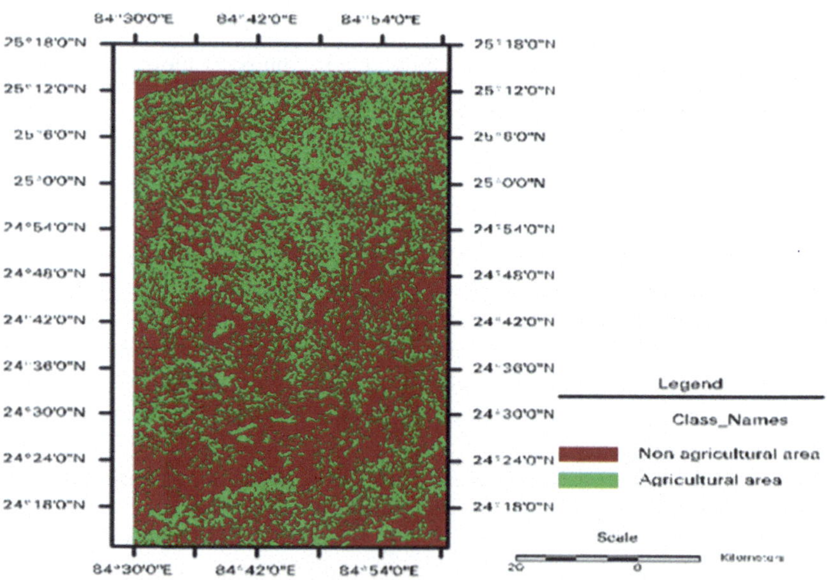

(b)

Fig. 4.9 **a** NDVI image of study area **b** NDVI image of agriculture and non-agricultural area

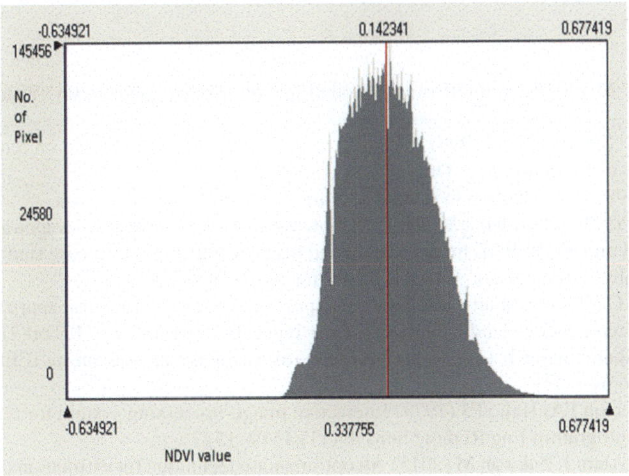

Fig. 4.10 NDVI value vs number of pixel

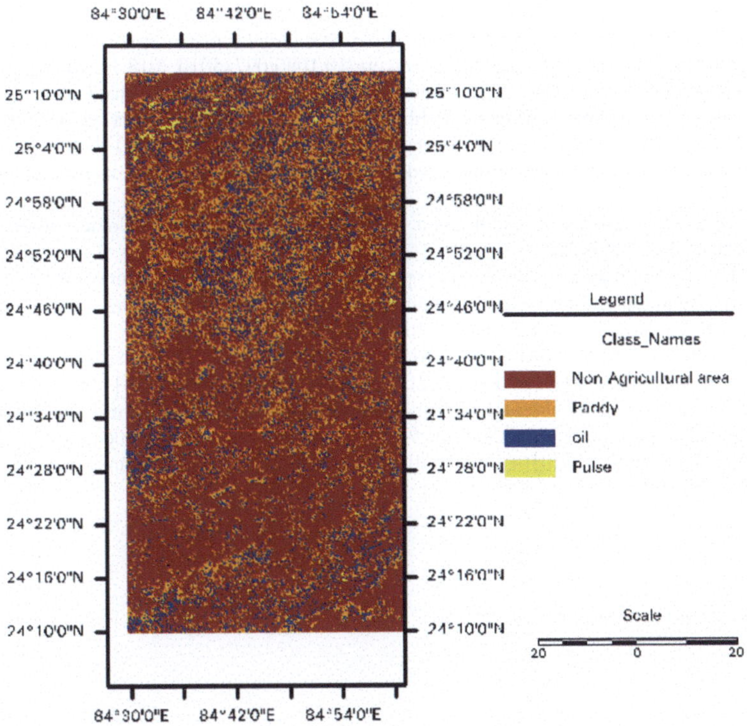

Fig. 4.11 NDVI image of non-agricultural area, paddy oil, and pulse

References

Ara Z, Zakwan M (2018) Rainfall runoff modelling for eastern canal basin. Water Energy Int 61(6):63–67
Bastiaanssen WGM, Bos MG (1999) Irrigation performance indicators based on remotely sensed data: a review of literature. Irrig Drainage Syst 13(4):291–311
Campbell JB (1987) Introduction to remote sensing. Guilford Press, New York
Chang Y, Ding Y, Zhao Q, Zhang S (2017) Remote estimation of terrestrial evapotranspiration by Landsat 5 TM and the SEBAL model in cold and high-altitude regions: a case study of the upper reach of the Shule River Basin. China Hydrol Proc 31(3):514–524
Fung T, Chan KC (1994) Spatial composition of spectral classes. A structural approach for image analysis of heterogeneous land-use & land-cover types. Photogram Engg RS 60(2)
Haralick RM, Shanmugam K (1973) Textural features for image classification. IEEE Trans Syst, Man Cybern 6:610–621
Jensen JR, Ennerson FA, Hajic EJ (1979) Interactive image-processing system for remote-sensing education. Photogramm Eng Remote Sens 45(11):1519–1527
Muzzammil M, Alam J, Zakwan M (2015) An optimization technique for estimation of rating curve parameters. Nation Symp Hydrol, New Delhi. 234–240
Muzzammil M, Alam J, Zakwan M (2018) A Spreadsheet approach for prediction of rating curve parameters. Hydrologic modeling. Springer, Singapore, pp 525–533
Pandey M, Zakwan M, Sharma PK, Ahmad Z (2018) Multiple linear regression and genetic algorithm approaches to predict temporal scour depth near circular pier in non-cohesive sediment. ISH J Hydraul Engg 26(1):96–103. https://doi.org/10.1080/09715010.2018.1457455
Reddy MA, Reddy KM (1996) Performance analysis of IRS—band for land use/land cover classification system using maximum Likelihood classifier., Int J Remote Sens 17(13):2505–2515
Sakthivadivel R, Thiruvengadachari S, Bastiaanssen UAW, Molden D (1999) Performance evaluation of the Bhakra irrigation system, India, using remote sensing and GIS techniques (Vol. 28)
Su Z (2000) Remote sensing of land use and vegetation for mesoscale hydrological studies. Int J Remote Sens 21(2):213–233
Zakwan M (2016) Estimation of runoff using optimization technique. Water Energy Int 59(8):42–44
Zakwan M (2018) Spreadsheet-based modelling of hysteresis-affected curves. Appl Water Sci 8(4):101–105. https://doi.org/10.1007/s13201-018-0745-3
Zakwan M, Ahmad Z, Sharief SMV (2018) Magnitude-frequency analysis for suspended sediment transport in ganga river. J Hydrol Eng 23(7). 10.1061/(ASCE)HE.1943-5584.0001671.
Zakwan M, Muzzammil M, Alam J (2016) Application of spreadsheet to estimate infiltration parameters. Perspect Sci 8:702–704. https://doi.org/10.1016/j.pisc.2016.06.064
Zakwan M, Muzzammil M (2016) Optimization approach for hydrologic channel routing. Water Energy Int 59(3):66–69

Chapter 5
Assessment of Multiple Satellite-Based Precipitation Estimates Over Muneru Watershed of India

Sushil Kumar Himanshu, Ashish Pandey, and Deen Dayal

Abstract Satellite-based precipitation estimates (SPEs) having the advantage of near-global coverage and high spatial-temporal resolution, can improve the present hydrological prediction potential for a river basin or watershed. In the present study, three popular SPEs i.e., Climate Prediction Centre Morphing algorithm (CMORPH), Precipitation Estimation from Remotely Sensed Information using Artificial Neural Networks (PERSIANN), and TRMM Multi-satellite Precipitation Analysis (TMPA) precipitation datasets were evaluated with gauge based India Meteorological Department (IMD) gridded dataset using contingency and statistical table methods over an agricultural Watershed of South India. The evaluation was carried out at a spatial resolution of 0.25° × 0.25° (~26.5 km × 26.5 km) on a daily, monthly, seasonal, and yearly time scales for the years 1998–2013. Similar characteristics were observed with different SPEs in detecting rainfall events on various time scales. Overall, statistical and contingency indices analysis revealed that the performance of the TRMM dataset in predicting rainfall was better than the PERSIANN and CMORPH datasets. It was observed that the correlation between watershed-wide monthly average precipitations of TRMM dataset against the IMD dataset was good ($R^2 = 0.79$) as compared to PERSIANN ($R^2 = 0.68$) and CMORPH ($R^2 = 0.57$) datasets. The occurrence frequencies of very light and very heavy rainfalls were observed maximum and minimum, respectively with the IMD dataset, and the same pattern was observed with all the SPEs. Although, the TRMM dataset performed better than PERSIANN and CMORPH datasets, considerable interannual and seasonal variations were observed due to bias in SPEs. The present analysis reveals that the application of SPEs could be a potential alternative approach for hydrologic simulation over an un-gauged or data-sparse regions after suitable bias corrections.

S. K. Himanshu (✉)
Texas A & M Agrilife Research, Texas A & M University System, Vernon, Texas, USA
e-mail: sushilkumarhimanshu@gmail.com

A. Pandey · D. Dayal
Department of Water Resources Development & Management, Indian Institute of Technology, Roorkee, India

A. Pandey et al. (eds.), *Water Management and Water Governance*,
Water Science and Technology Library 96,
https://doi.org/10.1007/978-3-030-58051-3_5

Keywords SPEs · IMD, evaluation · Precipitation · Muneru watershed

5.1 Introduction

Precipitation is a crucial parameter in the hydrologic cycle that affects the implications of water resources management strategies (Himanshu et al. 2018a, b). Knowledge of the spatial-temporal variations and quantity of regional rainfall is essential for hydrological applications (Pandey et al. 2016). An accurate near-global coverage and high spatio-temporal resolution of rainfall records are required to improve climate predictions and weather forecasts. The variations in rainfall distribution pattern provide a sign of climate change resulting from either anthropogenic activities or natural factors. High resolution watershed-wide hydrologic dataset (runoff, nutrient load, sediment yield, etc.) can be generated through hydrologic modeling forced with SPEs, which could be helpful in decision making on water resources development and management (Himanshu et al. 2017, 2018a, b). Despite the importance of rainfall, in general, rain gauge data are only available over the land. In a developing country, in general, rain gauge networks are sparsely/unevenly distributed because of climatic variability, the complexity of physical geography, and other restrictions (Meng et al. 2014).

Rainfall could be estimated remotely, either from satellite or from ground-based weather radars. Satellite-based measurements provide near-global coverage at higher spatial-temporal resolution. The Satellite-based precipitation estimates (SPEs) could be an excellent substitute for ground-based rain gauge data. SPEs have superiority over ground-based rainfall data in terms of temporal and spatial resolution, and areal coverage thus providing a substitute as a data source for un-gauged basins/watersheds regions (Jia et al. 2011; Meng et al. 2014). However, the implications of remotely sensed information for hydrological prediction could be a challenging task for the hydrology community.

Three popular SPEs, i.e., CMORPH (Joyce et al. 2004), PERSIANN (Hsu et al. 1997; Sorooshian et al. 2000), and TRMM (Huffman et al. 2007; Huffman and Bolvin 2011) were evaluated with gaugebased India Meteorological Department (IMD) datasets at a spatial resolution of 0.25° × 0.25° (~26.5 km × 26.5 km) over an agriculture-dominant Muneru Watershed of South India. Assessment of the SPEs was performed from January 1st, 1998 to December 31st, 2013 on a different time scale. Different continuous statistical and contingency table indices were analyzed considering their variability in the study area. Before qualitative and quantitative analysis of SPEs, pre-processing of SPEs datasets were performed. The preprocessing of the dataset included the computation of average precipitation over the watershed using the Thiessen Polygon Method and the extraction of data from grid points falling under the watershed. Figure 5.1 depicts the flow chart of the methodology adopted for the assessment of SPEs.

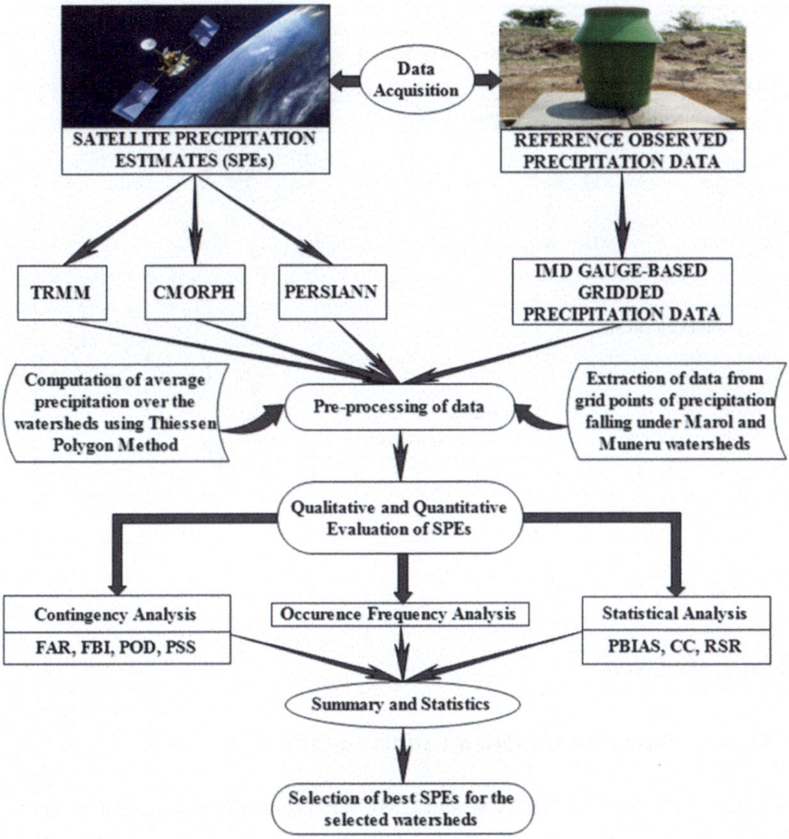

Fig. 5.1 Methodology adopted for evaluation of the SPEs

5.2 Materials and Methods

5.2.1 Study Area

The Muneru watershed is part of the lower Krishna River basin in the Andhra Pradesh state of India (Fig. 5.2). The watershed has an area of 10,260 km^2 lying between 79° 10′′ E to 80°50′ E longitude and 16° 40′ N to 18°10′ latitude. The climate is subtropical with an annual average rainfall of 1214 mm. The watershed has an elevation range of 20–795 m and an average land slope of 5.03%. The land use characterized by agriculture (75.02%), forest (13.71%), pasture (8.80%), water (1.18%) and built-up (1.28%). The major soil types in the study area are clay loam followed by the sandy clay loam. Heterogeneous land use and the absence of any large storage structures make this watershed suitable for the present study.

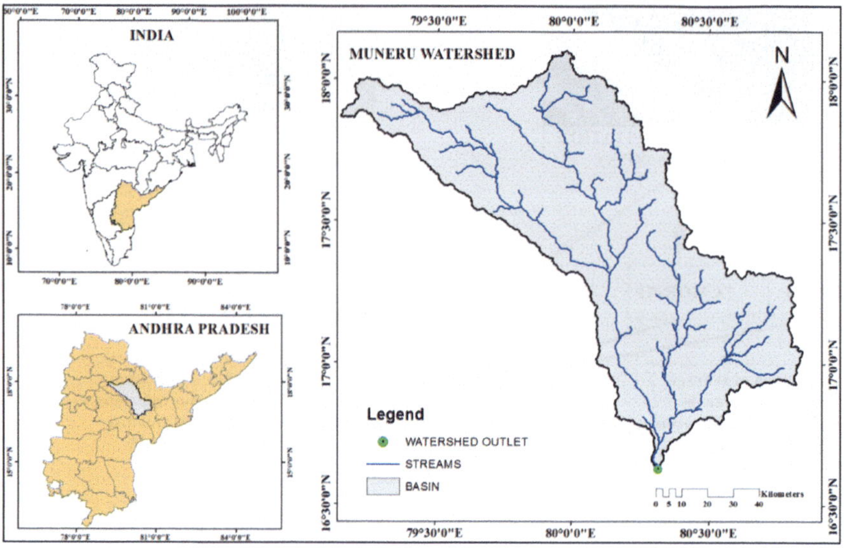

Fig. 5.2 The study area location map

5.2.2 Datasets Used

5.2.2.1 Satellite-Based Precipitation Estimates (SPEs)

Three SPEs, i.e., TRMM, PERSIANN, and CMORPH datasets available at a spatial resolution of 0.25° × 0.25° (~26.5 km × 26.5 km) on a 3-hourly time scale were used in the present study. The TMPA products are available in two different versions, 3B42 (research quality) and 3B42RT (near-real-time). Research quality version (3B42) has two successive versions, version 6 and the latest version 7. Research quality version 7 (3B42-V7) was evaluated in the present study. These SPEs were used globally for different hydro-meteorological applications (Su et al. 2008; Sapiano and Arkin 2009; Dinku et al. 2010; Sohn et al. 2010; Behrangi et al. 2011; Bitew and Gebremichael 2011a, b; Chen et al. 2011, 2013; Gourley et al. 2011; Vu et al. 2012; Xue et al. 2013; Qin et al. 2014; Qiao et al. 2014; Prakash et al. 2014; Tong et al. 2014; Yong et al. 2014; Bajracharya et al. 2015; Liu 2015a, b; Himanshu et al. 2017, 2018a, b). The Satellite-based precipitation estimates (SPEs) could be an excellent substitute for ground-based rain gauge data. SPEs have superiority over ground-based rainfall data in terms of temporal and spatial resolution and areal coverage. The details of the SPEs used in this study are presented in Table 5.1. Before qualitative and quantitative analysis of SPEs, pre-processing of SPEs datasets were performed. The preprocessing of the dataset included the computation of average precipitation over the watershed using the Thiessen Polygon Method and extraction of data from grid points falling under the watershed.

Table 5.1 Details of satellite precipitation estimates used in the study

Algorithm	Developer/Producer	Input Data	Spatial Resolution	Temporal Resolution	Areal coverage/start date	Update frequency	Latency	Reference
TRMM 3B42RP-V7 (Tropical Rainfall Measuring Mission)	National Aeronautics and Space Administration, Precipitation Processing System (NASA/PPS)	AMSR-E, AMSU, MHS, MW-VAR (IR), SSMI, SSMIS, TCI, TMI	0.25° × 0.25° (~ 26.5 km × 26.5 km)	3-hourly	50°N—50°S/1998	Monthly	2 months	Huffman et al. (2007)
CMORPH V1.0 RAW (Climate Prediction Center MORPHing technique)	National Oceanic and Atmospheric Administration, Climate Prediction Center (NOAA/CPC), U.S. Department of Commerce	AMSR-E, AMSU, IR vectors, MHS, SSMI, SSMIS, TMI	0.25° × 0.25° (~26.5 km × 26.5 km)	3-hourly	60°N - 60°S/1998	Daily	18 h	Joyce et al. (2004)
PERSIANN-CDR (Precipitation Estimation from Remote Sensing Information using Artificial Neural Network-Climate Data Record)	University of California, Irvine	GPCP Monthly Precipitation GRIDSAT-IRWIN,	0.25° × 0.25° (~ 26.5 km × 26.5 km)	6-hourly	60°N—60°S/1980	Monthly	2 months	Sorooshian et al. (2000)

Note AMSR-E: Advanced Microwave Scanning Radiometer for Earth Observing Systems; AMSU: Advanced Microwave Sounding Unit; GPCP: Global Precipitation Climatology Project; GRIDSAT-IRWIN: Gridded Satellite infrared window; MHS: Microwave Humidity Sounder; MW-VAR (IR): Microwave-adjusted variable rainfall product (Infrared); SSMI: Special Sensor Microwave Imager; SSMIS: Special Sensor Microwave Imager/Sounder; TCI: TRMM Combined Instrument; TMI: TRMM Microwave Imager

5.2.2.2 IMD Gauged Precipitation Dataset

The gauge-based observed gridded precipitation dataset computed by the IMD was used as standard reference data for the evaluation of the SPEs. The gauge-based IMD gridded dataset is available at 0.25° × 0.25° (~26.5 km × 26.5 km) grid on a daily time scale available since 1901(Pai et al. 2014, 2015). Shepard interpolation was adopted to interpolate the point observations to a regular grid (Shepard 1968). A total of 23 grid points of precipitation were available for the Muneru watershed area. Before the release of this dataset, the IMD gridded dataset was available at 0.5° × 0.5° grid (Rajeevan and Bhate 2009). The IMD data set reproduces the orographic precipitation's spatial gradient realistically due to finer spatial resolution (Prakash et al. 2014, 2016).

5.2.3 Performance Assessment of SPEs

To carry out quantitative performances, the SPEs were evaluated against the IMD dataset using the 2 × 2 contingency table indices (Schaefer 1990) and the continuous statistical indices (Tables 5.2 and 5.3). The detailed description of the contingency table indices and the statistical indices are presented in Tables 5.2 and 5.3, respectively.

5.3 Results and Discussion

To analyze the impact of rainfall inputs on hydrological models, there should be performance evaluation of SPEs against the IMD gridded dataset. A comparative analysis between SPEs and IMD dataset has been carried out for a period of January

Table 5.2 A 2 × 2 contingency table indices used

Statistics	Description	Formula	Perfect Value	Range
FAR	False Alarm Ratio	$\frac{F}{F+H}$	0	0 to 1
FBI	Frequency Bias Index	$\frac{F+H}{H+M}$	1	0 to ∞
POD	Probability of Detection	$\frac{H}{H+M}$	1	0 to 1
PSS	Peirce Skill Score	$(H/(H+M)) - (F/(F+R))$	1	−1 to 1

Note Observed IMD precipitation: no, SPEs estimated precipitation: yes = False alarms (F); Observed IMD precipitation: no, SPEs estimated precipitation: no = Correct rejects (R); Observed IMD precipitation: yes, SPEs estimated precipitation: no = Misses (M); Observed IMD precipitation: yes, SPEs estimated precipitation: yes = Hits (H)

Table 5.3 Statistics used for the evaluation of the SPEs

Statistics	Formula	Range	Perfect Value
Percent bias (PBIAS)	$\text{Pbias} = \left[\frac{\sum_{i=1}^{n}(Y_i^{obs} - Y_i^{sim}) * (100)}{\sum_{i=1}^{n} Y_i^{obs}}\right]$	$-\infty$ to $+\infty$	0
Correlation Coefficient (CC)	$CC = \left[\frac{\sum_{i=1}^{n}\left(Y_i^{obs} - \overline{Y^{obs}}\right)\left(Y_i^{sim} - \overline{Y^{sim}}\right)}{\sqrt{\sum_{i=1}^{n}\left(Y_i^{obs} - \overline{Y^{obs}}\right)^2}\sqrt{\sum_{i=1}^{n}\left(Y_i^{sim} - \overline{Y^{sim}}\right)^2}}\right]$	+1/−1	±1
RMSE observations standard deviation ratio (RSR)	$RSR = \left[\frac{\sqrt{\sum_{i=1}^{n}(Y_i^{obs} - Y_i^{sim})^2}}{\sqrt{\sum_{i=1}^{n}\left(Y_i^{obs} - \overline{Y^{obs}}\right)^2}}\right]$	0 to ∞	0

Note n = number of observations, Y_i^{obs} = observed value, $Y_i^{sim} = simulatedvalue$, $\overline{Y^{obs}}$ = average observed value and $\overline{Y^{sim}}$ = average simulated value in respective time steps i

1998–2013. The evaluation has been carried out on a grid-based spatial scale and watershed-wide average precipitation on different time scales. The evaluation of the SPEs were carried out against the IMD dataset using contingency and statistical table indices.

5.3.1 *Evaluation of Watershed-Wide Average Daily SPEs for Detection of Rainfall Events*

Detection capabilities of SPEs for rain events were evaluated using contingency table indices. Evaluation of daily SPEs on grid-based spatial scales as compared to grid-based IMD dataset for Muneru watershed are presented in Table 5.4 and Fig. 5.5 using contingency table indices for a period of 1998–2013 on seasonal and annual time steps. In general, the performance of SPEs in detecting rainfall events over the Muneru watershed has been found better during the monsoon period as compared to the non-monsoon period and on an annual time scale.

Similar characteristics were observed with different SPEs in detecting rainfall events on various time scales. The TRMM performed better in detecting rain events as compared to the PERSIANN and CMORPH SPEs. For the TRMM dataset, the values of the FAR were found close to 0, and the values of POD, FBI, and PSS were found close to 1 as compared to the PERSIANN and CMORPH datasets. FAR values for the TRMM dataset on monsoon, non-monsoon, and annual time scales were found as 0.12, 0.62, and 0.29, respectively. However, FAR values were found as 0.28, 0.72, and 0.37 on monsoon, non-monsoon, and annual time scales, respectively, for the PERSIANN dataset. Similarly, the FAR values were found as 0.24, 0.76, and 0.33 on monsoon, non-monsoon, and annual time scales, respectively for the CMORPH dataset. Values of FBI were found as 1.04, 1.58, and 1.18 on monsoon, non-monsoon,

Table 5.4 Evaluation of watershed-wide average daily SPEs with IMD rainfall in detecting rainfall events for the period of 1998–2013

Contingency TableMatrix	TRMM	PERSIANN	CMORPH	TRMM	PERSIANN	CMORPH	TRMM	PERSIANN	CMORPH	TRMM	PERSIANN	CMORPH
	FAR			FBI			POD			PSS		
Annual	0.29	0.37	0.33	1.18	0.82	0.84	0.83	0.61	0.66	0.53	0.51	0.49
Monsoon	0.12	0.28	0.24	1.04	0.86	0.82	0.91	0.74	0.72	0.30	0.34	0.27
Non-Monsoon	0.62	0.72	0.76	1.58	1.38	1.22	0.59	0.52	0.55	0.34	0.29	0.31

and annual time scale respectively for the TRMM dataset. In general, frequencies of rain events were overestimated during the non-monsoon season (FBI > 1) and underestimated on monsoon and annual time scales (FBI < 1), especially with the PERSIANN and CMORPH datasets. For the TRMM dataset, values of POD were found as 0.91, 0.59, and 0.83 on monsoon, non-monsoon, and annual time scale, respectively. Similarly, the values of PSS for the TRMM dataset were found as 0.30, 0.34, and 0.53 on monsoon, non-monsoon, and annual time scale, respectively. These results indicate that the TRMM dataset has better capabilities for the detection of rainfall events, however, a substantial bias was observed (Fig. 5.3).

5.3.2 Statistical Evaluation of Watershed-Wide Monthly Averaged SPEs

The statistical parameters, viz., mean, standard deviation, and sample variance were estimated for watershed-wide monthly averaged SPEs against IMD rainfall data, and the results are tabulated in Table 5.5. The coefficient of Skewness and coefficient of Kurtosis were estimated for watershed-wide monthly averaged SPEs against IMD rainfall data, and the results are tabulated in Table 5.6. Similarly, accuracy assessment (CC, PBIAS, and RSR) was also carried out for watershed-wide monthly averaged SPEs against IMD rainfall data (Table 5.7). The statistical analysis indicated that the TRMM rainfall dataset performed better as compared to PERSIANN and CMORPH datasets. During monsoon season, the values of CC for watershed-wide monthly averaged TRMM dataset and IMD dataset varied between 0.12 (June) and 0.56 (July), however, the values of CC for watershed-wide monthly averaged TRMM dataset and IMD dataset varied between 0.17 (February) and 0.83 (March) during non-monsoon season. The CC values for the watershed-wide monthly averaged PERSIANN dataset and IMD dataset varied between −0.14 (February) to 0.59 (December) and −0.27 (July) to 0.26 (October) during non-monsoon and monsoon season, respectively. Similarly, the CC values for watershed-wide monthly averaged CMORPH dataset and IMD dataset varied between −0.18 (February) to 0.89 (March) and 0.01 (October) to 0.72 (July) during non-monsoon and monsoon season, respectively. In general, the TRMM and PERSIANN datasets overestimated the watershed-wide monthly averaged precipitation during monsoon season and underestimated the watershed-wide monthly averaged precipitation during non-monsoon season. However, in general, the CMORPH dataset underestimated the watershed-wide monthly averaged precipitation during both monsoon and non-monsoon seasons. The value of RSR varies between 0.77 and 2.18 during monsoon and 0.37 to 1.11 during the non-monsoon season. The scatter plot between watershed-wide monthly averaged SPEs and the gauge-based IMD gridded dataset indicates that the prediction accuracy of the TRMM dataset was better as compared to PERSIANN and CMORPH datasets (Fig. 5.4). The coefficient of determination values for the TRMM, PERSIANN, and

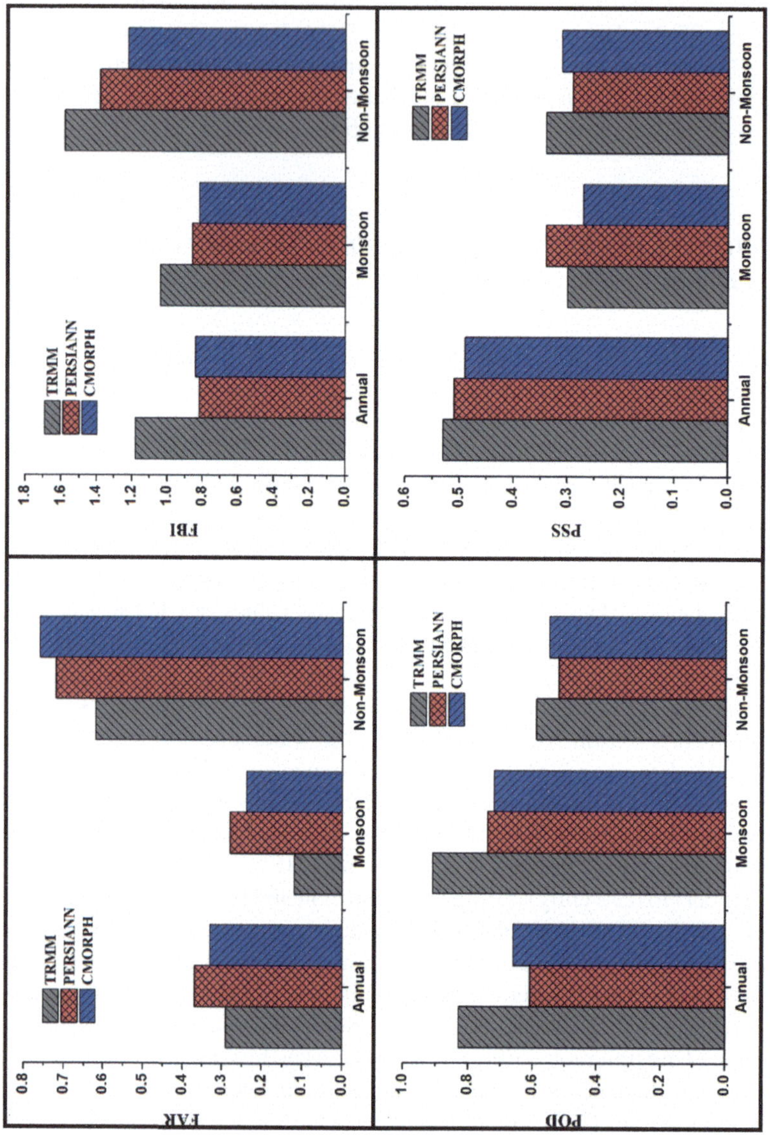

Fig. 5.3 Statistics of the SPEs with reference to IMD gridded dataset

Table 5.5 Statistical summary of watershed-wide monthly average SPEs against IMD rainfall

Month	Mean				Standard Deviation				Sample Variance			
	IMD	TRMM	PERSIANN	CMORPH	IMD	TRMM	PERSIANN	CMORPH	IMD	TRMM	PERSIANN	CMORPH
Jan	7.2	5.17	8.3	0.2	12.3	9.6	12.4	0.5	151.8	92.8	154.8	0.2
Feb	12.3	5.7	5.5	1.8	15.5	6.3	7.2	3.9	241.1	39.5	52.2	15.7
Mar	13.9	9.8	12.3	6.4	30.4	13.4	18.4	9.9	925.4	180.9	339.5	98.5
Apr	20.1	7.4	12.6	9.6	13.9	5.2	12.3	8.4	195.0	26.7	150.4	71.3
May	42.4	18.1	20.6	28.8	41.2	21.8	25.2	34.2	1694.1	476.5	636.4	1171.3
Jun	126.2	207.0	125.6	140.6	51.2	70.6	98.9	56.8	2622.7	4988.4	9774.6	3232.3
Jul	257.3	357.9	306.6	181.2	99.5	85.7	129.6	98.6	9908.2	7352.7	16797.9	6834.4
Aug	260.5	305.1	310.7	174.4	82.7	74.7	118.1	79.0	6834.4	5587.6	13959.9	6241.5
Sep	180.0	207.9	277.8	202.3	90.4	74.1	103.9	86.0	8166.6	5486.9	10810.4	7395.2
Oct	102.9	72.8	140.6	78.3	54.4	43.5	88.4	46.9	2958.5	1896.4	7808.6	2204.5
Nov	30.3	21.3	40.1	22.9	39.5	31.8	37.6	32.1	1560.9	1010.8	1414.2	1028.4
Dec	9.0	1.3	16.6	1.2	39.5	31.8	37.6	32.1	332.7	3.0	410.4	11.1

Table 5.6 Kurtosis and Skewness coefficients for watershed-wide monthly average SPEs with gauge-based IMD dataset

Month	Kurtosis				Skewness			
	IMD	TRMM	PERSIANN	CMORPH	IMD	TRMM	PERSIANN	CMORPH
Jan	0.61	10.83	2.01	3.58	1.51	3.13	1.69	2.09
Feb	−0.09	1.22	2.05	7.13	1.16	1.28	1.70	2.74
Mar	11.06	4.05	6.78	3.48	3.24	2.18	2.53	1.89
Apr	−0.66	−0.61	1.71	−1.13	0.46	0.57	1.30	0.58
May	1.45	2.11	2.40	7.51	1.36	1.66	1.83	2.53
Jun	1.02	0.27	4.24	−1.43	1.19	0.55	1.98	−0.07
Jul	0.95	3.75	−0.83	0.20	0.52	−1.89	−0.17	0.55
Aug	0.20	0.65	0.65	−0.91	0.55	0.80	−0.25	0.04
Sep	0.75	−1.06	−0.73	−0.84	1.03	0.15	−0.14	0.70
Oct	0.53	−1.01	−0.25	0.95	0.57	0.44	0.54	0.70
Nov	5.33	1.57	−1.54	1.55	2.20	1.68	0.42	1.59
Dec	3.22	−0.05	0.68	14.09	2.09	1.17	1.12	3.69

CMORPH datasets against the IMD dataset were observed as 0.79, 0.68, and 0.57, respectively for watershed-wide monthly averaged rainfalls.

5.3.3 *Evaluation of the Watershed-WideAnnual Average SPEs*

The watershed-wide annual average rainfall of the SPEs was analyzed against the IMD dataset (Fig. 5.5). On average, the watershed-wide annual average rainfall values with PERSIANN dataset were observed maximum followed by TRMM and CMORPH datasets. In general, the TRMM dataset was overestimated during all years except the years 2000, 2008, 2010, and 2012. Similarly, PERSIANN overestimated the watershed-wide annual average rainfall during all years except the years 2000, 2004, 2008, 2010, and 2012. Conversely, CMORPH underestimated the watershed-wide annual average rainfall during all years except years 1998, 2007, and 2009. The CMORPH dataset always underestimated the watershed-wide annual average rainfall in those years in which annual rainfall was higher than the average annual rainfall.

5.3.4 *Occurrence Frequency of SPEs*

Occurrence frequencies of different rainfall intensities were estimated on daily time scale for watershed-wide average rainfall (Fig. 5.6). The rainfall intensity threshold

Table 5.7 Evaluation of the watershed-wide monthly average SPEs against IMD dataset

Month	PBIAS			CC			RSR		
	TRMM	PERSIANN	CMORPH	TRMM	PERSIANN	CMORPH	TRMM	PERSIANN	CMORPH
Jan	−25.99	15.50	−90.44	0.39	0.29	−0.15	0.91	1.12	1.05
Feb	−50.37	−51.94	−80.17	0.17	−0.14	−0.18	1.06	1.16	1.19
Mar	−27.75	−10.94	−50.11	0.83	0.29	0.89	0.72	0.84	0.79
Apr	−59.38	−34.98	−49.13	0.45	0.32	0.25	1.22	1.09	1.28
May	−53.66	−48.15	−30.02	0.45	0.48	0.42	1.09	1.05	1.07
Jun	59.91	−0.45	10.64	0.12	0.24	0.08	2.18	1.96	1.47
Jul	36.62	17.92	−27.69	0.56	−0.27	0.72	1.39	1.96	1.00
Aug	16.01	18.04	−30.96	0.26	−0.06	0.40	1.33	1.91	1.39
Sep	14.53	50.86	11.59	0.73	0.13	0.43	0.77	1.85	1.06
Oct	−27.46	34.16	−22.39	0.38	0.26	0.01	1.05	1.84	1.16
Nov	−27.77	30.27	−22.62	0.40	−0.06	0.43	0.37	1.18	0.43
Dec	−79.57	78.08	−81.59	0.23	0.59	−0.11	1.11	1.02	1.16

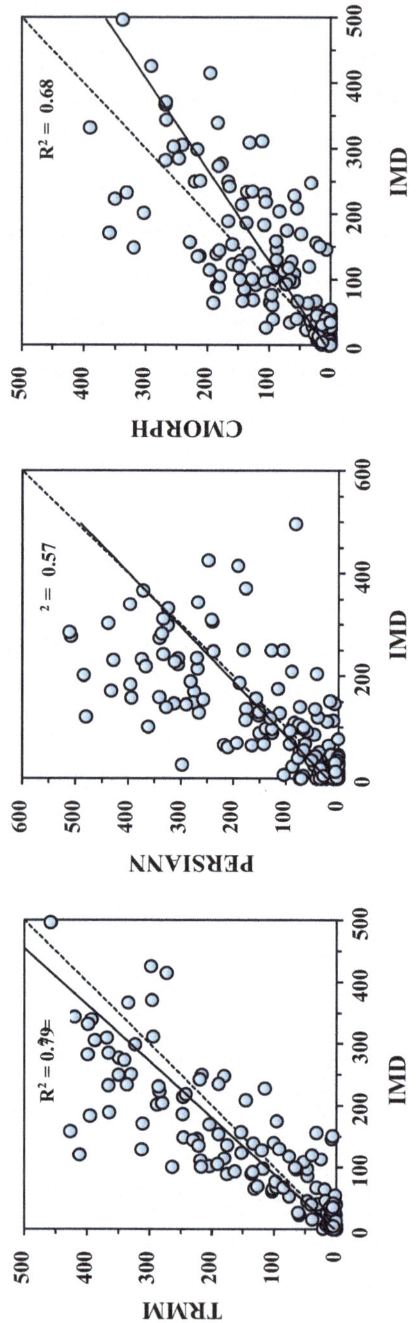

Fig. 5.4 Scatter plot between watershed-wide monthly average SPEs and the IMD datasets

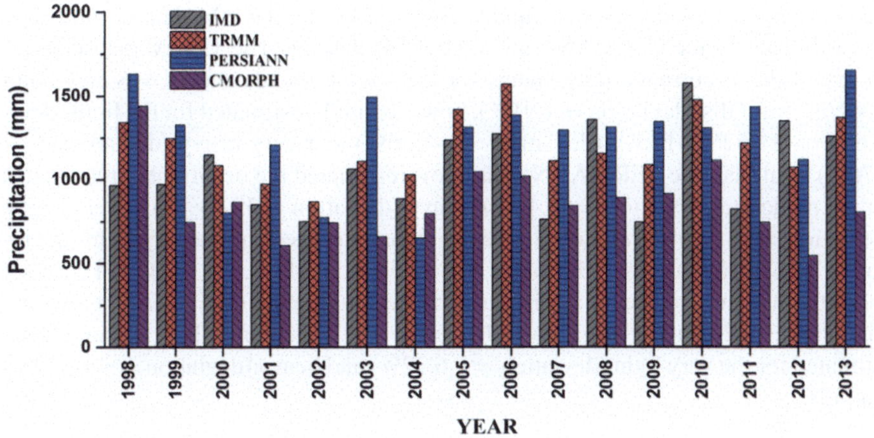

Fig. 5.5 Watershed-wide annual average precipitation of the SPES and the IMD datasets

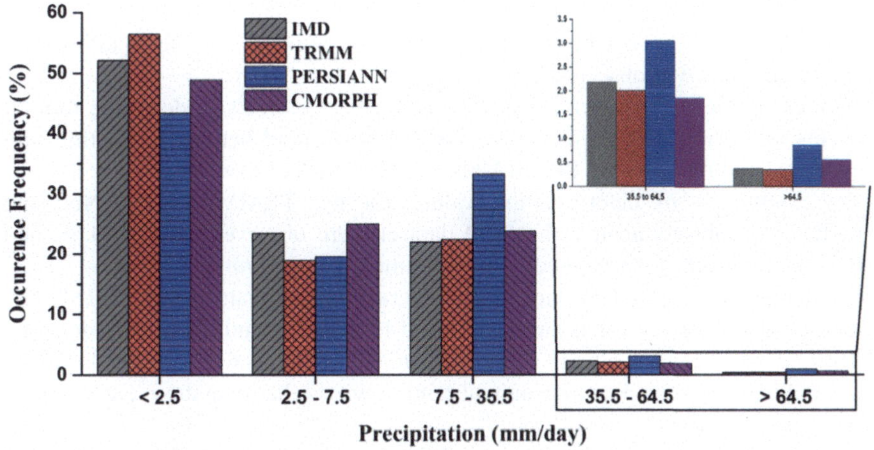

Fig. 5.6 Occurrence frequency of SPEs and IMD datasets for Muneru watershed

values were differentiated based on the IMD rainfall classification, i.e., very light, light, moderate, heavy, and very heavy rainfalls with rainfall intensities of 0.1–2.5, 2.5–7.5, 7.5–35.5, 35.5–64.5, and > 64.5 mm/day, respectively. The occurrence frequencies of very light rainfalls were observed maximum with the IMD dataset, and the same pattern was observed with all the SPEs. Similarly, the occurrence frequencies of very heavy rainfalls were observed minimum with the IMD dataset, and the same pattern was observed with all the SPEs. In case of watershed-wide very light average rainfalls, the occurrence frequencies were overestimated by the TRMM dataset, however, underestimated by the PERSIANN and CMORPH datasets. Conversely, in the case of watershed-wide very heavy average rainfalls, the occurrence frequencies were underestimated by the TRMM dataset, however, overestimated by the PERSIANN and CMORPH datasets.The TRMM dataset overestimated the occurrence frequencies for the very light (deviation = 8.36%) and moderate rainfalls (deviation = 1.91%), however, underestimated for the light (deviation = −19.52%), heavy (deviation = −8.06%) and very heavy (deviation = −6.76%) rainfalls. The PERSIANN dataset overestimated the occurrence frequencies for the moderate (deviation = 51.42%), heavy (deviation = 39.73%) and very heavy (deviation = 135.16%) rainfalls, however, underestimated for the very light (deviation = −16.84%) and light (deviation = −16.56%) rainfalls.The CMORPH dataset overestimated the occurrence frequencies for the light (deviation = 6.67%), moderate (deviation = 8.26%) and very heavy (deviation = 51.37%) rainfalls, however, underestimated for the very light (deviation = −6.22%) and heavy (deviation = −14.79%) rainfalls.

5.4 Conclusions

- SPEs performed good with the IMD dataset for Muneru watershed, although a significant bias was observed.
- Similar characteristics were observed with different SPEs in detecting rainfall events on various time scales. The TRMM performed better in detecting rain events as compared to the PERSIANN and CMORPH SPEs.
- The coefficient of determination values for the TRMM, PERSIANN, and CMORPH datasets against the IMD dataset were observed as 0.79, 0.68, and 0.57, respectively for watershed-wide monthly averaged rainfalls.
- On average, the watershed-wide annual average rainfall values with PERSIANN dataset were observed maximum followed by TRMM and CMORPH datasets. The CMORPH dataset always underestimated the watershed-wide annual average rainfall in those years in which annual rainfall was higher than the average annual rainfall.
- The occurrence frequencies of very light and very heavy rainfalls were observed maximum and minimum, respectively with the IMD dataset, and the same pattern was observed with all the SPEs.

References

Bajracharya SR, Palash W, Shrestha MS, Khadgi VR, Duo C, Das PJ, Dorji C (2015) Systematic evaluation of satellite-based rainfall products over the Brahmaputra basin for hydrological applications. Advances in Meteorology, Article ID-398687

Behrangi A, Khakbaz B, Jaw TC, AghaKouchak A, Hsu K, Sorooshian S (2011) Hydrologic evaluation of satellite precipitation products over a mid-size basin. J Hydrol 397(3):225–237

Bitew MM, Gebremichael M (2011a) Assessment of satellite rainfall products for streamflow simulation in medium watersheds of the Ethiopian highlands. Hydrol Earth Syst Sci 15(4):1147–1155
Bitew M M, Gebremichael M (2011b) Evaluation of satellite rainfall products through hydrologic simulation in a fully distributed hydrologic model. Water Resources Research, 47(6)
Chen C, Yu Z, Li L, Yang C (2011) Adaptability evaluation of TRMM satellite rainfall and its application in the Dongjiang River Basin. Procedia Environ Sci 10:396–402
Chen S, Hong Y, Cao Q, Gourley JJ, Kirstetter PE, Yong B, Tian Y, Zhang B, Shen Y, Hu J, Hardy J (2013) Similarity and difference of the two successive V6 and V7 TRMM multisatellite precipitation analysis performance over China. J Geophys Res Atm 118:13060–13074
Dinku T, Ruiz F, Connor SJ, Ceccato P (2010) Validation and intercomparison of satellite rainfall estimates over Colombia. J Appl Meteorol Climatol 49(5):1004–1014
Gourley JJ, Hong Y, Flamig ZL, Wang J, Vergara H, Anagnostou EN (2011) Hydrologic evaluation of rainfall estimates from radar, satellite, gauge, and combinations on Ft. Cobb basin, Oklahoma. J Hydrometeorol 12(5):973–988
Himanshu SK, Pandey A, Yadav B (2017) Assessing the applicability of TMPA-3B42V7 precipitation dataset in wavelet-support vector machine approach for suspended sediment load prediction. J Hydrol 550:103–117
Himanshu SK, Pandey A, Patil A (2018a) Hydrologic evaluation of the TMPA-3B42V7 precipitation data set over an agricultural watershed using the SWAT Model. J Hydrol Eng 23(4):05018003
Himanshu SK, Pandey A, Dayal D (2018b) Evaluation of satellite-based precipitation estimates over an agricultural watershed of India. In: World Environmental and Water Resources Congress 2018, ASCE, pp 308–320
Hsu KL, Gao X, Sorooshian S, Gupta HV (1997) Precipitation estimation from remotely sensed information using artificial neural networks. J Appl Meteorol 36(9):1176–1190
Huffman GJ, Bolvin DT (2011) TRMM and other data precipitation data set documentation. laboratory for atmospheres, NASA Goddard Space Flight Center and Science Systems and Applications Inc. Inc, Goddard, Maryland, United States (1968)
Huffman GJ, Bolvin DT, Nelkin EJ, Wolff DB, Adler RF, Gu G, Hong Y, Bowman KP, Stocker EF (2007) The TRMM multisatellite precipitation analysis (TMPA): Quasi-global, multiyear, combined-sensor precipitation estimates at fine scales. J Hydrometeorol 8(1):38–55
Jia S, Zhu W, Lű A, Yan T (2011) A statistical spatial downscaling algorithm of TRMM precipitation based on NDVI and DEM in the Qaidam Basin of China. Remote Sens Environ 115(12):3069–3079
Joyce RJ, Janowiak JE, Arkin PA, Xie P (2004) CMORPH: A method that produces global precipitation estimates from passive microwave and infrared data at high spatial and temporal resolution. J Hydrometeorol 5(3):487–503
Liu Z (2015a) Comparison of precipitation estimates between Version 7 3-hourly TRMM Multi-Satellite Precipitation Analysis (TMPA) near-real-time and research products. Atmos Res 153:119–133
Liu Z (2015b) Comparison of versions 6 and 7 3-hourly TRMM multi-satellite precipitation analysis (TMPA) research products. Atmos Res 163:91–101
Meng J, Li L, Hao Z, Wang J, Shao Q (2014) Suitability of TRMM satellite rainfall in driving a distributed hydrological model in the source region of Yellow River. J Hydrol 509:320–332
Pai DS, Sridhar L, Badwaik MR, Rajeevan M (2015) Analysis of the daily rainfall events over India using a new long period (1901–2010) high resolution (0.25 × 0.25) gridded rainfall data set. Climate Dyn 45(3–4):755–776
Pai DS, Sridhar L, Rajeevan M, Sreejith OP, Satbhai NS, Mukhopadhyay B (2014) Development of a new high spatial resolution (0.25 × 0.25) long period (1901–2010) daily gridded rainfall data set over India and its comparison with existing data sets over the region. Mausam, 65(1), 1–18 (1968)
Pandey A, Himanshu SK, Mishra SK, Singh VP (2016) Physically basedsoil erosion and sediment yield models revisited. CATENA 147:595–620

Prakash S, Mitra AK, Pai DS, AghaKouchak A (2016) From TRMM to GPM: How well can heavy rainfall be detected from space? Adv Water Resour 88:1–7
Prakash S, Sathiyamoorthy V, Mahesh C, Gairola RM (2014) An evaluation of high-resolution multisatellite rainfall products over the Indian monsoon region. Int J Remote Sens 35(9):3018–3035
Qiao L, Hong Y, Chen S, Zou CB, Gourley JJ, Yong B (2014) Performance assessment of the successive Version 6 and Version 7 TMPA products over the climate-transitional zone in the southern Great Plains, USA. J Hydrol 513:446–456
Qin Y, Chen Z, Shen Y, Zhang S, Shi R (2014) Evaluation of satellite rainfall estimates over the Chinese Mainland. Remote Sens 6(11):11649–11672
Rajeevan M, Bhate J (2009) A high resolution daily gridded rainfall dataset (1971–2005) for mesoscale meteorological studies. Curr Sci 96(4):558–562
Sapiano MRP, Arkin PA (2009) An intercomparison and validation of high-resolution satellite precipitation estimates with 3-hourly gauge data. J Hydrometeorol 10(1):149–166
Schaefer JT (1990) The critical success index as an indicator of warning skill. Weather and Forecasting 5(4):570–575
Shepard D (1968) A two-dimensional interpolation function for irregularly-spaced data. In: Proceedings of the 1968, 23rd ACM national conference, 517–524
Sohn BJ, Han HJ, Seo EK (2010) Validation of satellite-based high-resolution rainfall products over the Korean Peninsula using data from a dense rain gauge network. J Appl Meteorol Clim 49(4):701–714
Sorooshian S, Hsu KL, Gao X, Gupta HV, Imam B, Braithwaite D (2000) Evaluation of PERSIANN system satellite-based estimates of tropical rainfall. Bull Am Meteor Soc 81(9):2035–2046
Su F, Hong Y, Lettenmaier DP (2008) Evaluation of TRMM Multisatellite Precipitation Analysis (TMPA) and its utility in hydrologic prediction in the La Plata Basin. Journal of Hydrometeorology 9(4):622–640
Tong K, Su F, Yang D, Hao Z (2014) Evaluation ofsatellite precipitation retrievals and their potential utilities in hydrologic modeling over the Tibetan Plateau. J Hydrol 519:423–437
Vu MT, Raghavan SV, Liong SY (2012) SWAT use of gridded observations for simulating runoff-a Vietnam river basin study. Hydrol Earth Syst Sci 16(8):2801
Xue X, Hong Y, Limaye AS, Gourley JJ, Huffman GJ, Khan SI, Dorji C, Chen S (2013) Statistical and hydrological evaluation of TRMM-based Multi-satellite Precipitation Analysis over the Wangchu Basin of Bhutan: Are the latest satellite precipitation products 3B42V7 ready for use in ungauged basins? J Hydrol 499:91–99
Yong B, Chen B, Gourley JJ, Ren L, Hong Y, Chen X, Wang W, Chen S, Gong L (2014) Intercomparison of the Version-6 and Version-7 TMPA precipitation products over high and low latitudes basins with independent gauge networks: Is the newer version better in both real-time and post-real-time analysis for water resources and hydrologic extremes? J Hydrol 508:77–87

Chapter 6
Evaluation of Sentinel 2 Red Edge Channel for Enhancing Land Use Classification

Sucharita Pradhan, Kamlesh Narayan Tiwari, and Anirban Dhar

Abstract The region of maximum transition of spectral reflectance curve of a plant leaf also known as vegetation red edge is strongly associated with its biological characteristics such as chlorophyll contents and leaf area index (LAI). As the red edge is a sensitive indicator of several abiotic and climatic factors, incorporation of this band in multi-spectral satellite improves the potential for enhancing land use classification. This study evaluates the potential of three spectral bands of ESA's recently launched satellite Sentinel 2 which are centred at 0.705, 0.740, 0.783 μm in red-edge region. Sentinel 2 satellite images of March 2017 were collected for Damodar Left Bank Canal Irrigation System, West Bengal, India. The supervised classification was performed with maximum likelihood algorithm using various combinations of spectral bands, i.e., with and without red edge bands. Thereafter, the classified images were validated against the field data and high-resolution Google Earth images. The overall classification accuracy was found as 73.33% for classification excluding red-edge bands, while the incorporation of these bands improved the accuracy to 90%. Highest significant effect was observed for plantation class as producer's accuracy was found to increase from 20 to 80% which proved inclusion of red edge in image classification is efficient and can be used for application.

Keywords Sentinel 2 · Red edge bands · Land use classification · Canal Command Area · Canal Irrigation · Boro rice

---

S. Pradhan (✉) · K. N. Tiwari
Department of Agricultural and Food Engineering, Indian Institute of Technology Kharagpur, Kharagpur, West Bengal, India
e-mail: sucharita.pradhan20@gmail.com

K. N. Tiwari
e-mail: kamlesh@agfe.iitkgp.ac.in

A. Dhar
Department of Civil Engineering, Indian Institute of Technology Kharagpur, Kharagpur, West Bengal, India
e-mail: anirban@civil.iitkgp.ac.in

A. Pandey et al. (eds.), *Water Management and Water Governance*,
Water Science and Technology Library 96,
https://doi.org/10.1007/978-3-030-58051-3_6

6.1 Introduction

Red edge is the abrupt rise in the electromagnetic spectrum of 0.680–0.800 μm by plant leaves due to the combined effects of strong chlorophyll absorption which causes low red reflectance and high internal leaf scattering creating large NIR reflectance (Dawson and Curran 1998). It was first defined by Collins (1978) and is one of the most studied features on the vegetation spectral curve. High spectral difference among crop types, growth stages of vegetation, health conditions, and microclimate hinders the application of remote sensing to agriculture (Schuster et al. 2012). However, a combination of red and infrared radiance in red-edge region offers the basis for vegetation identification. Technical parameters of satellites along with classification procedures are constantly developing as there is always a strong requirement to advance the accuracy of classification results (Schuster et al. 2012). Several previous studies have evaluated the potential of red-edge band for assessment of plant chlorophyll content, biomass content (Filella and Penuelas 1994; Ju et al. 2010; Clevers and Gitelson 2013) plant stress (Carter and Miller 1994; Carter and Knapp 2001; Smith et al. 2004; Eitel et al. 2011) and in classification land use and land cover change (Schuster et al. 2012; Tigges et al. 2013).

ESA's recently launched satellite Sentinel 2 provides a resourceful set of 13 spectral bands covering from the visible and near-infrared to the shortwave infrared, introducing four bands at 10 m, six bands at 20 m, and three bands at 60 m spatial resolution. Frampton et al. (2013) evaluated the potential of Sentinel 2 multi-spectral instrument for estimating canopy chlorophyll content, leaf chlorophyll concentration, and LAI. Many researchers mainly aimed at the estimation of nitrogen content and vegetation state (Clevers and Gitelson 2012, 2013; Hill (2013); Delegido et al. 2012). Information to evaluate the influence of red edge channel of Sentinel 2 for land use classification in agriculture intensified canal command area is not available in the literature. Monitoring of the spatiotemporal distribution of agricultural areas is highly necessary for food security and water resources management. This study evaluates the accuracy of land use classification using red-edge spectral region and its effect on the classification accuracy of specific land use classes.

6.2 Material and Methods

The study area lies between latitude 220 33′ 42.05″ N to 230 40′ 14.06″ N and longitude 870 18′ 20.44″ E to 880 24′ 01.67″ E of Damodar canal command, West Bengal, India. The Damodar Valley Project Development is amongst the first few multipurpose water resource projects that were completed in the post-independence era of India. The main purpose of the project was to moderate the floods of the river Damodar by constructing a series of dams on the upper catchment of the river and utilizing the stored water for regulated release throughout the year. In the lower reaches farmlands are being irrigated by diversion barrage and vast network of canal systems. The major portion of the command area (80%) lies in the left bank of Damodar River and spreads over the districts of Bardhaman, Howrah, and Hooghly of West Bengal, India. The irrigation system originally envisaged an irrigation potential of 393927 ha for *Kharif* (July–October) and 40486 ha for *Rabi* (October–March) Irrigation. In addition to *Kharif* and *Rabi* irrigation, *Boro*(Mid February–Mid April) cultivation which was not a part of the original proposal, has also been extended depending upon the availability of surplus water after considering all the committed requirements. The study site comprises six land use classes and is therefore appropriate for the objective of this study. The data used for this study are four images of Sentinel 2 satellite from March 16th, 2017 which falls during the period of intensive cultivation of Boro rice. The delivered scenes were free from clouds or haze (Fig. 6.1). For the analysis of red-edge effect on classification results, two different spectral feature stacks were used. Details of feature stack combinations are given in Table 6.1. Thereafter supervised land use classifications with Maximum Likelihood (ML) algorithm were performed using these combinations of spectral band as input. Feature set 1 represents the primary analysis whereas feature set 2 is supplemented by three vegetation red edge bands to check the change in accuracies in the classification process. Topographic maps and data from field visits were employed for the generation of classification training and validation samples. The classification comprised six different classes including brushwood, habitation, plantation, fallow land, rice cultivation, and water body.

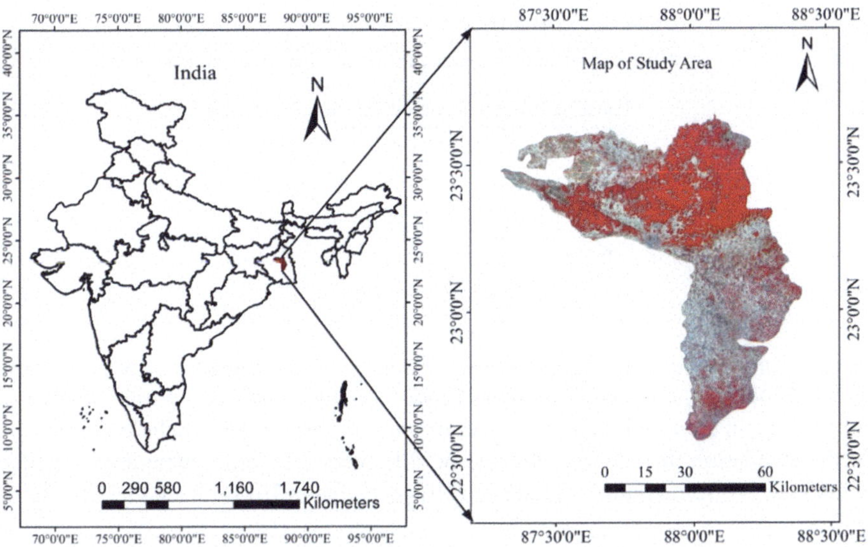

Fig. 6.1 Location of the study area and Sentinel 2 scene from March 16th 2017 displayed in the standard false color composite (FCC)

Table 6.1 Combination of spectral feature for analysis

Spectral feature	Spectrum (μm)	Feature set 1	Feature set 2
Blue	0.490	✓	✓
Green	0.560	✓	✓
Red	0.665	✓	✓
Veg. Red Edge	0.705	×	✓
Veg. Red Edge	0.740	×	✓
Veg Red Edge	0.783	×	✓
NIR	0.842	✓	✓

(✓—Inclusion of band, ×—Exclusion of band)

Several polygons distributed over the study area were selected to represent each class. For depiction and illustration of results, accuracy assessments were accomplished by taking 10 to 15 number of validation points for each classified feature sets which were collected during field visits and using high-resolution Google Earth images. The accuracy of the results was evaluated in terms of users accuracy (A_u), producer accuracy (A_p), overall accuracy (A_o), and kappa coefficient ($\hat{K}$) as follows (Story and Congalton 1986; Congalton 1991):

$$A_u = \frac{\text{Number of correctly classified pixels in each category}}{\text{Total number of pixels classified in that category}} \quad (6.1)$$

$$A_p = \frac{\text{Number of correctly classified pixels in each category}}{\text{Number of training set pixels used in that category}} \quad (6.2)$$

$$A_o = \frac{\text{Total number of corrected classified pixels}}{\text{Total number of reference pixels}} \quad (6.3)$$

$$\hat{K} = \frac{N \sum_{i=1}^{r} x_{ii} - \sum_{i=1}^{r} (x_{i+} \times x_{+i})}{N^2 - \sum_{i=1}^{r} (x_{i+} \times x_{+i})} \quad (6.4)$$

where r = number of rows in the confusion matrix, x_{ii} = number of observations in row i and column i, x_{i+} = total number of observations in row i, x_{+i} = total number of observations in column i, and N = total number of observations included in matrix. For the evaluation of red edge effects, the difference of results including red edge channel (R_{FS2}) from results excluding it (R_{FS1}) is quantified for each class by using Eq. 6.5:

$$\text{Percentage Deviation} = \frac{R_{FS2} - R_{FS1}}{R_{FS1}} \quad (6.5)$$

The percent deviation is evaluated for interdependency with the accuracy of each classification.

6.3 Results and Discussions

The incorporation of red edge leads to significant improvement in classification results. Figs. 6.2 and 6.3 show the classified images of feature set 1 (excluding red edge) and feature set 2 (including red edge). Accuracy assessment was carried out from sixty validation points which were distributed all over the study area (Fig. 6.4). Results from the analysis showed that overall accuracy is 16.67% higher due to inclusion of red edge channel in feature stack. The accuracy results indicated that the

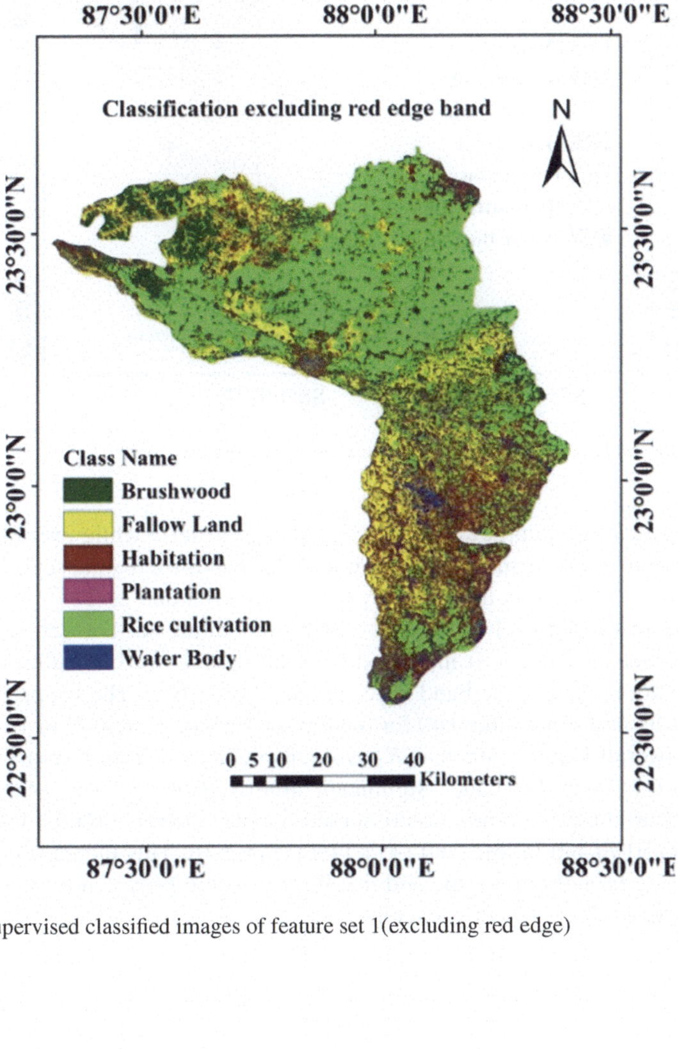

Fig. 6.2 Supervised classified images of feature set 1(excluding red edge)

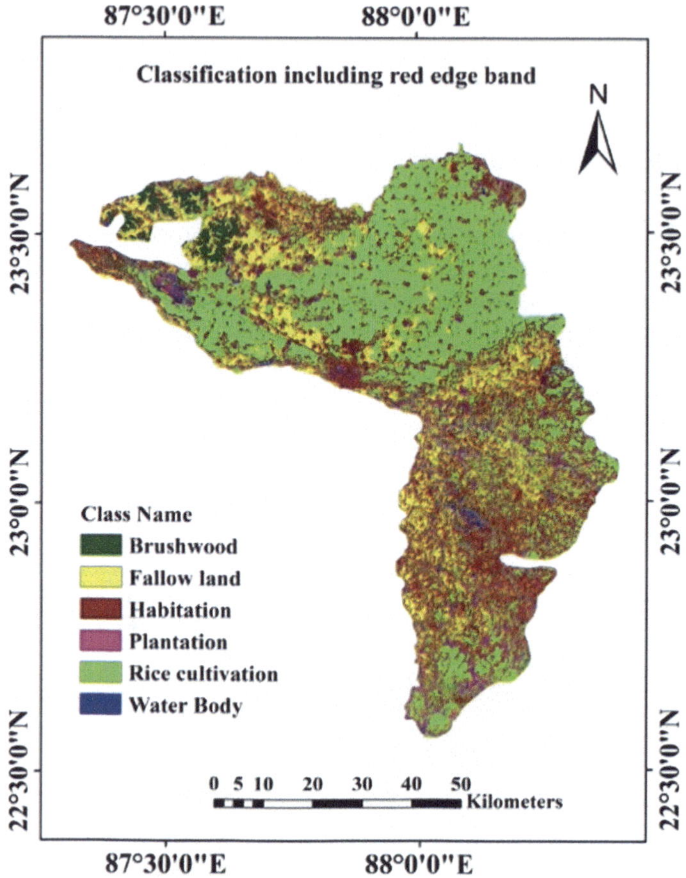

Fig. 6.3 Supervised classified images of feature set 2 (including red edge)

overall classification accuracy was 73.33% for feature set 1 while overall accuracy increased to 90% with the incorporation of red edge bands for feature set 2. Tables 6.2 and 6.3 show the class-specific accuracy for feature set 1 and 2.

Producer accuracy for plantation was 20% with feature set 1 which increased to 80% for feature set 2 while it increased from 60 to 90% for habitation class with the incorporation of red edge band in feature set 2 (Fig. 6.5). The kappa coefficient ($\hat{K}$) for brushwood was found 0.40 for feature set 1 which increased to 1 for feature set 2. The overall kappa statistics ($\hat{K}$) were also changed from 0.68 to 0.88 with the inclusion of red-edge band. Significant improvement in terms of percentage deviation of producer accuracy was observed for plantation (+ 300%) followed by habitation (+50%) and fallow land (+11.11%) (Fig. 6.6). This increase in accuracy results directly brought up by the additional information provided by the inclusion of red-edge band during feature stacking.

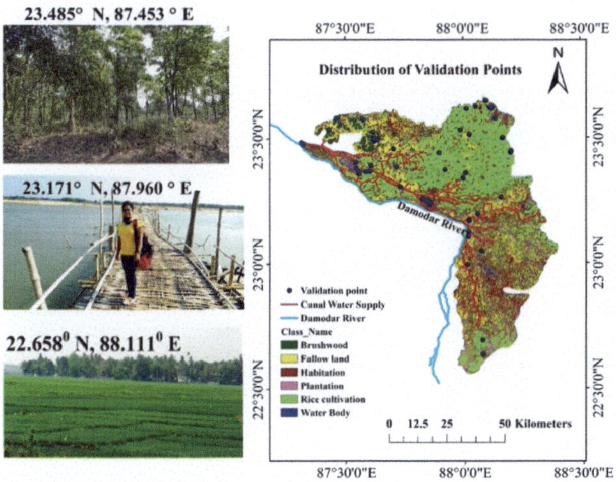

Fig. 6.4 Spatial distribution of validation point, field photos obtained during field survey with canal water supply length

These red-edge effects depended upon land use classes. Land use classes like plantation, habitation, and brushwood were affected by rise in accuracy, while accuracy in other classes like rice cultivated area, waterbody, fallow land did not appear specifically sensitive. This was due to the fact that 58% of the study area was under rice cultivation while features like water body and fallow land were distinctive features with different spectral reflectance which made these features easier to identify and less sensitive for improvement by the inclusion of red edge bands. This analysis was also verified from the results of the regression plot (Fig. 6.6) which explained the relationship between percent deviations of each class to respective classification accuracies. The linear regression coefficient was 0.88 which indicated that the scope for improvements is less for easily classified land use classes.

The sensitivity difference can be explained such as land use classes that contained vegetation signal showed higher classification accuracies with incorporation of red edge bands. Accuracy improvement in case of habitation can be explained as a combined effect of an increase in overall accuracy.

6.4 Conclusions

The incorporation of red edge channel brought significant effects on land use classification based on Sentinel 2 data. The results were influenced by investigated land use classes, area coverage, and major difference in spectral reflectance. However, it was confirmed that Sentinel 2 red edge channel has the capability to increase classi-

Table 6.2 Classification accuracy of output classified map of feature set 1

Class name	Water body	Rice cultivation	Fallow Land	Plantation	Habitation	Brushwood	Producer accuracy	Users accuracy	Conditional Kappa ($\hat{K}$)
Water Body	10	0	0	0	0	1	100%	90.91	0.891
Rice Cultivation	0	9	0	1	0	0	90	90	0.88
Fallow Land	0	0	9	0	3	0	90	75	0.70
Plantation	0	0	0	2	0	0	20	100	1
Habitation	0	0	1	1	6	1	60	66.67	0.60
Brushwood	0	1	0	6	1	8	80	50	0.40

Overall Classification Accuracy = 73.33%
Overall Kappa Statistics = 0.68

Table 6.3 Classification accuracy of output classified map of feature set 2

Class name	Water body	Rice cultivation	Fallow land	Plantation	Habitation	Brushwood	Producer accuracy	Users accuracy	Conditional Kappa ($\hat{K}$)
Water body	10	0	0	0	1	1	100%	83.33	0.80
Rice cultivation	0	9	0	2	0	0	90	81.82	0.782
Fallow land	0	0	10	0	0	0	100	100	1
Plantation	0	1	0	8	0	0	80	88.89	0.867
Habitation	0	0	0	0	9	1	90	90	0.88
Brushwood	0	0	0	0	0	8	80	100	1

Overall classification accuracy = 90.00%

Overall Kappa Statistics ($\hat{K}$) = 0.88

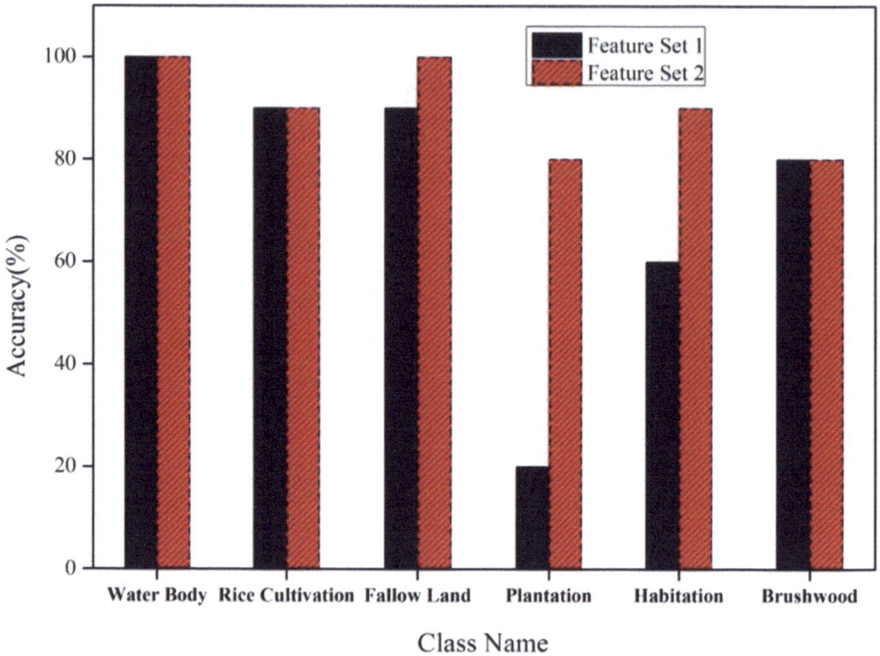

Fig. 6.5 Classification results of feature set 1 and 2

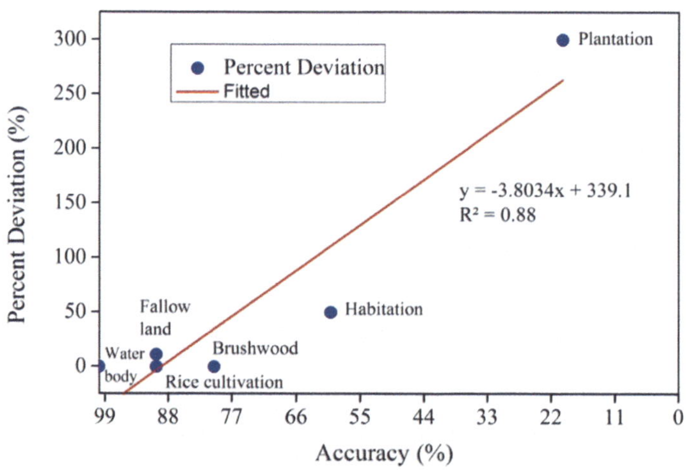

Fig. 6.6 Comparison between percentage deviation and respective classification accuracy

fication accuracy for land use classes that include vegetation signals. Furthermore, the interpretation of results for the habitation class invites advanced research in great detail.

References

Carter GA, Knapp AK (2001) Leaf optical properties in higher plants: linking spectral characteristics to stress and Chlorophyll concentration. American J Bot 88(4):677

Carter GA, Miller RL (1994) Early detection of plant stress by digital imaging within narrow stress-sensitive wavebands. Remote Sens Environ 50:295–302

Clevers JGPW, Gitelson AA (2012) Using the red-edge bands on Sentinel-2 for retrieving canopy chlorophyll and nitrogen content Proceedings of the First Sentinel-2 Preparatory Symposium, 23–27 April, Frascati, Italy: ESA-8p

Clevers JGPW, Gitelson AA (2013) Remote estimation of crop and Grass Chlorophyll and Nitrogen content using red-edge Bands on Sentinel-2 and -3. Int J Appl Earth Observ Geoinf 23:344–351

Collins W (1978) Remote sensing of crop type and maturity. Photogrammetric Eng Remote Sens 44(1):43–55

Congalton RG (1991) A review of assessing the accuracy of classifications of remotely sensed data. Remote Sens Environ 37(1):35–46

Dawson TP, Curran PJ (1998) Technical note A new technique for inter-polating the reflectance red edge position. Int J Remote Sens 19(11):2133–2139

Delegido J, Verrelst J, Alonso L, Moreno J (2012) Evaluation of sentinel-2 red-edge bands for empirical estimation of Green LAI and Chlorophyll Content. Sensors 11(7):7063–7081

Eitel JUH, Vierling LA, Litvak ME, Long DS, Schulthess U, Ager A, Krofcheck DJ, Stoscheck L (2011) Broadband, red-edge information from satellites improves early stress detection in a New Mexico Conifer Woodland. Remote Sens Environ 115:3640–3646

Filella I, Penuelas J (1994) The red edge position and shape as indicators of Plant Chlorophyll content, biomass and hydric status. Int J Remote Sens 15(7):1459–1470

Frampton WJ, Dash J, Watmough G, Milton EJ (2013) Evaluating the capabilities of Sentinel-2 for quantitative estimation of biophysical variables in vegetation. ISPRS J Photogrammetry Remote Sens 82:83–92

Hill MJ (2013) Vegetation index suites as indicators of vegetation state in grassland and savanna: An analysis with simulated SENTINEL 2 data for a North American transect. Remote Sens Environ 137:94–111

Ju CH, Tian YC, Yao X, Cao WX, Zhu Y, Hannaway D (2010) Estimating Leaf Chlorophyll Content Using Red Edge Parameters. Pedosphere 20(5):633–644

Schuster C, Förster M, Kleinschmit B (2012) Testing the red edge channel for improving land-use classifications based on high-resolution multi-spectral satellite data. Int J Remote Sens 33(17):5583–5599

Smith KL, Steven MD, Colls JJ (2004) Use of Hyperspectral derivative ratios in the red-edge region to identify plant stress responses to gas leaks. Remote Sens Environ 92:207–217

Story M, Congalton RG (1986) Accuracy assessment: a user's perspective. Photogrammetric Eng Remote Sens 52(3):397–399

Tigges J, Lakes T, Hostert P (2013) Urban vegetation classification: benefits of multi-temporal rapideye satellite data. Remote Sens Environ 136:66–75

Chapter 7
Reference Crop Evapotranspiration Estimation Using Remote Sensing Technique

Samuel Malou Mukpuou, Ashish Pandey, and V. M. Chowdary

Abstract Reference crop evapotranspiration (ET_0) is a basic parameter for irrigation water management, and its estimation is crucial, and therefore, there exists a number of empirical and physically based approaches for its estimation. However, lack of conventional ground weather station data, required by these approaches, is a big challenge in a developing country like South Sudan, Eastern Africa, the population of which mainly depend on rainfed agriculture and their production has diminished tremendously in recent years. Irrigation is being taken up as a remedy for diminished food production but due to non-availability of data for reference crop evapotranspiration estimation, irrigation planning and management are seriously affected. Thus, this study proposes a simple remote sensing technique for estimating monthly reference crop evapotranspiration without ground weather station data, using high spatial resolution remote sensing data, in dry season of South Sudan. The study evaluated the use of land surface temperature retrieved from Landsat 8 data as an alternative input for three commonly used temperature-based models (Blaney-Criddle, Thornthwaite and Hargreaves) in Juba County of South Sudan. The proposed methodology has also been compared with the previous studies carried out by other researchers (Maeda et al. in Appl Geogr 31:251–258, 2011) that use moderate resolution imaging spectroradiometer (MODIS) land surface temperature. Further, the study also proposes an automatic satellite image processing algorithm for each of the temperature-based ET_0 methods for reasons of simplicity of image processing and calculations involved. Weather data obtained from Juba County was used in FAO Penman–Montieth model, for calibration and validation. The results of the evaluated methods based on the

S. M. Mukpuou (✉)
Department of Agricultural Engineering, University of Juba, Juba, South Sudan
e-mail: mukpuou@gmail.com

A. Pandey
Department of Water Resources Development and Management, IIT Roorkee, Roorkee, India
e-mail: ashish.pandey@wr.iitr.ac.in

V. M. Chowdary
NRSC, ISRO, DOS Branch Secretariat, Loknayak Bhawan, New Delhi, India
e-mail: muthayya.chowdary@gmail.com

A. Pandey et al. (eds.), *Water Management and Water Governance*, Water Science and Technology Library 96,
https://doi.org/10.1007/978-3-030-58051-3_7

proposed methodology, with FAO Penman–Montieth, show that Hargreaves method performed well in this specific study area, with overall RMSE of 0.106 mm/day, MAE of 0.016 mm/day and coefficient of determination (R^2) of 0.928, indicating the satisfactory performance. Thus, the proposed methodology may be used for irrigation planning in Juba County as well as South Sudan at large if no ground station data is available.

Keywords Reference crop evapotranspiration · Land surface temperature · Remote sensing technique

7.1 Introduction

Water is a crucial asset for all forms of life on this earth. It is used for agriculture and other purposes. Wisser et al. (2008) indicated that agriculture is a major water-consuming sector accounting for 70% of the world water consumption. Hence, irrigation water management is very important in the world over. One of the basic parameters in irrigation management is reference crop evapotranspiration (ET_0), which indicates the water lost through evaporation and transpiration from hypothetical grass surface not short of moisture (Allen et al. 1998). Its estimation is of great value in planning, design and management of irrigation systems (Valipour 2015; Trajkovic 2005; Maeda et al. 2011; Hargreaves 1994; Azhar and Perera 2011; Droogers and Allen 2002).

Many empirical- and physical-based approaches are there for ET_0 estimation (Allen et al. 1998). However, the lack of ground weather station data required by these approaches is a big drawback in a developing country like South Sudan, Eastern Africa. South Sudan's population depend mainly on rainfed agriculture, whose production diminished tremendously in recent years and irrigation is being taken up as a remedy for diminished food production, but due to non-availability of reference crop evapotranspiration data, irrigation planning and management are seriously affected. However, in such data scarce places, like South Sudan, satellite remote sensing can provide an alternate means for acquiring data needed (Wagner et al. 2009). According to Maeda et al. (2011), a blend of ET methods with remotely sensed data can provide an alternative for ET_0 estimation. Remote sensing data had been used to estimate evapotranspiration by many researchers (Ray and Dadhwal 2001; Bois et al. 2008; El-Shirbeny 2016; Zheng and Zhu 2015).

Looking to the aforementioned, the main aim of this paper is to estimate ET_0 without ground station data. The paper proposes a simple remote sensing technique for estimation of monthly ET_0 using high spatial resolution remote sensing data, in dry season of Juba County, South Sudan. The study evaluates the use of land surface temperature retrieved from Landsat 8 data as an alternative input for three commonly used temperature-based ET_0 models (Blaney-Criddle, Thornwaite and Hargreaves). Further, the proposed methodology has been also compared with the similar procedure, proposed by others (Maeda et al. 2011), which use moderate resolution imaging spectroradiometer (MODIS) land surface temperature (LST). The study also proposes an automatic satellite image processing algorithm for each of the temperature-based ET_0 methods for reasons of simplicity of image processing and calculations involved.

7.2 Material and Methods

7.2.1 Study Area

The study area, Juba County, is situated in South Sudan, the newest nation in Africa that became independent from Sudan on July 9, 2011. It is a landlocked country, bounded on the north by Sudan, on the West by Central Africa Republic, Democratic Republic of Congo on Southwest, Uganda on the South, Kenya on southeast and Ethiopia on the East. Geographically, South Sudan is situated in the tropical region between latitudes 3° N and 13° N and longitudes 24° E and 36° E and covers nearly 647,000 km^2 geographical area (Fig. 7.1). South Sudan is divided into six agro-climatic zones, namely flood plains, greenbelt, iron stone plateau, hills and mountains, Nile and Sobat and arid (Tizikara et al. 2015) (Fig. 7.1). Major occupation in South Sudan is agriculture, which is mostly rainfed. And rainfall distribution in this area is erratic in nature. This made farmers and government to devise strategies for protective irrigation as dry season prevails for almost six months. The weather data used in this study was collected from Juba International Airport Station, situated at approximately 4.86° N latitude and 31.6° E longitude, representative of flood plains agro-climatic zone (Fig. 7.1). Flood plains are the biggest agro-climatic zone in South Sudan covering an area of nearly 18,789 km^2.

7.2.2 Climatological ET_0 Methods

Several empirical- and physical-based climatological ET_0 models exist, and these models differ in complexity and data requirements. The complex ones need more climatological data and result in good performance for variety of climates while models with limited data are applicable for specific climatic regions. Particularly, the availability of climatic data required by the models is difficult for developing country like South Sudan. Even simple air temperature data required by temperature-based ET_0 methods can be difficult to obtain in South Sudan. In order to overcome data problem, temperature-based empirical ET_0 models were evaluated by replacing the air temperature with satellite-derived land surface temperature (LST) for ET_0 computations. Details of the temperature-based models are presented below.

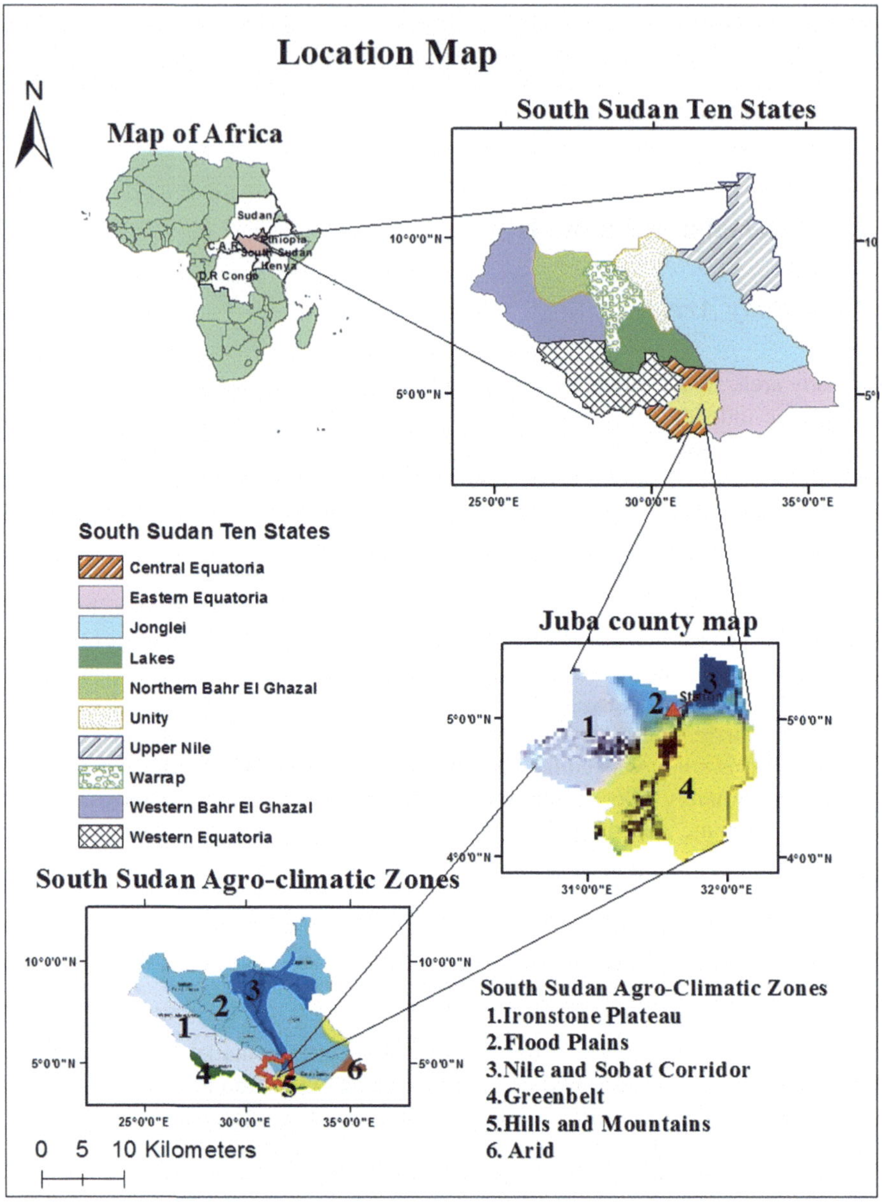

Fig. 7.1 Location map of the study area

7.2.2.1 Blaney–Criddle Method

Blaney and Criddle method of ET_o estimation is based on the mean air temperature and monthly percentage of daylight hours of the period under consideration. It is expressed as (Doorenbos and Pruitt 1977)

$$ET_o = p(0.46T + 8) \tag{7.1}$$

where ET_o is reference evapotranspiration, mm/day, T is average daily temperature, and °C, p daylight per cent.

7.2.2.2 Thornthwaite Method

Thornthwaite (1948) equation related a reference crop evapotranspiration with average monthly temperature as follows (Subedi and Chávez 2015):

$$ET_o = 16\left(10\frac{T}{I}\right)^a \tag{7.2}$$

where ET_o is reference evapotranspiration of standard month of 30 days and 12 h in mm/month, T is the mean monthly temperature in °C, I is the annual thermal/heat index, i.e., sum of monthly heat index $\left(i = \left(\frac{T}{5}\right)^{1.514}\right)$, and $\boldsymbol{a}$ is constant which depends on annual heat index and is expressed as follows: $a = 6.75 * 10^{-7}I^3 - 7.71 * 10^{-5}I^2 + 1.792 * 10^{-2}I + 0.49239$.

7.2.2.3 Hargreaves Method

Hargreaves and Samani (1985) equation (Hosseinzadeh Talaee 2014) for computing reference evapotranspiration from air temperature is as follows:

$$ET_o = 0.0023 * R_a(T_{mean} + 17.8)(T_{max} - T_{min})^{0.5} \tag{7.3}$$

where ET_o is reference evapotranspiration in mm/day, R_a is extraterrestrial radiation in mm/day which can be obtained from standard tables available in most literatures, T_{mean} is the average daily temperature, and T_{max} and T_{min} are maximum and minimum daily temperature, respectively.

7.2.3 Input Data

Traditionally, weather data used in the ET_0 models is obtained from established ground weather stations. However, in South Sudan as well as in other developing nations, getting such data and establishing weather stations to record such a data are cumbersome. Even recording of air temperature is very difficult.

Thus, in this study, remotely sensed data (satellite land surface temperature) is tested as an alternative input for temperature-based ET_0 methods. Satellite land surface temperature refers to radiometric temperature emitted from the earth surface observed by a sensor on the satellite. Data from two satellites one with low spatial resolution (Terra data) and one with high spatial resolution (Landsat 8 data) has been evaluated in this study.

7.2.3.1 Landsat 8 Data

Operation Land Imager (OLI) and Thermal Infrared Sensor (TIRS) images, onboard of Landsat 8, were used for computing land surface temperature (LST). Further, LST was used to compute reference crop evapotranspiration using temperature-based ET_0 methods. Landsat 8 imageries were downloaded from the United States Geological Survey (USGS) Web site (https://earthexplorer.usgs.gov/). Landsat 8 acquires data from 11 spectral bands, nine of which are in shortwave, captures by OLI sensor, with 30 m spatial resolution except band 8; panchromatic which have 15 m. And two thermal bands (10 and 11) are captured by TIRS, with 100 m resolution. In this study, the bands used were red band (R), near infrared (NIR) band and one TIR band, i.e., 10. R and NIR bands were used for computing land surface emissivity (LSE). Although there are two thermal bands (10 and 11), available in the Landsat 8 images for retrieval of radiometric temperature, use of band 11 is not recommended by USGS due to large calibration uncertainty. Thus, band 10 was used for calculating brightness temperature which in turn was used together with land surface emissivity obtained from R and NIR band to estimate actual land surface temperature. Band 10 has spatial resolution of 100 m and is resampled to 30 m in this study. Study location, Juba County, is covered by four scenes, and a total of 28 images representing 7 months, in two dry seasons, were employed in this study. Acquisition details of the satellite data used in this study are presented in Table 7.2.

Since Landsat 8 satellite acquires data at a time, which may not correspond to maximum, mean or minimum temperature acquisition times, spatial LST values obtained from Landsat 8 were scaled using historical long-term maximum, minimum air temperatures as per ET_0 model needs. For scaling the LST, historical temperature (maximum and minimum) data pertaining to identified gridded stations at 31°33′45″ E longitude and 4°50′22″ N latitude for Juba county obtained from National Center for Environmental prediction (NCEP) Web site (https://globalweather.tamu.edu/) was used. Maximum and minimum temperature values from NCEP data corresponding to image acquisition months were averaged for 35 years (1979–2013),

Table 7.1 Juba County Landsat 8 LST Scaling data

2014–2015: Dry season scaling data (all in °C except ratios)							
Month	Max	Min	Mean	LST	Max: LST	Min: LST	Mean: LST
November	36	22	29	34	1.05	0.64	0.85
December	36	21	28	38	0.95	0.54	0.74
January	37	19	28	35	1.06	0.55	0.81
February	39	21	30	37	1.05	0.55	0.80
Average scaling ratio					**1.03**	**0.57**	**0.80**

NB: Max, Min and Mean are averages of maximum, minimum and mean temperature values, respectively, obtained from NCEP data, and LST is the land surface temperature obtained from Landsat 8 imagery.

and their means were also computed. Then scaling ratios were computed, and their averages were used. Scaling was done for the data pertaining to one season, while other season has been used in validation (Table 7.1).

7.2.3.2 MODIS Data

MODIS sensor is onboard of two satellites, namely Terra and Aqua, and provides images in 36 bands. The MODIS land surface temperature was downloaded from Land Processes Distributed Active Achieve Center (LPDAAC) Web site (https://lpdaac.usgs.gov/data_access/reverb). Its spatial resolution is 1 km, and temporal resolution is 1–2 days. In this study, MODIS MOD11A2 product, which provides the average values of clear sky LSTs during eight-day period (day and night) on 1 km resolution, in sinusoidal grid was used.

The daytime LST corresponds to one obtained by a sensor approximately between 10:00 am and 12:00 pm, and nighttime LST corresponds to one obtained by sensor approximately between 22:00 pm and 23:00 pm. A total of 138 MODIS images (tile h21v08) of the study location were downloaded from LPDAAC Web site for the years, 2013, 2014 and 2015. Re-projection of images to geographic coordinate system (Datum WGS84) was carried out in ENVI 5.3 software and projected to UTM zone 36 N in ArcGIS 10.4. Monthly LST was computed by aggregating eight-day LST images and, one that corresponds to months in which Landsat 8 LST was derived, was used. The LST values originally in kelvin were converted to degree Celsius as per ET_0 models requirement. The daytime MODIS LST was taken as maximum temperature and nighttime LST as minimum temperature and their average as mean temperature for application in ET_0 models (Table 7.2).

Table 7.2 Satellite data acquisition for the Juba County

Path-row	Season 1 (November 2014–February 2015)		Season 2 (November 2013–February 2014)	
	Acquisition date	Local solar time	Acquisition date	Local solar time
172-56	30-11-2014	0806	29-12-2013	0807
172-57	30-11-2014	0806	29-12-2013	0807
173-56	05-11-2014	0812	20-12-2013	0813
173-57	05-11-2014	0812	20-12-2013	0813
172-56	16-12-2014	0806	30-01-2014	0806
172-57	16-12-2014	0806	30-01-2014	0806
173-56	23-12-2014	0812	05-01-2014	0813
173-57	23-12-2014	0812	05-01-2014	0813
172-56	17-01-2015	0806	15-02-2014	0806
172-57	17-01-2015	0806	15-02-2014	0806
173-56	08-01-2015	0812	22-02-2014	0812
173-57	08-01-2015	0812	22-02-2014	0812
172-56	02-02-2015	0806		
172-57	02-02-2015	0806		
173-56	09-02-2015	0812		
173-57	09-02-2015	0812		

7.2.4 Methodology

7.2.4.1 Estimation of ET_0 from LANDSAT 8 Data

The general framework for computing reference crop evapotranspiration from Landsat 8 data is presented in Fig. 7.2. Initially, null data pixels from the images, downloaded from USGS website, were removed using ArcGIS 10.4. And subsequently, these images were mosaicked, subset and extracted band wise corresponding to study location in ERDAS IMAGINE 2015. Once the data is acquired, preprocessing of satellite images (bands 4, 5 and 10) was carried out for retrieval of brightness temperature (BT) and land surface emissivity (LSE). Further, both BT and LSE are used to compute land surface temperature (LST). LST is then refined by masking out extreme values after which it is scaled to maximum, mean and minimum values using NCEP data as mentioned earlier. The scaled LST is then used for computation of ET_0 using temperature-based methods.

The specific procedures for each of the three temperature-based ET_0 methods are presented below.

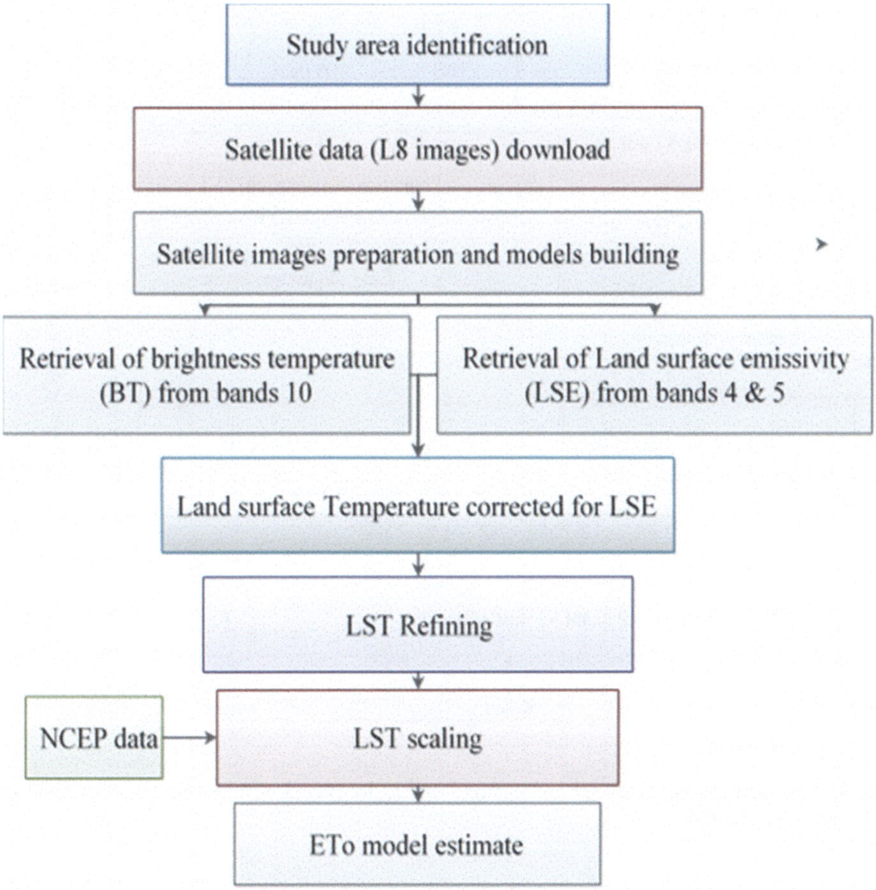

Fig. 7.2 Schematic framework for computation of ET_0 using satellite-derived LST methods

Blaney–Criddle ET_0 Model

The algorithm was developed in ArcGIS 10.4 to automatically derive ET_0 from Landsat 8 data-based Blaney–Criddle method. The input data for this model includes band 10, NDVI, maximum NDVI and minimum NDVI, satellite images rescaling factors and Blaney–Criddle ET_0 equation constant (P). Algorithm uses raster resampling tool to resample band 10, which is originally of 100 m spatial resolution to the same resolution as NDVI, i.e., 30 m and raster calculator tool for all computations. Steps used by the mentioned algorithm for ET_0 calculation from Landsat 8 data are presented in Fig. 7.3. With group of steps in the box (**GENERAL FOR ALL MODELS**) as common for all temperature-based ET_0 models considered in this study. Condition is set in the model to ignore LST values less than 0 °C and more than 46 °C.

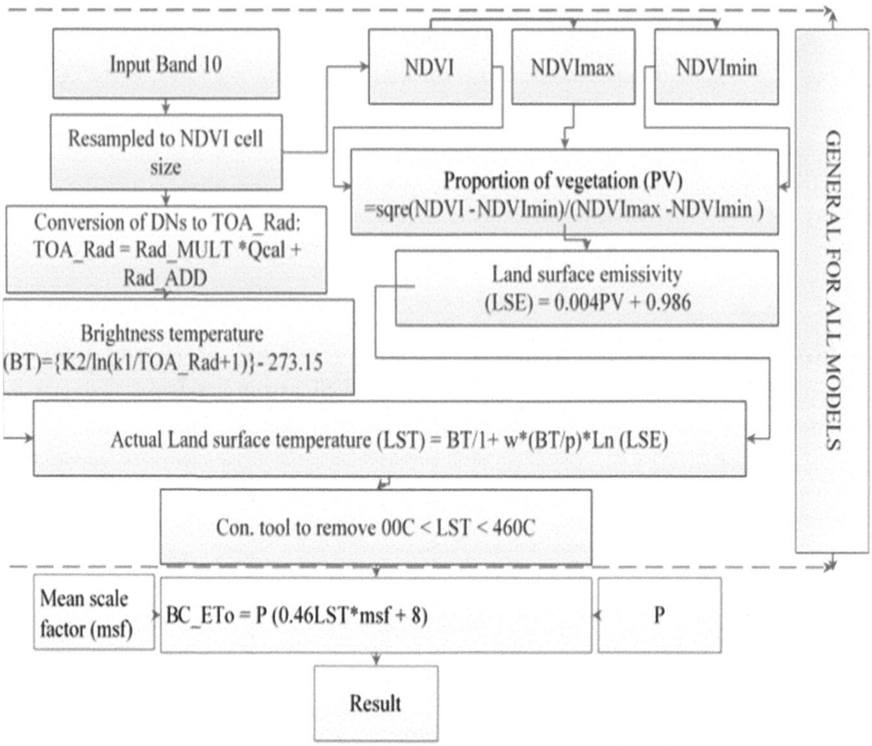

Fig. 7.3 Schematic framework for computation of ET_0 by Blaney–Criddle model using Landsat 8-derived LST method

Top of Atmosphere Spectral Radiance (L_λ): In order to account atmospheric effect, digital numbers (DNs) of band 10 are converted into top of atmosphere spectral radiance using the formula obtained from USGS Web site (https://landsat.usgs.gov/using-usgs-landsat-8-product) and are given below:

$$L_\lambda = M_L * Q_{cal} + A_L \tag{7.4}$$

where L_λ is radiance in w sr^{-1} m^{-2}, M_L is the band-specific multiplicative rescaling factor, Q_{cal} is the band digital number (DN), i.e., raw image itself, and A_L is the band-specific additive rescaling factor. M_L and A_L are provided in file metadata. The values of M_L and A_L are **0.0003342** and **0.1**, respectively. They are relatively constant for all images.

Conversion of radiance to brightness temperature: Once DNs are converted to top of the atmospheric radiance, i.e., corrected for atmospheric effect, satellite or brightness temperature is computed from the L_λ using the algorithm as given below (https://landsat.usgs.gov/using-usgs-landsat-8-product).

$$\mathrm{BT} = \left(\frac{K_2}{\mathrm{In}\left(\frac{K_1}{L_\lambda} + 1\right)} \right) - 273.15 \quad (7.5)$$

In which BT = brightness temperature in °C, K_1 and K_2 are band thermal constants available in metadata of the image's file. And 273.15 is the conversion factor from kelvins to degree Celsius. For band 10 of the chosen area images, K_1 is 774.8853 and K_2 is 1321.0789.

Normalized difference vegetation index (NDVI) is computed using the following relation:

$$\mathrm{NDVI} = \frac{\mathrm{NIR(band\,5)} - R(\mathrm{band\,4})}{\mathrm{NIR(band\,5)} + R(\mathrm{band\,4})} \quad (7.6)$$

where NIR is near infrared band (band 5) and R is red band (band 4). But prior to using Eq. 7.6, atmospheric and solar angle correction of the reflectance were done using the following equation, given in USGS page (https://landsat.usgs.gov/using-usgs-landsat-8-product) to get top of atmosphere reflectance (TOA_Ref).

$$\mathrm{TOA\_\ Ref} = \frac{(\mathrm{Ref\_\ MULT} * \mathrm{band\,DNs} + \mathrm{Read})}{\sin(\theta)} \quad (7.7)$$

where θ is sun elevation angle, Ref_MULT is band's reflectance multiplicative rescaling factor, and Ref_ADD is band's reflectance additive rescaling factor, both given in file metadata. For the images of the chosen areas, Ref_MULT = 0.00002 and Ref_ADD = −0.1. In calculation of NDVI, the parameter sin (θ) in the above relation cancels out for two bands and does not need to be involved in computation.

The proportion of vegetation (P_V) is calculated by using simple equation given by Carlson and Ripley (1997)

$$P_V = \mathrm{square}\left(\frac{\mathrm{NDVI} - \mathrm{NDVI_{min}}}{\mathrm{NDVI_{max}} - \mathrm{NDVI_{min}}} \right) \quad (7.8)$$

where $\mathrm{NDIV_{max}}$ and $\mathrm{NDVI_{min}}$ are maximum and minimum NDVI values, respectively.

Lands surface emissivity (LSE): LSE is computed by using the equation proposed by Giannini et al. (2015).

$$\mathrm{LSE} = 0.004 P_V + 0.986 \quad (7.9)$$

where P_v is proportion of vegetation.

Land surface temperature with different land cover considered is calculated using the Eq. 7.10 proposed by Stathopoulou and Cartalis (2007).

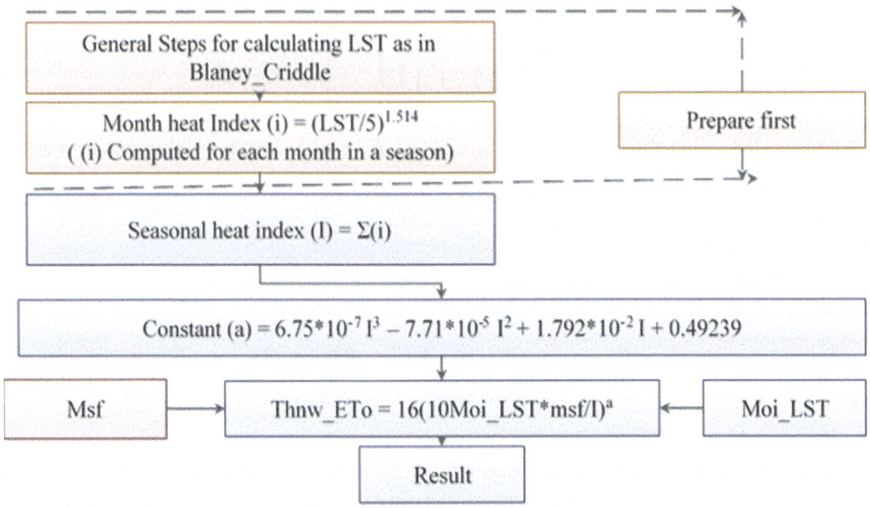

Fig. 7.4 Schematic framework for computation of ET_o by Thornwaite model using Landsat 8-derived LST method

$$\mathrm{LST} = \mathrm{BT}/1 + w * (\mathrm{BT}/p) * \mathrm{Ln}(\mathrm{LSE}) \tag{7.10}$$

where LST is land surface temperature, ***w*** is band wavelength, P is constant = hc/σ (14,380), h is Planck's constant (6.626×10^{-34} J × s), σ is Boltzmann constant ($1.38*10^{-23}$ J/K), and c is velocity of light ($2.998*10^{8}$ m/s) and other variables as defined previously.

Lastly, the model uses the following relation to get ET_o by Blaney–Criddle method (BC_ET_o), in mm/day (average monthly)

$$\mathrm{BC\_\ ET_{o}}\text{-} = P(0.46\mathrm{LST} * \mathrm{msf} + 8) \tag{7.11}$$

where LST is land surface temperature, msf is mean scaling factor obtained from Table 7.1 (Sect. 7.2.3.1), and ***P*** is per cent of daylight hours. P is function of latitude of a place and month of the year. Thus, standard values of P are adopted from literature.

Thornwaite ET_o Model

The algorithm was created in ArcGIS 10.4 to compute ET_o based on Thronwaite method, from Landsat 8 imageries. The model inputs are month of interest LST (moi_LST) and land surface heat index (i) of different months in a season. The procedures used by algorithm are presented in Fig. 7.4.

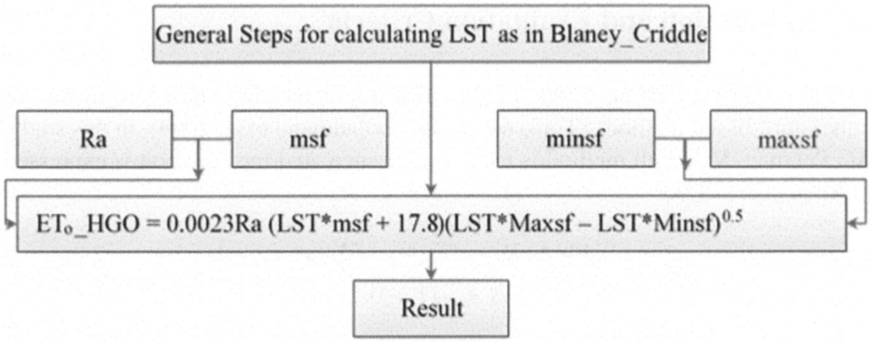

Fig. 7.5 Schematic framework for computation of ET_o by Hargreaves model using Landsat 8-derived LST method

Hargreaves (1985) ET_o Model

The algorithm for computation of ET_o by Hargreaves original equation is given in Fig. 7.5. The model inputs are band 10, NDVI, $NDVI_{max}$, $NDVI_{min}$ andextraterrestrial radiation value (Ra). Maximum, minimum and mean LST scaling factors are obtained as mentioned in Sect. 7.2.3.1. Ra is a function of latitude and period of year, and it is obtained from standard values available in literature.

Note that in the flowchart, msf, minsf and maxsf refer to mean, minimum and maximum LST scaling factors, respectively.

7.2.4.2 Assessment of ET_o Using MODIS LST

Eight-day MODIS land surface temperature images (MOD11A2) were downloaded and projected to geographic coordinate system (datum WGS84) in ENVI 5.3. Images with no data pixels were first converted to points and then interpolated using nearest neighborhood interpolation technique in ArcGIS spatial analyst tool. Further, eight-day LST products were aggregated into monthly scale and re-projected to UTM coordinate system (datum WGS84) and subsetted to the area of interest. Rescaling was done using the factor 0.002, as per the metadata file of the LST images and converted to degree Celsius from kelvin. LST extracted from the images acquired during daytime and nighttime was considered as maximum and minimum, respectively. The appropriate finished product(s) of the above processes were then used in computation of ET_o in the appropriate method.

7.3 Calibration and Evaluation Criteria

Several researchers recommended calibration for enhancing the performance of temperature-based methods (Trajkovic 2005; Fooladmand et al. 2008). In this study, FAO Penman–Monteith method is used as reference, and the equation is expressed as (Allen et al. 1998).

$$\mathrm{ET_o} = \frac{0.408\Delta(R_n - G) + \gamma \frac{900}{T+273} u_2(\mathrm{es} - \mathrm{ea})}{\Delta + \gamma(1 + 0.34u_2)} \tag{7.12}$$

where $\mathrm{ET_o}$ is reference evapotranspiration in mm/day; R_n is net radiation in MJ/m^2/day; G is soil heat flux in MJ/m^2/day; T is mean daily air temperature in °C at reference height (2 m); u_2 is wind speed in m/s at reference height; $e_s - e_a$ is saturation vapor pressure deficit in kPa; e_s is saturation vapor pressure in kPa; e_a is the actual vapor pressure in kPa; Δ is slope of vapor pressure curve in kPa/°C, and γ is psychometric constant kPa/°C.

Juba County Station Data

The monthly averages of temperatures (maximum and minimum; Table 7.3) were collected from Juba International Airport Station, located at an altitude of 457 m above mean sea level, 4.86° N latitude and 31.6° longitude.

ET_o estimation with limited meteorological data: $\mathrm{ET_o}$ was computed using CROPWAT 8.0 software, as this particular software has capability to estimate $\mathrm{ET_o}$ with limited weather data. The software has an option of computing reference

Table 7.3 Station data of Juba County

Year	2013		2014		2015	
Month	Monthly Max. *T*	Monthly Min. *T*	Monthly Max. *T*	Monthly Min. *T*	Monthly Max. *T*	Monthly Min. *T*
January	37.5	19.8	33.3	18.5	35.9	17.9
February	39.0	20.4	33.4	18.3	37.9	22.1
March	38.1	22.5	27.6	22.4	37.5	23.8
April	35.6	22.6	35.1	23.1	33.4	21.9
May	34.1	22.9	33.3	23.0	34.9	22.3
June	33.1	22.6	31.8	21.8	32.7	23.2
July	31.6	22.0	31.5	21.9	33.5	21.7
August	32.3	21.3	31.2	21.7	33.8	21.0
September	32.9	23.2	32.3	21.6	34.4	21.3
October	33.0	22.9	32.3	21.9	33.5	20.3
November	33.6	22.0	34.3	21.6	33.4	21.0
December	35.7	20.6	35.8	20.3	34.7	20.2

crop evapotranspiration using maximum and minimum temperatures, based on FAO Penman–Montieth method. Methodology for estimation of parameters required for estimation of ET_o is adopted from FAO Irrigation and Drainage paper-56 (Allen et al. 1998). It is indicated that errors produced by FAO-PM method based on temperature are low for monthly or higher periods (López-Moreno et al. 2009), therefore, adopted in this study as standard, because ET_o estimation is for longer period.

Further, spatial ET_o estimated using remote sensing data is calibrated with reference to ET_o from meteorological station data, using pixel values corresponding to same meteorological station's location (latitude/longitude). Following linear regression equation is used.

$$ET_{o}\_cal = a + bET_{o}\_sat \tag{7.13}$$

where ET_o_cal represents the calibrated ET_o and ET_o_sat is ET_o estimated from satellite data. "a" and "b" are obtained by regression analysis between Penman–Montieth equation as standard with ET_o_sat.

Comparison of models was done using standard statistics and linear regression analysis (Douglas et al. 2009; Maeda et al. 2011). Root mean squared error (RMSE), mean absolute error (MAE), percentage error (PE) and coefficient of determination (R^2) were computed using the equations described below:

$$\text{RMSE} = \left(\frac{1}{n}\Sigma(ET_{o}\_cal - ET_{o}\_PM)^2\right)^{0.5} \tag{7.14}$$

$$\text{MAE} = \frac{1}{n}\Sigma|ET_{o}\_cal - ET_{o}\_PM| \tag{7.15}$$

$$\text{PE} = \frac{(ET_{o}\_cal - ET_{o^-}\_PM)}{ET_{o}\_\ PM} * 100 \tag{7.16}$$

$$R^2 = \frac{\Sigma(ET_{o}\ \_cal - ET_{o^-}\_cal\_avg)^2}{\Sigma\left(ET_{o\_PM} - ET_{o\_PM\_avg}\right)^2} \tag{7.17}$$

7.4 Results and Discussion

The error and regression analysis results are presented in Table 7.4. It is inferred that models parameterized with Landsat 8 LST show best fit results than models parameterized with MODIS LST, a performance which may be attributed to coarse spatial resolution of MODIS data. Using Landsat 8 LST data, the seasonal RMSE values for Blaney–Criddle, Hargreaves and Thornthwaite were 0.1123 mm/day, 0.1064 mm/day and 0.1117 mm/day, respectively, and MAE were 0.0954 mm/day, 0.0163 mm/day

Table 7.4 Summary of error and linear regression analysis results

Method/evaluation Item	Blaney–Criddle		Hargreaves		Thornthwaite	
Satellite data used	Landsat 8	MODIS	Landsat 8	MODIS	Landsat 8	MODIS
R-sqaure	0.9193	0.5583	0.9275	0.3949	0.9202	0.5709
R	0.9588	0.7472	0.9631	0.6284	0.9593	0.7556
RSME	0.1123	0.2627	0.1064	0.3075	0.1117	0.2689
MAE	0.0954	0.2102	0.0163	0.1320	0.0944	0.2237
Calibration parameter (a)	−3.8274	−21.236	1.1727	−0.3784	1.0836	−2.4869
Calibration parameter (b)	1.7184	4.4913	0.7997	0.9943	1.0357	1.4107

and 0.0944 mm/day, respectively. While using MODIS LST data, RMSE values were 0.2627 mm/day, 0.3075 mm/day and 0.2689 mm/day, respectively, and the values of MAE were 0.2102 mm/day, 0.1320 mm/day and 0.2237 mm/day for Blaney–Criddle, Hargreaves and Thornthwaite method, respectively. The calibrated models were tested on training data (i.e., data that was used in calibration), and error resulted in per cent is presented in Fig. 7.6. Although all the modeled ET_o values are close to the ET_o values of FAO Penman–Montieth method as shown by percentage deviation errors in Fig. 7.6. It is equally noticed that only the ones in which Landsat 8 LST has been used are relatively very close. Further, out-of-time validation was done (for December 2013–February 2014, months whose Landsat 8 images were less contaminated by clouds, in this specific dry season), and the errors of prediction, in per cent,

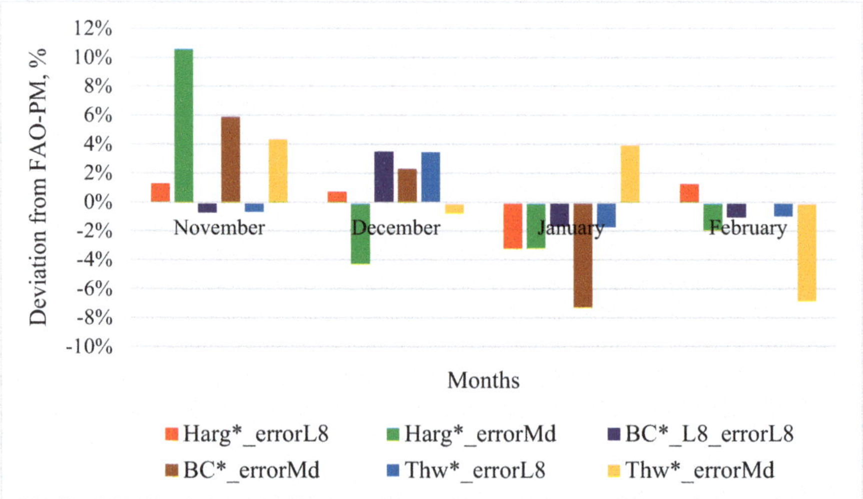

Fig. 7.6 Errors after calibration: dry season of 2014–2015

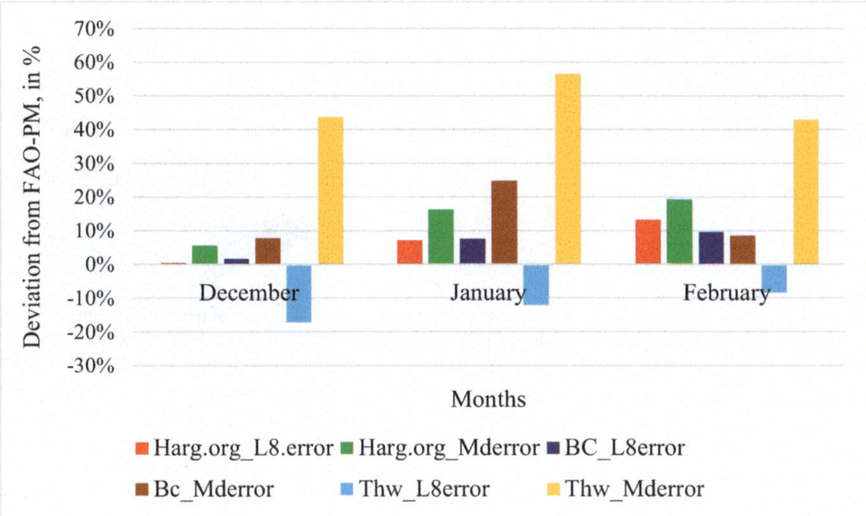

Fig. 7.7 Errors of out-of-time validation: dry season of 2013–2014

are presented in Fig. 7.7. And it can be noticed from Fig. 7.7 that the errors, when models were parameterized with Landsat 8 LST, are minimal; all models produced errors less or equal 17%. While with MODIS LST, the models' errors were high.

Based on regression and error observation, Hargreaves model based on Landsat 8 data appears to be the best in the study area, in the period considered, it is relatively consistent, with overall MAE values of 0.0163 mm/day, RMSE values of 0.1064 mm/day and R^2 values of 0.9275 and produces small monthly errors as presented in Figs. 7.6 and 7.7, except in February (season 2013–2014) where its error (13%) is slightly higher than Blaney–Criddle error (9%), but still reasonable. And based on minimizing errors of prediction, Blaney–Criddle, Landsat 8-based, is the second best after Hargreaves as manifested in error as shown in Figs. 7.6 and 7.7. Blaney–Criddle was giving negative ET_o values on some pixel for the images which were somehow contaminated by clouds, and this shows that it is more sensitive to temperature difference and, hence, cannot be applicable for the proposed methodology, for any variation that exists between LST and air temperature would greatly affect its results. Therefore, Hargreaves Landsat 8-based model is only selected, as relatively best in this study. ET_o maps of the study area obtained by Hargreaves based on Landsat 8 LST data for two dry seasons (2013–2014 and 2014–2015 seasons) is presented in Fig. 7.8.

Linear regression results of the satellite data parameterized ET_o models showed that Landsat ET_o models values have close relationship with FAO-PM values than MODIS ET_o methods (Table 7.4); a thing which may be accounted to spatial resolution of the data two satellites. Because coarse spatial resolution of MODIS data does not consider land surface heterogeneity, this have been found in many researches, for instance (Li et al. 2017; Kustas et al. 2003). Some studies like (Xiaozhou et al.

Fig. 7.8 Maps showing average ET_0 for 3 months in dry season, using Hargreaves model parameterized with satellite-derived LST

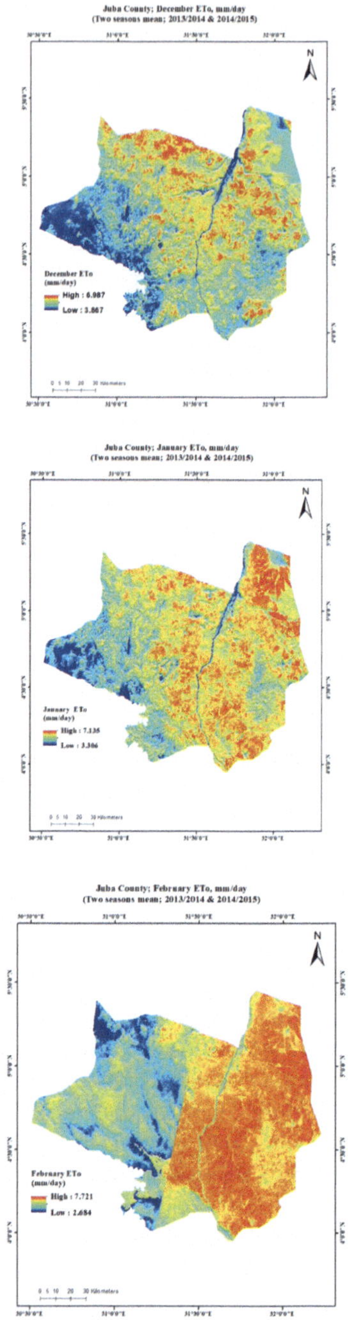

2005) indicated that MODIS data causes major errors in estimating any of the land surface energy fluxes like evapotranspiration. However, to assure the relative confident in the prediction of ET_0 using high spatial resolution satellite-derived LST, the calibrated satellite LST ET_0 models were validated by predicting other season's ET_0 values and were found that Hargreaves based on Landsat showed good results with less monthly error as shown in Fig. 7.7. The authors in this study assumed that the variation of different ET_0 models even when parameterized with the same satellite data is attributed to the empirical nature of the original models. It can be seen from the ET_0 maps prepared by best method (Landsat 8 Hargreaves), showing average of ET_0 values of three months in two dry seasons that ET_0 have its highest maximum in February (Fig. 7.8), and this may be due to increase in both LST and air temperature in February as shown in Table 7.1. Based on the findings in the current study, the selected satellite model may be used for ET_0 estimation in the study area when no ground station data input for conventional ET_0 models is available. Besides being simple procedure for ET_0 estimation, it has a benefit of obtaining ET_0 values for heterogeneous area spatially.

7.5 Summary and Conclusion

The main objective of this study was to estimate ET_0 values in Juba County of South Sudan, using high spatial resolution satellite data, without involving ground station data. The study evaluated the performance of three commonly used temperature-based ET_0 models (Thornthwaite, Blaney-criddle and Hargreaves methods), when taking satellite remote sensing-derived LST as input instead of air temperature. Also, satellite image processing algorithms were built in ArcGIS 10.4, based on the considered empirical temperature-based ET_0 methods to automatically extract pixel by pixel ET_0 values from satellite images. The monthly ET_0, in dry season of Juba County, South Sudan, was estimated by three temperature-based ET_0 models, based on the proposed methodology. And the proposed methodology was compared with analogous procedure suggested by (Maeda et al. 2011) that uses low spatial resolution satellite data (MODIS LST). Empirical models based on satellite LST were calibrated, validated and analyzed using simple statistical test parameters of coefficient of determination (R^2), root mean squared error (RMSE), mean absolute (MAE) and percentage error (PE) against FAO Penman–Montieth model parameterized with ground station data as standard. Average ET_0 maps of the study area (Juba County, South Sudan) were prepared. The following conclusions can be drawn from this study:

- Landsat ET_0 models performed better than MODIS ET_0 models with R^2 value of >0.9.
- Hargreaves ET_0 model (Landsat 8 based) was the best, with R^2 of 0.9275 and produces errors less than any other model, in both in-time and out-of-time validation.

- The satellite image processing models developed in this study can be helpful in quickening of the complex processing steps in computing ET_0 from satellite images.
- Spatial ET_0 maps may help planners, in showing how ET_0 varies in different months of the season.

Landsat 8 data can be freely downloaded from Internet with high spatial resolution of 30 m and 16-day temporal resolution; hence, the proposed procedure can be helpful in estimating ET_0 even at smaller field scale, considering the resolution and can assist, when no station data available, in planning and management of irrigation systems and other water management sectors.

References

Allen RG, Pereira LS, Raes D, Smith M (1998) Crop evapotranspiration—guidelines for computing crop water requirements—FAO Irrigation and drainage paper 56. Irrig Drainage, 1–15

Azhar AH, Perera BJC (2011) Evaluation of reference evapotranspiration estimation methods under Southeast Australian conditions. Irrig Drainage Eng 137(May):268–279

Bois B, Pieri P, Van Leeuwen C, Wald L, Huard F, Gaudillere JP, Saur E (2008) Using remotely sensed solar radiation data for reference evapotranspiration estimation at a daily time step. Agric Meteorol 148(4):619–630

Carlson TC, Ripley DA (1997) On the relationship between NDVI, fractional vegetation cover, and leaf area index. Remote Sens Environ 62:241–252

Doorenbos J, Pruitt WO (1977) Guidelines for predicting crop water requirements. FAO Irrig Drainage Paper 24:144

Douglas EM, Jacobs JM, Sumner DM, Ray RL (2009) A comparison of models for estimating potential evapotranspiration for Florida land cover types. J Hydrol 373(3–4):366–376

Droogers P, Allen RG (2002) Estimating reference evapotranspiration under. Irrig Drainage Syst 16:33–45

El-Shirbeny MA (2016) Evaluation of Hargreaves based on remote sensing method to estimate potential crop evapotranspiration. Int J Geomate 11(23):2143–2149

Fooladmand HR, Zandilak H, Ravanan MH (2008) Comparison of different types of Hargreaves equation for estimating monthly evapotranspiration in the south of Iran. Arch Agron Soil Sci 54(3):321–330

Giannini MB, Belfiore OR, Parente C, Santamaria R (2015) Land surface temperature from Landsat 5 TM images: comparison of different methods using airborne thermal data. J Eng Sci Technol Rev 8(3):83–90

Hargreaves BGH (1994) Reference evapotranspiration by George H. Hargreaves, 1 Fellow, ASCE. J Irrig Drainage Eng 120(6):1132–1139

Hargreaves GH, Samani ZA (1985) Reference crop evapotranspiration from temperature. Appl Eng Agric 1(2):96–99

Hosseinzadeh Talaee P (2014) Performance evaluation of modified versions of Hargreaves equation across a wide range of Iranian climates. Meteorol Atmos Phys 126(1–2):65–70

Kustas WP, Norman JM, Anderson MC, French AN (2003) Estimating subpixel surface temperatures and energy fluxes from the vegetation index—radiometric temperature relationship. Remote Sens Environ 85:429–440

Li X, Xin X, Jiao J, Peng Z, Zhang H, Shao S, Liu Q (2017) Estimating subpixel surface heat fluxes through applying temperature-sharpening methods to MODIS data. Remote Sens 9(8):836

López-Moreno JI, Hess TM, White SM (2009) Estimation of reference evapotranspiration in a mountainous mediterranean site using the Penman-Monteith equation with limited meteorological data. Pirineos 164:7–31 (JACA)

Maeda EE, Wiberg DA, Pellikka PKE (2011) Estimating reference evapotranspiration using remote sensing and empirical models in a region with limited ground data availability in Kenya. Appl Geogr 31(1):251–258

Ray S, Dadhwal V (2001) Estimation of crop evapotranspiration of irrigation command area using remote sensing and GIS. Agric Water Manag 49:239–249

Stathopoulou M, Cartalis C (2007) Daytime urban heat islands from Landsat ETM+ and Corine land cover data: an application to major cities in Greece. Sol Energy 81(3):358–368

Subedi A, Chávez JL (2015) Crop Evapotranspiration (ET) estimation models: a review and discussion of the applicability and limitations of ET methods. J Agric Sci 7(6):50–68

Tizikara C, George L, Lugor L (2015) Post-conflict development of agriculture in South Sudan: perspective on approaches to capacity strengthening. https://europa.eu/capacity4dev/sorudev/document/post-conflict-development-agriculture-south-sudan-approaches-capacity-strengthening

Trajkovic S (2005) Temperature-based approaches for estimating reference evapotranspiration. J Irrig Drainage Eng 131(August):316–323

Valipour M (2015) Temperature analysis of reference evapotranspiration models. Meteorol Appl 22(3):385–394

Wagner S, Kunstmann H, Bárdossy A, Conrad C, Colditz RR (2009) Water balance estimation of a poorly gauged catchment in West Africa using dynamically downscaled meteorological fields and remote sensing information. Phys Chem Earth 34(4–5):225–235

Wisser D, Frolking S, Douglas EM, Fekete BM, Vörösmarty CJ, Schumann AH (2008) Global irrigation water demand: variability and uncertainties arising from agricultural and climate data sets. Geophys Res Lett 35(24)

Xin X, Liu Q, Tang Y, Tian G, Gu X (2005) Estimating surface evapotranspiration using combined MODIS and CBERS-02 data. Sci China Ser E Eng Mater Sci 48:145–160

Zheng X, Zhu J (2015) Temperature-based approaches for estimating monthly reference evapotranspiration based on MODIS data over North China. Theoret Appl Climatol 121(3–4):695–711

Chapter 8
Assessing Irrigation Water Requirement and Its Trend for Betwa River Basin, India

Ashish Pandey, Reetesh Kumar Pyasi, and Santosh S. Palmate

Abstract In this study, Remote Sensing (RS) and Geographical Information System (GIS) are used to estimate the crop coefficient (Kc) and crop evapotranspiration (ETc) of the Betwa basin (Area = 43,500 km^2) of Central India. Reference crop evapotranspiration (ETo) was calculated for 18 stations of the Betwa basin using CROPWAT 8.0 software for November to April. Crop coefficients were estimated by developing its relationship with RS derived Soil Adjusted Vegetation Index (SAVI). The Kc and ETo maps for respective months were integrated to obtain spatial variation of crop evapotranspiration (ETc) in the Betwa basin. Analysis revealed that irrigation water requirements in the Betwa basin from November to April were 54.57, 85.74, 74.54, 63.82, 71.62, and 23.14 Mm^3. Further, trend analysis of irrigation water requirement using Mann–Kendall test and Sen's slope estimator showed a negative trend in Rabi season except in the month of December.

Keywords Crop evapotranspiration · Soil adjusted vegetation index (SAVI) · Crop coefficient · Evapotranspiration · Remote sensing

8.1 Introduction

Crop evapotranspiration (ETc) is the process of water loss in the atmosphere from the cropped area, which includes evaporation from the soil surface and transpiration from the plants body. Its estimation is essential for efficient irrigation planning and water

A. Pandey (✉) · R. K. Pyasi
Department of Water Resources Development and Management, Indian Institute of Technology Roorkee, Roorkee, Uttarakhand 247667, India
e-mail: ashish.pandey@wr.iitr.ac.in

R. K. Pyasi
e-mail: reetesh.pyasi@gmail.com

S. S. Palmate
Wetlands International South Asia, New Delhi, Delhi 110024, India
e-mail: santoshpalmate@gmail.com

A. Pandey et al. (eds.), *Water Management and Water Governance*, Water Science and Technology Library 96,
https://doi.org/10.1007/978-3-030-58051-3_8

resource allocation (Michael 1978). Mathematically, ETc is the product of reference evapotranspiration (ETo) and crop coefficient (Kc). Reference evapotranspiration (ETo) can be treated as a complex climatic parameter as it depends on whether parameters only (Xu et al. 2006). Crop coefficient depends on crop type and their growth stages.

FAO recommended Penman–Monteith equation has been considered as the best method for estimation of reference evapotranspiration (ETo) as compared to other methods and worldwide accepted by scientists (Allen et al. 2005; Cai et al. 2007). Penman–Monteith equation uses number of weather parameters, which has high spatial distribution like wind speed and temperature, which can vary significantly within few kilometers Ray and Dadhwal (2001). Crop type and their growth vary from field to field, hence crop coefficient. There will be a considerable error to derive spatial distribution of crop evapotranspiration directly from crop coefficient available in literature due to the empirical nature of Kc values (Jagtap and Jones 1989). Therefore, Vegetation Indices (VI) are used to obtain a single value correlating to physical vegetation parameters (such as biomass, productivity, leaf area index).

In the past, many researchers have used VI in agriculture and water resources application studies (Bausch 1993; Seevers and Ottmann 1994; Bausch 1995; Carlson and Ripley 1997; Bastiaanssen et al. 2000; Calera et al. 2001; Heinemann et al. 2002; Agrawal et al. 2003; Courault et al. 2005; Duchemin et al. 2006; Jayanthi et al. 2007; Raki et al. 2007; Tasumi and Allen 2007; Ozdogan 2010; Maeda et al. 2011; Salifu and Agyare 2012; Jamali et al. 2014). Ray and Dadhwal (2001) developed a relationship between the Soil Adjusted Vegetation Index (SAVI) and crop coefficient for the Mahi Right Bank Canal (MRBC) command area of Gujarat, India. Mishra et al. (2005) applied a similar technique for estimating ETc for paddy field in the Tarafeni South Main Canal command of West Bengal. Casa et al. (2009) estimated the crop water requirement (CWR) using only four Landsat ETM + images and some meteorological and geographical vector layers. Aghdasi (2010) developed a relationship between NDVI and Kc for assessment of crop water requirements in Iran. Gontia and Tiwari (2010) used remote sensing and GIS techniques for the estimation of Kc and ET of wheat crop. The application of Remote sensing data responds effectively to real crop conditions and can detect the spatial and temporal variability in the area. Basic hypothesis behind usefulness of remote sensing data for estimating crop coefficient is that both crop coefficients and VI (derived from reflectance) are influenced by leaf area index and a fraction of ground cover (Bausch 1995).

As evapotranspiration depends on the climatic factor, which may affect irrigation water requirement and water resource planning due to climate change. Tabari et al. (2011) examined trends in the Penman–Monteith (P-M) method based ETo estimates using the Mann–Kendall (MK) test, the Sen's slope estimator. Their study indicated that magnitude of positive trends in annual ETo varied up to 11.28 mm/year. Duhan et al. (2013) studied the spatial and temporal change in T_{MAX}, $T_{MIN,}$ and T_{MEAN} for Madhya Pradesh (MP). The annual T_{MEAN}, $T_{MAX,}$ and T_{MIN} increases by 0.60, 0.60, and 0.62 °C, respectively, over 102 years. Jhajharia et al. (2012) computed ETo by

Penman–Monteith method and studied its trend using MK test after removing significant lag-1 serial correlation from the time series by pre-whitening. Based on 22 years of data they found decreasing trends of annual and seasonal ETo for northeast region of India. Chakraborty et al. (2013) studied the effect of different weather variables on reference evapotranspiration and Irrigation Water Requirement (IWR). They used MK test and Theil-Sen's slope estimator for trend analysis of various climatological parameters and computed ETo using the CROPWAT 8.0. Overall, the results show an increase in irrigation water requirements. Looking to the aforementioned, the present study has been planned with the specific objectives of (1) estimation of crop water requirement using remote sensing and GIS techniques, and (2) trend analysis of irrigation water requirement for the Betwa River Basin (BRB).

8.2 Study Area

BRB is located in Central India and lies between 77°10′ E to 80°20′ E longitude and 22°54′ N to 26°05′ N latitude, covers 43,500 km^2(Fig. 8.1). Total length of the river from its origin to its confluence with the Yamuna River is 590 km, in which 232 km lies in Madhya Pradesh (MP) and the rest 358 km in Uttar Pradesh (UP). The climate of the BRB is characterized as moderate, mostly dry except during the southwest monsoon (Chaube et al. 2011) with an average annual rainfall of 1138 mm. The daily mean temperature and mean relative humidity varies from 8.1 to 42.3 °C and 18% (April and May) to a maximum of 90% (August), respectively (Suryavanshi et al. 2013).

8.3 Data Acquisition

8.3.1 Climate Data

In the study area, Indian Meteorological Department (IMD) has installed 18 meteorological stations for measurement of various weather parameters, i.e., maximum and minimum temperatures, rainfall, wind speed, relative humidity, and sunshine hour. 29 years monthly data was obtained from the IMD. Further, weather data was also downloaded from other online sources or website of Indian Water portal (www.indiawaterportal.org) and website of NASA atmospheric science data Centre. The missing weather data was filled with the corresponding long-term historical mean values. Some district wise weather data was available on website. Therefore, same weather data was considered for the stations within any particular district.

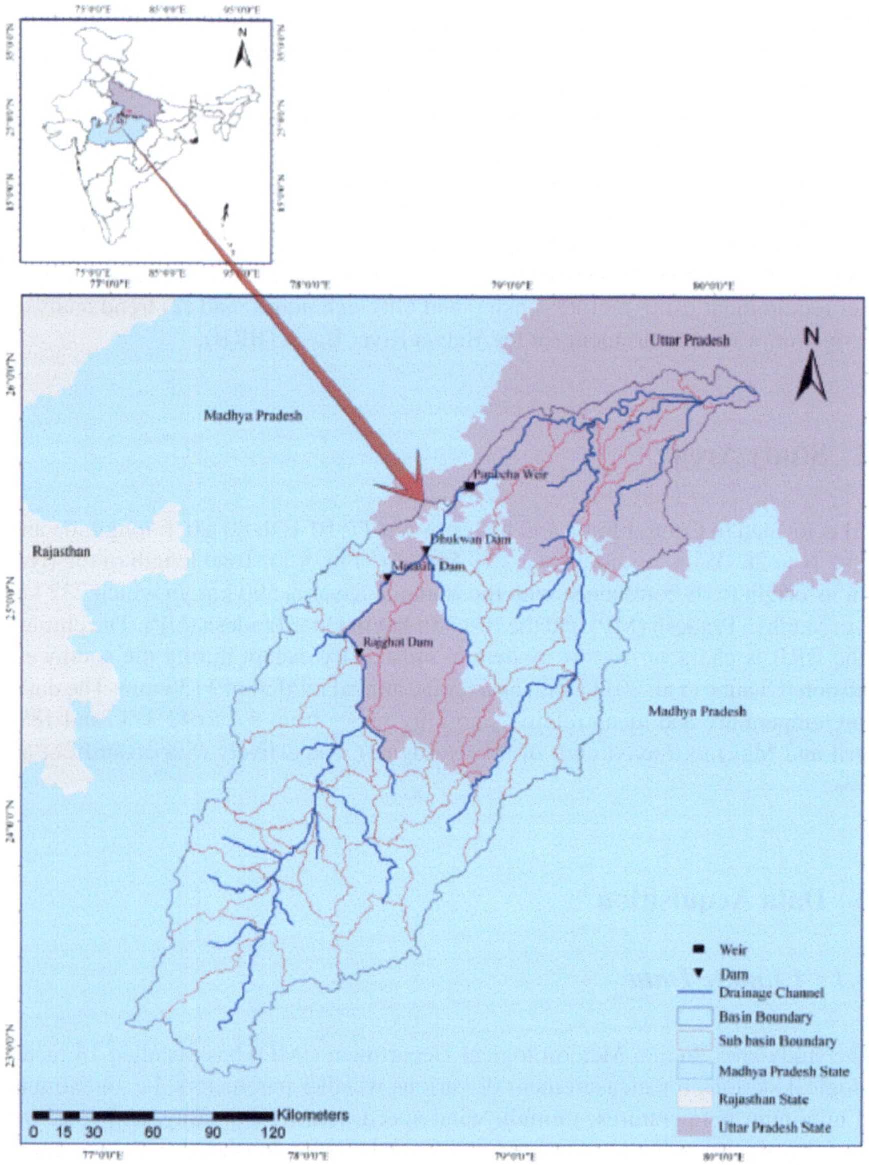

Fig. 8.1 Location map of the Betwa River Basin

8.3.2 Satellite Data

Digital data of Landsat-5 TM satellite have been used for the land use/land cover (LULC) classification and SAVI calculation. Imageries of 30 m × 30 m spatial

Table 8.1 Agricultural pattern in the study area

Rabi		Kharif	
Crop name	Agricultural area (km^2)	Crop name	Agricultural area (km^2)
Gram	6685	Soybean	11,726
Wheat	8904	Urad	2523
Lentil	1491	Sesasum	1320
Mustard	2520	Groundnut	1108
Linseed	700	Paddy	931
Others	3756	others	3456

resolution were downloaded from USGS website (http://glovis.usgs.gov/) for each month from November-2012 to April-2013. Total basin area falls under six satellite scenes. Hence 36 satellite imageries were used in this study.

In the study area, more than 70% of population depends on agriculture. Wheat and soybean are prevalent crops in Rabi and Kharif seasons, respectively. Cropping pattern information of the Betwa Basin collected from the website of Department of Farmer Welfare and Agriculture Development Madhya Pradesh (www.mpkrishi.org/) and Uttar Pradesh agriculture department is presented in Table 8.1.

8.4 Methodology

Betwa River Basin encompasses a large agricultural area (24,360 km^2) and different crop varieties in Rabi season. Since sowing of crop starts from the first week of November to second week of December. Therefore, for estimation of crop water requirement in Rabi season, November to April is considered.

8.4.1 Estimation of Reference Crop Evapotranspiration (ETo) and Spatial Mapping

The ETo may be considered as a complex climatic parameter and can be computed from weather data (Xu et al. 2006). In this study, CROPWAT 8.0 software was used for estimation of monthly ETo(Clarke et al. 1992; Chatterjee et al. 2012) using various input parameters, i.e., monthly minimum and maximum temperature, wind speed, sunshine hour, and relative humidity. CROPWAT software uses FAO recommended Penman–Monteith method for estimation of ETo. As suggested by Espadafor et al. (2011) the Penman–Monteith method considers the most significant variables, and the influence of each of them can be analyzed. The ETo derived by Allen et al. (1998) can be estimated by Eq. 8.1:

$$ET_O = \frac{0.408\Delta(R_n - G) + \gamma \frac{900}{T+273} U_2(e_s - e_a)}{\Delta + \gamma(1 + 0.34U_2)} \quad (8.1)$$

where ETo = Reference evapotranspiration [mmday^{-1}]; R_n= Net radiation at the crop surface [MJ m^{-2} day^{-1}]; G = Soil heat flux density [MJ m^{-2} day]; γ = Psychrometric constant (kPa °C^{-1}); T = Daily mean air temperature at 2 m height; U_2 = Wind speed at 2 m height [m s^{-1}]; e_s= Saturation vapor pressure [kPa]; e_a= Actual vapor pressure [kPa]; Δ = Slope of vapor pressure and temperature relationship [kPa °C-1].

To study the spatial variability of ETo, Inverse Distance Weighted (IDW) method has been employed and ArcGIS 9.3 software was used. A similar methodology was adopted by Jiang et al. (2009); Aguilar and Polo (2011); Sheikh and Mohammadi (2013); Mancosu et al. (2014); Talaee et al. (2014)studied the spatial distribution of weather parameters using IDW method. Tong et al. (2007) used IDW method to interpolate seasonal ETc for spring wheat. Ahmadi and Fooladmand (2008) applied IDW method for interpolation of monthly ETo values.

8.4.2 Crop Coefficient and Crop Evapotranspiration

Figure 8.2 shows the methodology adopted for the derivation of monthly crop evapotranspiration (ETc). Multi-date remotely sensed Landsat-5 TM imagery data were used for mapping of the Kc. Six satellite images for each month from November to April were processed for the study area. Land use/land Cover (LULC) map of the basin was prepared with the help of ERDAS IMAGINE software. Google earth application and subsequent ground truth verifications were carried out twice during the seasons. The masking tool has been used for the extraction of the Agricultural area from the classified image of the basin. Soil Adjusted Vegetation Index (SAVI) maps were generated for the agricultural area by using ArcGIS 9.3 software.

8.4.3 Soil Adjusted Vegetation Index (SAVI)

In agricultural land, the soil background affects satellite observed reflectance, which influences the NDVI values (Gontia and Tiwari 2010). To avoid soil reflectance, Huete (1988) proposed the concept of SAVI using the following relationship (Eq. 8.2):

$$SAVI = \frac{NIR - R}{NIR + R + L} \times (1 + L) \quad (8.2)$$

where NIR and R are the reflectance in Near-Infrared and Red wavelength respectively and L is an adjustment factor.

Huete (1988) suggested that soil background effect on canopy reflectance was adequately minimized for canopy cover varies from sparse to dense with a value

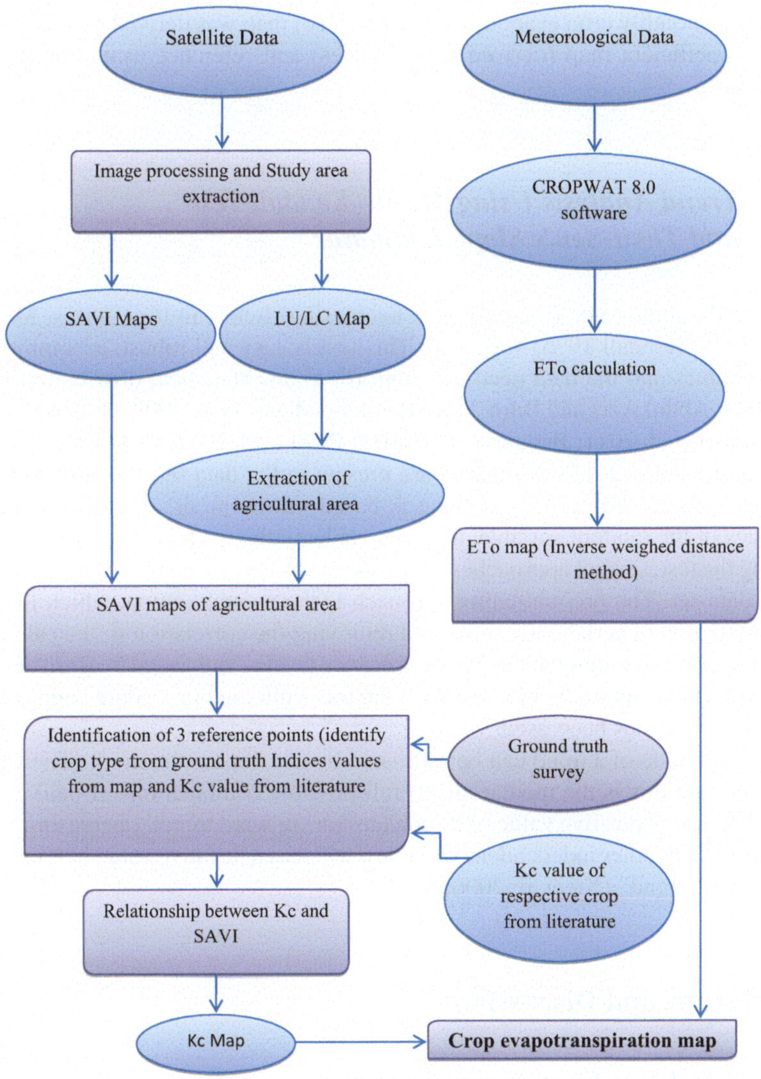

Fig. 8.2 Methodology adopted for derivation of monthly ETc map

of L as 0.5. SAVI values of three reference points with different crop types were recorded for each month of Rabi season. Crop types of all three sites were identified by ground truth verification and crop coefficient values of respective crops for various developing stages were adopted from Doorenbos and Pruitt (1977). Linear regression method was used to develop relationship between SAVI and crop coefficient values. Crop coefficient maps were developed for each month using the regression equation.

Finally, the monthly crop evapotranspiration (ETc) map was derived by multiplying the crop coefficient map (derived from indices) and reference evapotranspiration (ETo) map.

8.4.4 Trend Analysis Using Mann–Kendall Test and Theil-Sen's Slope Estimator

To detect the monotonic (increasing or decreasing) trends in time series, MK test (Mann 1945; Kendall 1975) was used. This test is easy and robust, accommodates missing values and the data need not conform to any statistical distribution (Burn et al. 2004; Abdul Aziz and Burn 2006; Bandyopadhyay et al. 2009; Dinpashoh et al. 2011; Tabari et al. 2011; Jhajharia et al. 2012; Patra et al. 2012; Croitoru et al. 2013; Duhan and Pandey 2013). If outliers are present in the data set, the nonparametric MK test is useful because its statistic is based on the (+ or –) signs rather than the values of the random variable, and trend is less affected by the outliers. Prior to applying the test, several approaches are proposed for removing the serial correlation from a data set. The pre-whitening approach is the most common, which involves the computation of serial correlation and removing the correlation if, the calculated serial correlation is significant at 5% significance level as suggested by Burn and Hag Elnur (2002). In this study, MK test with the pre-whitening procedure suggested by Yue et al. (2002) has been used.

The magnitude of a trend can be estimated by the slope estimator β (Theil 1950; Sen 1968) and this is the median of overall possible combinations of pairs for the whole data set. A positive value of β indicates an "upward trend" (increasing values with time or any other independent variable), whereas a negative value of β indicates a "downward trend" (Xu et al. 2006).

8.5 Results and Discussion

8.5.1 Land Use/Land Cover Analysis

Landsat imagery of the year 2013 was downloaded to extract LULC information of the BRB and supervised classification method was applied to determine LULC of the study area (Table 8.2 and Fig. 8.3). LULC analysis shows that 52%, 19%, and 12% of the study area falls under agriculture forest cover and barren land, respectively. Accuracy assessment results with commission error, omission error, producer's accuracy, and user's accuracy which shows class by class comparison between ground truth data and classification results are also presented in Table 8.3. Overall classification accuracy and Kappa statistics of LULC were 79% and 0.7051, respectively.

Table 8.2 Land use/land cover classes of the study area

Land use/land coverclass	Area (%)	Area (km^2)
Waterbody	5	2175
Settlement	8	3480
Riverbed	4	1740
Forest	19	8265
Barren land	12	5220
Agricultural land	52	22,620

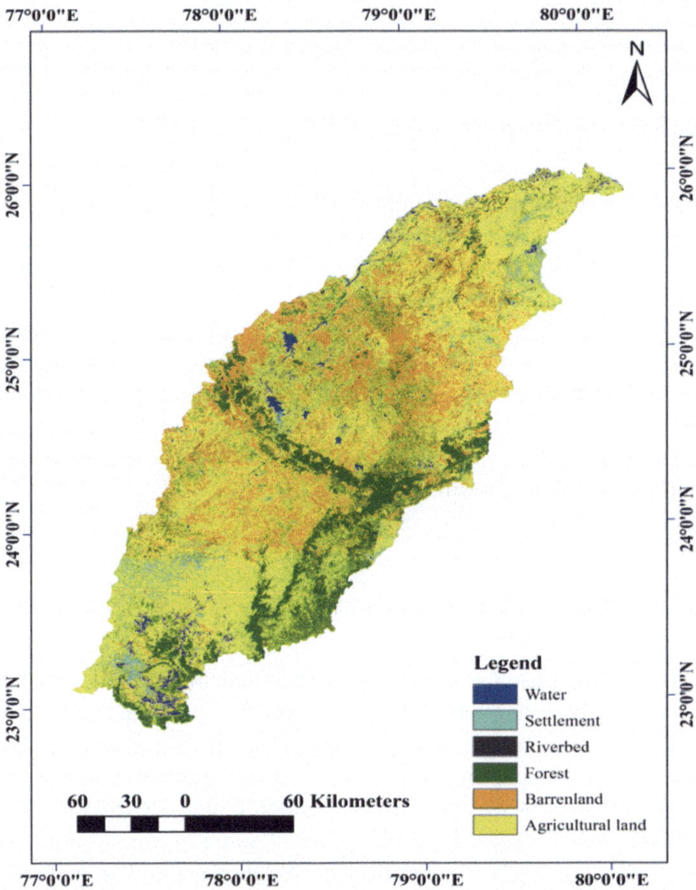

Fig. 8.3 Land use/land cover classification of using 2013 Landsat 5 TM imagery

Table 8.3 Classification accuracy assessment

Class name	Commission (%)	Omission (%)	Producers Accuracy	Users Accuracy
Settlement	50	16.6	83.33	50
Riverbed	25	0	100	75
Forest	19.23	0	100	80.77
Waterbody	60	20	80	40
Barren land	20	38.46	61.54	80
Agricultural	5	26.92	73.08	95

Overall Classification Accuracy = 79%, Kappa Statistics = 0.7051

8.5.2 Reference Evapotranspiration (ETo) Maps

In this study, Inverse Distance Weighted (IDW) method has been employed for interpolation and to generate ETo maps for November to April using point data of eighteen stations. Ray and Dadhwal (2001) and Darshana et al. (2013) also applied this method to interpolate reference ETo values.

The spatial distribution of ETo varies from 3.83 to 4.52 mm/day for February month (Fig. 8.4). The value of ETo is high (4.32–4.52 mm/day) in the South-Western part (Upper Betwa basin) and low (3.83–4.02) in the North-Eastern part (lower Betwa basin) of the study area. In the central part of the Betwa basin, ETo value ranges from 4.13 to 4.22 mm/day. Thus, more ETo was obtained in the Upper Betwa basin and less ETo in the lower Betwa basin.

8.5.3 Generation of SAVI Maps and Regression Equation

Agricultural area was separated from the classified map of the Betwa basin, and 52% of the basin is mostly falling under the agriculture area (Fig. 8.3). SAVI values for three points from agricultural areas were obtained from the six scenes representing six months. The crops sowing in three points were wheat, gram, and mustard. However, mustard and gram sowing was done in November, hence, the SAVI values for the April month was not considered. Similarly, SAVI value of a point representing the wheat area is neglected for November because of late sowing. For analyzing overall crop water requirement in the basin for each month of Rabi season, mean of all values was estimated. The minimum, maximum and mean value of SAVI, derived Kc and ETc values for the Betwa River Basin are given in Table 8.4.

On a seasonal time scale, in Rabi season maximum (1.47) and minimum (0.007) values of SAVI were found during the months of February and April, respectively. The mean monthly value of SAVI was maximum (0.47) and minimum (0.06) in December and April, respectively. The High SAVI value was in December due to the

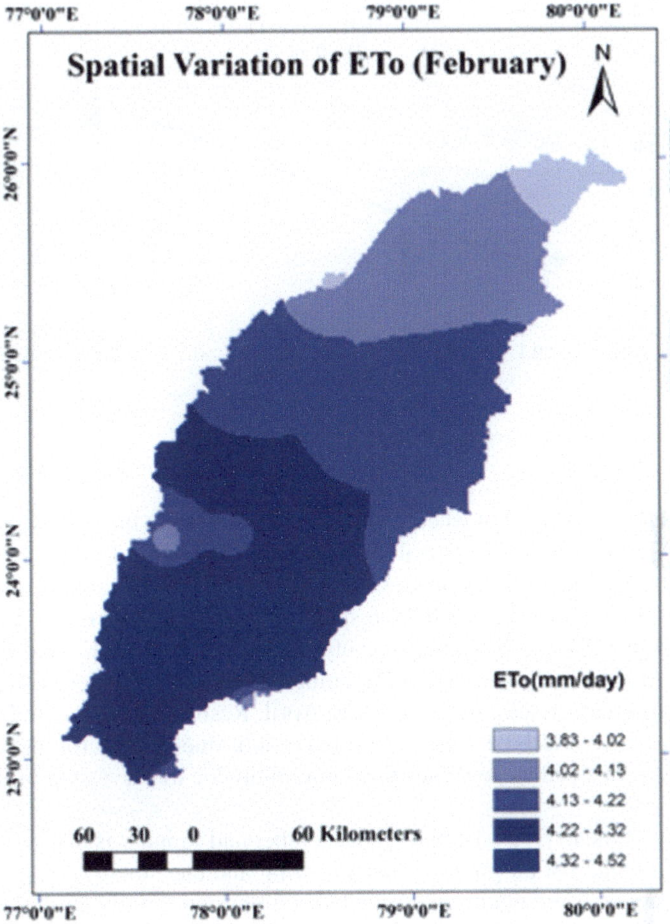

Fig. 8.4 Spatial distribution of ETo in February

Table 8.4 SAVI (minimum, maximum, and mean), and SAVI derived values of Kc and ETc

Month	SAVI (min)	SAVI (max)	SAVI (mean)	Kc (SAVI mean)	ETc (SAVI mean)
November	0.012	1.41	0.21	0.44	2.24
December	0.010	1.42	0.47	0.81	3.52
January	0.019	1.42	0.44	0.78	3.06
February	0.014	1.47	0.34	0.63	2.62
March	0.015	1.45	0.28	0.54	2.94
April	0.007	1.43	0.06	0.22	0.95

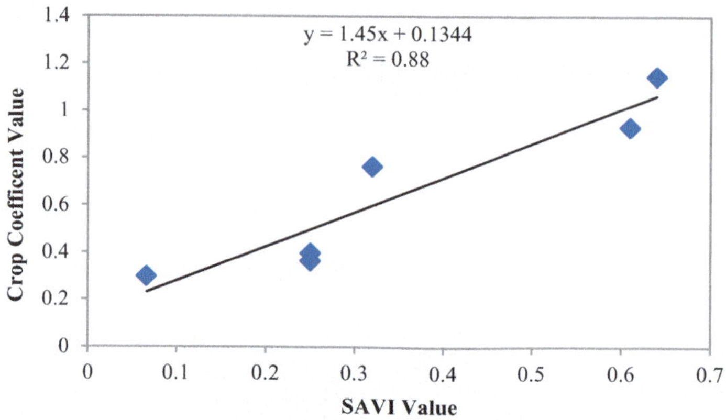

Fig. 8.5 Comparison between crop coefficient (Kc) and SAVI value

full crop development stage of wheat crop. On the other hand, the harvesting of Rabi crop before April month was responsible for the low SAVI value. Further, Kc and ETc values were derived from SAVI mean values (Table 8.4). SAVI estimated minimum Kc and ETc values were 0.22 and 0.95, respectively for April, and maximum Kc and ETc values were 0.81 and 3.52, respectively for December. The Kc and ETc value rapidly increases from November to December then decreases from December to March and rapidly decreases from March to April. Results reveal that in December SAVI estimated Kc and ETc values were maximum due to the crop development stage. However, in April these values were minimum due to harvesting of Rabi crop in the BRB.

High value of mean ETc especially in December and January is mainly attributed to development stage of all growing (early sowing and late sowing) crops resulting in more irrigation water requirement for basin area, particularly in these months. Average value of SAVI and corresponding FAO-56 crop coefficient values of respective crops for each month are used for developing linear relationships (Fig. 8.5). Coefficient of determination between SAVI value and Kc is 0.88 which, shows a positive response and exhibited a strong correlation.

8.5.4 Crop Coefficient and Crop Evapotranspiration Maps

Once the relationship between SAVI and Kc were developed, Kc maps were generated using ArcGIS 9.3 software for November to April. Crop coefficient maps derived from SAVIis presented in Fig. 8.6(a and f). It shows that, derived Kc values are less than 0.4 for November and April months in most of the BRB area, because of the initial stage and end of the late-season stage of Rabi crops. The maximum area under low (Kc < 0.4) and high (Kc > 1) Kc values were 18,129.60 km^2 (in April) and 6330.79 km^2 (in November) respectively (Table 8.5). Further, month-wise crop

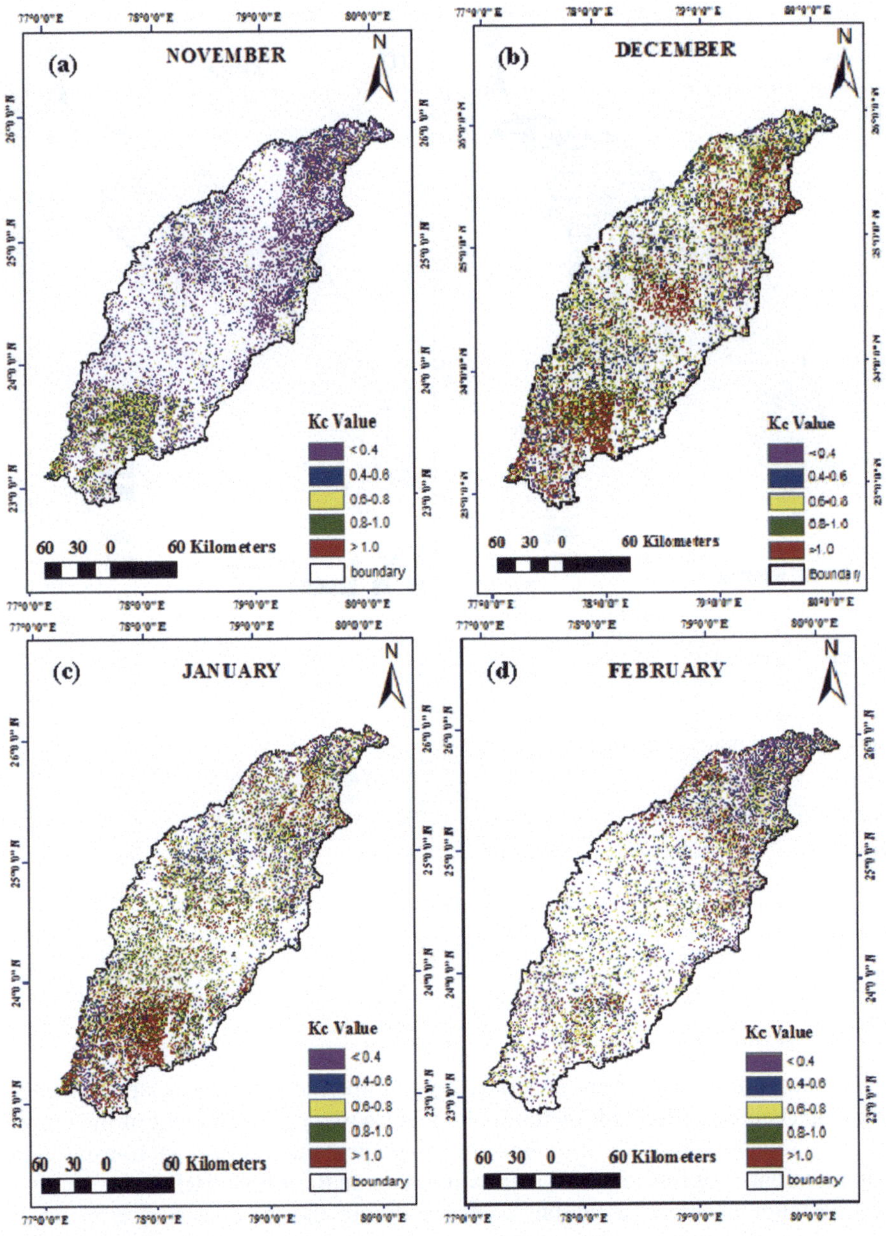

Fig. 8.6 Spatial distribution of crop coefficient for **a** November, **b** December, **c** January, **d** February, **e** March, and **f** April

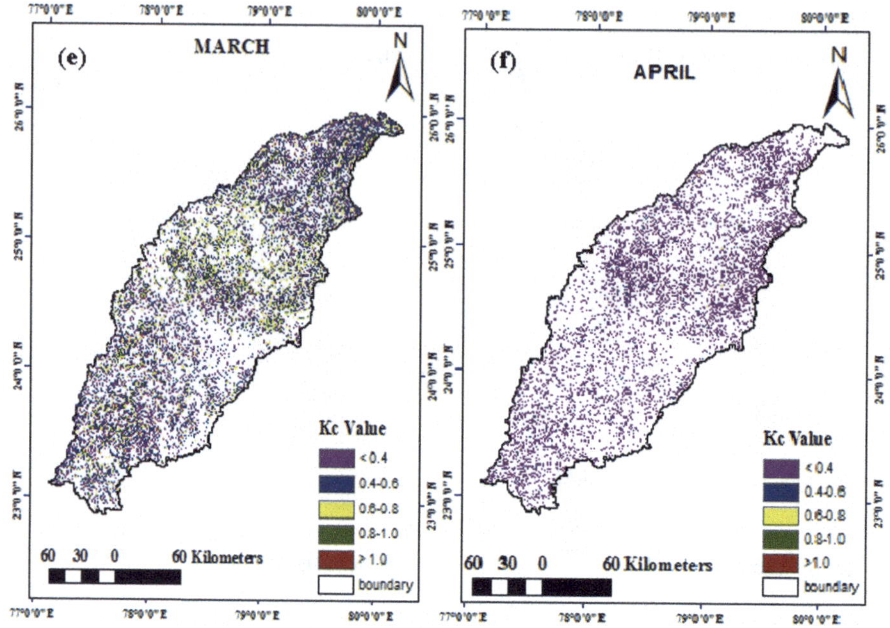

Fig. 8.6 (continued)

Table 8.5 Area (km^2) under Kc value

Month	<0.4	0.4–0.6	0.6–0.8	0.8–1.0	>1
November	6517.43	5776.97	3479.57	2255.24	6330.79
December	3480.19	5565.05	6504.12	6499.39	2311.25
January	2591.61	4816.70	5722.22	6073.84	5155.63
February	6195.24	2766.40	1770.88	8824.09	4803.39
March	8349.96	10,177.80	3690.92	2133.01	8.31
April	18,129.61	5883.95	331.11	14.86	0.47

evapotranspiration maps were generated by multiplying reference evapotranspiration map (Penman–Monteith method) and pixel-wise crop coefficient of the corresponding months (Fig. 8.7). Similar to crop coefficient map, crop evapotranspiration map represents pixel-wise crop water requirement in the whole BRB area.

As negligible effective rainfall occurred during the Rabi season, crop water requirement (CWR) has been considered as an irrigation water requirement (IWR) of the area. Spatial distribution of CWR/IWR derived from SAVI in the basin is presented in Fig. 8.7 for all the six months from November to April for the Betwa basin. Figure 8.7(a—c) shows ETc distribution in the month of November, December, and January. Low ETc values (<2 mm/day) has been observed in most of the agricultural area of the basin in the month of November. While moderate ETc (2–4 mm/day)

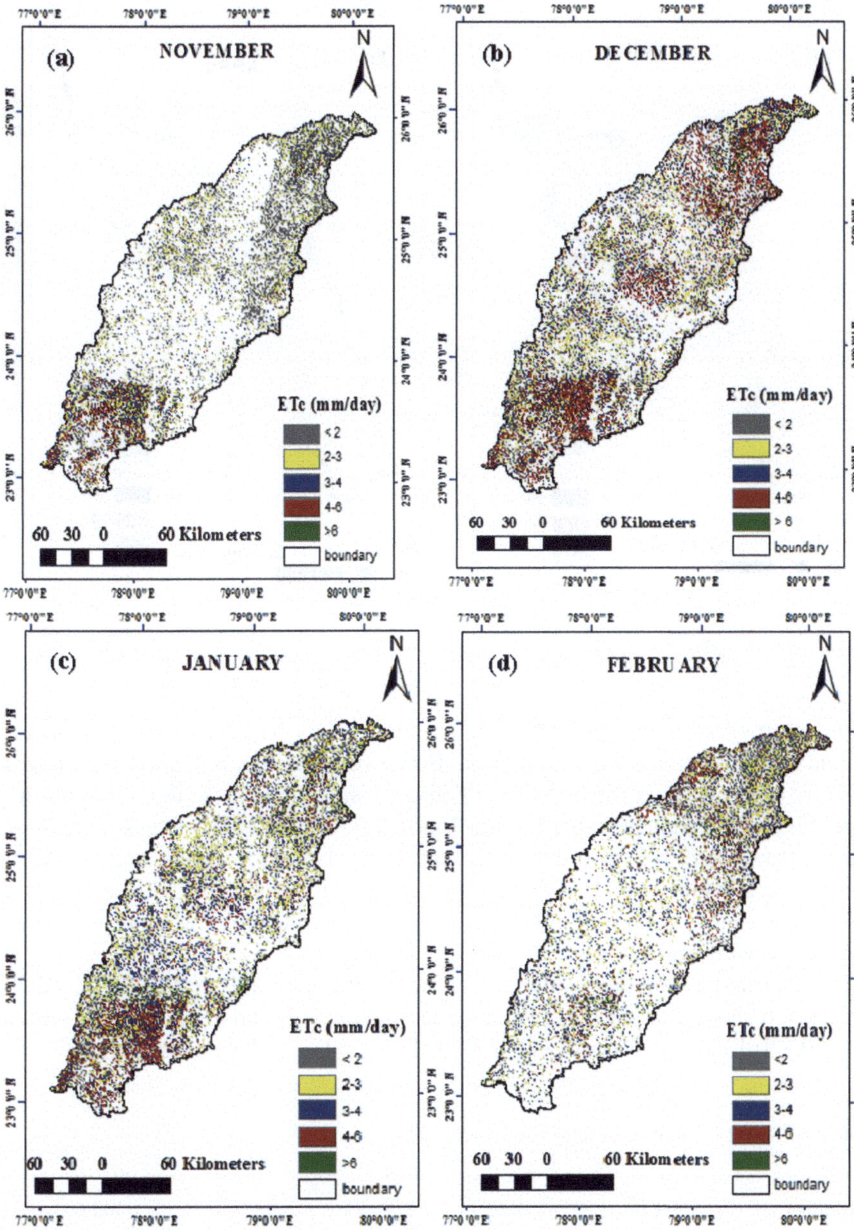

Fig. 8.7 Spatially distributed map of actual crop evapotranspiration for **a** November, **b** December, **c** January, **d** February, **e** March, and **f** April

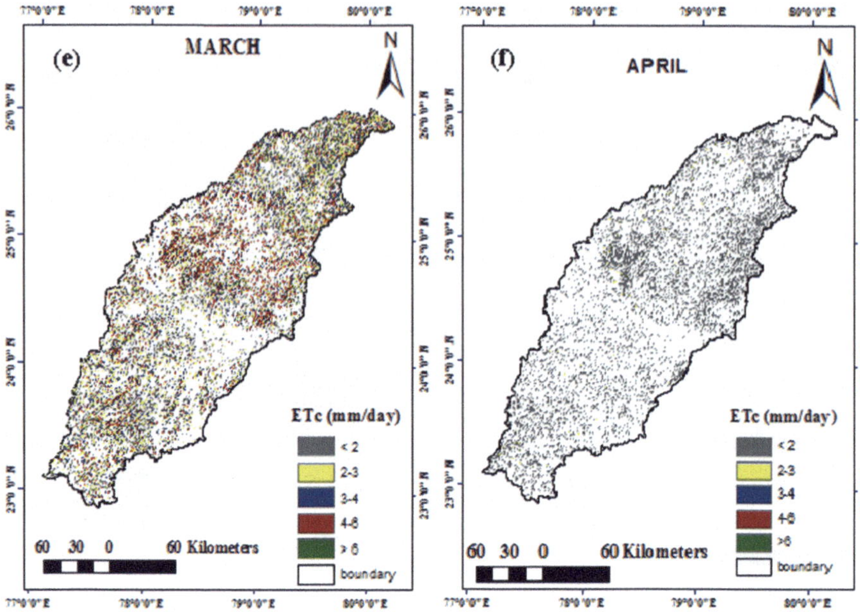

Fig. 8.7 (continued)

has been observed in months of December and January because of the crop development stage in most of the region of the Betwa basin. Figure 8.7(e and f) indicates high ETc (>4 mm/day) in the months of February and March due to the crop maturity stage. Figure 8.7f indicates that for April month ETo value is less than 0.2 mm/day in the Betwa basin. It is also due to the end of the late Rabi season stage in April month, which exhibits less ETo (<0.2 mm/day) from the study area. Hence, the crop water requirement of April month was found to be significantly lower because most of the crops were harvested.

The distribution of the monthly area under different ETc values is given in Table 8.6. It shows that maximum area under the low (<2) and high (>6) ETc values were 10,730.13 km^2 (in April) and 4658.09 km^2 (in December), respectively.

Table 8.6 Distribution of Area (km^2) under ETc value

Month	<2	2–3	3–4	4–6	>6
November	10,599.12	5291.65	3272.59	1826.98	3369.66
December	2348.97	4980.98	6112.93	6259.03	4658.09
January	4152.78	5888.34	6279.24	3476.26	4563.38
February	8006.30	6021.44	4692.56	2814.17	2825.53
March	6366.72	4843.17	2917.19	3797.77	6435.15
April	10,730.13	2893.89	9449.45	1286.23	0.30

From LULC analysis, it was estimated that the total agricultural land which needs irrigation in Rabi season is 24,360 km^2. Therefore, the monthly volume of irrigation water requirement for the Betwa basin has been estimated by multiplying mean irrigation water requirement depth by agricultural area within the basin. Irrigation water requirements during November, December, January, February, March, and April were 54.57, 85.74, 74.54, 63.82, 71.62, and 23.14 Mm^3, respectively for the BRB. Therefore, total crop water requirement was found to be 373.43 Mm^3. It reveals that irrigation water requirement is low in November, maximum in December and reduced in later months as the crop is approaching its maturity.

8.5.5 *Trend Analysis of Monthly IWR*

Assuming a similar cropping pattern for 29 years, trend of irrigation water requirement (mm) was estimated using MK test and Theil-Sen's slope estimator. In all months except December, a decreasing trend in irrigation water requirement was noticed. However, a significant decreasing trend was noticed only in months of April (0.18 mm/day) and March (0.31 mm/day) at 0.01 and 0.1 level of significance respectively. Table 8.7 shows the trend results of IWR for each month of Rabi season.

The trend results of irrigation water requirement per unit area are given in Table 8.7. Change in volume of irrigation water requirement over a basin depends on the change in agricultural land and change in the climate. It can be assumed that the agricultural area is reducing or constant during study period. Thus, it can be concluded that volume of irrigation water requirement will not increase in future, and planning of irrigation scheme and design of irrigation structures can be carried out without considering the impact of climate change in the Betwa River Basin.

Table 8.7 Irrigation water requirement trends by MK test and Theil-Sen's slope estimator

Month	Z Statistics	Significance	Q values
January	−0.43		−0.19
February	−1.11		−0.36
March	−1.89	+	−0.31
April	−2.98	**	−0.18
November	−0.28		−0.09
December	0.69		0.33

'**' if trend at 0.01 significance level, '+' if trend at 0.1 significance level

8.6 Conclusion

In this study, estimation of crop coefficient (Kc), evapotranspiration (ET), and irrigation water requirement in Rabi season for the Betwa basin were carried out using Remote sensing derived SAVI and GIS techniques. The following conclusions are drawn:

- The relationship between estimated SAVI and crop coefficient exhibits strong correlation ($R^2 = 0.88$) in Rabi season.
- The mean value of ETc derived from vegetation indices vary from 0.95 mm/day to 3.52 mm/day.
- The high value of crop evapotranspiration was obtained from December to March due to the development and maturity stages of the crops. However, crop water requirement was significantly low in April due to the harvesting of most of the crops.
- The classification accuracy can influence the accuracy of ET estimation and Kc mapping. Furthermore, the source of errors may be due to FAO recommended standard crop coefficient values used in the study, which maybe affected due to local climatic, crop, and soil conditions.
- The Crop coefficient based on vegetation indices provided accurate and spatially distributed crop water demand for agricultural areas.
- The irrigation water requirement showed a decreasing trend in most of the months of the Rabi season. Therefore, long-term irrigation planning of the area can be carried out considering the present status of crop water demand, as irrigation water requirement will not increase in the near future.

References

Abdul Aziz OI, Burn DH (2006) Trends and variability in the hydrological regime of the Mackenzie River Basin. J Hydrol 319(1):282–294

Aghdasi F (2010) Crop water requirement assessment and annual planning of water allocation, M.Sc., Faculty ITC, Twente University

Agrawal S, Joshi P, Shukla Y, Roy P (2003) SPOT vegetation multi temporal data for classifying vegetation in south. Curr Sci 84(11)

Aguilar C, Polo MJ (2011) Generating reference evapotranspiration surfaces from the Hargreaves equation at watershed scale. Hydrol Earth Syst Sci 15:2495–2508

Ahmadi SH, Fooladmand HR (2008) Spatially distributed monthly reference evapotranspiration derived from the calibration of Thornthwaite equation: a case study. South of Iran, Irrigation Science 26(4):303–312

Allen RG, Pereira L, Raes D, Smith M (1998) FAO Irrigation and drainage paper No. 56. Food and Agriculture Organization of the United Nations, Rome, 26–40

Allen RG, Tasumi M, Morse A, Trezza R (2005) A landsat-based energy balance and evapotranspiration model in Western US water rights regulation and planning. Irrig Drain Syst 19:251–268

Bandyopadhyay A, Bhadra A, Raghuwanshi N, Singh R (2009) Temporal trends in estimates of reference evapotranspiration over India. J Hydrol Eng 14(5):508–515

Bastiaanssen WG, Molden DJ, Makin IW (2000) Remote sensing for irrigated agriculture: examples from research and possible applications. Agric Water Manag 46(2):137–155

Bausch WC (1993) Soil background effects on reflectance-based crop coefficients for corn. Remote Sens Environ 46(2):213–222

Bausch WC (1995) Remote sensing of crop coefficients for improving the irrigation scheduling of corn. Agric Water Manag 27(1):55–68

Burn DH, Hag Elnur MA (2002) Detection of hydrologic trends and variability. J Hydrol 255:107–122

Burn DH, Cunderlik JM, Pietroniro A (2004) Hydrological trends and variability in the Liard River basin. Hydrol Sci J 49(1):53–67

Cai J, Liu Y, Lei T, Pereira LS (2007) Estimating reference evapotranspiration with the FAO Penman-Monteith equation using daily weather forecast messages. Agric For Meteorol 145(1):22–35

Calera A, Martínez C, Melia J (2001) A procedure for obtaining green plant cover: relation to NDVI in a case study for barley. Int J Remote Sens 22(17):3357–3362

Carlson TN, Ripley DA (1997) On the relation between NDVI, fractional vegetation cover, and leaf area index. Remote Sens Environ 62(3):241–252

Casa R, Rossi M, Sappa G, Trotta A (2009) Assessing crop water demand by remote sensing and GIS for the Pontina Plain, Central Italy. Water Resour Manag 23(9):1685–1712

Chakraborty S, Mishra SK, Pandey R, Chaube U (2013) Long-term changes of irrigation water requirement in the context of climatic variability. ISH J Hydraul Eng 19(3):257–266

Chatterjee SK, Banerjee S, Bose M (2012) Climate change impact on crop water requirement in Ganga River Basin, West Bengal, India. In: 3rd international conference on biology, environment and chemistry IPCBEE, vol 46

Chaube UC, Suryavanshi S, Nurzaman L, Pandey A (2011) Synthesis of flow series of tributaries in Upper Betwa basin. Int J Environ Sci 1(7):59–75

Clarke D, Smith M, El-Askari K (1992) CROPWAT—a computer program for irrigation planning and management. FAO Irrigation and Drainage paper 46, Rome, Italy

Courault D, Seguin B, Olioso A (2005) Review on estimation of evapotranspiration from remote sensing data: from empirical to numerical modeling approaches. Irrigat Drain Syst 19(3–4):223–249

Croitoru AE, Piticar A, Dragotă CS, Burada DC (2013) Recent changes in reference evapotranspiration in Romania. Glob Planet Change 111:127–132

Darshana, Pandey A, Pandey R (2013) Analysing trends in reference evapotranspiration and weather variables in the Tons River Basin in Central India. Stoch Environ Res Risk Assess 1–15

Dinpashoh Y, Jhajharia D, Fakheri-Fard A, Singh VP, Kahya E (2011) Trends in reference crop evapotranspiration over Iran. J Hydrol 399(3):422–433

Doorenbos J, Pruitt W (1977) Crop water requirements, FAO irrigation and drainage paper 24, land and water development division. FAO, Rome

Duchemin B, Hadria R, Erraki S, Boulet G, Maisongrande P, Chehbouni A, Escadafal R, Ezzahar J, Hoedjes J, Kharrou M (2006) Monitoring wheat phenology and irrigation in Central Morocco: on the use of relationships between evapotranspiration, crops coefficients, leaf area index and remotely-sensed vegetation indices. Agric Water Manag 79(1):1–27

Duhan D, Pandey A (2013) Statistical analysis of long term spatial and temporal trends of precipitation during 1901–2002 at Madhya Pradesh, India. Atmos Res 122(2013):136–149

Duhan D, Pandey A, Gahalaut KPS, Pandey RP (2013) Spatial and temporal variability in maximum, minimum and mean air temperatures at Madhya Pradesh in central India. CR Geosci 345(1):3–21

Espadafor M, Lorite I, Gavilán P, Berengena J (2011) An analysis of the tendency of reference evapotranspiration estimates and other climate variables during the last 45 years in Southern Spain. Agric Water Manag 98(6):1045–1061

Gontia NK, Tiwari KN (2010) Estimation of crop coefficient and evapotranspiration of wheat (Triticum aestivum) in an irrigation command using remote sensing and GIS. Water Resour Manage 24(7):1399–1414
Heinemann A, Hoogenboom G, De Faria R (2002) Determination of spatial water requirements at county and regional levels using crop models and GIS: An example for the State of Parana, Brazil. Agric Water Manag 52(3):177–196
Huete A (1988) A soil-adjusted vegetation index (SAVI). Remote Sens Environ 25(3):295–309
Jagtap S, Jones J (1989) Stability of crop coefficients under different climate and irrigation management practices. Irrig Sci 10(3):231–244
Jamali S, Seaquist J, Eklundh L, Ardö J (2014) Automated mapping of vegetation trends with polynomials using NDVI imagery over the Sahel. Remote Sens Environ 141:79–89
Jayanthi H, Neale CM, Wright JL (2007) Development and validation of canopy reflectance-based crop coefficient for potato. Agric Water Manag 88(1):235–246
Jhajharia D, Dinpashoh Y, Kahya E, Singh VP, Fakheri-Fard A (2012) Trends in reference evapotranspiration in the humid region of northeast India. Hydrol Process 26(3):421–435
Jiang L, Islam S, Guo W, Singh Jutla A, Senarath SU, Ramsay BH, Eltahir E (2009) A satellite-based daily actual evapotranspiration estimation algorithm over South Florida. Glob Planet Change 67(1):62–77
Kendall MG (1975) Rank correlation methods. Griffin, London, UK
Maeda EE, Wiberg DA, Pellikka PK (2011) Estimating reference evapotranspiration using remote sensing and empirical models in a region with limited ground data availability in Kenya. Appl Geogr 31(1):251–258
Mancosu N, Snyder RL, Spano D (2014) Procedures to develop a standardized reference evapotranspiration zone map. J Irrig Drainage Eng
Mann HB (1945) Non-parametric tests against trend. Econometrica 13:245–259
Michael AM (1978) Irrigation theory and practice. Vikas Publishing House, India
Mishra P, Tiwari KN, Chowdary VM, Gontia NK (2005) Irrigation water demand and supply analysis in the command area using remote sensing and GIS. Hydrol J IAH 28(1–2):59–69
Ozdogan M (2010) The spatial distribution of crop types from MODIS data: temporal immixing using independent component analysis. Remote Sens Environ 114(6):1190–1204
Patra JP, Mishra A, Singh R, Raghuwanshi N (2012) Detecting rainfall trends in twentieth century (1871–2006) over Orissa State, India. Clim Change 111(3–4):801–817
Raki S, Chehbouni A, Guemouria N, Duchemin B, Ezzahar J, Hadria R (2007) Combining FAO-56 model and ground-based remote sensing to estimate water consumptions of wheat crops in a semi-arid region. Agric Water Manag 87(1):41–54
Ray S, Dadhwal V (2001) Estimation of crop evapotranspiration of irrigation command area using remote sensing and GIS. Agric Water Manag 49(3):239–249
Salifu T, Agyare WA (2012) Distinguishing land use types using surface albedo and normalized difference vegetation index derived from the SEBAL model for the Atankwidi and Afram subcatchments in Ghana. J Eng Appl Sci 7(1)
Seevers P, Ottmann R (1994) Evapotranspiration estimation using a normalized difference vegetation index transformation of satellite data. Hydrol Sci J 39(4):333–345
Sen PK (1968) Estimates of the regression coefficient based on Kendall's tau. J Am Stastitics Assoc 63:1379–1389
Sheikh V, Mohammadi M (2013) Evaluation of reference evapotranspiration equations in semi-arid regions of Northeast of Iran. Int J Agric Crop Sci 5(5):450–456
Suryavanshi S, Pandey A, Chaube UC, Joshi N (2013) Long-term historic changes in climatic variables of Betwa Basin, India. Theor Appl Climatol 1–16
Tabari H, Marofi S, Aeini A, Talaee PH, Mohammad K (2011) Trend analysis of reference evapotranspiration in the western half of Iran. Agric For Meteorol 151(2):128–136
Talaee PH, Tabari H, Abghari H (2014) Pan evaporation and reference evapotranspiration trend detection in western Iran with consideration of data persistence. Hydrol Res 45(2):213–225

Tasumi M, Allen RG (2007) Satellite-based ET mapping to assess variation in ET with timing of crop development. Agric Water Manag 88(1):54–62

Theil H (1950) A rank invariant method of linear and polynomial regression analysis, Part 3. Netherlands Akademie van Wettenschappen, Proceeding 53:1397–1412

Tong L, Kang S, Zhang L (2007) Temporal and spatial variations of evapotranspiration for spring wheat in the Shiyang river basin in northwest China. Agric Water Manag 87(3):241–250

Xu CY, Gong L, Jiang T, Chen D, Singh V (2006) Analysis of spatial distribution and temporal trend of reference evapotranspiration and pan evaporation in Changjiang (Yangtze River) catchment. J Hydrol 327(1):81–93

Yue S, Pilon P, Cavadias G (2002) Power of the Mann-Kendall and Spearman's rho test for detecting monotonic trends in hydrological series. J Hydrol 259:254–271

Chapter 9
Comparison of FAO56 and NDVI-Derived K_c Curves for Major Crops in Pare Basin of Arunachal Pradesh

Waikhom Rahul Singh, Khyoda Meema, Tana Matthew, A. Bhadra, and A. Bandyopadhyay

Abstract This study compared the crop coefficients of major crops of the Pare basin acquired from the FAO56 with the NDVI-generated crop coefficients. The FAO56's K_c was modified using average monthly RH and wind speed of the study area and compared with NDVI-generated K_c. The coefficient of determination (R^2) was determined to evaluate the NDVI-generated K_c values. Results showed that sweet potato, mustard, and sugarcane results were satisfactory; cucumber and maize were moderate and paddy was not satisfactory which may be attributed to the presence of standing water for the most part of its life. The uncertainty involved with the results may be because of not incorporating the effects of parameters like crop height, crop canopy, and water management in the estimation of K_c. This study illustrated the usefulness of the NDVI images to generate K_c curves for various crops in a region where conducting field experiments are not feasible.

Keywords Crop coefficient · NDVI-derived K_c · FAO56 · Major crops of arunachal pradesh

9.1 Introduction

The accurate calculation of daily crop evapotranspiration (ET_o) is a necessary requirement within the irrigation scheduling program. The crop coefficient values of different crops are important for estimating accurate crop evapotranspiration (ETc), water transfer planning, efficient irrigation management, and hydrological studies

---

W. R. Singh (✉)
Center for Flood Management Studies, National Institute of Hydrology, Dispur, Guwahati 781006, Assam, India
e-mail: rahulwaikhom0@gmail.com

W. R. Singh · K. Meema · T. Matthew · A. Bhadra · A. Bandyopadhyay
Department of Agricultural Engineering, North Eastern Regional Institute of Science and Technology, 791109 Nirjuli (Itanagar), Arunachal Pradesh, India

A. Pandey et al. (eds.), *Water Management and Water Governance*, Water Science and Technology Library 96,
https://doi.org/10.1007/978-3-030-58051-3_9

(Petillo and Castel 2007; Jyothy et al. 2011; Montazar et al. 2017). Estimating evapotranspiration by direct measurement using lysimeter or indirect measurements of soil moisture can help in finding out the net irrigation requirement. In the recent past, lysimeter has been improved greatly and errors involved in the measurements can be reduced by de-signing lysimeter to meet specific requirements, by proper lysimeter operation, and by managing the lysimeter site according to design requirements (Howell et al. 1991). But this process is a difficult process and also tedious. Therefore, general practice is employed where reference evapotranspiration is first estimated from a standard surface and then to apply an appropriate empirical crop coefficient (K_c) for other crops, which accounts for the difference between the standard surface and crop ET (ET_0) (Allen et al. 1994). The methodology of estimating crop ET is the use of a potential or reference ET and a crop coefficient. The crop coefficient is an empirical ratio of crop ET to some reference ET and is generally derived from experimental data. In addition, by using ET_0, it is easier to select consistent K_c values and to make reliable ET estimates in new areas. The use of ET_0-K_c approach has been enormously successful in obviating the need to calibrate a separate ET equation for each crop and its stage of growth (Wright 1982; Jensen et al. 1990). The crop coefficient is specific to each crop and to the method used for ET_0 and reflects the canopy development and water management practices over the course of the growing season (Kamble et al. 2013). Carefully developed K_c values, along with ET_0 values, can provide robust and accurate ET estimates. This method relies heavily on published K_c values for different crops. While there are significant assumptions imbedded in the K_c values, which may not apply to many locations/conditions, the simplicity of this method has been one of the main reasons for its widespread adoption. K_c values for the same crop can show significant variation between locations due to differences in crop variety, soil properties, irrigation method and frequency, climate, and perhaps more importantly, crop management practices. Changes in the vegetation and ground cover mean that the crop coefficient varies during the growing season.

Various methods for estimating crop evapotranspiration were presented in Subedi and Chavez (2015) from the simple empirical method which uses temperature data only to the complex Shuttleworth model. A method for estimating reference crop evapotranspiration, ET_0 from measured temperature alone was presented by Hargreaves and Samani (1985) which provide a reasonable ET_0 for areas having limited meteorological data. American Society of Civil Engineers, ASCE has now recognized the FAO-56 PM method (Allen et al. 1998) as the sole international standard method for indirect estimation of ET_0 when comprehensive datasets are available. It is also the most widely recommended equation for daily reference grass evapotranspiration (ET_0) computations from meteorological data (ASCE-EWRI 2005). However, under non-ideal data conditions, other methods can also play a major role in estimating accurate ET_0 depending on data availability and site conditions. FAO-56 PM method requires several meteorological data to be measured with sufficient accuracy to produce accurate estimates. The crop coefficient curve requires knowledge of only three values of K_c defining the initial stage ($K_{c\,ini}$), mid-season stage ($K_{c\,mid}$), and the end of late-season stage ($K_{c\,end}$). The tabulated data of K_c values given in FAO-56 for various agricultural crops were used as the reference dataset for generalized K_c values for this study. The coefficients presented are organized by group type (i.e., small vegetables, legumes, cereals, etc.) to assist in locating the crop in the table and to aid in comparing crops within the same group. The weekly average K_c values for groundnut, paddy, tobacco, and sugarcane crops measured using lysimeters were compared with those recommended in FAO 56 K_c values in Jyothy et al. (2011). A similar study was also carried out by Ko et al. (2009) in which they found some crops corresponded to K_c values with those using FAO 56 method and some did not. The relationship between crop evapotranspiration estimated from the FAO-56 method and those measured using the weighing type lysimeter is linear (Kumar et al. 2011). In this study, the K_c curves were developed based on the NDVI-generated images and then compared with the calculation-based K_c curves. The main objective here was to focus on the behavior of crop coefficient curves of major crops of the Pare Basin in Arunachal Pradesh.

9.2 Study Area

Pare river basin is situated at the south-western part of Arunachal Pradesh, India. It lies at the upstream of the Dikrong river basin extending from 93° 13' 9.39" E to 93° 47' 8.99" E longitude and 27° 9' 32.5" N to 27° 22' 11.08" N latitude. The total area of the region is 773.5 sq. km. Papumpare district, in which the Pare basin situated, falls under the mid-tropical hill zones. Climate is wet and humid in most parts of the basin. Monsoon season starts from May and continues up to September/October. The average annual rainfall of the area is 2542.8 mm and the temperature ranges between 10 and 32°C (Dabral et al. 2008) while it experiences an average relative humidity of about 82% throughout the year. The topography is undulated with elevation ranges from 255 m to 2895 m above mean sea level (msl). The study area, Pare river basin is shown in Fig. 9.1.

9.3 Methodology

9.3.1 *Data Acquisition and Determination of Growth Stage*

Six crops were identified as per the survey conducted on farmers at different locations of the study area, additionally with details of the sowing and harvesting dates for the crops. A comprehensive dataset of corresponding lengths of crop development stages for the three growth stage (initial, mid and late stages) was prepared with reference to the standard values given in Table 9.1, obtained from FAO-56, for the identified crops, which will be used as the main dataset for the calculation of local crop coefficients after correction for local climatic parameters. The crop calendar for the study area is shown in Table 9.2.

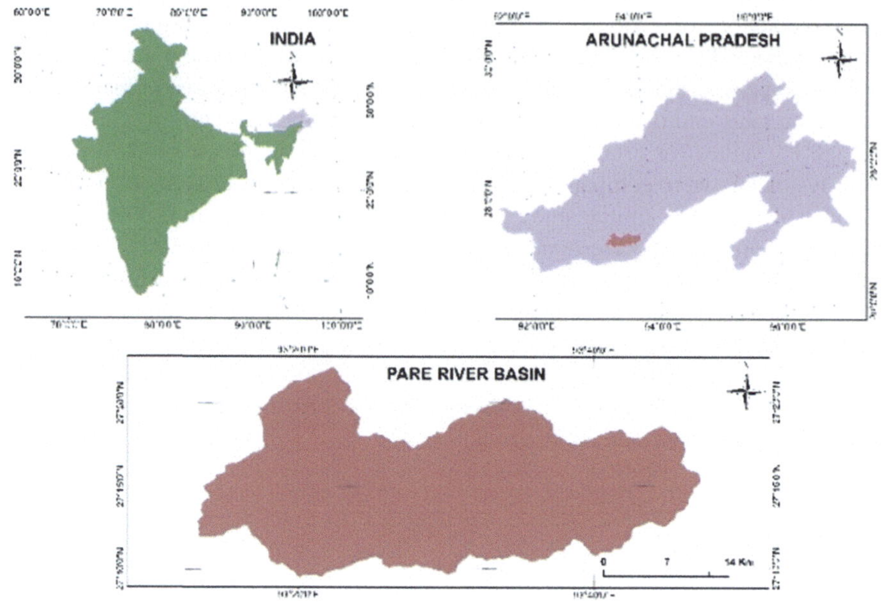

Fig. 9.1 Pare River Basin, Arunachal Pradesh, India

9.3.2 Determination of RH and Wind Speed of the Region

RH and wind speed values of the study area were obtained from www.mosdac.in. The values obtained were in an hourly basis for a span of almost 8 years in Greenwich Mean Time, GMT timeline. Therefore, the first step was to convert the timeline into Indian standard Time, IST format. Accordingly, the values of daily RH and wind speed were converted into the IST timeline, and then a comprehensive dataset of average monthly RH and wind speeds were prepared for the span of 8 years (2008–2015). Table 9.3 shows the average monthly wind speed and RH of the study area.

9.3.3 Estimation of Crop Coefficient ($\mathbf{K}_c$)

The trends in K_c during the growing period are represented by the crop coefficient curve. Only three values for K_c are required to construct the crop coefficient curve: those during the initial stage ($K_{c\ ini}$), the mid-season stage ($K_{c\ mid}$), and at the end of the late-season stage ($K_{c\ end}$). Tabulated K_c values in Table 9.4 are typical values for $K_{c\ ini}$, $K_{c\ mid}$, and $K_{c\ end}$ for various agricultural crops. The values of crop coefficients in Table 9.4 are for typical values expected for average K_c under a standard climatic condition, which is defined as a sub-humid climate with average day-time minimum

Table 9.1 (FAO56): Lengths of crop development stages* for various planting periods and climatic regions (days)

Crop	Initial (L_{ini})	Dev (L_{dev})	Mid (L_{mid})	Late (L_{late})	Total	Plant date	Region
Rice	30 30	30 30	60 80	30 40	150 180	Dec.; May Dec.	Tropics; Mediterranean tropics
Maize (sweet)	20 20 20 30 20	20 25 30 30 40	30 25 50/30 30 70	10 10 10 103 10	80 80 90 110 140	Mar. May./Jun. Oct./Dec. Apr. Jan.	Philippines Mediterranean Arid climate Idaho, USA Calif. desert, USA
Cucumber	20 25	30 35	40 50	15 20	105 130	Jun./Aug. Nov.; Feb.	Arid region Arid region
Sweet potato	20 15	30 30	60 50	40 30	150 125	April rainy season	Mediterranean Tropical regions
Mustard	30	30	30	30	120	Jan. Sept.	–
Sugarcane	25 30 35	70 50 105	135 180 210	50 60 70	280 320 420		Low latitudes Tropics Hawaii, USA

*Lengths of crop development stages provided in this table are indicative of general conditions but may vary substantially from region to region, with climate and cropping conditions, and with crop variety. The user is strongly encouraged to obtain appropriate local information

relative humidity (RH_{min}) ≈ 45% and having calm to moderate wind speeds averaging 2 m/s (Allen et al. 1998).

The crop coefficients depend on various climatic factors other than relative humidity and wind speed, however, most of the various climatic parameters were already incorporated in the estimation of reference evapotranspiration (ET_o). Hence, K_c varies predominately with the crop characteristics and only to a limited extent with climate. The transferring of these K_c values to another location and climate were carried out in this study as described in Allen et al. (1998); Allen et al. (2005) and Allen and Pereira (2009). The initial stage $K_{c\,ini}$ of the crops was not adjusted as in this stage, wind and RH are not significant for their growth. For specific adjustments in climates, RH_{min} differs from 45% or where u2 is larger or smaller than 2.0 m/s (Table 9.4).

$$\text{Adjusted } K_{c\,mid} = K_{c\,mid} + \left[0.04(u2 - 2) - 0.04\left(RH_{min} - 45\right)\right](h/3)^{0.3} \quad (9.1)$$

$$\text{Adjusted } K_{c\,end} = K_{c\,end} + \left[0.04(u2 - 2) - 0.04\left(RH_{min} - 45\right)\right](h/3)^{0.3} \quad (9.2)$$

Table 9.2 Crop calendar of the crops in the study area

Sl.no.	Crops	Jan	Feb	Mar	Apr	May	Jun	Jul	Aug	Sep	Oct	Nov	Dec
1	Paddy					Initial	Dev	Mid	Mid	Late	Late		
2	Maize			Initial	Dev	Dev	Mid	Mid	Late	Late			
3	Cucumber				Initial	Dev	Mid	Late					
4	Sweet potato						Initial	Dev	Dev	Mid	Mid	Late	
5	Mustard	Initial	Dev	Mid	Late					Initial	Dev	Mid	Late
6	Sugar Cane			Initial	Dev	Dev	Dev	Mid	Mid	Mid	Mid	Mid	Mid
		Late	Late	Late	Late								

Table 9.3 Average monthly wind speed and RH of the study area

Months	Jan	Feb	Mar	Apr	May	Jun	Jul	Aug	Sep	Oct	Nov	Dec
Avg. Wind Speed (m/s)	0.692	0.464	0.553	0.639	0.438	0.345	0.311	0.263	0.249	0.32	0.257	0.324
Avg. RH (%)	74.49	69.850	71.017	72.145	84.956	79.831	87.886	90.968	86.969	83.771	76.859	73.358

Table 9.4 (FAO56): Single (time-averaged) crop coefficients, K_c, and mean maximum plant height for non-stressed, well-managed crops in sub-humid climates ($RH_{min} \approx 45\%$, $u2 \approx 2$ m/s) for use with the FAO Penman-Monteith, ET_0

Crop	$K_{c\ ini}$	$K_{c\ mid}$	$K_{c\ late}$	Maximum crop height (h) (m)
Rice (cereal)	1.05	1.2	0.9	1
Maize (cereal), sweet	0.55	1.15	1.05	1.5
Cucumber – Fresh Market – Machine harvest	0.6 0.5	1.00 1.00	0.75 0.90	0.3 0.3
Sweet potato (roots and tubers)	0.5	1.15	0.65	0.4
Mustard	0.299	1.2	0.6	–
Sugarcane	0.40	1.25	0.75	3

where u2 is the monthly average wind speed [ms^{-1}]; RH_{min} is the monthly average minimum relative humidity [%] and h is the mean plant height during the stage [m]. The modified K_c curves for all the crops considered for this study are shown in Table 9.5. The variety of mustard grown in the study area is grown twice in a year; one is sown as a Rabi crop early in the month of December–January and the other as a Kharif in the month of September. Therefore, for the purpose of evaluation, both the varieties of mustard have been considered separately, and accordingly, the crop coefficient curves have also been developed separately.

9.3.4 Generation of K_c_NDVI

The K_c value for the study can be estimated using the regression equation developed by Kamble et al. (2013), which is given below:

$$K_{c\_NDVI} = 1.457\,\text{NDVI} - 0.1725 \tag{9.3}$$

The values of K_{c_NDVI} for each crop obtained from the above equation are compared to the values obtained by Eqs. (9.1) and (9.2). MOD13Q1 16_Day tile of resolution 250 m for Arunachal Pradesh, India was downloaded for eight years (2008–2015) to obtain the NDVI layer of vegetation area. MODIS Re-Projection Tool (MRT) (version 4.1) was used to convert the downloaded image from HDF-EOS to GeoTIFF. The raster images were then clipped to the necessary boundary conditions of the study area and time series of the sample crops were extracted for a period of 8 years using ArcGIS. The step-wise illustration of the procedure is given in Fig. 9.2. For further evaluation, few pixels were selected which showed the best values of the coefficient of determination (R^2) while comparing the conventional K_c and generated K_{c_NDVI} curves for each crop. Coefficient of determination is defined as a number that indicates the proportion of the variance in the dependent variable

Table 9.5 Modified K_c for respective crops

Sl.no.	Crop	K_{in}	K_{mid}	K_{late}	Plant date	Harvest date	Crop height	Revised K_{mid}	Revised K_{late}
1	Paddy	1.05	1.2	0.90-0.6	May, Jun	Oct- Nov	1	1.14	0.72–0.42
2	Maize	0.55	1.15	1.05	Mar, Apr	Sep	1.5	0.97	0.85
3	Cucumber	0.6	1.15	0.75	Apr, May	Jun-Jul-Aug	0.3	1.05	0.63
4	Sweet potato	0.5	1.15	0.65	Mar, Apr	Nov-Dec	0.4	0.97	0.54
5	Mustard	0.3	1.2	0.6	Sep	Jan	0.13	0.62	0.53
6	Sugarcane	0.4	1.25	0.7	Mar	May	3	1.03	0.54

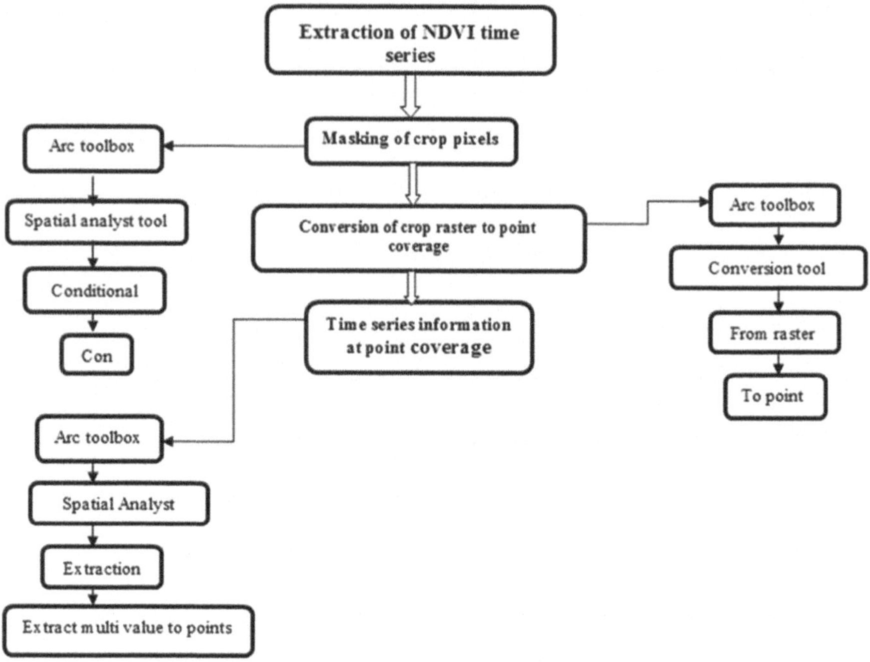

Fig. 9.2 Procedure for extraction of K_{c_NDVI} time series

that is predictable from the independent variable(s). It provides a measure of how well-observed outcomes are replicated by the model, based on the proportion of total variation of outcomes explained by the model.

9.4 Results and Discussion

9.4.1 Conventional Crop Coefficient Curves

The conventional K_c curves for different crops after being adjusted for wind speed and relative humidity are shown in Figs. 9.3, 9.4, 9.5, 9.6, 9.7, 9.8 and 9.9. These curves were prepared by averaging the adjusted crop coefficient values over the eight-years period. It was observed from Fig. 9.3 that the values of late conventional K_c of maize were higher than that of initial K_c and slightly smaller than that of mid-K_c. This is because during the initial stages the plants have less leaf area and plant height but during the later stage, the full development of leaves and stems increases the metabolic activities for the maturation of the corn. For rice growing in paddy fields with a water depth of 0.10-0.20 m, the ET during the initial stage mainly consists of evaporation from the standing water. Since the study area has a more humid climate

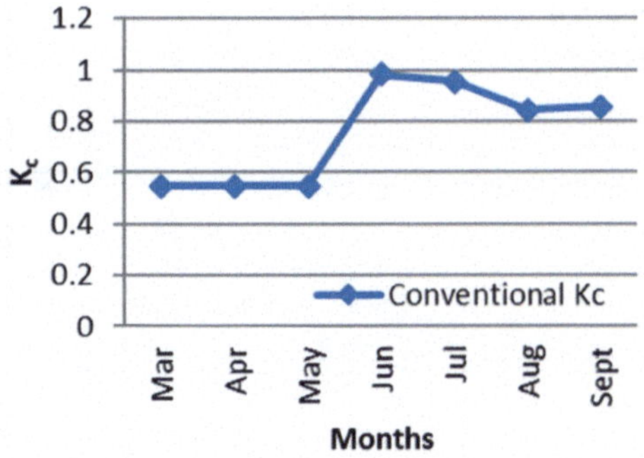

Fig. 9.3 Conventional K_c of maize

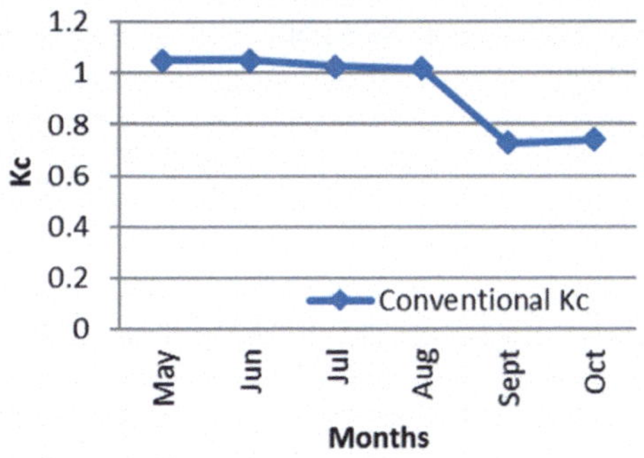

Fig. 9.4 Conventional K_c of paddy

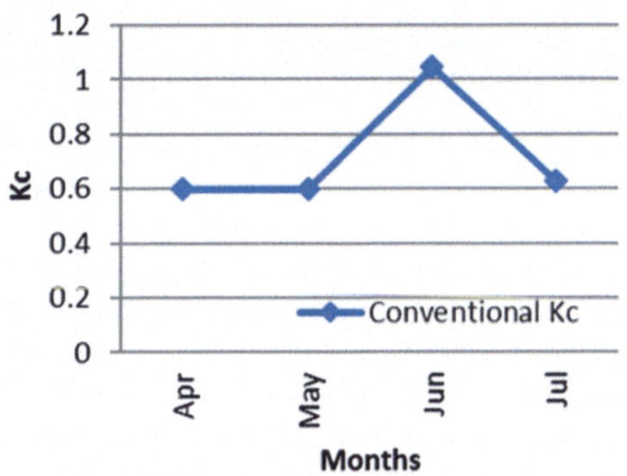

Fig. 9.5 Conventional K_c of cucumber

and conditions of lower wind speeds, the K_c mid values of paddy are somewhat lower than that of the $K_{c\,ini}$. The K_c curve of the paddy for the study area is shown in Fig. 9.4. Consistent decreasing trend of the K_c curves was observed until the end of the mid-stage. The sudden decrease in the curve indicates the onset of the late stage of crop life.

For cucumber, the full development of the crop completes in approximately 120–130 days. So, the trend of the K_c curve cannot be understood elaborately as each stage of growth is assigned as a period of one month. The curve was observed to be

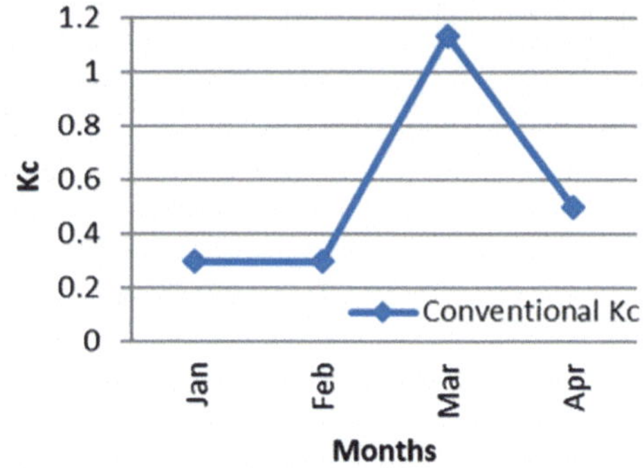

Fig. 9.6 Conventional K_c of mustard

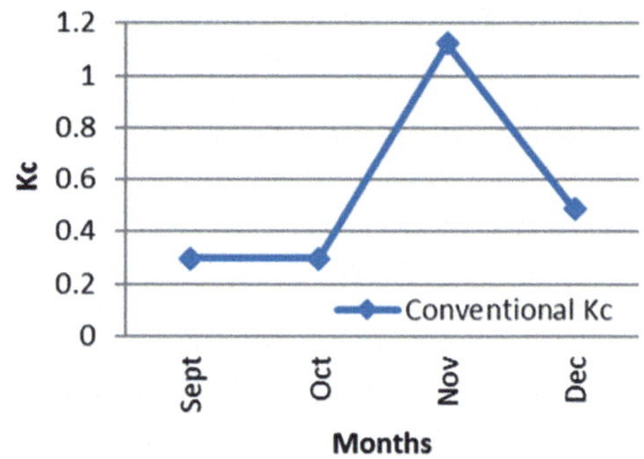

Fig. 9.7 Conventional K_c of mustard

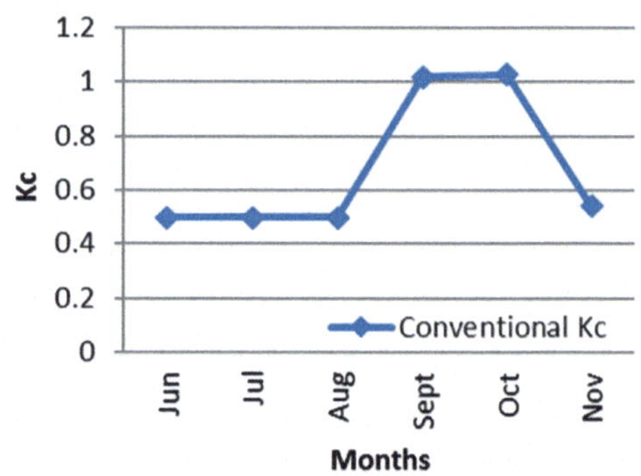

Fig. 9.8 Conventional K_c of sweet potato

triangular in shape as shown in Fig. 9.5. The initial stage and late stage had similar K_c values (0.6) while the mid-season stage had a value of more than one. In case of mustard, the shape of the K_c curves was observed as triangular, similar to the curve of cucumber, and are shown in Figs. 9.6 and 9.7. This is because only one month was assigned for $K_{c\,mid}$ as per the crop calendar developed for the sample crops of the study area. For mustard, values of $K_{c\,late}$ are higher than initial K_c by a margin of about 0.2 for both the versions of mustard. Same trend was observed in the case of maize. Also, the curves were found similar in both the cropping seasons.

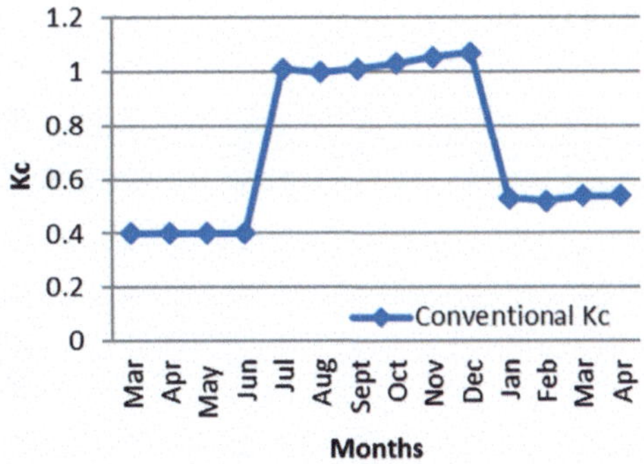

Fig. 9.9 Conventional K_c of sugarcane

In Fig. 9.8, the K_c curve for sweet potato is shown. Unlike cucumber and mustard which gives a very vague idea of the behavior of the K_c curve throughout the different stages of their growth, the data on sweet potato provides a fairly elaborate idea of the nature of the K_c curve. In fact, sweet potato shows an ideal and agreeable trend throughout the growth stages and can be considered as a reference for further discussions. Figure 9.8 shows the K_c curve for sweet potato. The trend analysis of the crop coefficient of sugarcane shows a very promising nature as shown in Fig. 9.9. The behavior of the curve suggests that the K_c value is higher by 0.1 during the later stage in comparison to that in the initial stage, which in turn indicates that the crop ET is higher during the late stage than the initial stage. This trend affirms the claim that was put forward in case of the trend analysis of the K_c curves of maize as well as mustard. The value of K_c witnesses a gradual increase during the mid-season as the crop approaches the later stage.

9.4.2 NDVI Crop Coefficient Curve

The generated K_c curves from NDVI data using Eq. (9.3) and those conventional K_c curves using Eqs. (9.1) and (9.2) were compared by plotting K_c on the Y-axis against dates as abscissa. These graphs are illustrated in the figures from Fig. 9.10,

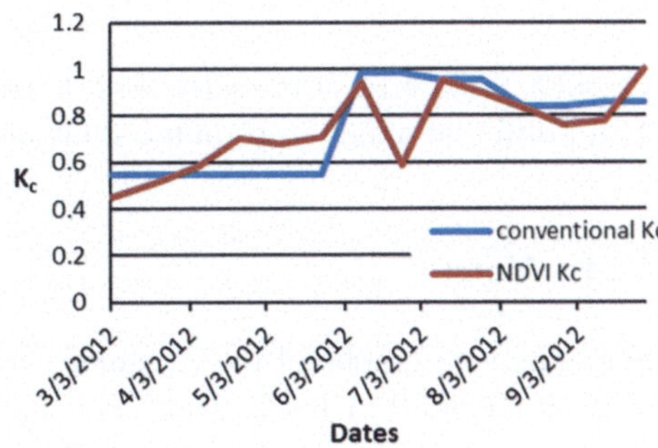

Fig. 9.10 Maize K_c curve

Fig. 9.11 Maize K_c curve

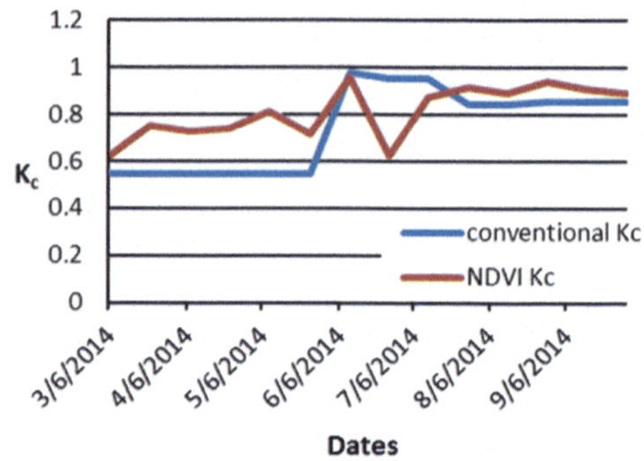

Fig. 9.12 Paddy K_c curve

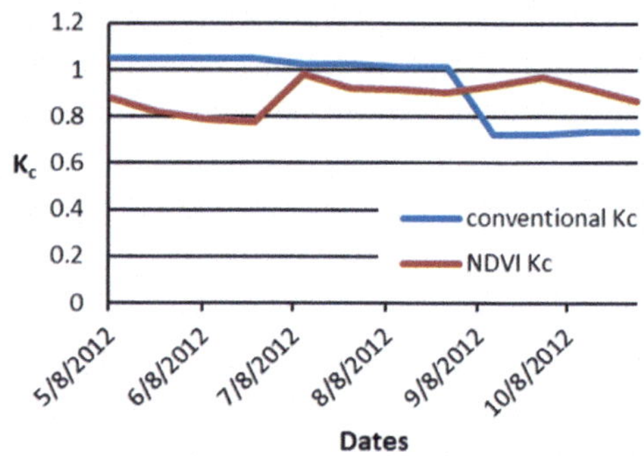

Fig. 9.13 Paddy K_c curve

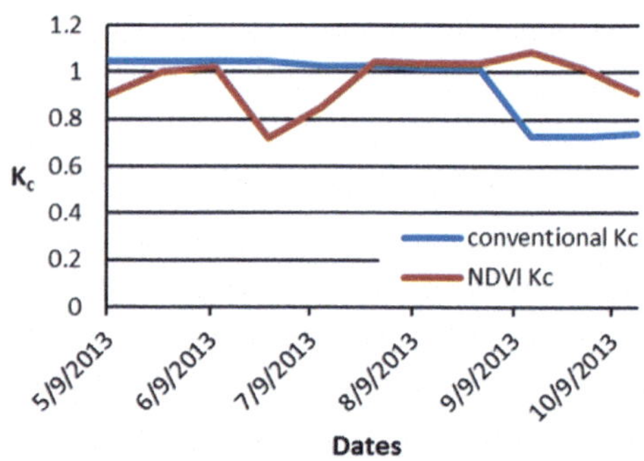

9.11, 9.12, 9.13, 9.14, 9.15, 9.16, 9.17, 9.18, 9.19, 9.20, 9.21, 9.22, 9.23, 9.24 and 9.25 for different major crops in the study area.

9.4.2.1 Maize

The comparison graphs of the K_c curves for the years 2012 and 2014 are shown in Figs. 9.10 and 9.11 respectively. For maize, pixel 13,516 gave a good value of R^2 which is around 0.475 for the year 2012 and the next best value of R^2 obtained

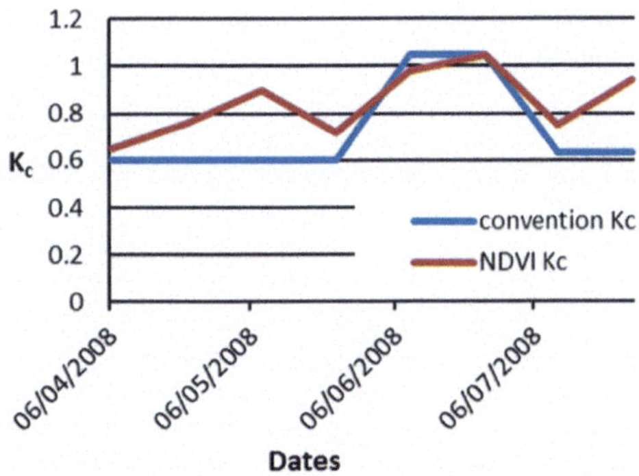

Fig. 9.14 Cucumber K_c curve

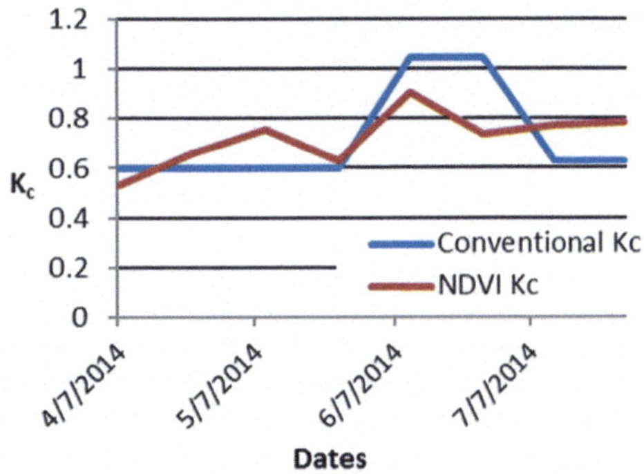

Fig. 9.15 Cucumber K_c curve

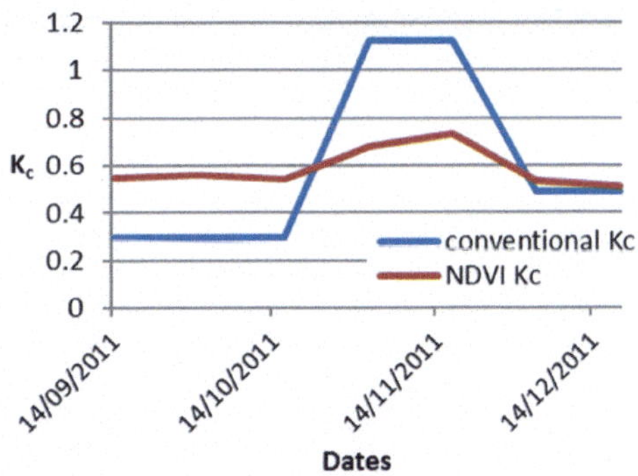

Fig. 9.16 Mustard (Winter) K_c curve

was 0.345 for the year 2014. Therefore, it can be suggested that in this case, since the results were moderate, the procedure showed nearly 50% chances of success in the determination of K_c using NDVI which is less than the expected level. With its high values of correlation between NDVI and K_c, as per Kamble et al. (2013), the coefficients of determination values were supposed to be better than the ones obtained. This shortcoming could be because of the difference in the species of maize, water management, growth conditions in the study area.

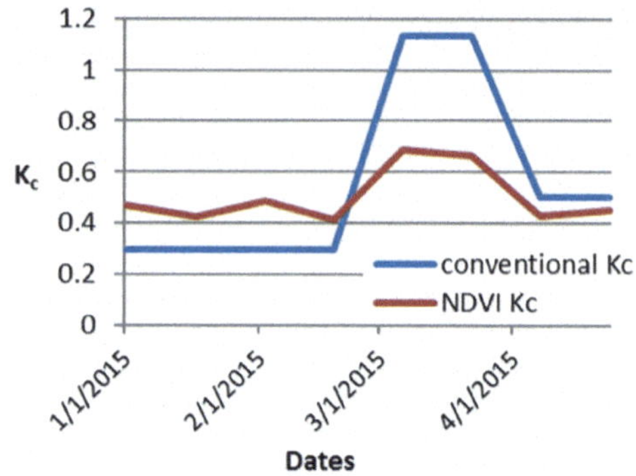

Fig. 9.17 Mustard (Summer) K_c curve

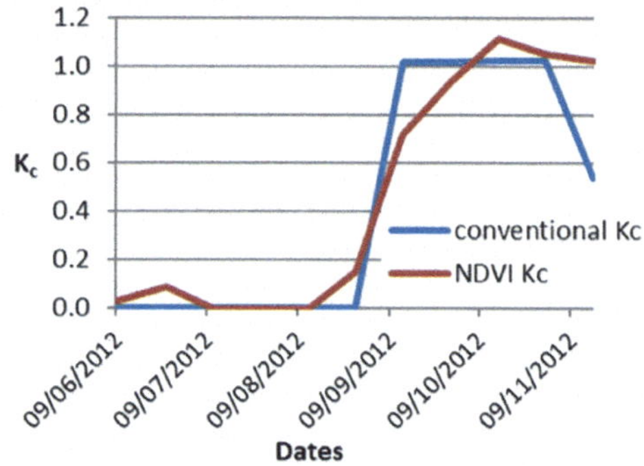

Fig. 9.18 Sweet potato K_c curve

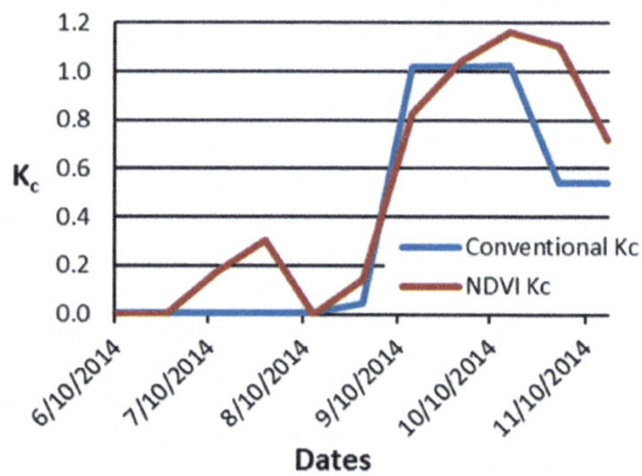

Fig. 9.19 Sweet potato K_c curve

9.4.2.2 Paddy

In the case of paddy, the results were not satisfactory as the values of R^2 were found to be very small. Though the best among the K_c comparison graphs have been shown in Figs. 9.12 and 9.13 for the year 2012 and 2013 respectively. The R^2 values are highly disappointing to be considered a success. Hence, this procedure was a complete failure in the case of paddy. This can, however, be justified by the fact that paddy is

Fig. 9.20 Sugarcane 1 K_c curve

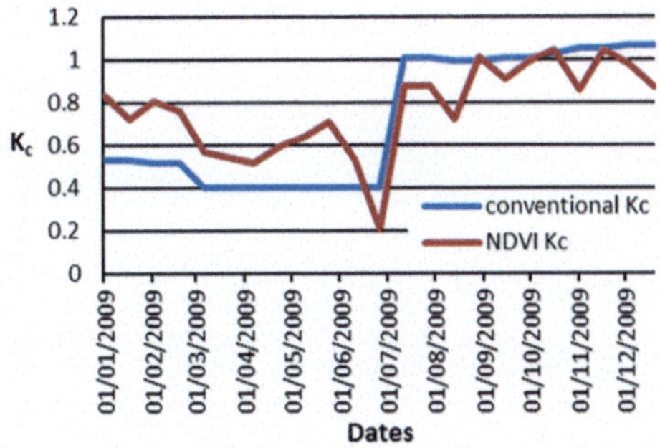

Fig. 9.21 Sugarcane 1 K_c curve

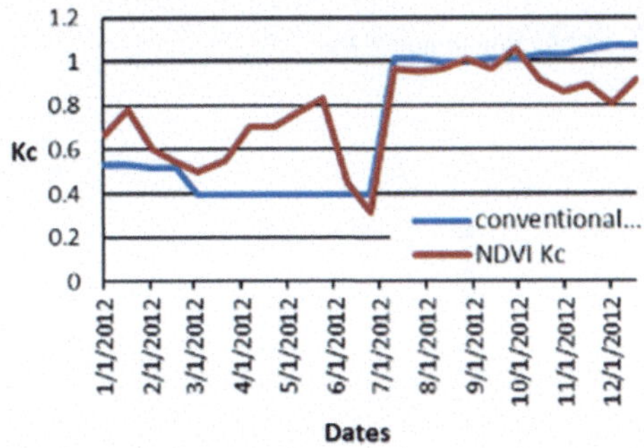

Fig. 9.22 Sugarcane 1 K_c curve

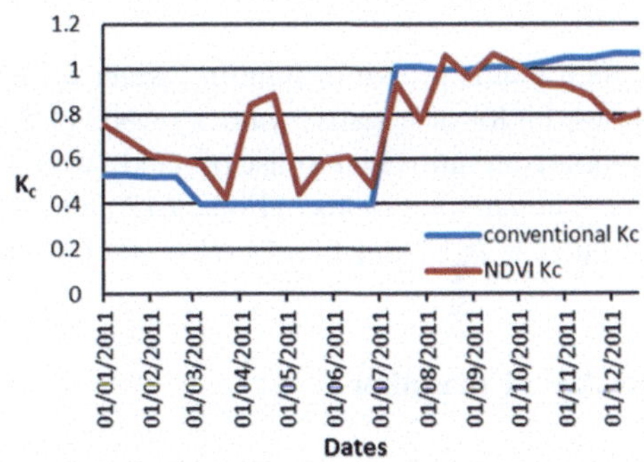

Fig. 9.23 Sugarcane 2 K_c curve

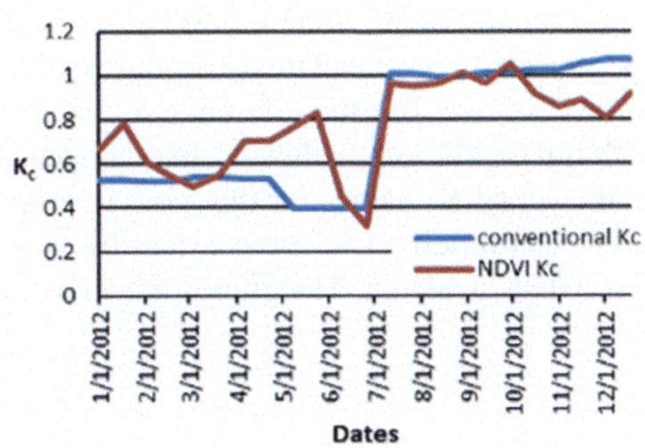

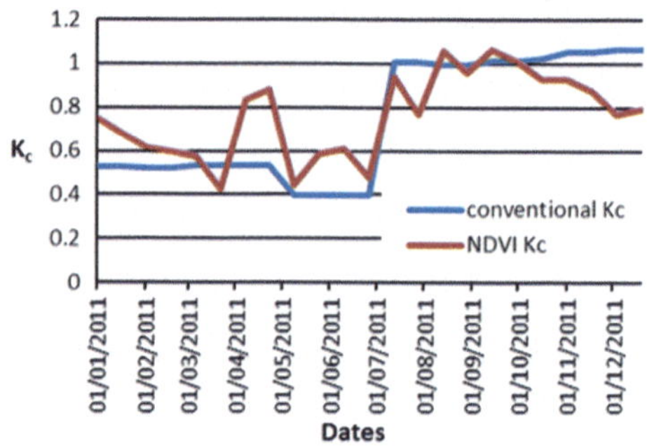

Fig. 9.24 Sugarcane 2 K_c curve

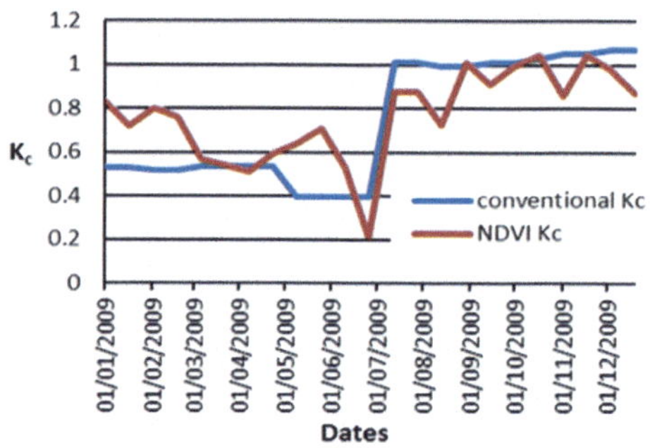

Fig. 9.25 Sugarcane 2 K_c curve

a plant that is grown in standing water for most of its life. The time series of paddy showed a lot of negative values of NDVI K_c, as a result of which the values of the comparison ratios in the case of paddy were very low compared to all the other crops. The comparison graphs in Figs. 9.12 and 9.13 are for the pixels 14,994 ($R^2 = 0.188$) in the year 2012 and 14,743 ($R^2 = 0.068$) in 2013 respectively.

9.4.2.3 Cucumber

For cucumber, the pixels 13,435 of 2008 and 15,429 of the year 2014 showed the highest value of R^2, 0.576, and 0.332, respectively. The values of the correlation coefficients suggest that this procedure of development of K_c curve had a 50% chance of success rate. But the reference K_c values obtained from FAO56 cannot be taken with full confidence as those values are generalized values and might not necessarily be the actual K_c values for the types of crops grown in the study area although they have been modified to some extent by using the local relative humidity and wind speed data available. The graphs are shown in Figs. 9.14 and 9.15 respectively.

9.4.2.4 Mustard

In this part of the region mustard is planted twice in a year, summer mustard, and winter mustard and therefore, for this study, it was necessary to calculate both the seasonal mustards separately. Interestingly, the behavior of the comparison graphs of the NDVI K_c and the conventional K_c for both the types of mustard were very promising. The winter mustard which is planted in September and harvested in December has an overwhelming value of R^2 for the pixel 13,761 (0.819) in 2010 which is exemplary which strengthens the notion that during that particular period of time that area was used for the cultivation of mustard. The highest value of the correlation coefficients for the summer mustard which is grown from January to April is seen for the pixels 15,673 (0.873) in the year 2015 and the graphs are shown in Figs. 9.16 and 9.17 respectively.

9.4.2.5 Sweet Potato

Sweet potato showed the best comparison trend between the conventional K_c and the NDVI- generated K_c. Even the values of R^2 were the highest amongst all the other sample crops, which suggests that this procedure of development of K_c is also successful in the case of sweet potato. The pixels identified as sweet potato grown areas are 5020 ($R^2 = 0.864$) in the year 2012 as well as 22,200 ($R^2 = 0.832$) in the year 2014 and their comparison graphs with the conventional K_c values are shown in Figs. 9.18 and 9.19, respectively.

9.4.2.6 Sugarcane

In and around the study area, sugarcane is sown in the month of March and the crop takes more than a year to be fully matured. The values of the pixels were available not on the basis of growth periods but over a single year. Since the growth period of sugarcane exceeded a period of one year, the procedure was carried on by considering sugarcane as two different types of crops, and K_c values for the overlapping months were replaced in the K_c dataset of the first sugarcane to prepare a comprehensive dataset for the second sugarcane crop. All the other calculations were preceded by considering the two sugarcane crops separately. Additionally, the trends of the comparison graphs for both the versions of sugarcane are a point of interest in this study. Basically, this procedure showed good results for sugarcane as the value of the correlation coefficient was overwhelming and the trend observed in the graphs was very promising. The values of R^2 were observed to be better in the case of the first sugarcane. The highest values were observed in the pixel-13,138 ($R^2 = 0.651$) for the year 2009, pixel-12,891 ($R^2 = 0.646$) in 2012 and pixel-17,563 ($R^2 = 0.579$) in 2011. However, the values of R^2 were equally good for the second sugarcane in pixel-12,891 ($R^2 = 0.638$) in 2012, pixel-17,563 ($R^2 = 0.614$) in 2011 and pixel-13,138 ($R^2 = 0.611$) in 2009. The trends observed in the comparison graphs of sugarcane could be regarded as one of the best amongst the other sample crops. The graphs of the above-mentioned pixels are shown in Fig. 9.20, 9.21, 9.22, 9.23, 9.24, and 9.25, respectively.

9.5 Conclusion

According to the trend analysis of the K_c curves generated by modifying the FAO56 K_c values to fit the local time and climatic conditions of the study area, interesting behavior was seen in paddy which had higher values of $K_{c\,ini}$ than that of $K_{c\,mid}$ and $K_{c\,end}$, which in turn ascertains the fact that the ET was maximum due to the surface evaporation from the surface of standing water in the paddy fields. The curve of sweet potato showed an ideal trend whereas in the case of mustard, maize, and sugarcane the $K_{c\,end}$ values were higher than that of the $K_{c\,ini}$ which maybe be the result of the limitations of this procedure as other parameters of K_c determination such as leaf area, crop canopy, plant height, water management and among others, were not examined. As per the trend analysis of the comparison graphs of the NDVI-generated and the conventional $K_{c,}$ the NDVI approach for K_c determination was a success in the case of sweet potato, mustard, and sugarcane. The high values of the coefficient of determination (R^2) for these three crops hereby further validate the correlation of NDVI with K_c. In the case of paddy, the R^2 values were small which suggests the less likeliness of K_c determination using NDVI values for paddy due to the low reflectance of the Red and NIR bands off the standing water of paddy fields. The moderate values of R^2 for cucumber and maize suggests a near 50% chance of certainty for the success of the application of this approach for K_c calculation. But the lack of examination of the effects of other variables of the crop grown in this procedure could be the reason for the partial success in some of these crops.

References

Allen RG, Pereira A (2009) Estimating crop coefficients from fraction of ground cover and height. Irrig Sci 28:177–34

Allen RG, Pereira LS, Raes D, Smith M (1998) Crop evapotranspiration. Guidelines for computing crop water requirements, Irrigation and drainage paper 56, food and agric. Organization of the United Nations, Rome, Italy

Allen RG, Smith M, Perrier A, Pereira LS (1994) An update for the definition of reference evapotranspiration. ICID Bull. 43(2):1–34

Allen RG, Walter IA, Elliott RL, Howell TA, Itenfisu D, Jensen ME, Snyder RL (2005) The ASCE Standardized Reference Evapotranspiration Equation. Am Soc Civil Eng, Reston, p 59

ASCE-EWRI (2005) The ASCE standardised reference evapotranspiration equation. Environmental and water resources institute of the ASCE, Report by the task committee on standardisation of reference evapotranspiration

Dabral PP, Baithuri N, Pandey A (2008) Soil erosion assessment in a hilly catchment of north eastern India using usle, gis and remote sensing. Water Resour Manage 22(12):1783–1798

Hargreaves GH, Samani ZA (1985) Reference crop evapotranspiration from temperature. Am Soc Agric Eng, Paper No. 85–2517, https://doi.org/10.13031/2013.26773

Howell TA, Schneider AD, Jensen ME (1991) History of lysimeter design and use for evapotranspiration measurement. In: Allen RG, Howell TA, Pruitt WO, Walter IA, Jensen ME (eds), Proceeding of international symposium on lysimetry, July 23–25, 1991, Honolulu, Hawaii, pp 1–9

Jensen ME, Burman RD, Allen RG (1990) Evapotranspiration and irrigation water requirements. Am Soc Civ Eng, Engg. Pract. Manual No. 70

Jyothy SA, Reddy KCS, Mallikarjuna P (2011) Development of crop coefficient models for weekly crop evapotranspiration estimation, Proc of Int. Conf. on Recent Trends in Transportation, Environmental and Civil Engineering

Kamble B, Kilic A, Hubbard K (2013) Estimating crop coefficients using remote sensing-based vegetation index. Remote Sensing 5:1588–1602

Ko J, Piccinni G, Marek T, Howell T (2009) Determination of growth-stage-specific crop coefficients (K_c) of cotton and wheat. Agric Water Manage 96:1691–1697

Kumar R, Shankar V, Kumar M (2011) Development of crop coefficients for precise estimation of evapotranspiration for mustard in mid hill zone—India. Univers J Environ Res Technol 1(4):531–538

Montazar A, Rejmanek H, Tindula G, Little C, Shapland T, Anderson F, Inglese G, Mutters R, Linquist B, Greer CA, Hill J, Synder RL (2017) Crop coefficient curve for paddy rice from residual energy balance calculations. J Irrig Drain Eng 143(2)

Petillo MG, Castel JR (2007) Water balance and crop coefficient estimation of a citrus orchard in Uruguay. Spanish J Agric Res 5(2):232–243

Subedi A, Chavez JL (2015) Crop evapotranspiration (ET) estimation models: a review and discussion of the applicability and limitations of ET methods. J Agric Sci 7(6):50–68

Wright JL (1982) New evapotranspiration crop coefficients. J Irrig Drainage Div ASCE 108:57–74

Chapter 10
Monitoring of Soil Moisture Variability and Establishing the Correlation with Topography by Remotely Sensed GLDAS Data

Kaushlendra Verma and Y. B. Katpatal

Abstract Monitoring of temporal and spatial soil moisture variability is important to understand the surface hydrology of the region. The objective of this study is to analyze the spatial variability of soil moisture with the topography of the region. The study has been performed for the Maharashtra region of India. The monthly data of soil moisture has been taken from Global Land Data Assimilation System (GLDAS) NOAH model of 0.25 spatial resolution. For the topographic analysis of the terrain, CARTOSAT DEM data of 20 m spatial resolution have been used. The soil moisture anomalies were calculated for available monthly data sets and correlated with the topography of the study area. The Pearson Product-Moment Correlation and Spearman's rank correlation coefficient have been used for establishing the spatial correlation between soil moisture and topography of the region. The results of the study indicate that the correlation of soil moisture and topography of the study area harmonizes well with each other.

Keywords Remote sensing · GIS · Soil moisture · GLDAS NOAH · CARTOSAT DEM

10.1 Introduction

Soil moisture is a vital hydrological component that helps in understanding the hydrological process such as flood, drought, and interaction between land and atmosphere but induced variability in the spatial and temporal domain (Owe et al. 1982; Grayson et al. 1997). Soil moisture available in surface and subsurface affects each other i.e., the fully saturated subsoil moisture triggers the surface runoff (Burt and Butcher

K. Verma (✉)
Water Resource Engineering, Department of Civil Engineering, VNIT, Nagpur, India
e-mail: kaushlendraverma@students.vnit.ac.in

Y. B. Katpatal
Department of Civil Engineering, VNIT, Nagpur, India
e-mail: ybkatpatal@rediffmail.com

A. Pandey et al. (eds.), *Water Management and Water Governance*, Water Science and Technology Library 96,
https://doi.org/10.1007/978-3-030-58051-3_10

1985). Previous scientific efforts have been made to understand the soil moisture variability in horizontal and temporal scale but (Hawley et al. 1983; Nyberg 1996; Famiglietti et al. 1998) but very few studies have been done to explore the soil variability related to profile features (Loague1992). The previous studies have shown that soil moisture variability is greatly influenced by soil properties (Reynolds 1970) precipitation (Famiglietti et al. 1998), solar radiation (Western et al. 1999), slope position, gradient, relative elevation and curvature (Crave and Gascuel-odux 1997; Moore et al. 1988; Western et al. 1999). Soil moisture and its variability in rootzone affect plant growth and productivity. So, the continuous measurements are needed for agricultural practices, irrigation planning, and management. The soil moisture in the root zone changes with the variation in the topography of the catchment area spatially as well as temporally. Attempts have been made in the earlier studies for characterization of the spatial distribution of soil moisture, which varies with seasonal time-varying as a pattern according to the dominance of a particular process in relation to the wetness of the catchment. Variation in the relief of a surface affects the soil moisture by affecting the lateral flow of the groundwater and runoff processes (Grayson et al. 1997). The ready availability of digital terrain models has to lead to terrain indices being one of the most common predictors used to estimate spatial moisture patterns, even in situations where they may not represent the vital controlling processes (Grayson et al. 1997). Remote sensing and ground-based measurement studies have been done in the past for the spatial distribution of soil moisture. Generally, the ground-based measurements have focused on relatively small scales, whereas remote sensing studies for large scale. The calibration of ground measurement data is relatively easy but limited to small areas and the number of correlation measurements (Western et al. 2003, Brocca et al. 2010, Verma et al. 2020). The current study shows the prediction of soil moisture variability and its correlation with surface topography using remotely sensed satellite data.

10.2 Study Area

Maharashtra state of India is considered as study area (Fig. 10.1). Maharashtra is the second-most populous and third largest state by area of India. The study area is spread over 307,713 km^2, which lies between longitudes of 72°43′19.70″ to 80°53′49.59″ E and latitudes of 21°58′52.64″ to 15°35′57.66″ N. The annual precipitation of the study area is in the range of 991 mm with an annual temperature variation between a minimum of 14°C to a maximum of 48°C.

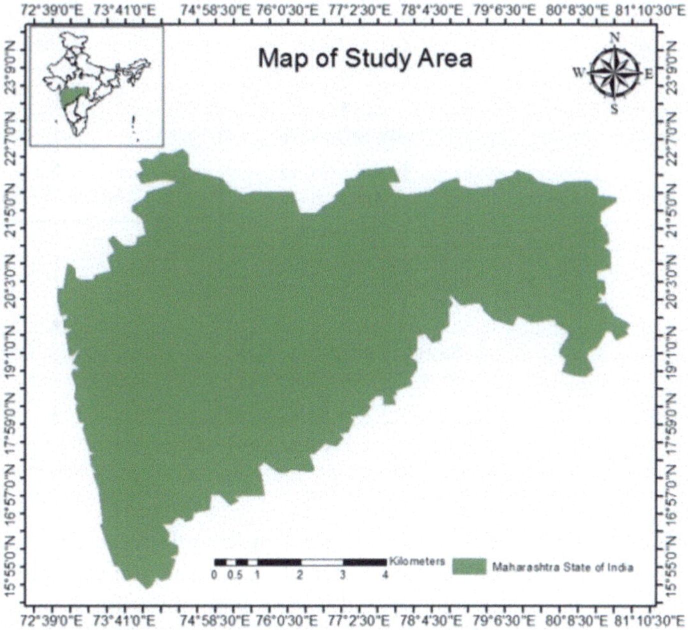

Fig. 10.1 Location map of the study area

10.3 Data Used and Methodology

10.3.1 Data Used

The monthly data of soil moisture has been taken from Global Land Data Assimilation System (GLDAS) NOAH model of 0.25° × 0.25° spatial resolution, The average of hydrological outputs from the Noah, Mosaic, Common Land Model (CLM), and Variable Infiltration Capacity (VIC) land surface models (LSM) were used to calculate the soil moisture anomaly (SMA). For the topographic analysis of the terrain, CARTOSAT DEM data of 20 m spatial resolution have been used.

10.3.2 Methodology

Global Land Data Assimilation System data has been extracted from my study area in the raster format of having a spatial resolution of 0.25° × 0.25° for different depths of 0–10, 10–40, 40–100, 100–200 cm and root zone depth which is further being clubbed as 0–40, 40–200 cm and root zone depth for the clear visualization. CARTOSAT Digital Elevation Model of 20 m resolution has been taken for the analysis. The raster to point conversion has been done using ArcMap software for

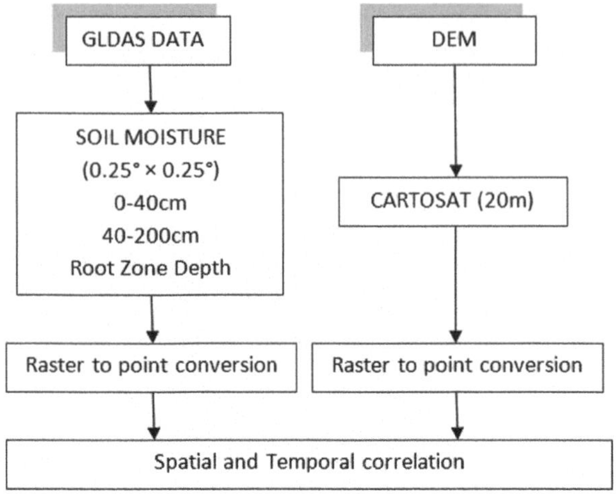

Fig. 10.2 Methodology flowchart of study

both GLDAS and DEM. The data have been taken from the year 2000 to 2015 for the soil moisture. After the calculation of DEM values and soil moisture values at different depth have been correlated spatially and temporally using Pearson Product Moment and Spearman correlation methods (Fig. 10.2). The following equations have been used for correlations (Eqs. 10.1 and 10.2).

$$r = \frac{N\sum xy - \sum(x)(y)}{\sqrt{[N\sum x^2 - \sum(x)^2][N\sum y^2 - \sum(y)^2]}} \tag{10.1}$$

r Pearson r correlation coefficient,
N number of observations,
x DEM values,
y soil moisture values.

$$\rho = 1 - \frac{6\sum \mathrm{d}i^2}{n(n^2 - 1)} \tag{10.2}$$

ρ Spearman rank correlation,
di the difference between the ranks of the corresponding variables,
n no of observations.

10.4 Results and Discussions

Digital Elevation Model profile has been plotted in 2-D by using the surface profile tool in QGIS (Fig. 10.3). This profile shows the elevation values and at different section at definite interval. The elevation has been extracted at different section and exported to excel.

The soil moisture values have also been extracted for the same interval as DEM. The soil moisture profile has been plotted for the same section as DEM (Fig. 10.3).

The soil moisture at different depths of 0–40, 40–200 cm and root zone has been plotted with the DEM at same interval and section. This plot shows the relationship of DEM and soil moisture for different seasons. The soil moisture values are the average values of three months, representing for pre-monsoon, monsoon, post-monsoon Rabi and Kharif seasons (Figs. 10.4, 10.5 and 10.6).

The plot between DEM and soil moisture at the depth 0–40 cm shows the less fluctuation. Only the monsoon season shows the relatively more fluctuation than other seasons. The soil moisture at the depth 40–200 cm shows relatively more fluctuation in comparison to 0–40 cm depth. The plot of correlation between soil moisture and DEM shows the linear correlation at most of the section of DEM.

The values of DEM and soil moisture at different section have been correlated with two different methods. The Pearson Product Moment Correlation Coefficient

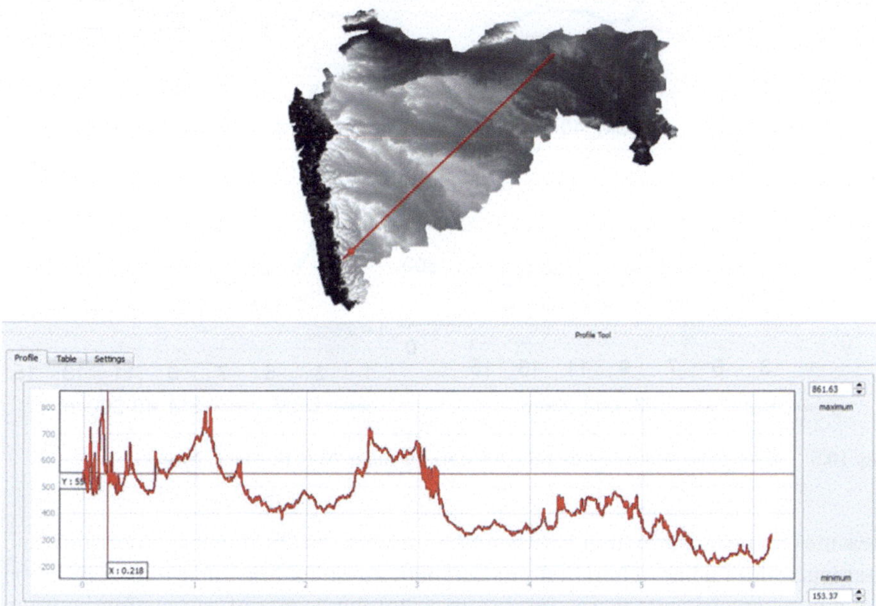

Fig. 10.3 DEM plotted in 2D by profile tool in QGIS software

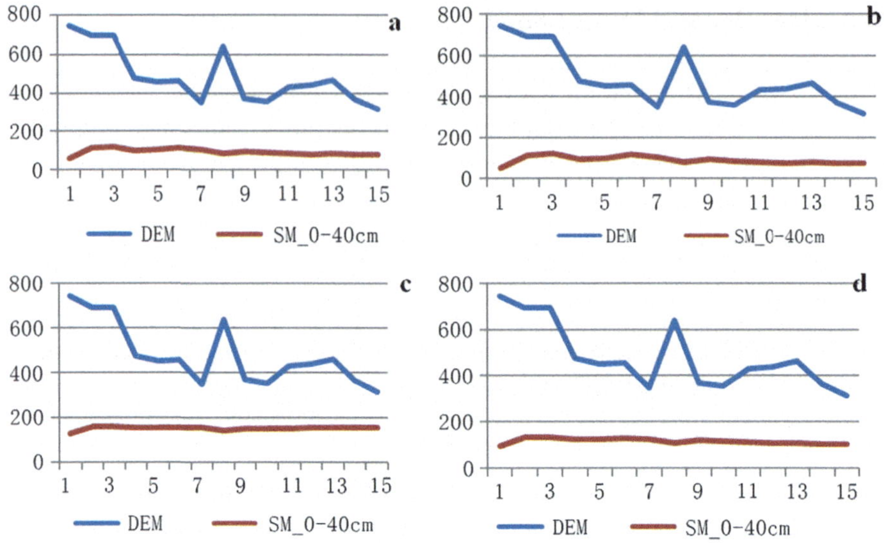

Fig. 10.4 DEM versus soil moisture (0–40 cm) for Jan **a**, May **b**, Aug **c**, Nov **d**

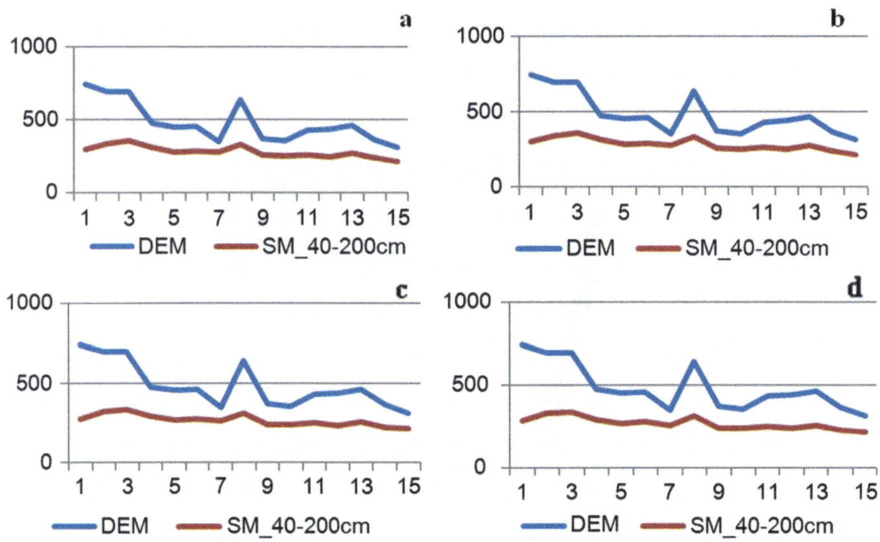

Fig. 10.5 DEM versus soil moisture (40–200 cm) for Jan **a**, May **b**, Aug **c**, Nov **d**

measures a linear association between two variables as DEM and soil moisture and attempts to draw a line of best fit the data of both variables. Spearman rank correlation is a non-parametric test that is used to measure the degree of association between two variables (Table 10.1). The output from the GLDAS soil moisture NOAH model

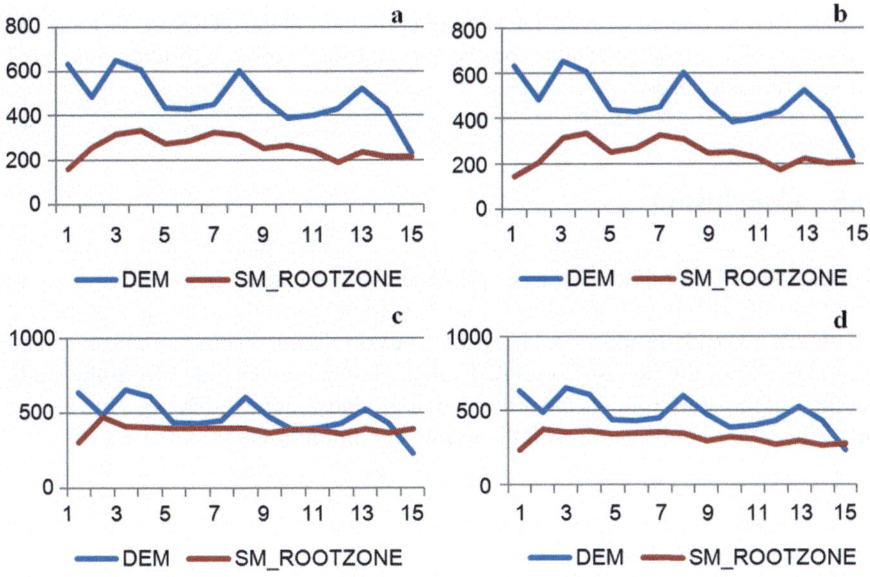

Fig. 10.6 DEM versus soil moisture (root zone) for Jan **a**, May **b**, Aug **c**, Nov **d**

Table 10.1 Spearman and Pearson correlation between DEM and soil moisture (a) S.M. 0–40 cm (b) S.M. 40–200 cm. (c) S.M. at root zone

(a)	*DEM*	
SM_0–40 cm	*Pearson*	*Spearman*
JAN	0.07	0.21
MAY	0.12	0.2
AUG	0.14	0.11
NOV	0.07	0.25
(b)	*DEM*	
SM_40–200 cm	*Pearson*	*Spearman*
JAN	0.06	0.22
MAY	0.06	0.22
AUG	0.06	0.22
NOV	0.04	0.2
(c)	*DEM*	
Root zone depth	*Pearson*	*Spearman*
JAN	0.28	0.34
MAY	0.28	0.27
AUG	0.09	0.23
NOV	0.14	0.32

data has been correlated with the DEM output profile from the QGIS profile tool. The Pearson and Spearman correlation has been established between different depths of soil moisture and DEM.

10.5 Conclusions

The response of correlation between DEM and soil moisture is found harmony in nature for different depth of soil moisture. It is found from the analysis that the soil moisture is inversely proportional to DEM. The correlation results from Pearson and Spearman shows the inverse correlated values. It can be concluded from the study that the soil moisture follows the reflection of ground water availability. However, it shows the inverse correlation with the terrain condition or DEM of the area.

10.6 Annexure 10. A

10.6.1 Input Images for Calculating Soil Moisture Anomalies:

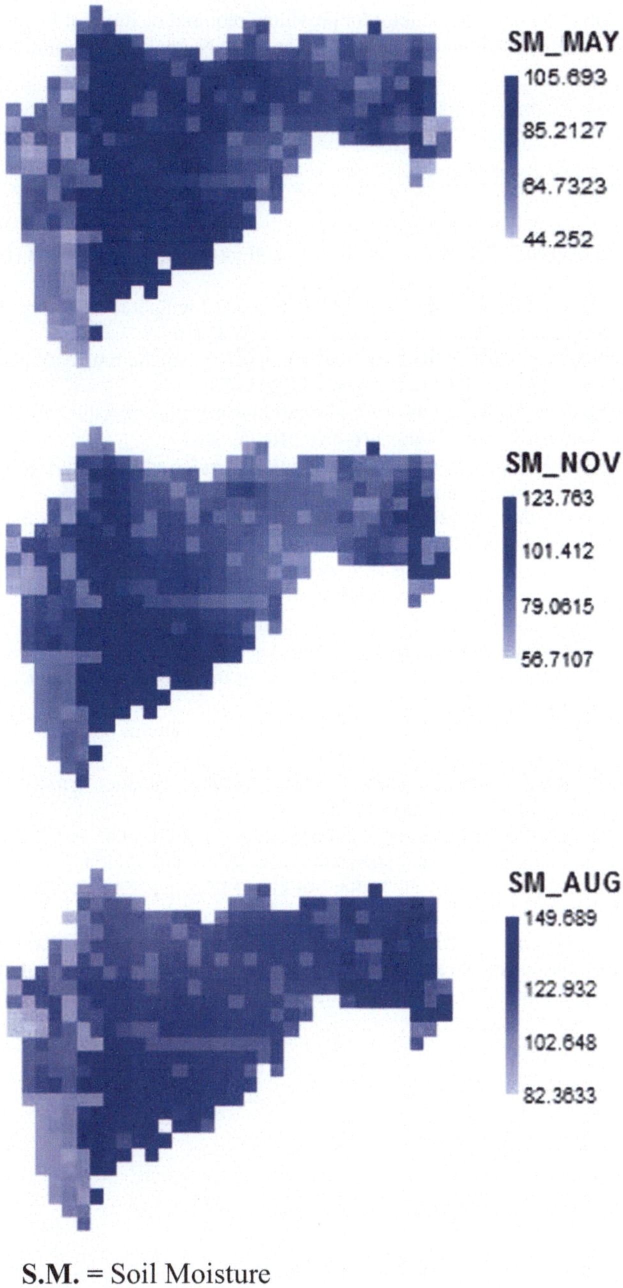

S.M. = Soil Moisture.

Acknowledgements Authors are thankful for providing required facilities of Department of Civil Engineering, Visvesvaraya National Institute of Technology Nagpur, Maharashtra India.

References

Burt TP, Butcher DP (1985) Topographic controls of soil moisture distributions. J Hydrol 36:469–486

Brocca L, Melone F, Moramarco T, Morbidelli R (2010) Spatial-temporal variability of soil moisture and its estimation across scales. Water Resour Res 46:W02516

Crave A, Gascuel-odux C, (1997) The influence of topography on the time and space distribution of soil surface water content. Hydrol Processes 11:203–210

Famiglietti JS, Rudnicki JW, Rodell M (1998) Variability in surface moisture content along a hill slope transect: Rattlesnake Hill, Texas. J Hydrol 210:259–281

Grayson RB, Western AW, Chiew FHS, Bloschl G (1997) Preferred states in spatial soil moisture patterns: local and non local controls. Water Resour Res 33(12):2897–2908

Hawley ME, Jackson TJ, McCuen RH (1983) Surface soil moisture variation on small agricultural watersheds. J Hydrol 62:179–200

Loague K (1992) Soil water content at R-5. Part 1. Spatial and temporal variability. J Hydrol 139:233–251

Moore ID, Burch GJ, MacKenzie DH (1988) Topographic effects on the distribution of surface soil water and the location of ephemeral gullies. Trans Am Soc Agric Eng 31(4):1098–1107

Nyberg L (1996) Spatial variability of soil water content in the covered catchment of Gardsjon, Sweden. Hydrol Processes 10:89–103

Owe M, Jones EB, Schmugge TJ (1982) Soil moisture variation patterns observed in Hand Country, South Dakota. Water Resour Bull 18:949–954

Reynolds SG (1970) The gravimetric method of soil moisture determination. I: a study of equipment, and methodological problems. J Hydrol 11:258–273

Verma K, Katpatal YB (2020) Groundwater monitoring using GRACE and GLDAS data after downscaling within basaltic aquifer system. Groundwater 58(1):143–151

Western AW, Grayson RB, Bloschl G, Willgoose GR, McMahon TA (1999) Observed spatial organization of soil moisture and its relation to terrain indices. Water Resour Res 35(3):797–810

Western AW, Zhou SL, Grayson RB, McMahon TA, Bloschl G, Wilson DJ (2003) Spatial correlation of soil moisture in small catchment sand its relationship to dominant spatial hydrological processes. J Hydrol 286(1–4):113–134

Chapter 11
Agricultural Crop Mapping Using MODIS Time-Series Data in DVC Command Area

Dibyandu Roy, Sneha Murmu, and Sujata Biswas

Abstract Crop mapping in a command area assumes significant importance in the context of understanding agricultural water use for different crop growing seasons. In this paper, an endeavor has been made to segregate and estimate the areal extent of the major crops of Rabi and Kharif seasons in the DVC command area, West Bengal. Time-series imagery acquired from MODIS has great potential for crop monitoring. In this study, MOD09Q1 250 m, 8-day composite surface reflectance data over the period of 2014–2017 have been used to derive Normalized Difference Vegetation Index (NDVI) maps. Subsequently, monthly maps were also generated using maximum value composite (MVC) technique. Seasonal NDVI profiles for each of the classes from ground-truth data have been analyzed, and crop types for each pixel were identified using supervised classification technique. Dominant crop areas for both the seasons have been estimated. The accuracy of classification was assessed based on ground-truth data. Overall accuracy has been achieved above 85% and 80% for Kharif and Rabi seasons, respectively, for the above-mentioned period.

Keywords Crop mapping · MODIS · NDVI

11.1 Introduction

Identification of various crop types of an area is one of the crucial information for assessment and evaluation of that agricultural system. Remote sensing images from space have always been an indispensable and evident source of information for

D. Roy
Department of Civil Engineering, Indian Institute of Technology, Kharagpur, West Bengal 721302, India
e-mail: d2roy.1992@gmail.com

S. Murmu (✉) · S. Biswas
Department of Civil Engineering, Indian Institute of Engineering Science and Technology, Shibpur, Howrah, West Bengal 711103, India
e-mail: sneha@civil.iiests.ac.in

A. Pandey et al. (eds.), *Water Management and Water Governance*, Water Science and Technology Library 96,
https://doi.org/10.1007/978-3-030-58051-3_11

deriving crop phenological information. These images have also been increasingly used for irrigation planning and management. This is mainly due to its capabilities of timely acquiring of images by the sensors that provide repeatable, continuous, human independent measurements for large territories. Agricultural statistics data gives an idea of how cropped areas change, whereas satellite imagery can provide spatially detailed maps of varying cropping patterns over a specific time period. Multi-temporal remote sensing imagery is more effective compared to the single images for determining the crop structure as they contain phenological information. The Moderate Resolution Imaging Spectroradiometer (MODIS), a high-temporal resolution sensor offers a good potential for assessing crop area mapping as well as provide crop condition parameter that can be used to assess crop yields. MODIS data are considered as highly suitable for crop related LULC classifications due to its spatial, temporal, and spectral specifications.

There are several methods of remote sensing image analysis to determine plant conditions and vegetation characteristics developed by many scientists in the last few decades. Indices such as the Normalized Difference Vegetation Index (NDVI) relate reflectance values to vegetation cover by making use of the specific relationship between the ratio of red and near-infrared reflectance and plant status (Jensen 2014). NDVI have successfully been applied for time-series analysis to gather agricultural information at different scales. After filtering noise from the NDVI time-series data for crop mapping using parametric and nonparametric classification algorithms, the maximum likelihood classifier (MLC) and SVMs yearly estimates of rice growing areas and delineation of spatial distribution of rice crops have been performed (Pringle et al. 2012). In addition with NDVI, many other indices have been used to derive crop maps such as Normalized Difference Tillage Index (NDTI), Enhanced Vegetation Index (EVI), Land Surface Water Index (LSWI), Normalize Difference Water Index (NDWI), Normalize Difference Soil Index (NDSI) which can also represent land condition (Potgieter et al. 2010; Liaqat et al. 2017). Gumma et al. (2014) described an approach to accurately separate out crops and quantify crop acreage in the major command area in the Krishna River Basin using IRS-P6 and MODIS 8-day time-series images. Chen et al. (2011) evaluated the MODIS and Landsat multi-band vegetation indices, e.g., Soil Adjusted Vegetation Index (SAVI), Modified Soil Adjusted Vegetation Index (MSAVI), NDVI and Enhanced Vegetation Index (EVI), to evaluate the most appropriate index that performs better in cropping area of irrigated Indus Basin.

In this study, an attempt has been made to segregate and estimate the areal extent of the major crops of Rabi and Kharif seasons in the DVC command area, West Bengal, for the period of 2014–2017. MODIS 250 m 8-day composite data have been used here to generate NDVI time-series images. Detailed ground-truth data were collected for sampling the various crops and other lands covers present in the study area.

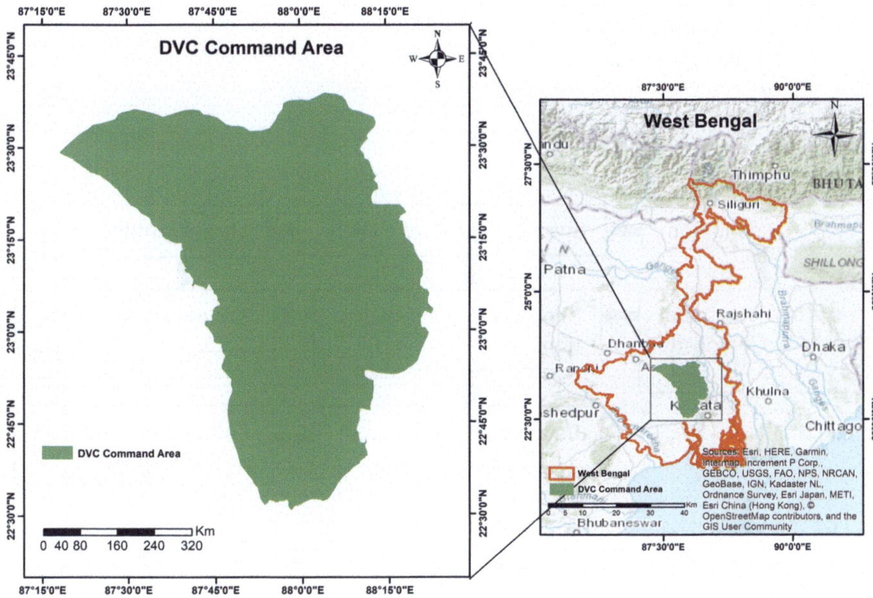

Fig. 11.1 Study area map

11.2 Study Area

The study area (DVC command area) located under Hoogly, Howrah, and Burdwan District, West Bengal, India, is represented in Fig. 11.1. The areal extent of the study area is 7004 km^2. Its located between 88°4.4′–87°57.16′ E and 23°26.71′–23°3.55′ N. In the study area, there are two dominant cropping seasons, the summer season which is commonly called as Kharif that starts from July and ends by October and winter season, i.e., Rabi, starts from October and ends by March. Potato and rice are the major Rabi crop, while rice is the major Kharif crop. Onion and mustard are being sown in various tracts in Rabi season. Sugarcane, lentil, vegetables, jute are shown in summer season which starts from March and ends by June. Annual average rainfall in the study area ranges from 220 to 265 mm in the months of June to September and mean daily maximum temperature rarely exceeding 35 °C in summer.

11.3 Materials and Methods

The steps involved in image processing technique and image classification procedure are illustrated by Fig. 11.2.

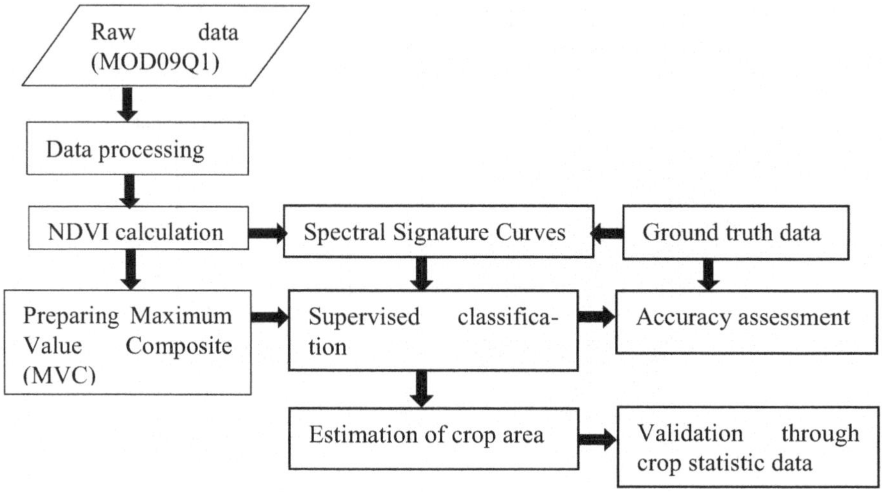

Fig. 11.2 Methodological flowchart of crop mapping

11.3.1 Data Acquisition and Data Processing

The MODIS/terra 8-day composite surface reflectance data (MOD09Q1) with spatial resolution of 250-m have been acquired from NASA Distributed Active Archive Center for the period May 1, 2014, to April 30, 2017. This product has red and near-infrared spectral band as Band1 and Band2. So, 46 images of each year of each band (total 276 images) were acquired. The projection of the images was converted into geographical coordinate system; the datum in World Geodetic System 84 (WGS84) and the scale were also changed (by multiplying 0.0001). Finally, the time-series dataset was subset to the study area boundary which is further used for classification of crop types.

11.3.2 NDVI Calculation

The Normalized Difference Vegetation Index (NDVI) is a graphical indicator which is widely used to analyze remote sensing measurements. Rouse et al. (1974) developed the generic NDVI to following equation:

$$\text{NDVI} = (\text{Band2} - \text{Band1})/(\text{Band2} + \text{Band1}) \tag{11.1}$$

where Band2 and Band1 are the reflectance values for the near infrared and red ranges, respectively. So, finally we got 46 NDVI-time-series images for each year. Each pixel of NDVI images gives the values ranging from -1 to $+1$.

11.3.3 Preparing MVC

After calculation of NDVI, image denoising was applied to some of the images which have pixels with value beyond the range (−1 to + 1) by setting "Null" value. Due to the recurrent unstable weather conditions, many data points within the study area are found which are not useful for vegetation monitoring. It was therefore necessary to regenerate the time series of NDVI data. The maximum-value iteration filter was used to remove the bulk of useless pixel data and prepare monthly maximum value composite (MVC) from 46 images of each year of NDVI images followed by Julian day calendar. The maximum values of each pixel of monthly NDVI images are composed in MVC images of each month.

11.4 Ground-Truth Datasets

Ground-truth data were collected during October–November, 2016 and February 2017 for 219 sample sites covering major crop classes in the study area. The precise locations of the samples were recorded by a Garmin GPS unit in latitude/longitude coordinate system with a common datum of WGS84. The sample size per class varied from 5 to 26. Each class was assigned with a label in the field. Classes have the flexibility to merge to a higher class or break into a distinct class based on the land cover percentages observed at each location. At each location, some of the land use–land cover classes such as forest, water, fallow lands were recorded. For the major crops, the crop types, pattern, and crop calendar for different seasons have been noted. Also, the source such as irrigated, rain-fed, supplemental irrigation of water for each of the field has been noted.

11.5 Spectral Signature Curves

The variability in the phenology of each crop type is reflected in its NDVI profile. The range of NDVI values and its change over time tells us about the crop type including crop intensity. NDVI signatures were extracted in some of the observation points from the MODIS time-series data. The points selected to generate the NDVI signatures indicates a definitive crop type and/or cropping system. From these curves, the values of NDVI for each crop for a specific time can be calculated. Some crop signatures of MODIS MVC of March, 2017 are shown in Fig. 11.3.

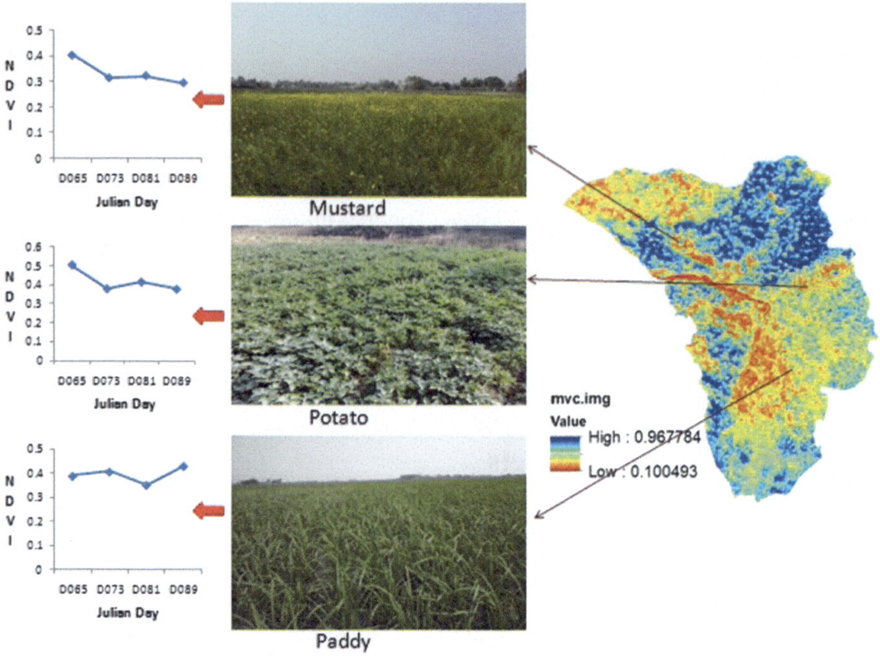

Fig. 11.3 Spectral signature for different crops in the study area, March, 2017

11.6 Image Classification

The monthly NDVI-MVC images of each season of each year (image of March for Rabi and September for Kharif) were classified using supervised classification technique. Specific training areas are identified for each informational category. The training areas were identified based on NDVI time-series plots, spectral signature curves, and ground-truth data. Training areas were distributed throughout the image so that they provide a basis for representation of the diversity present within the scene. Various supervised classification methods may be used to assign an unknown pixel to one of a number of classes. Among the most frequently used classification algorithms, the maximum likelihood algorithm (MLC) was used in this study. Accuracy assessment was performed to using the classification outputs to determine the classification accuracy.

11.7 Results and Discussion

11.7.1 Spectral Signatures of Various Crop Types

Temporal NDVI series have been analyzed to study crop phenology, crop calendar and to estimate cropping intensities (single or double crop). The temporal class signature allows the segregation between various crop classes. Figures 11.4, 11.5, 11.6 and 11.7 show the spectral signatures of all the crops in the area, which has been derived from the MVCs of the NDVI images.

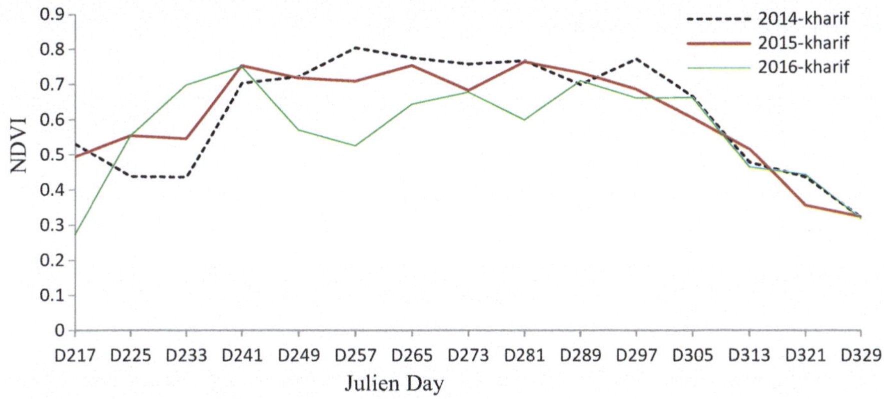

Fig. 11.4 NDVI-time series, Kharif paddy

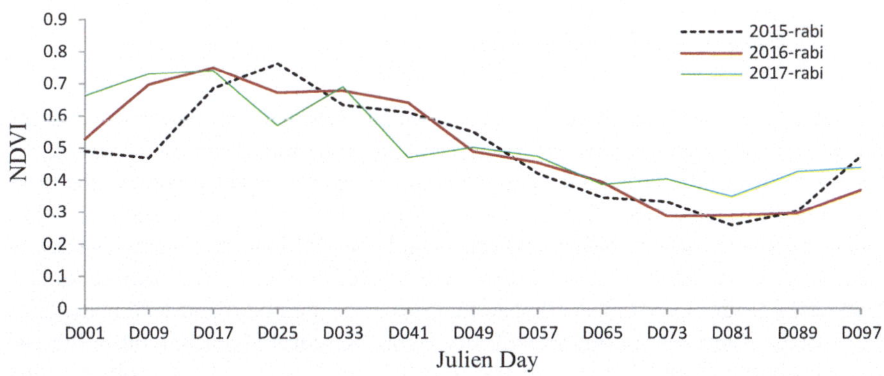

Fig. 11.5 NDVI-time series, Rabi paddy

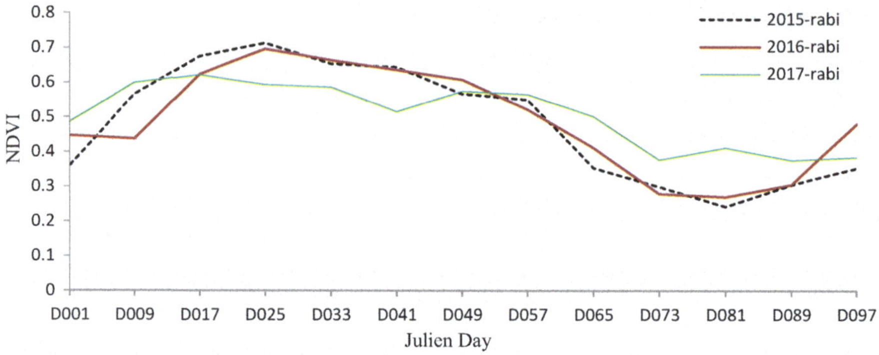

Fig. 11.6 NDVI-time series, Potato

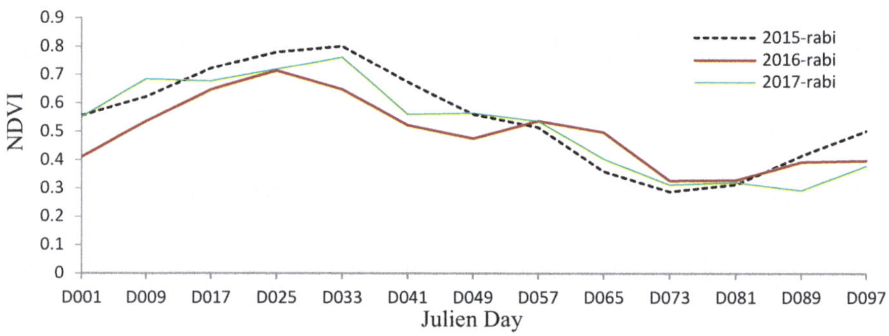

Fig. 11.7 NDVI-time series, Mustard

11.7.2 Crop Classification Maps

The images of both Rabi and Kharif season of each year were classified using maximum likelihood classifier of supervised classification technique. Crop classes were identified using the MODIS-derived NDVI images based on ground truth, information from farmers, and NDVI signatures. Four to six major crops such as paddy, potato, mustered, winter paddy, and other land use and land cover classes such as waterbody, field, forest were identified and leveled for each crop seasons during 2014–2017 (Figs. 11.8 and 11.9). The major part of the command area is irrigated with surface water and ground water. Most of the agricultural lands in the command area are double-cropped area with paddy in Kharif season and some other winter crops like potato, winter paddy, etc.

Accuracy assessment based on ground-truth data was carried out for the classification results of 2014–2017 images of each season shown in the above figures. The results of accuracy assessment of the classified images for the year 2015 are shown in Tables 11.1 and11.2.

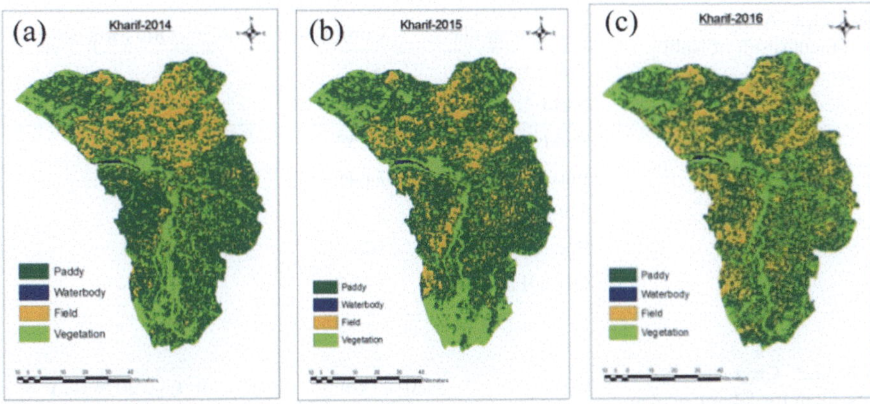

Fig. 11.8 Crop classes of DVC command area, Kharif season, **a** 2014, **b** 2015, **c** 2016 based on MODIS NDVI images

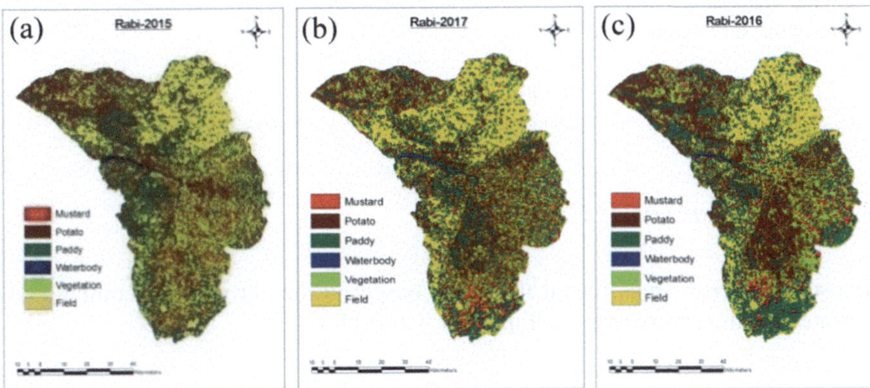

Fig. 11.9 Crop classes of DVC command area, Rabi season, **a** 2015, **b** 2016, **c** 2017 based on MODIS NDVI images

Table 11.1 Accuracy assessment result of Kharif season 2015

Land cover class	Commission error	Omission error
Waterbody	0	0
Field	0	0
Paddy	5.56	5.56
Forest	8.7	8.7
Overall accuracy	97.64	

Table 11.2 Accuracy assessment result of Rabi season 2015

Land cover class	Commission error	Omission error
Water body	0	5
Field	0	0
Paddy	28.57	28.57
Forest	34.48	0
Mustard	33.33	60
Potato	100	0
Overall accuracy	83.72	

Table 11.3 Crop area assessment for Kharif season

Crop	2014 (km^2)	2015 (km^2)	2016 (km^2)
Kharif paddy	4130.6	4165	3596.2

Table 11.4 Crop area assessment for Rabi season

Crop	2015 (km^2)	2016 (km^2)	2017 (km^2)
Rabi paddy	1824.1	2061.8	1923.5
Mustard	799.2	576	774.5
Potato	1574.5	1893.4	1597

11.7.3 Assessment of Crop Area

The estimated crop areas for all the crops based on the classification outputs are illustrated for the two seasons in Tables 11.3 and 11.4.

11.7.4 Validation Using Crop Statistics Data

Crop statistics data have been collected from the agriculture divisions of DVC of the three districts Burdwan, Hoogly, and Howrah. There are total 33 blocks under the study area. The crop acreage of the whole study area for Kharif and Rabi season is shown in Tables 11.5 and 11.6, respectively.

The classification results have been compared with these datasets provided by Agriculture Department of DVC command area. The estimated values from the classification results indicate higher values which may be due to many factors. The

Table 11.5 Crop acreage for Kharif season

Crop	2014 (km^2)	2015 (km^2)	2016 (km^2)
Kharif paddy	3710.3927	3482.6127	3474.9627

Table 11.6 Crop acreage for Rabi season

Crop	2015 (km^2)	2016 (km^2)	2017 (km^2)
Rabi paddy	1659.82	1371.41	1605.16
Mustard	322.1	356.34	297.87
Potato	1275.95	1384.86	1171.9

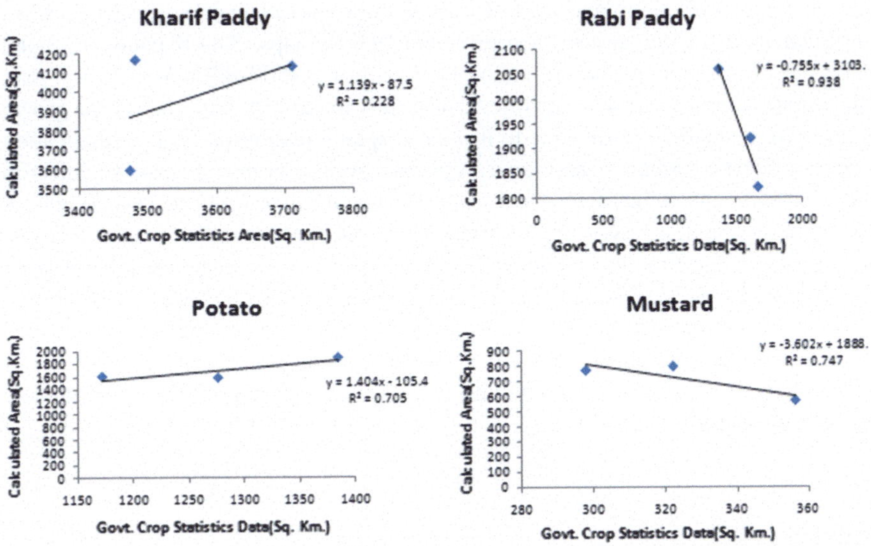

Fig. 11.10 Comparison of crop area derived from MODIS NDVI and statistical data

correlation between time-series MODIS-derived area and statistical area of the all the crops is shown in Fig. 11.10. The results show a reasonable consistency between estimated and observed data, with correlation coefficient(r) values 0.48, 0.97, 0.84, and 0.86, respectively, for Kharif Paddy, Rabi Paddy, Potato, and Mustard. It can be observed that for Rabi paddy, Potato, and Mustard there is relatively more significant correlation between the estimated areas derived from satellite data and the statistical area. Moreover, the crop statistics data for Howrah District were not available for the study area for the year 2016, which has resulted in the variation of the estimated data with it. The difference between estimated and observed values are acceptable, considering the 250 m coarse resolution pixel size, climatic variation as well as sowing period in the parts of the command area and uncertainty included in field measurements.

11.8 Conclusion

This study demonstrates the potential of multi-temporal medium resolution satellite imagery to differentiate crop types and cropping pattern in regions of high spatial and temporal complexity. MODIS 250-m time-series data in conjunction with detail ground-truth data have been used here to classify the major crops and other LULC classes in the DVC command area. Relatively high overall accuracies (above 84%) were achieved for the series of crop-related LULC maps. Kharif paddy is found to be the major crop in the command area covering 3500–4200 km^2 every year. The classification result demonstrates that it fails to classify the other crops grown in the area in smaller regions due to its coarse spatial resolution. This classification procedure can be applied to large-area cropland mapping, crop yield assessment, crop water requirement studies at regional scale. The result can be more accurate if images having higher spatial resolution are used. However, there is some inconsistency in the crop acreage estimation seen in different areas of the command area that depend on non-imagery factors such as variation in sowing time, lack of water availability, climatic factors, and in effective sowing methods. These factors were neither reflected in satellite images nor considered in developing classification images in this study. It is assumed that these results can be improved further if crop production parameters remain consistent.

References

Chen CF, Son NT, Chen CR, Chang LY (2011) Wavelet filtering of time-series moderate resolution imaging spectroradiometer data for rice crop mapping using support vector machines and maximum likelihood classifier. J Appl Remote Sens 5(1):3525–3540

Gumma MK, Pyla KR, Thenkabail PS, Reddy VM, Gundapaka N, Ahmed MI, Rafi IMD (2014) Crop dominance mapping with IRS-P6 and MODIS 250-m time series data. J Agric 4:113–131

Jensen JR (2014) Remote sensing of the environment, 2nd edn, Dorling Kindersley (India) Pvt. Ltd. New Delhi, India

Liaqat U, Cheema MJM, Huang W, Mahmood T, Zaman M, Khan MM (2017) Evaluation of MODIS and Landsat multiband vegetation indices used for wheat yield estimation in irrigated Indus Basin. J Comput Electron Agric 138:39–47

Potgieter AB, Apan A, Hammer G, Dunn P (2010) Early-season crop area estimates for winter crops in NE Australia using MODIS satellite imagery. J Photogrammetry Remote Sens 65(4):380–387

Pringle MJ, Denham RJ, Devadas R (2012) Identification of cropping activity in central and southern Queensland, Australia, with the aid of MODIS MOD13Q1 imagery. J Appl Earth Observ Geoinf 19:276–285

Rouse JW, Haas RH, Schell JA, Deering DW (1974) Monitoring vegetation systems in the great plains with ERTS. In: Third ERTS symposium, NASA SP-351 I, pp 309–317

Chapter 12
Groundwater Modeling in a Semi-confined Aquifer Using GIS and MODFLOW

Y. B. Katpatal, Priyanka Thakare, and S. N. S. Maunika

Abstract Groundwater modeling is an effective tool used to understand the behavior of the aquifer system under different hydrological stresses and to prevent excessive withdrawal of water. The Chauras region located in Bhandara District in Maharashtra, India, is a typical semi-confined aquifer in the thick alluvium deposits which is depleting very fast due to overexploitation. The agriculture in the region is very good, and three crops are grown in the area. The main objective of this research is the simulation of future groundwater scenarios using remote sensing and GIS, and groundwater modeling to find a solution to drastically decreasing groundwater levels. The study also focuses on methods to recharge the semi-confined aquifer, which is overlain by a 12-m-thick clay layer. A suitability analysis concludes that recharge using the direct technique can convert the total fallow land to agricultural land with crop, thus further increasing the productivity of the region.

Keywords Groundwater modeling · Remote sensing · GIS · Recharge

12.1 Introduction

In this digital and industrial era, rapid industrialization, urbanization, pollution, uneven rainfall, increased agricultural needs, etc., have led to a shortage of freshwater in many parts of the world. Continuous and excessive use of surface water alone may lead to many other problems like water logging, which may make the fertile lands unproductive. Usage of groundwater may solve these problems, whereas excessive

Y. B. Katpatal
Department of Civil Engineering, VNIT, Nagpur, India
e-mail: ybkatpatal@rediffmail.com

P. Thakare (✉) · S. N. S. Maunika
Water Resource Engineering, Department of Civil Engineering, VNIT, Nagpur, India
e-mail: pthakare73@gmail.com

S. N. S. Maunika
e-mail: samhi.sns@gmail.com

A. Pandey et al. (eds.), *Water Management and Water Governance*,
Water Science and Technology Library 96,
https://doi.org/10.1007/978-3-030-58051-3_12

withdrawals lead to lowering of water tables, drying up of wells, and an increase in the heads for pumping water. Hence, there is a need to keep a special emphasis on sustainable development of water resources and management. Surface water can be estimated by various methods, whereas groundwater needs modeling. The infiltration parameters and the hydraulic conductivity were calibrated in the MODFLOW, by using steady-state approach to match with the study area field conditions. The parameters were verified and optimized for a transient model (Michl 1996). Numerical models provide flexibility for analysis of groundwater application, and these methods generally involve approximating continuous partial differential equations (Faust and Mercer 1980). The numerical models which represent the hydro-geological condition of the area often require large data set, which is difficult to manage. GIS can facilitate the data automation, and thus, to reduce the efforts to link GIS with groundwater model, many interface programs have been developed (Watkins et al. 1996). MODFLOW and GIS interface can also be used to assess changes in groundwater dynamics after applying various water saving measures (Xu et al. 2011). Before building a numerical model of any site to be modeled, development of a conceptual model is necessary (Singhal and Goyal 2011). To obtain a general overview of groundwater condition and to minimize model assumption, GIS-based groundwater flow model can be used (Jani 2012).

12.1.1 Study Area

The study area, Chauras is located in Pauni and Lakhandur districts, situated in the northeastern side of Maharashtra, India, is 340 km^2 which lies between longitudes of 79°37′37.58″–79°49′52.31″ E and latitudes of 20°51′53.32″–20°40′35.72″ N. It is bounded by the Wainganga River from north to south on the ends and Chulband River from south to east. The mean minimum temperature is 6 °C, and the mean maximum temperature is 45 °C (Fig. 12.1).

12.2 Data Used and Methodology

12.2.1 Data Used

See Tables 12.1 and 12.2.

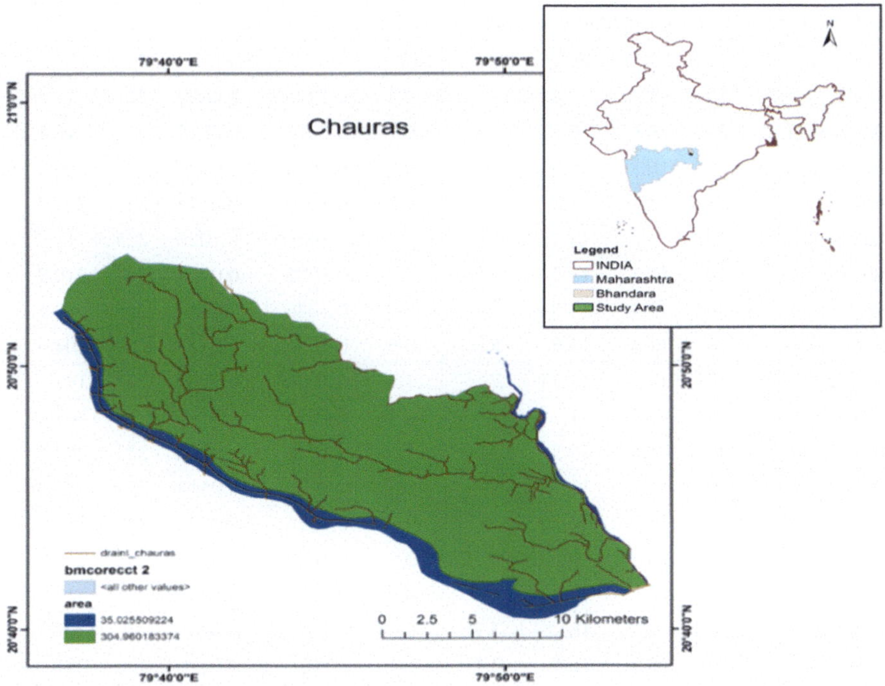

Fig. 12.1 Location map of the study area

Table 12.1 List of the data used in the study area

S. no.	Description	Source
1	Satellite image of study area	Google earth pro
2	Map showing well location, river streams etc.	Generated in Lab
3	Contour maps of heads in pre- and post monsoon	Generated in Lab
4	Land use/Land cover map of study area	Generated in Lab
5	TIN of the area	Generated in Lab
6	Aquifer system and strata geometry	GSDA
7	Data of hydrogeological parameters	GSDA
8	Ground water heads	GSDA

12.2.2 Methodology

To make a conceptual model, input data, such as LULC map, geological data, and soil data, is used by overlying them with DEM using ArcMap as input for MODFLOW, and by using GIS maps and other hydro-geological data such as K, n, S_y, groundwater modeling has been done.

Table 12.2 Data used for calculating groundwater recharge

Well	TD	PRD	POD	IH	RH	*Sy*	WTF	*E*	*H*
Asola	20	9.3	8.45	10.7	0.85	0.1026	0.0872	221	211.7
Dambevirli	20	8.6	8	11.4	0.6	0.1026	0.0615	221	212.4
Dhoke-sarandi	20	16.3	15.5	3.7	0.8	0.1026	0.0820	223	206.7
Itan	20	8.3	8.4	11.7	−0.1	0.1026	−0.0102	222	213.7
Karandala	20	15	14	5	1	0.1026	0.1026	225	210
Khairna	20	14.1	13.2	5.9	0.9	0.1026	0.0923	225	210.9
Kirmati	20	12.9	12.5	7.1	0.4	0.1026	0.0410	215	202.1
Kudegaon	20	14.4	13.9	5.6	0.5	0.1026	0.0513	219	204.6
Manded	20	13	11.75	7	1.25	0.1026	0.1282	212	199
Moharna	20	12.2	11.7	7.8	0.5	0.1026	0.0513	220	207.8
Nanded	20	14.5	13.5	5.5	1	0.1026	0.1026	223	208.5
Opara	20	11.5	11	8.5	0.5	0.1026	0.0513	222	210.5
Rajani	20	17	16.1	3	0.9	0.1026	0.0923	230	213
Rohani	20	10.5	10	9.5	0.5	0.1026	0.0513	226	215.5
Tembri	20	11	10.2	9	0.8	0.1026	0.0820	221	210
Virli BK	20	16.3	16	3.7	0.3	0.1026	0.0307	222	205.7
Virli Kh	20	16.9	16.5	3.1	0.4	0.1026	0.0410	230	213.1
Bhendala	20	16.2	14	3.8	2.2	0.1026	0.2257	230	213.8
Dhamni	20	18.2	17.3	1.8	0.9	0.1026	0.0923	223	204.8
Junona	20	15.8	16	4.2	−0.2	0.1026	−0.0205	227	211.2
Khairi Diwan	20	19.15	17	0.85	2.15	0.1026	0.2205	232	212.85
Mohari	20	17.3	17	2.7	0.3	0.1026	0.0307	224	206.7
Mokhara	20	16.5	16.2	3.5	0.3	0.1026	0.0307	231	214.5
Palora	20	17.2	16.7	2.8	0.5	0.1026	0.0513	234	216.8
Pauna BK	20	14	13.4	6	0.6	0.1026	0.0615	226	212
Sendri Bh	20	16.7	16.2	3.3	0.5	0.1026	0.0513	235	218.3
Sindpuri	20	17.9	17	2.1	0.9	0.1026	0.0923	230	212.1
Wadegaon	20	16.7	16	3.3	0.7	0.1026	0.0718	230	213.3

12.3 Model Calibration and Validation

Model calibration is the process of changing the values of input parameters to obtain or to match field conditions within some acceptable criteria. For MODFLOW, the inputs required can be categorized into aquifer geometry, hydro-geological parameters, i.e., hydraulic conductivity, specific storage, specific yield, and the main measured variable; groundwater heads. The steady-state calibration is done in the pre- and post-monsoon conditions for the year 2014 (Fig. 12.2; Table 12.3).

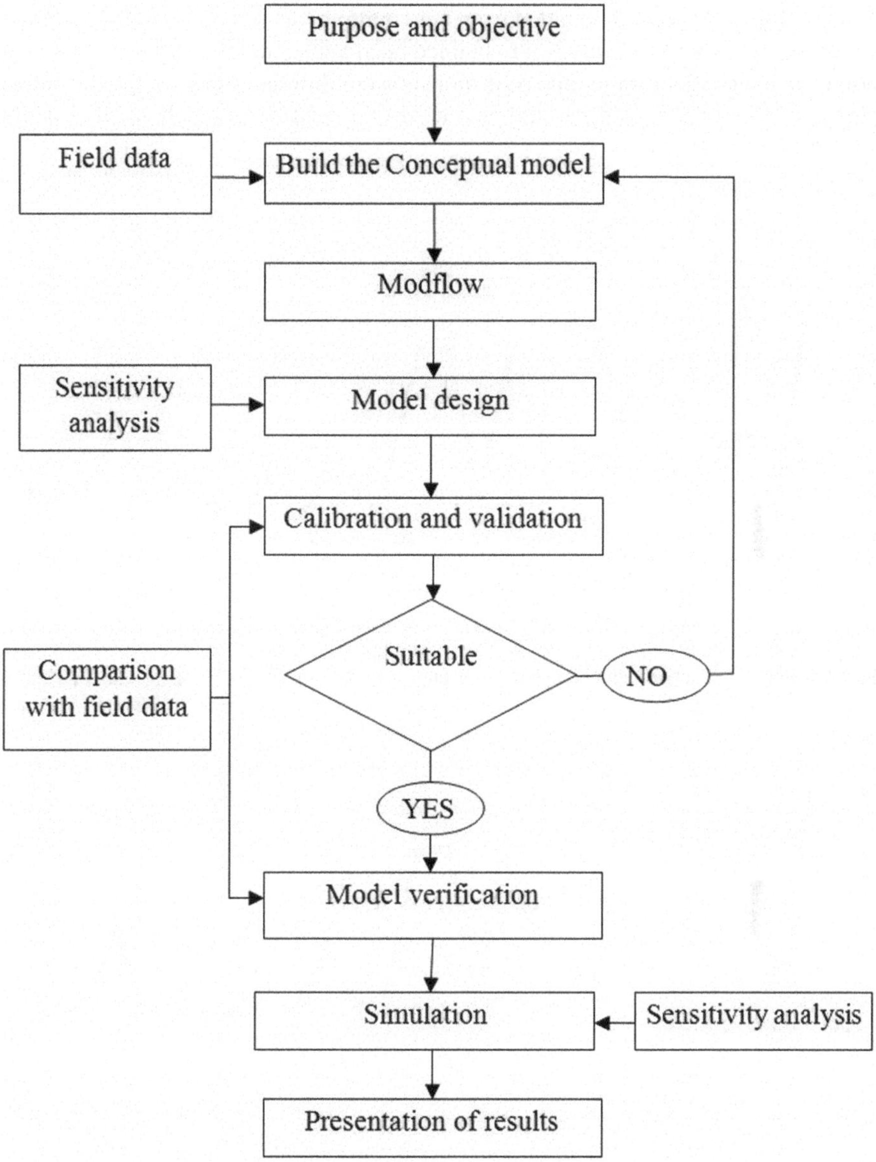

Fig. 12.2 Methodology adopted in MODFLOW

Table 12.3 Hydro-geological parameters of layer 1 and layer 2, respectively

	Hydraulic conductivity	Specific storage	Specific yield
Layer 1	0.000203	0.0001	0.1025
Layer 2	6.25	0.0001	0.25

The results (viz., values obtained from field pumping tests, i.e., heads) are verified by plotting the input head value and obtained head value, and the variation is shown in Figs. 12.3 and 12.4 for pre- and post-monsoon conditions, respectively. The correlation coefficient values for pre- and post-monsoon conditions are obtained as 0.989 and 0.906, respectively.

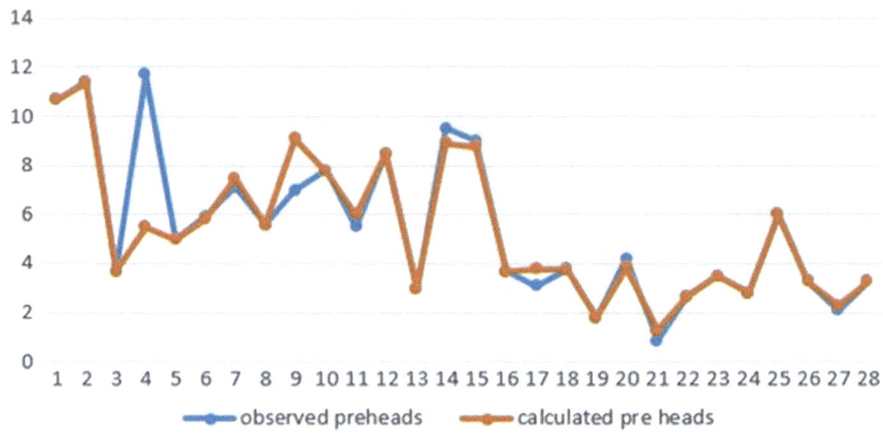

Fig. 12.3 Observed versus calculated heads for pre-monsoon conditions

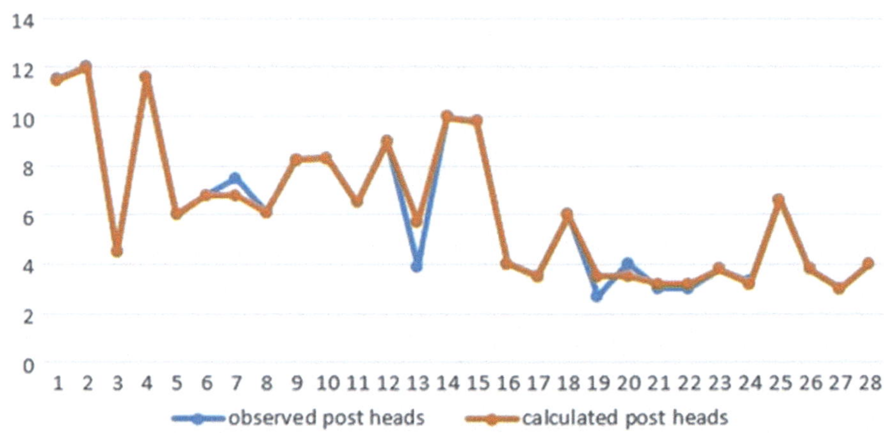

Fig. 12.4 Observed versus calculated heads for post-monsoon conditions

12.4 Results and Discussion

For the year 2020, it is observed that the groundwater level is decreased by 2 m. The water heads map for 2020 is shown in Fig. 12.5.

Also, the study area in the month of May for the year 2014 is obtained and land use classification is done, which conveys that 42% of the total area is in agricultural practice in that month and 30.6% is fallow land which can be utilized for cropping if water is available (Figs. 12.6 and 12.7).

From the above results obtained, it is clear that there is a need for adopting water saving and conservation practices. A recharge structure is proposed which conserves the rainwater which consists of three wells each having more depth than the earlier with gravel beds of increasing size for filtering water. The recharge capacity from each structure is calculated by the formula as follows:

For filter, the rate of filtration is 4000–5000 L/m^2/h.

Area of the filter well $= 3.14\text{m}^2$

$$\begin{aligned} \text{Rate of filtration of a well} &= 3.14 \times 5000 \times 24\,\text{h} \\ &= 376,800\,\text{L/day} = 377\,\text{m}^3\text{/day} \end{aligned}$$

$$\begin{aligned} \text{Recharge} &= \text{Rate of Filtration of single well} \times \text{No. of Days} \\ &= 377 * 90 = 33.93\,\text{Ha - m/90 days}. \end{aligned}$$

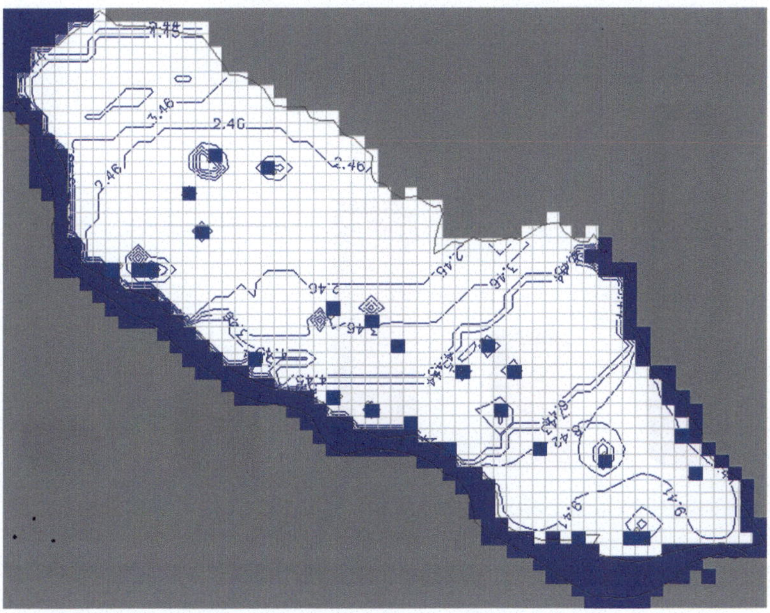

Fig. 12.5 Contour map of forecasted groundwater heads for the year 2020

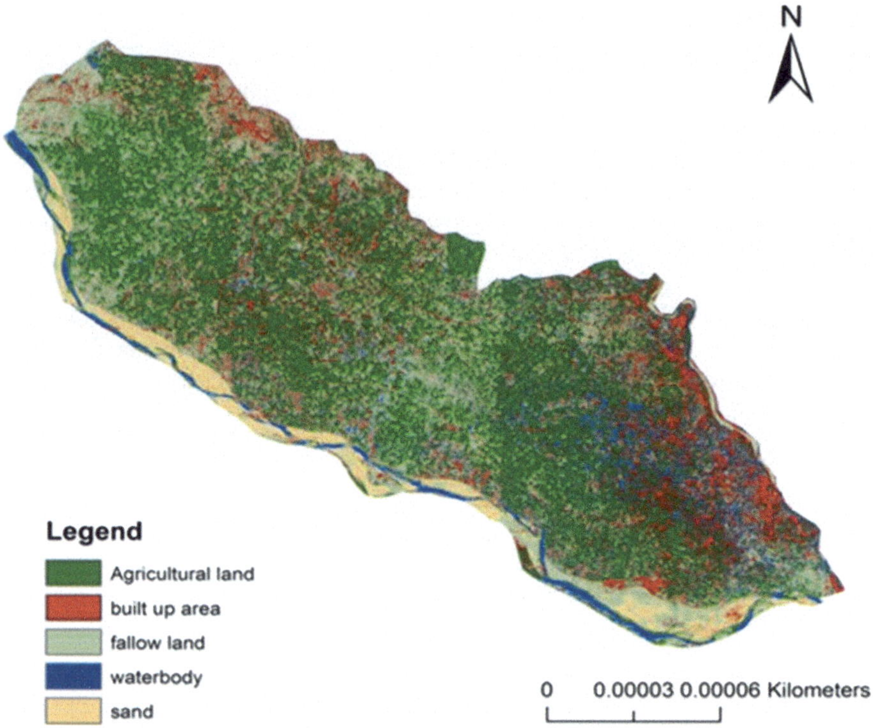

Fig. 12.6 Land use/Land cover classification

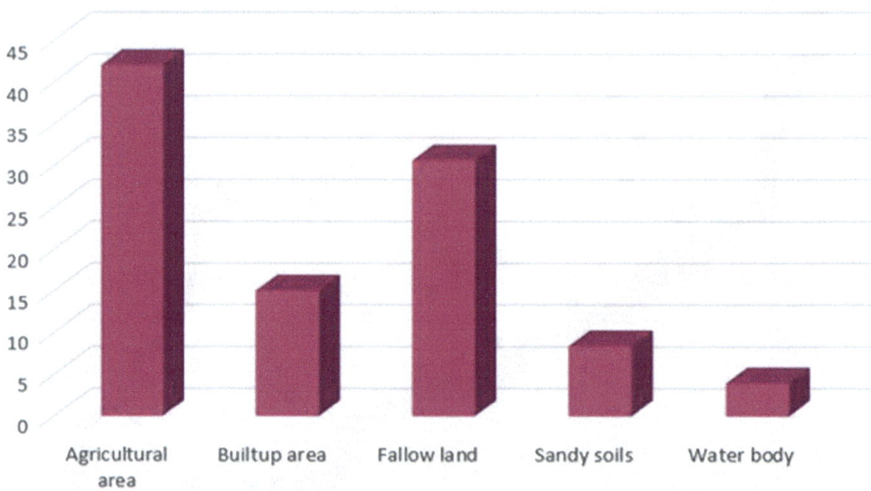

Fig. 12.7 Percentage LULC area

It is estimated that 33.93 Ha-m quantity of water will recharge the over-exploited aquifer beneath. The amount of water required for making the 30.6% fallow land into agricultural land is calculated, taking the crop water requirement of paddy as 1.2 m in arid climates and obtained as 12,505 Ha-m, which shows that 3200 structures have to be built. From the observations in heads made, the groundwater level in the northern and northeastern part is very low. Hence, the distribution of the recharge structures should be more in that area.

12.5 Conclusions

The continued depletion of groundwater and its changing trend has been discussed in the present paper. The groundwater condition has been analyzed with the help of groundwater modeling tool MODFLOW. Correlation coefficient values for pre- and post-monsoon conditions are also calculated for checking the accuracy of the results. It is also found that the groundwater decreased by 1.85 m in two years span, i.e., from 2014 to 2016, and by 2 m by the year 2020. GIS and MODFLOW interface made analysis more simpler.

References

Faust CR, Mercer JW (1980) Ground water modeling: numerical methods. Ground Water 18(4):395–409

Jani J (2012) GIS as a tool for modelling groundwater flow. IIEEE symposium on business, engineering and industrial applications, pp 513–517

Michl C (1996) Using GIS, MODFLOW and MODPATH for groundwater management of an alluvial aquifer of The River Sieg, Germany. Hydro GIS96: application of geographic information systems in hydrology and water resources management IAHS Publ. no. 235, pp 551–557

Singhal V, Goyal R (2011) Development of conceptual groundwater flow model for Pali area, India. Afr J Environ Sci Technol 5(12):1085–1092

Watkins DW, McKinney DC, Maidment DR, Lin M (1996) Use of geographic information system in ground water flow modelling. J Water Resour Plann Manage 122(2):88–96

Xu X, Huang G, Qu Z, Pereira LS (2011) Using MODFLOW and GIS to assess changes in groundwater dynamics in response to water saving measures in irrigation districts of the Upper Yellow River Basin. Water Resour Manage 25:2035–2059

Chapter 13
Effect of Monthly Variation of Near-Surface Lapse Rate on Snowmelt Runoff Simulation in Eastern Himalayas

Minotshing Maza, Liza G. Kiba, Arnab Bandyopadhyay, and Aditi Bhadra

Abstract Most of the temperature-based studies in the Himalayan basins used a constant value of 0.65 °C/100 m as the near-surface lapse rate parameter. However, studies on lapse rate have established that it varies with location as well as with the time of the year. In this study, three scenarios were tested: (i) constant lapse rate of 0.65 °C/100 m; (ii) constant lapse rate specific to Eastern Himalayas; and (iii) monthly varying lapse rate specific to Eastern Himalayas. It was found that the snowmelt model WinSRM captured the observed hydrograph more efficiently when constant lapse rate of 0.5 °C/100 m (specifically determined for Eastern Himalayas from observed temperature data) was used compared to monthly varied lapse rate or constant lapse rate of 0.65 °C/100 m. Based on the results obtained, a constant near-surface lapse rate of 0.5 °C/100 m is recommended for the temperature-index based snowmelt models in the Eastern Himalayas.

Keywords WinSRM · Lapse rate · Snowmelt runoff modelling · Eastern Himalayas

13.1 Introduction

The Himalaya, which is in the northern part of the Indian Peninsula, is the abode of eternal snows, from which the rivers of the Ganga, the Indus and the Brahmaputra originate. These river systems, besides supplying water for urban and rural household uses, generate millions of kilowatts of power and are also extensively used for irrigating millions of hectares of agricultural land. Therefore, monitoring the distribution of snow in mountainous basins is necessary for accurate streamflow simulation which is of great importance for optimal water resources planning and management, and climate change studies. Significant amount of runoff is generated from both

M. Maza (✉) · L. G. Kiba · A. Bandyopadhyay · A. Bhadra
Department of Agricultural Engineering, North Eastern Regional Institute of Science and Technology, Nirjuli (Itanagar), Arunachal Pradesh 791109, India
e-mail: minotshing@gmail.com

A. Pandey et al. (eds.), *Water Management and Water Governance*, Water Science and Technology Library 96,
https://doi.org/10.1007/978-3-030-58051-3_13

melting of ice and snow but due to lack of detailed ground truth observations, their relative contribution to runoff is difficult to quantify (Wulf et al. 2016).

Many conceptual and physically based hydrological models are used to explore, quantify and understand the processes associated with the discharge. The runoff generated from ice and snow is commonly estimated using either surface energy balance (Anderson 1976) or temperature-index models (Rango and Martinec 1981; Lang and Braun 1990; Rulin et al. 2008). Temperature-index models are commonly used for snowmelt modelling studies owing to easy accessibility to air temperature data and good performance in spite of their simplicity (Hock 2003). Temperature-index models are developed based on empirical relationships between ablation and air temperature. Many studies have established that a high degree of correlation exists between snowmelt and air temperature. Surface air temperature is important to many fields of research as it is one of the main controlling factors of many environmental processes. Near-surface temperature typically decreases with increase in elevation. However, the opposite effect may occur under certain conditions. Temperature lapse rates vary regularly and seasonally (Blandford et al. 2008). The variation in lapse rates is found to be more during the warmer months as compared to colder months and during daytime than at nighttime. Frequent temperature inversions during colder months partly describe the seasonality of temperature lapse rate (Whiteman et al. 1999). Factors such as location and time of the year affect the atmospheric lapse rate. Quantity of water vapour present in the air also strongly affects the lapse rate (Bandyopadhyay et al. 2014).

Many researches in the Himalayan basins used an annual constant lapse rate of 0.65 °C/100 m (Singh and Jain 2003; Pandey et al. 2014; Romshoo et al. 2015). Blandford et al. (2008) found that taking the lapse rate as 0.65 °C/100 m, which is the average value of environmental lapse rate, incurred some errors which led to lower accuracy of the modelling results.

Based on the existing hydro-meteorological information and the basin hydrological characteristics, the Snow Runoff Model (WinSRM) was used to simulate the discharge for the study area. The main objective of this study is to evaluate the impact of employing seasonally varied and annual constant near-surface lapse rates on hydrological simulations for the Nuranang watershed in the Eastern Himalayas.

13.2 Methodology

13.2.1 Description of WinSRM

WinSRM developed by Martinec et al. (2008) was used for this study to evaluate the effect of variation of near-surface lapse rate on snowmelt modelling. WinSRM is a deterministic, degree-day hydrological model developed to simulate and forecast daily runoff in hilly catchments where snowmelt is one of the dominant runoff contributing components. The model can be used for mountainous basins of any size and of any elevation range with a maximum simulation period of 366 days. Computed daily runoff from rainfall and snowmelt when superimposed on the recession flow gives daily streamflow as follows:

$$Q_{n+1} = [c_{Sn}a_n(T_n + \Delta T_n)S_n + c_{Rn}P_n]\frac{A_Z 10{,}000}{86{,}400}(1 - k_{n+1}) + Q_n k_{n+1} \quad (13.1)$$

where Q is mean daily runoff, m^3 s^{-1}; a represents degree-day factor, cm °C^{-1} d^{-1}; c_S express snowmelt runoff coefficient; ΔT is temperature lapse rate adjustment, °C d; T means number of degree-days, °C d; S represents ratio of snow-covered area to basin area; c_R is rainfall runoff coefficient; P is runoff contributing precipitation, cm; A_Z is catchment area or zonal area, km^2; n represents the number of days for runoff calculation; k represents recession coefficient. Figure 13.1 shows the flowchart of WinSRM.

13.2.2 Study Area

The study area, Nuranang watershed which is dominated by seasonal snow, is situated at Tawang district of Arunachal Pradesh (Fig. 13.2). Nuranang River which is one of the tributaries of Tawang River originates from Sela Lake and meets Tawang River at Jang. The total area of the watershed is 53.45 km^2 and extends from 27°30′ N to 27°35′ N latitude and from 92° E to 92°7′ E longitude. Elevation of the study area ranges from 3465 to 4835 m above mean sea level (MSL) with a mean slope of 51%. The study area receives an average annual precipitation of 1139 mm between the summer monsoon months of May and September. Snowfall begins from late October and lasts till March, and the ablation period begins from February and ends in early June.

13.2.3 Acquisition of Data

13.2.3.1 Hydro-meteorological Data

Hydro-meteorological data for the Nuranang watershed was acquired from Central Water Commission (CWC), Jang. The daily data for discharge, precipitation and temperature for three years, i.e. 2004, 2005, and 2008, was used for this study. The Nuranang watershed was delineated at Jang, which is the CWC discharge monitoring site. The watershed outlet is located at 27°33′01″ N latitude and 92°01′13″ E longitude with an elevation of 3465 m above MSL. The watershed was further subdivided into three zones with 456 m elevation difference between each one of them (Table 13.1).

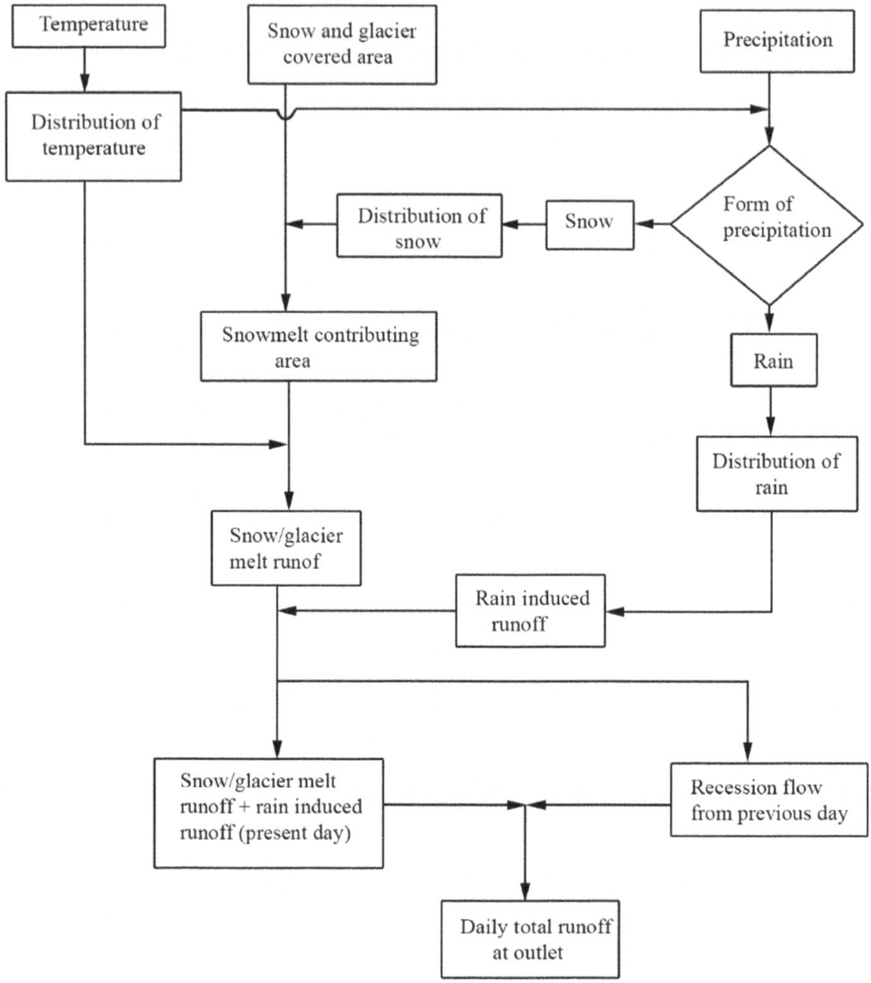

Fig. 13.1 Flowchart of WinSRM (*Source* Singh and Bengtsson 2005)

13.2.3.2 Moderate Resolution Imaging Spectroradiometer (MODIS) Snow Cover

MODIS/Terra snow cover product (MOD10A1) for the years 2004, 2005 and 2008 was obtained from https://reverb.echo.nasa.gov/ and was converted from HDF-EOS format to GeoTIFF format with the help of MODIS Reprojection Tool (MRT). The images obtained after preprocessing were used to get the snow cover area (SCA) information about the study area.

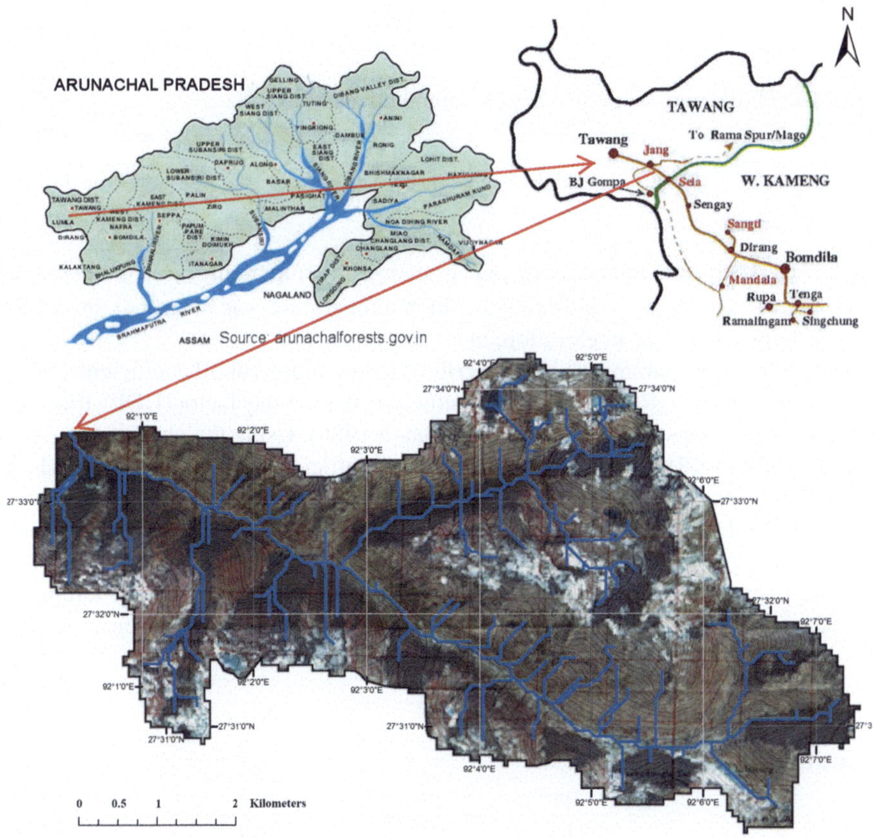

Fig. 13.2 Nuranang River basin

Table 13.1 Zonal elevation area and mean hypsometric elevations

Zone	LE (m)	HE (m)	HME (m)	Area (km^2)
Z1	3465	3921	3777.50	5.90
Z2	3922	4378	4230.75	33.16
Z3	4379	4835	4479.00	14.39

13.2.4 Input Data and Model Parameters

Precipitation, temperature and SCA were supplied as the input data to WinSRM. Temperature data was preprocessed to get the maximum, minimum and mean temperatures on a daily time scale. Daily degree days were then calculated using the mean temperature. Altitude adjustment of degree days was applied for all the elevation zones as follows:

$$T_{adj} = T - \Delta T \quad (13.2)$$

where the altitude adjustment ΔT was computed as:

$$\Delta T = \gamma\left(h_{st} - \bar{h}\right)\frac{1}{100} \quad (13.3)$$

where T_{adj} is adjusted temperature; ΔT is change in temperature; γ is temperature lapse rate, °C/100 m; h_{st} is elevation at which temperature was measured, m; and $\bar{h}$ is zonal hypsometric mean elevation, m.

The SRM model parameters include critical temperature, runoff coefficients, rainfall contribution area (RCA), recession coefficient, degree-day factor (DDF), time lag and temperature lapse rate (Martinec and Rango 1986). Using the observed streamflow of Nuranang River at the outlet, recession coefficients of Nuranang watershed were calculated as Senzeba et al. (2015). In this paper, time lag (L) of 2 h was adopted for all the calibration years (Senzeba et al. 2015). Values of DDF for each month were taken as per the recommendation given by WMO (1964). Table 13.2 shows the monthly variation of DDF value considered for this study. Yearly average temperature lapse rate of 0.5 °C/100 m (Table 13.3) was used for model calibration as per the

Table 13.2 Adopted degree-day factor (*a*) cm/°C (WMO 1964)

Month	DDF
March	0.1
April	0.3
May	0.4
June	0.6
July	0.5
August	0.5
September	0.5

Table 13.3 Monthly variation of lapse rate

Month	Lapse rate (°C/100 m)
March	0.49
April	0.49
May	0.50
June	0.46
July	0.46
August	0.48
September	0.52
Yearly average	0.50

Table 13.4 Starting sets of parameters for calibration years

Parameter	Range	2004	2005
T_{CRIT}, °C	0.0–5.0	1.0	1.0
c_R	0.1–1.0	0.5	0.5
c_S	0.1–1.0	0.6	0.6

finding of Bandyopadhyay et al. (2014) over the region. Detailed description about the model parameters can be found in Martinec and Rango (1986).

13.2.5 Calibration and Validation of WinSRM

The model was calibrated by matching the observed runoff with the simulated runoff for the depletion period of 2004 (13 April–21 August) and 2005 (17 April–4 September) and was validated for the depletion period of 2008 (17 April–20 September). Out of the seven model parameters, critical temperature T_{CRIT}, runoff coefficient for rain, c_R and runoff coefficient for snow, c_S, were calibrated, while the others were taken from the previous research findings (Senzeba et al. 2015). Calibration of the model parameters was done by varying one parameter at a time while keeping the rest of the parameters at their initial values. Table 13.4 shows the initial sets of the calibration parameters used for the calibration period.

The average values of calibration parameters obtained during the calibration years were used for validating the model.

13.2.6 Accuracy Assessment

The accuracy assessment of the simulation was done by comparing the simulated streamflow with the observed streamflow. Two dimensionless statistical performance criteria were used as follows:

$$\text{Modelling Efficiency, ME} = \frac{\left[\sum_{i=1}^{n_d}\left(O_{v,i} - \overline{O}_v\right)^2 - \sum_{i=1}^{n_d}\left(P_{v,i} - O_{v,i}\right)^2\right]}{\sum_{i=1}^{n_d}\left(O_{v,i} - \overline{O}_v\right)^2} \tag{13.4}$$

$$\text{Coefficient of Residual Mass, CRM} = \frac{\left[\sum_{i=1}^{n_d} O_{v,i} - \sum_{i=1}^{n_d} P_{v,i}\right]}{\sum_{i=1}^{n_d} O_{v,i}} \tag{13.5}$$

where $P_{v,i}$ = simulated streamflow; $O_{v,i}$ = observed streamflow; $\overline{O}_v$ = mean observed streamflow and n_d = number of days. Lower value of ME indicates poor model performance, while a value close to 1 or equal to 1 indicates a good model performance.

CRM value of zero indicates a perfect model, whereas a positive value of CRM indicates underestimation and a negative value of CRM indicates an overestimation.

13.3 Results and Discussion

The model was calibrated for the years 2004 and 2005 by matching the model simulated runoff with the observed runoff at the basin outlet. Depending on the values of the model performance indicators, while comparing the simulated runoff with the observed runoff, the best possible values of T_{CRIT} were obtained as 1 °C for 2004 and 3 °C for 2005 as shown in Table 13.5. Better matches were obtained for c_R as 0.6 and c_S as 1.0 for both the calibration years (Table 13.5). The average values of calibration parameters obtained during the calibration years were used for validating the model.

Figures 13.3, 13.5 and 13.7 show the time series plots of observed versus simulated daily runoff from WinSRM, precipitation and mean temperature for the calibration years 2004 and 2005, and validation year 2008. Figures 13.4, 13.6 and 13.8 show the

Table 13.5 Calibrated parameters for the years 2004 and 2005

Parameter	Calibration years		Validation year
	2004	2005	2008
T_{CRIT}, °C	1.0	3.0	2.0
c_R	0.6	0.6	0.6
c_S	1.0	1.0	1.0

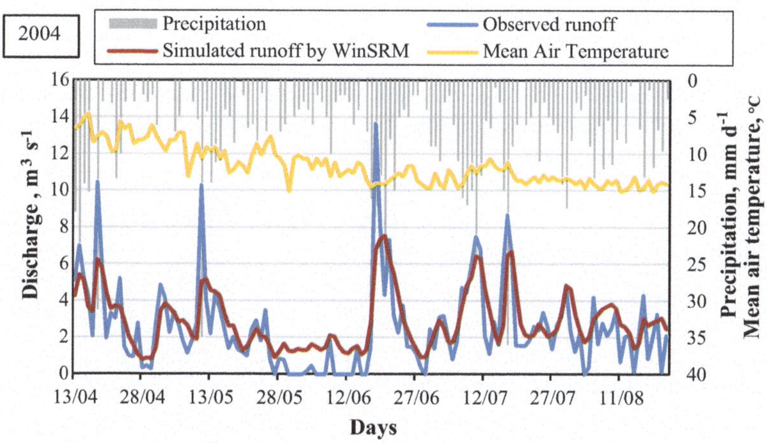

Fig. 13.3 Time series plot of observed versus simulated daily runoff, precipitation and mean temperature for the year 2004

scatter plots of observed and simulated runoff for the calibration years and validation year.

During the calibration years 2004 and 2005, ME values of 0.610 and 0.608 were obtained, respectively, while ME value of 0.611 was obtained during the validation

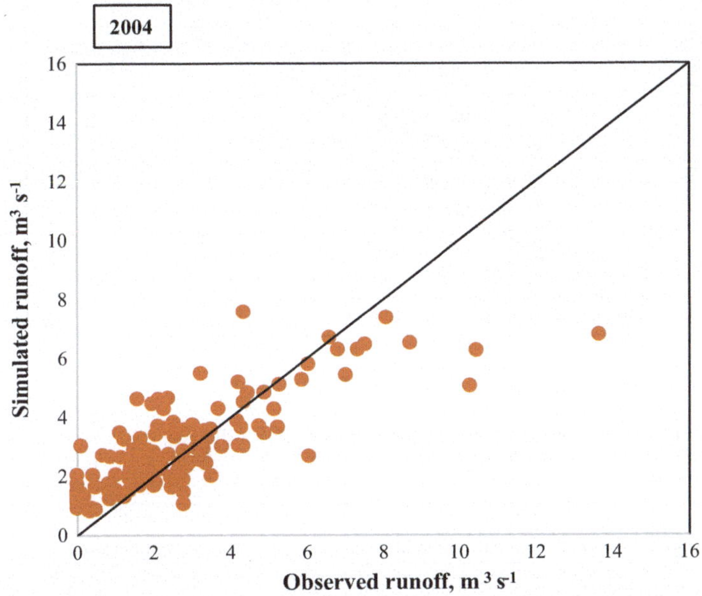

Fig. 13.4 Scatter plot of observed versus simulated runoff for the year 2004

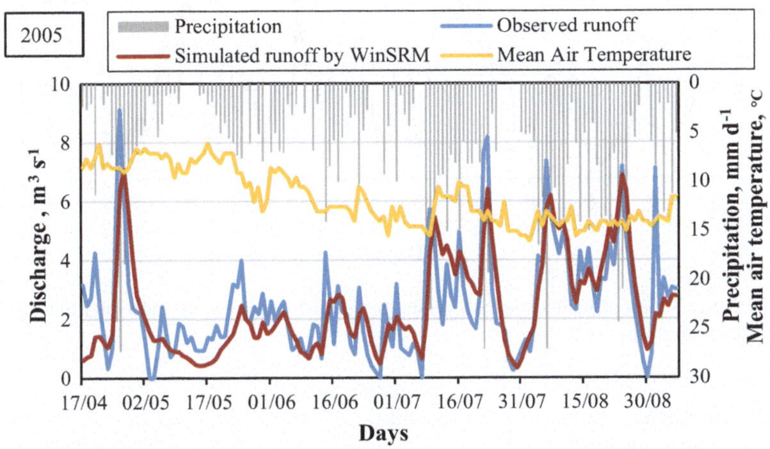

Fig. 13.5 Time series plot of observed versus simulated daily runoff, precipitation and mean temperature for the year 2005

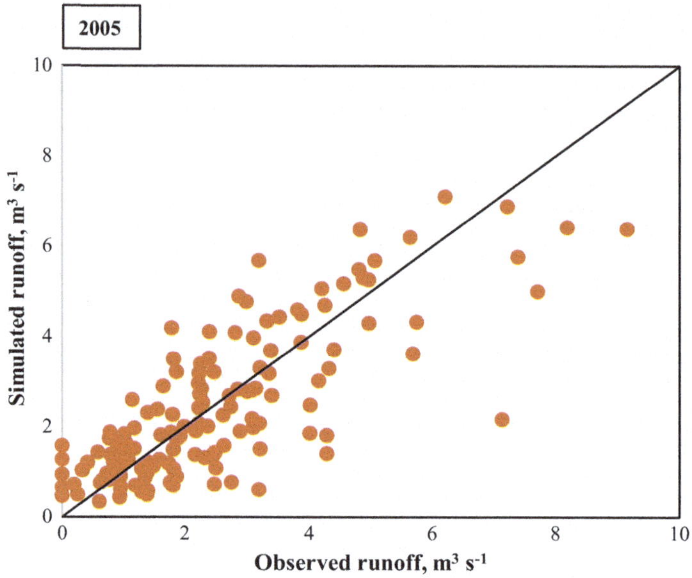

Fig. 13.6 Scatter plot of observed versus simulated runoff for the year 2005

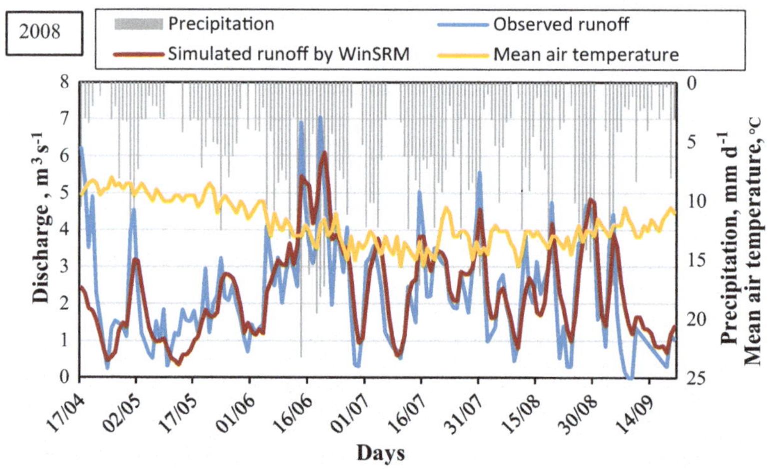

Fig. 13.7 Time series plot of observed versus simulated daily runoff, precipitation and mean temperature for the year 2008

period 2008. CRM values of −0.148 and 0.027 were obtained during the calibration period 2004 and 2005, respectively, while during the validation period 2008, CRM value of −0.0256 was obtained. The positive value of CRM in the year 2005 shows model underestimation in the simulated runoff and negative values of CRM show model overestimation in the simulated runoff during the years 2004 and 2008.

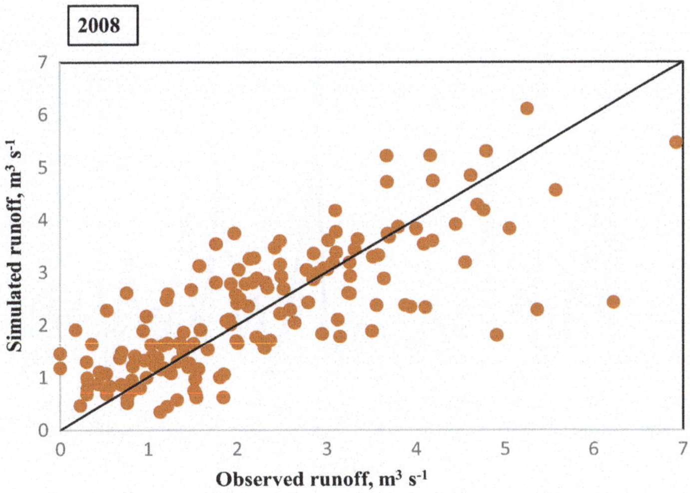

Fig. 13.8 Scatter plot of observed versus simulated runoff for the year 2008

Overall, the statistical performance indicators show acceptable modelling results during calibration and validation for the data scarce Himalayan watershed.

13.3.1 Effect of Monthly Variation of Near-Surface Lapse Rate (LR) on Runoff Simulation

While constant lapse rate of 0.5 °C/100 m was employed for calibration and validation, to observe the effect of employing different lapse rates on model simulation, the model was re-simulated for three years 2004, 2005 and 2008 with different lapse rates as follows: (i) constant lapse rate of 0.65 °C/100 m; (ii) constant lapse rate as determined for Eastern Himalayas (0.5 °C/100 m); and (iii) monthly varying lapse rate specific to Eastern Himalayas (Table 13.3) (Bandyopadhyay et al. 2014).

Figures 13.9, 13.10 and 13.11 show the hydrographs generated by the model for the years considered for this study. The hydrographs represent the observed runoff, runoff generated by using annually constant lapse rate of 0.65 °C/100 m, runoff generated by using a constant lapse rate of 0.5 °C/100 m and runoff generated by using the monthly varying lapse rates. From the given figures, we can observe that all the hydrographs generated with different lapse rates follow a similar trend with the observed hydrograph.

Table 13.6 shows the hydrograph characteristics of the observed and simulated stream flows. From the table, it was observed that in the year 2004, the average runoff generated by taking a constant lapse rate of 0.5 °C/100 m when compared with the observed runoff was found to be better than the average runoff from the other two conditions where the lapse rates were taken as an annual constant of 0.65 °C/100 m

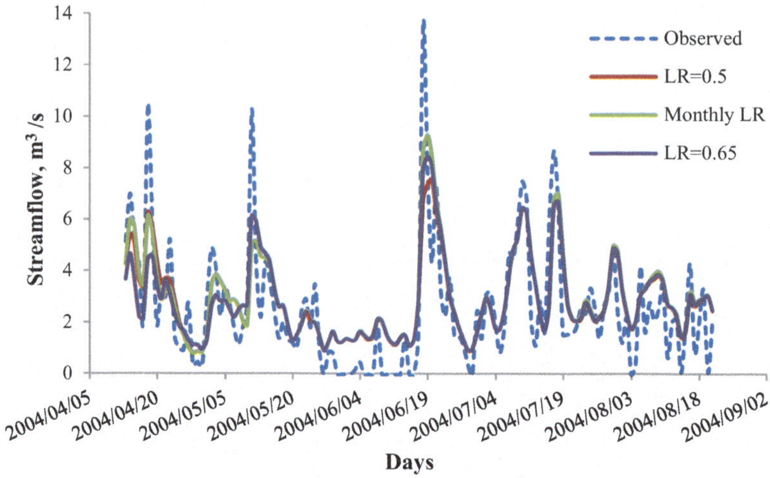

Fig. 13.9 Daily time series plot of observed and simulated streamflow for the year 2004

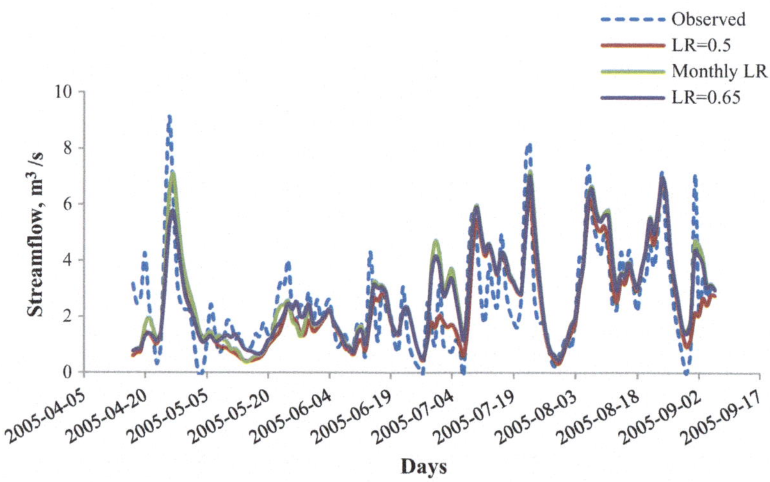

Fig. 13.10 Daily time series plot of observed and simulated streamflow for the year 2005

and a monthly varying lapse rate. For the year 2005, the average runoff generated by taking the monthly varying lapse rate was found to give better results than the other two conditions, while, for the year 2008, the average runoff generated by using a constant LR of 0.5 °C/100 m was found to give better results than the other two conditions.

For the year 2004, from Table 13.7, it can be seen that the monthly varying LR-based simulation gave an ME value of 0.621 which was found to be the best when compared with the ME values from the other two simulations. The ME value from the

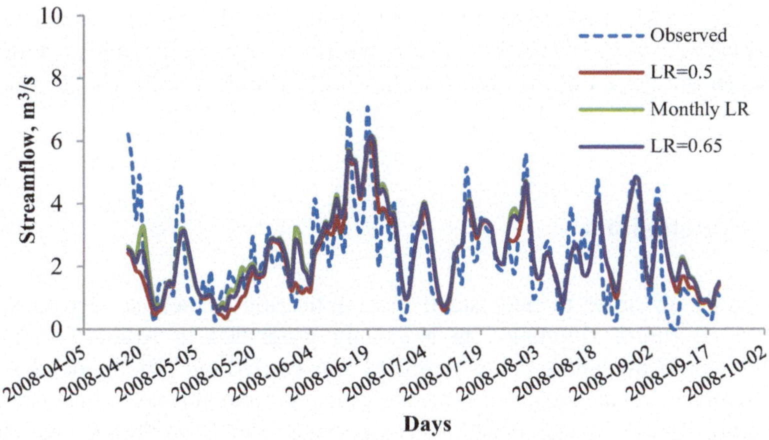

Fig. 13.11 Daily time series plot of observed and simulated streamflow for the year 2008

Table 13.6 Hydrograph characteristic of observed and simulated runoffs

Year	Observed		Computed (with a constant LR of 0.5 °C/100 m)		Computed (with annual LR, 0.65 °C/100 m)		Computed (with monthly LR)	
	Runoff volume (10^6 m^3)	Avg. runoff (m^3/s)	Runoff volume (10^6 m^3)	Avg. runoff (m^3/s)	Runoff volume (10^6 m^3)	Avg. runoff (m^3/s)	Runoff volume (10^6 m^3)	Avg. runoff (m^3/s)
2004	44.52	2.41	44.19	2.39	42.14	2.28	44.07	2.32
2005	41.28	2.23	34.18	1.85	37.7	2.04	40.43	2.19
2008	38.61	2.09	37.4	2.02	39.36	2.13	42.32	2.29

Table 13.7 Statistics of the hydrographs

Year	Modelling efficiency (ME)			Coefficient of residual mass (CRM)		
	LR of 0.5 °C/100 m	Annual LR (0.65 °C/100 m)	Monthly LR	LR of 0.5 °C/100 m	Annual LR (0.65 °C/100 m)	Monthly LR
2004	0.610	0.588	0.621	−0.148	−0.130	−0.174
2005	0.608	0.553	0.525	0.027	−0.089	−0.122
2008	0.611	0.560	0.539	−0.0256	−0.105	−0.134

simulation based on constant LR of 0.5 °C/100 m gave the second best value of 0.610 which is close to 0.621. It is worth noting that although using a constant lapse rate of 0.5 °C/100 m resulted in better volumetric water balance for the year 2004 than using monthly varying lapse rate, trends in high and low flows are being captured better by the latter. For the other two years, i.e. 2005 and 2008, the simulation based on constant lapse rate of 0.5 °C/100 m gave the best ME value of 0.608 and 0.611, respectively. CRM value shows an overestimation for all the years when the lapse rate was taken as 0.65 °C/100 m and also when it was varied monthly. In the year 2005, when the lapse rate was taken as 0.5 °C/100 m, a positive CRM value was obtained which indicates an underestimation in the runoff by the model.

13.4 Conclusions

In this study, in order to understand the implication of near-surface lapse rate on snowmelt runoff simulation in Nuranang watershed in Arunachal Pradesh, snowmelt runoff modelling using WinSRM was carried out. The model was calibrated by matching the observed runoff with the simulated runoff at the outlet for the depletion period of 2004 and 2005. The average values of calibration parameters obtained during the calibration period were used for validating the model. Acceptable modelling results were obtained during calibration and validation for the data scarce Himalayan watershed. After obtaining satisfactory model performance for calibration and validation, runoff simulation was carried out for all the three years 2004, 2005 and 2008 using different lapse rates as follows: (i) constant lapse rate of 0.65 °C/100 m, (ii) constant lapse rate of 0.5 °C/100 m (specific to Eastern Himalayas) and (iii) monthly varying lapse rate specific to Eastern Himalayas.

For the years 2004 and 2008, the average runoff generated by taking a constant lapse rate of 0.5 °C/100 m when compared with the observed runoff was found to give better results than the average runoff from the other two conditions where the lapse rates were taken as an annual constant value of 0.65 °C/100 m and a monthly varying lapse rate, while, for the year 2005, the average runoff generated by taking the monthly varying lapse rate was found to give better results than the other two conditions. For the years 2005 and 2008, simulations based on constant lapse rate of 0.5 °C/100 m gave the best ME values of 0.608 and 0.611, respectively, while the best ME value of 0.621 was obtained using monthly varying lapse rate in the year 2004. In all the scenarios, simulations based on a constant lapse rate of 0.65 °C/100 m perform the worst. The model overestimates runoff for all the periods considered when monthly varying lapse rate and constant lapse rate value of 0.65 °C/100 m were used. Overall, using a constant lapse rate of 0.5 °C/100 m gave better simulations results. Instead of using average environmental lapse rate, specific lapse rate designed for a particular region must be used for better hydrological modelling. Based on the results obtained, a constant near-surface lapse rate of 0.5 °C/100 m is recommended for the temperature-index based snowmelt models in the Eastern Himalayas.

References

Anderson A (1976) A point energy and mass balance model of a snow cover. NOAA Tech. Rep. NWS 19, U.S. Dept. of Commerce, National Oceanic and Atmosphere Administration, National Weather Service. Silver Spring, MD

Bandyopadhyay A, Bhadra A, Maza M, Shelina RK (2014) Monthly variations of air temperature lapse rates in Arunachal Himalaya. J Indian Water Resour Soc 34(3):16–25

Blandford TR, Humes KS, Harshburger BJ, Moore BC, Walden VP, Ye H (2008) Seasonal and synoptic variations in near-surface air temperature lapse rates in a mountainous basin. J Appl Meteorol Climatol 47(1):249–261

Hock R (2003) Temperature index melt modelling in mountain areas. J Hydrol 282(1–4):104–115

Lang H, Braun L (1990) On the information content of air temperature in the context of snow melt estimation. In: Molnar L (ed) Hydrology of Mountainous Areas, Proceedings of the Strbske Pleso Symposium 1990, IAHS Publ. no. 190, pp 347–354

Martinec J, Rango A (1986) Parameter values for snowmelt runoff modelling. J Hydrol 84(3–4):197–219

Martinec J, Rango A, Roberts R (2008) Snowmelt Runoff Model (SRM) user's manual. College of Agriculture and Home Economic, Las Cruces, New Mexico, USA

Pandey PK, Williams CA, Frey KE, Brown ME (2014) Application and evaluation of a snowmelt runoff model in the Tamor River basin, Eastern Himalaya using a Markov Chain Monte Carlo (MCMC) data assimilation approach. Hydrol Process 28(21):5337–5353

Rango A, Martinec J (1981) Accuracy of snowmelt runoff simulation. Nord Hydrol 12:265–274

Romshoo SA, Dar RA, Rashid I, Marazi A, Ali N, Zaz SN (2015) Implications of shrinking cryosphere under changing climate on the streamflows in the Lidder catchment in the Upper Indus Basin, India. Arct Antarct Alp Res 47(4):627–644

Rulin O, Liliang R, Weiming C, Zhongbo Y (2008) Application of hydrological models in a snowmelt region of Aksu River Basin. Water Sci Eng 1(4):1–13

Senzeba KT, Bhadra A, Bandyopadhyay A (2015) Snowmelt runoff modelling in data scarce Nuranang catchment of eastern Himalayan region. Remote Sens Appl Soc Environ 1:20–35

Singh P, Jain SK (2003) Modelling of streamflow and its components for a large Himalayan basin with predominant snowmelt yields. Hydrol Sci J 48(2):257–276

Singh P, Bengtsson L (2005) Impact of warmer climate on melt and evaporation for the rain-fed, snow-fed and glacier-fed basins in the Himalayan region. J Hydrol 300(1–4):140–154

Whiteman CD, Bian X, Zhong S (1999) Wintertime evolution of the temperature inversion in the Colorado Plateau Basin. J Appl Meteorol 38(8):1103–1117

WMO: Guide for hydrometeorological practices. World Meteorological Organisation (WMO). Geneva, Switzerland (1964)

Wulf H, Bookhagen B, Scherler D (2016) Differentiating between rain, snow, and glacier contributions to river discharge in the western Himalaya using remote-sensing data and distributed hydrological modelling. Adv Water Resour 88:152–169

Chapter 14
Estimation of Surface Runoff Using SCS Curve Number Method Coupled with GIS: A Case Study of Vadodara City

Lakhwinder Singh and Deepak Khare

Abstract The curve number method was developed by the USDA Natural Resources Conservation Service, which was formerly called the Soil Conservation Service or SCS. It is widely used in hydrological investigations. The present study is an attempt to estimate surface runoff using SCS curve number method coupled with remote sensing and GIS tools. SCS is an empirical formula based on curve numbers that are defined using landuse and hydrological soil groups. In 2016, Muche derived relation between curve number and NDVI considering soil classes. His study utilized 12 years of rainfall and runoff data from four small watersheds in Northeast Kansas on the Konza Prairie Long-Term Ecological Research (LTER). A relationship is developed based on practical tests that used here; for our study area, we have calculated runoff using both methods, i.e., using direct value of curve number and derived NDVI. Nasiri and Alipur (Int J Environ Chem Ecol Geol Geophys Eng 8(5):342–345, 2014) determined curve number using NDVI. Our result shows similar pattern of runoff using both the methods. Therefore, his equation can be used for any soil type. Because natural vegetation type also depends on soil type and urban landuse classes, so it is possible to link CN with NDVI that used in this study. In this study, we used pre-defined soils by FAO. The study area lies in single soil of hydrological soil group C. Usable range of curve number is between 30 and 98 where lower curve number value represents more infiltration and less runoff. Higher value represents more runoff and less infiltration (Fan et al. in Remote Sens 5:1425–1438, 2013). This study attempts to link these values with NDVI. This study used to drive runoff using SCS curve number method using GIS model. The results from both the methods are also validated by taking 400 random sample points and compared runoff value with the observed. It is found that runoff is based on NDVI and without NDVI having relation of $r^2 = 0.80$. This is acceptable for any area for estimation of surface runoff.

L. Singh (✉) · D. Khare
Department of Water Resources Development and Management, Indian Institute of Technology Roorkee, Roorkee, Uttarakhand 247667, India
e-mail: lsingh@wr.iitr.ac.in

D. Khare
e-mail: kharefwt@iitr.ac.in

A. Pandey et al. (eds.), *Water Management and Water Governance*, Water Science and Technology Library 96,
https://doi.org/10.1007/978-3-030-58051-3_14

Keywords Landuse · Runoff · SCS curve number · GIS model · Hydrology

14.1 Study Area

Vadodara is city of Gujarat state of India, located in limits of 22°12′40′ N–22°25′23″ N and 73°04″37″ E–73°16″40″ E city area as per city boundary which is considered as 250 km^2 shown in Fig. 14.1.

14.1.1 Data Sources

Rainfall data is used of India Meteorological Department for average rainfall of last 35 years. Landsat satellite image is used of January month for calculation of landuse and NDVI. Another option of MODIS NDVI is also here but not used because urban area required a fine resolution image, but MODIS has 250-m resolution, so this MODIS NDVI is not suitable for city-level analysis works.

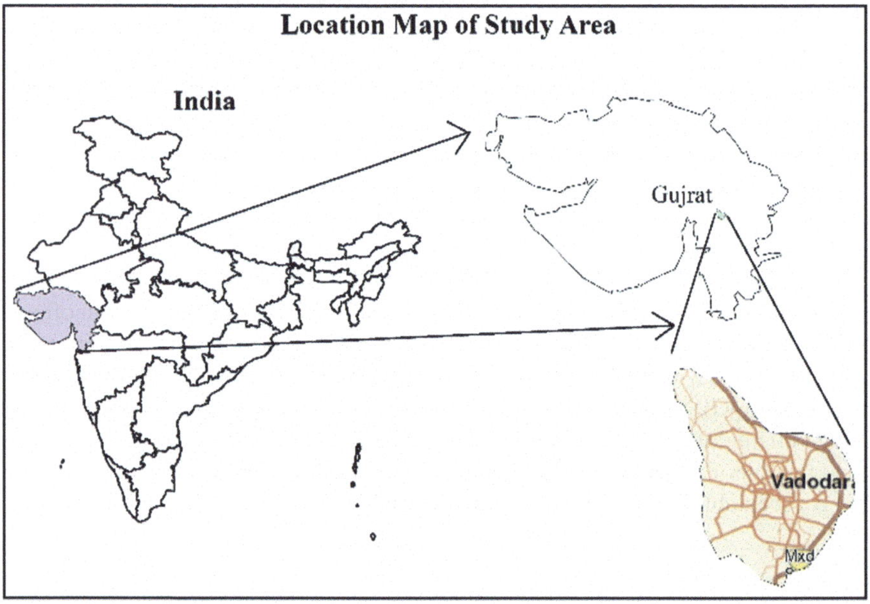

Fig. 14.1 Study area

14.2 Methodology

Methodology is divided in two parts: The first part is SCS curve number method as it is, and the second part is to develop a model with NDVI method. SCS-CN method, developed by Soil Conservation Services (SCS) of USA in 1969, is a simple, predictable, and stable conceptual method for estimation direct runoff depth based on storm rainfall depth. It relies on only one parameter CN. Currently, it is a well-established method, having been widely accepted for use in USA and many other countries. The SCS-CN method is based on the water balance equation of the rainfall in a known interval of time (Subrahmanyam 2014). The SCS-CN model expressed as (Eq. 14.1):

$$Q = \frac{(P - I_a)^2}{P - I_a + S} \tag{14.1}$$

where P is rainfall, I_a is initial abstraction, and S represents potential maximum retention.

S is expressed as (Eq. 14.2):

$$S = \frac{25{,}400}{\mathrm{CN}} - 254 \tag{14.2}$$

The constant 254 is used to express S in mm. Now, curve number is related to S.

14.2.1 CN and NDVI

NDVI is normalized difference vegetation index. Its value lies between −1 and 1 where 1 represents more healthy vegetation and −1 represents hard surface like urban area. So, it is possible to link landuse with NDVI. On the other side, forest type and agriculture also depend on soil type. The vegetation above any different soil has different signatures and different soil which are not covered by vegetation having different NDVI. So, once derived curve number by linking to NDVI, no need to consider about soil because soil type also impacts on NDVI. Accuracy of the results will increase if we link it perfectly and put in model. In this study for digital calculation, we have re-scaled actual NDVI value to 8-bit raster in 0–255 range where 0 represents −1 and 255 represents 1. The study ranges between 86 and 155 DN's.

NDVI is calculated as (Eq. 14.3):

$$\mathrm{NDVI} = \frac{(\mathrm{NIR} - \mathrm{Red})}{(\mathrm{NIR} + \mathrm{Red})} \tag{14.3}$$

where NIR is near infrared and red bands of satellite image. Now, CN is in function of NDVI (Eq. 14.4):

$$CN = a\text{NDVI} + c \tag{14.4}$$

where a is −0.117 and c is 97.3 and NDVI is value between 0 and 255. Figure 14.2 shows developed integrated tool for ArcGIS and graphic user interface. Flowchart of methodology is shown in Fig. 14.3

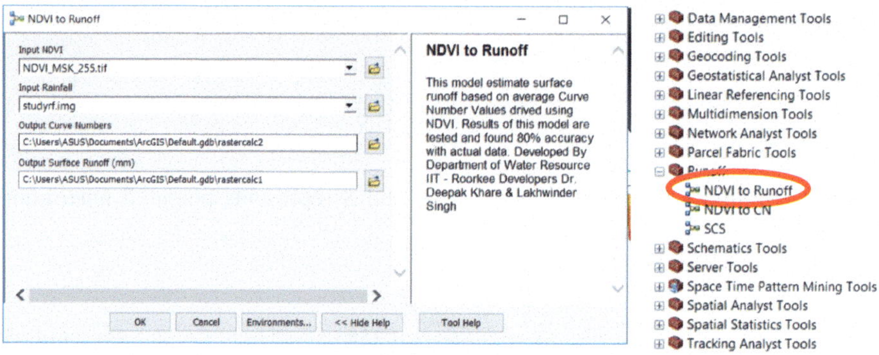

Fig. 14.2 Model GUI and ArcGIS integration

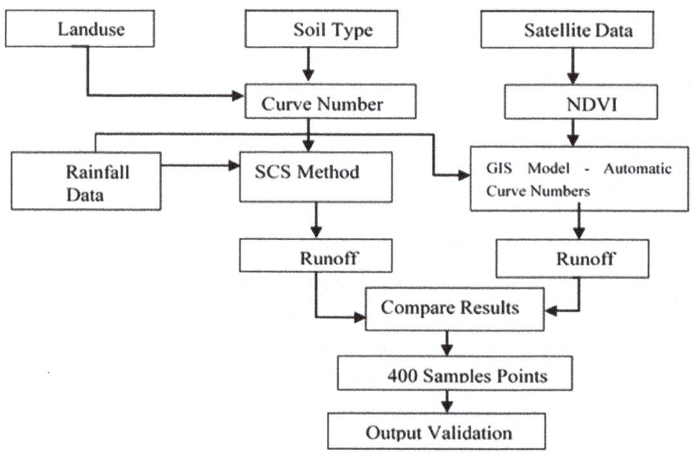

Fig. 14.3 Flowchart of methodology

Table 14.1 Landuse 2017

Class	Area (km^2)
Water	1.45
Forest	25.91
Barren	17.24
Agriculture	90.40
Settlements/urban	115.07
Total	250.07

14.3 Results and Discussion

14.3.1 Landuse

Landuse of 2017 is derived from Landsat 8 image. Satellite image was preprocessed; then, advance approach of landuse preparation is used by mixing supervised and unsupervised classification. Accuracy of classification 84% is achieved after manual correction of pixels. The details of landuse are shown in Table 14.1. Landuse classification is shown in Fig. 14.4.

14.3.2 GIS Layers and Results

The study area lies in hydrological soil group C as per FAO soil classification. Surrounding soil of area is shown in Fig. 14.5. Curve number values decided from the literature are shown in Fig. 14.6. The study area lies in single soil zone. So, manual curve number value as per landuse class. As shown, urban area is very dense in this region, so curve number 90 is used for dense urban area and hydrological soil group C of soil. Similarly, 82 is for agriculture, 79 for pasture land or barren land with grass, and 70 for forest is used where this area has few water bodies, so CN value of 0 is always used for water body (Subrahmanyam 2014). Figure 14.7 represents NDVI values which range from −0.14 to 0.54. Lower NDVI value represents hard surface. So, NDVI value is easily linked with curve numbers. Figure 14.8 shows curve number derived from NDVI using Eq. 14.4. Corresponding satellite image is also shown in Fig. 14.9. For calculation of rainfall, an average of 35 years of rainfall is used in this study to generate throughout average trend of runoff in urban area, whereas trend of rainfall over small urban area is negotiable using IMD data (Pai et al. 2014). It shows 735–912 mm of rainfall in area as shown in Fig. 14.10. It is a very small variation derived using long-term data. This trend shows that the south part of city is always getting more rainfall, but the north part of city is getting comparatively less rainfall. The amount and spatial pattern of rainfall also impact on surface runoff (Khare et al. 2015). In this study, we have generated two curve number maps to validate accuracy of the results. Runoff is calculated two times with both values

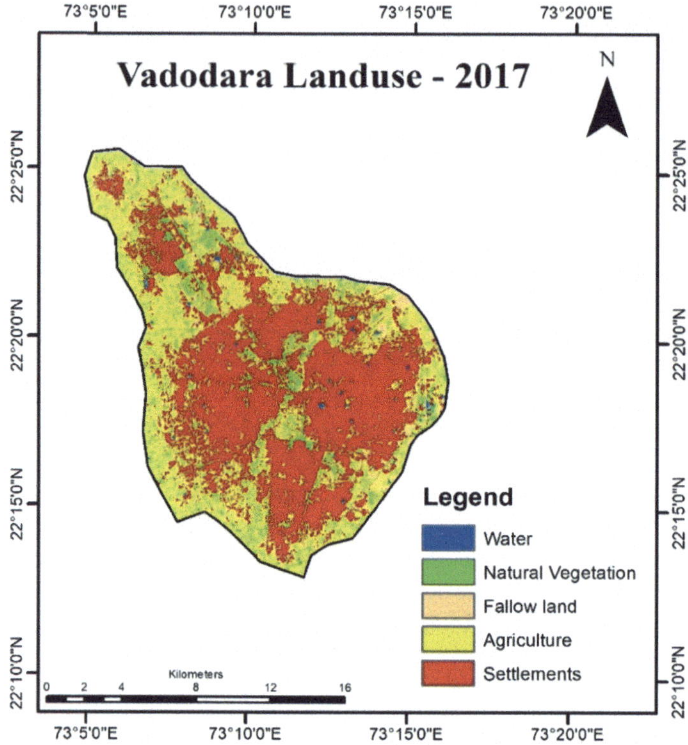

Fig. 14.4 Landuse classification

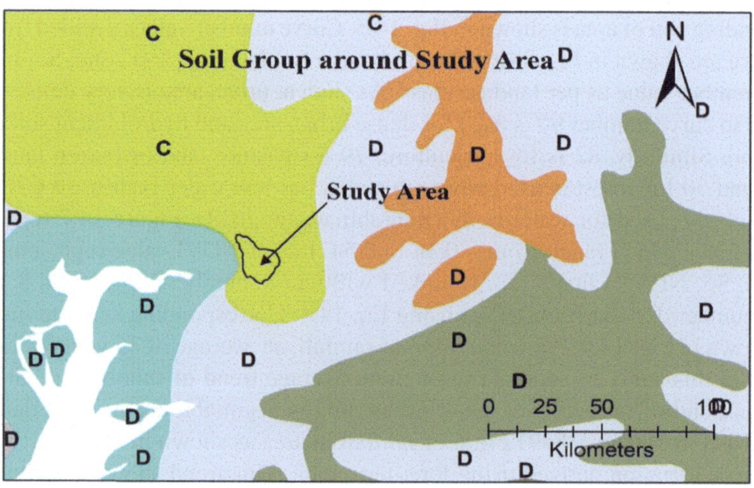

Fig. 14.5 FAO hydrological soil group

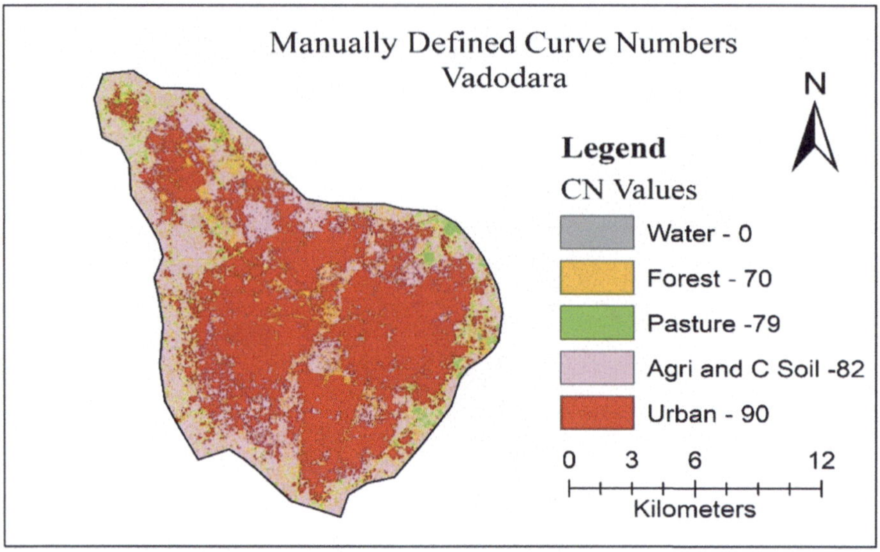

Fig. 14.6 Manually derived curve numbers

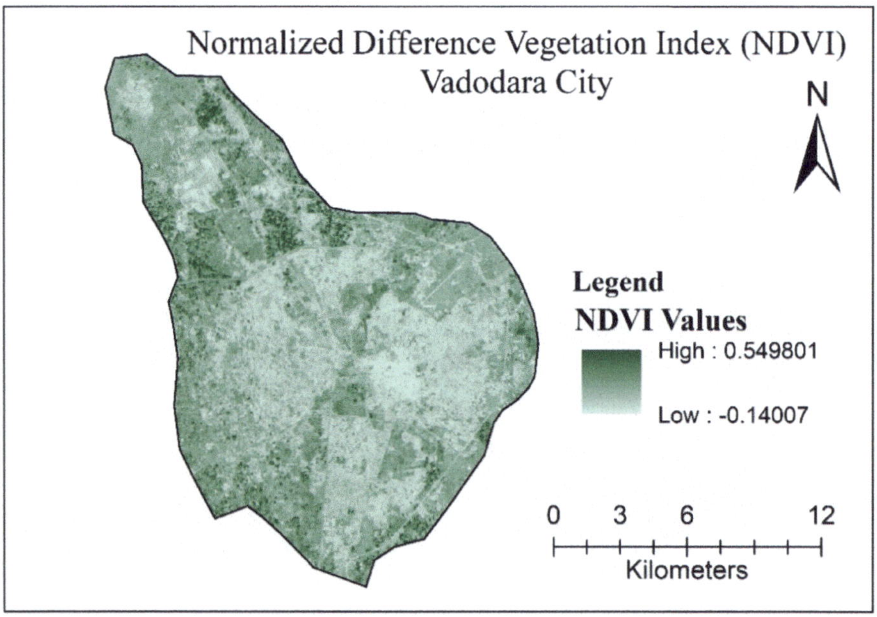

Fig. 14.7 NDVI values

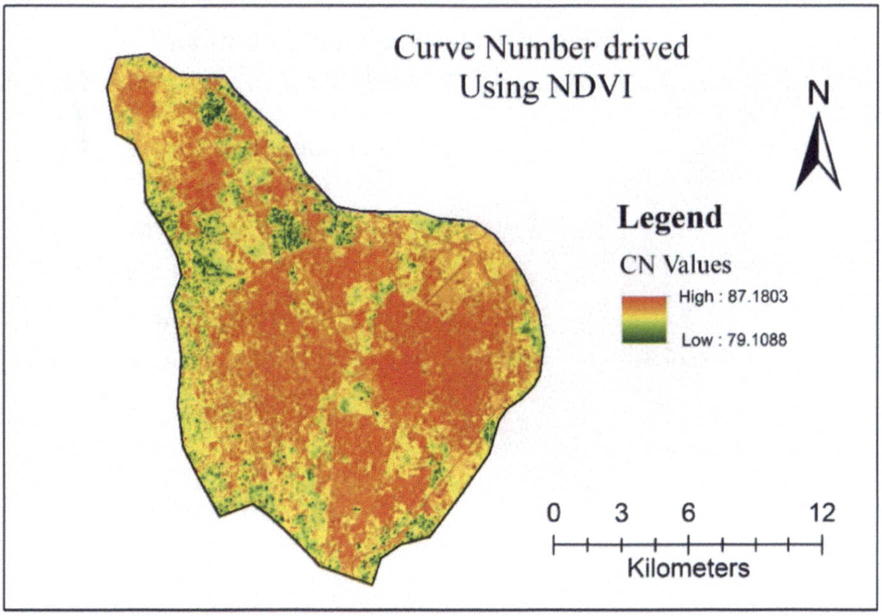

Fig. 14.8 CN values derived from NDVI

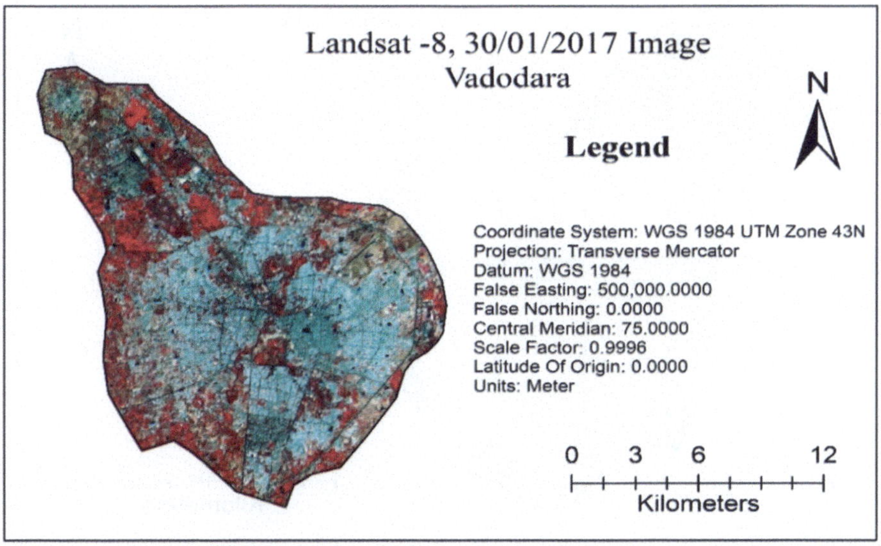

Fig. 14.9 Landsat image

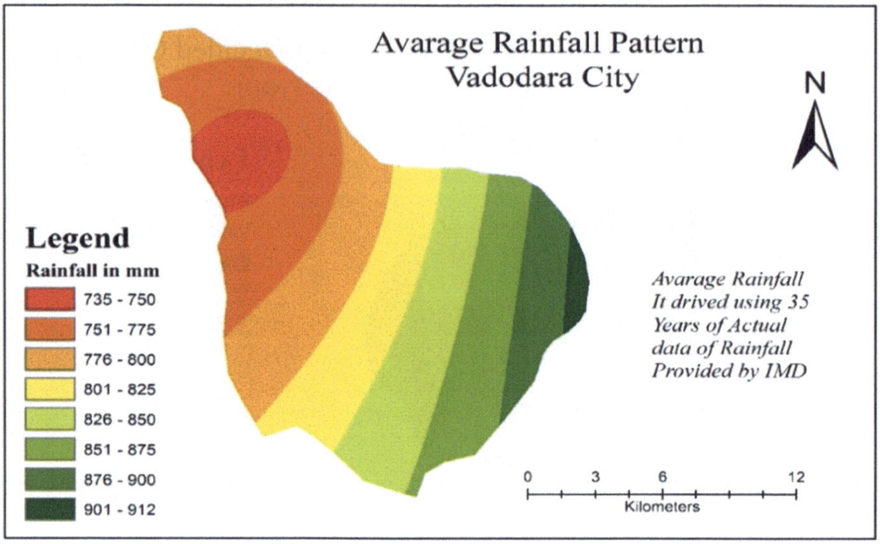

Fig. 14.10 Rainfall distribution

of curve numbers. Both maps show almost the same trend in surface runoff. One is based on landuse and other is based on NDVI. Our study reveals that NDVI can be used as function of curve number to faster the process of runoff estimation. But some points need to care while calculating runoff using NDVI. A satellite image must be of session when crops are growing in fields and representing green signature not in time of harvesting. Therefore, selection of time is most important for estimation of runoff. On the other side, if area is large, then the average MODIS and ISRO NDVI product can be used, for similar set of datasets.

In this study, NDVI is calculated using both methods—direct curve numbers and NDVI-based curve number; both the results are shown in Figs. 14.11 and 14.12. The results are almost similar following the same pattern with little variation in values. But at acceptable range, as in landuse and urban area settlement has the same value in digital data. When we drive NDVI of the same area, it has variation based on surface hardness, due to which different wavelengths are reflected. We can say it is a much better approach than landuse because it uses actual data to drive CN automatically. There are more possibilities of research present in this direction beyond this study. In the present study, the results are validated by performing calculation based on model-based NDVI and traditional CN-based runoff model. For testing, random sampling method is used to get output result of both methods and validate values of the same location. Thus, 400 random sample points are generated in the same study area, and later, value of 400 points is taken from both the results for validation.

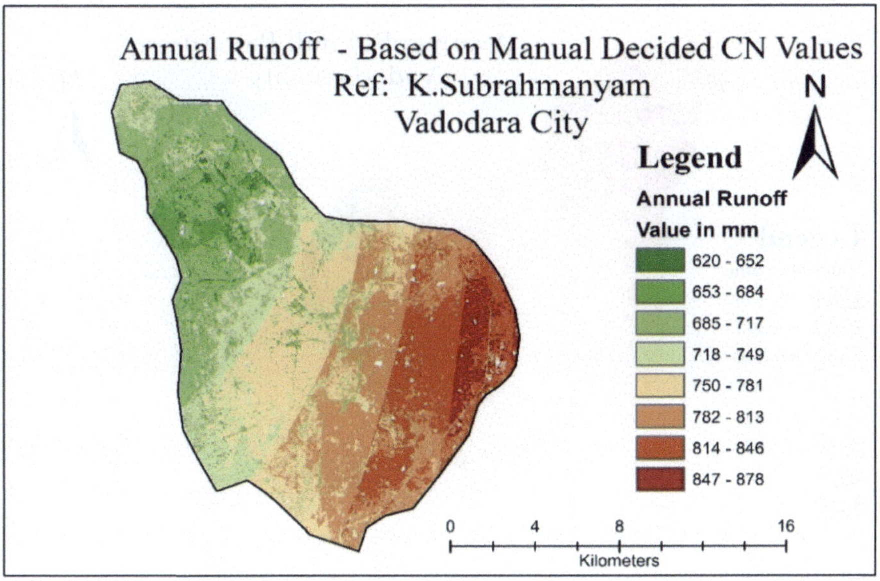

Fig. 14.11 Runoff—literature-based CN

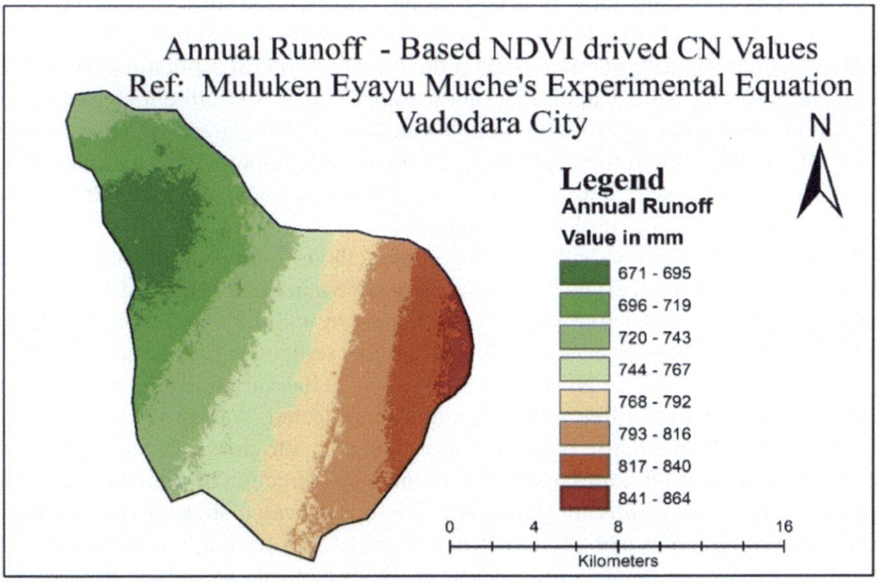

Fig. 14.12 Runoff—NDVI-based CN

14.3.3 Validation

While validating, it is found that both the results follow the same trend with $r^2 = 0.80$. Sample points are shown in Fig. 14.13, and relationship in Fig. 14.14. Thus, calculation of runoff by NDVI is faster than traditional methods. Spatial-based model provides quick and fast results if integrated in models and tools.

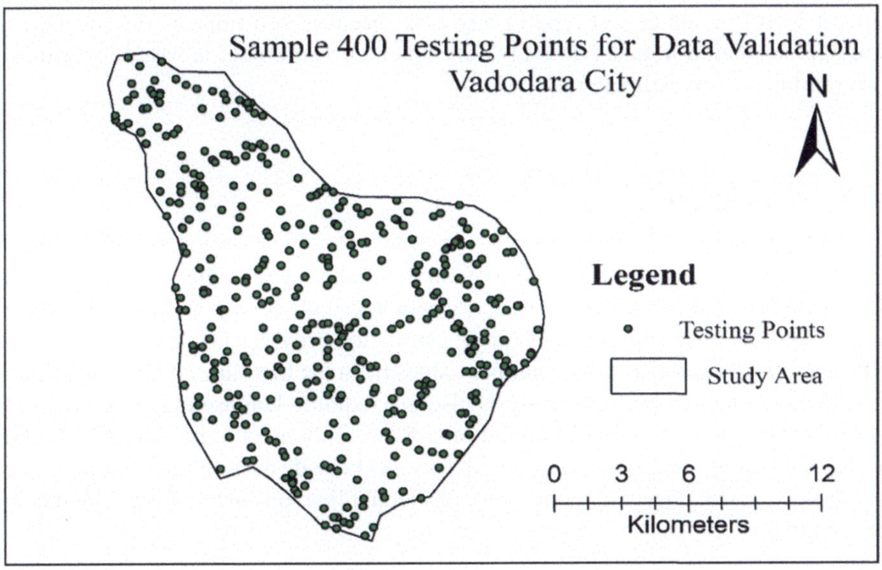

Fig. 14.13 Test points

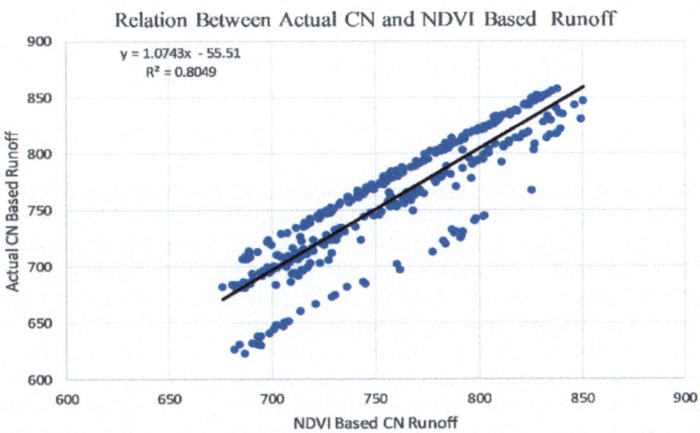

Fig. 14.14 NDVI and direct CN-based runoff

As shown in Figs. 14.11 and 14.12, runoff from both methods follows the same trend. It is easily seen 4 landuse having and same slope of angle this tends to similarity between runoff of both methods. In Fig. 14.14, lines appear representing four types of landuse. But while manual CN given value is fixed, but when CN derived using NDVI value is near to that actual values but not far from actual values, due to this reason, it represents lines following the same trend. Trend line represents correlation of 0.80 which is on acceptable limit. After validated, this approach is also usable to determine actual curve number based on NDVI, providing near accurate result without worrying about soil type in any area, because soil impacts on vegetation type and vegetation impacts on NDVI on which curve number is based, later used with rainfall data to estimate runoff.

14.4 Conclusion

NDVI can also be used for estimation of curve number. After validation, it provides 80% accurate results, and this is under acceptable limit of any analysis. By using this, GIS model-based approach of runoff estimation provides fast results because it saves time from exploring curve number values from the literature. It also saves time from making landuse, and this analysis also not required landuse map to be built in GIS. After fitting, everything in model in GIS, we only need to provide NDVI and rainfall data for any location. As output this model provide curve numbers and runoff for whole region at once. So, it has many further applications in handling digital data for Hydrology.

Acknowledgements The authors would like thanks for the support of the sponsor project "Assessment of Climate change and its impact on urban hydrology: Indian Perspective [Project Number = AIC-566-WRC]."

References

Fan F, Deng Y, Hu X, Weng Q (2013) Estimating composite curve number using an improved SCS-CN method with remotely sensed variables in Guangzhou, China. Remote Sens 5(3):1425–1438

Khare D, Singh R, Mishra N, Shukla R (2015) Assessment of surface runoff in a Barinallah watershed using distributed parameter model (SWAT Model). J Water Resour Environ Eng 1(1):31–38

Muche ME (2016) Surface water hydrologic modeling using remote sensing data for natural and disturbed lands. Kansas State University, Manhattan, Kansas

Nasiri A, Alipur H (2014) Determination the curve number catchment by using GIS and remote sensing. Int J Environ Chem Ecol Geol Geophys Eng 8(5):342–345

Pai DS, Sridhar L, Rajeevan M, Sreejith OP, Satbhai NS, Mukhopadyay B (2014) Development of a new high spatial resolution (0.25 × 0.25) long period (1901–2010) daily gridded rainfall data set over India and its comparison with existing data sets over the region. Mausam 65(1):1–18

Subrahmanyam K (2014) Engineering hydrology. McGraw-Hill Professional

Chapter 15
Determination of ERA-INTERIM Proficiency for Rainfall–Runoff Modeling

Tanmoyee Bhattacharya, Deepak Khare, and Manohar Arora

Abstract In case of ungauged or poorly gauged basins, such as data-scarce Beas River basin, the main challenge is determining the utility of reanalysis data for simulation of river discharge. According to the researchers, the availability of large number of rain gauge station can improve the result of rainfall–runoff models. But the drawback is difficulty of collecting data, missing data and a point measurement. In this study, we have examined the adequacy of ERA-INTERIM dataset for estimating rainfall–runoff in data-scarce Beas River basin using ArcSWAT software for modeling purpose.

Keywords ERA-INTERIM · BEAS · ArcSWAT · Rainfall-runoff model

15.1 Introduction

Hydrometeorological data is needed for driving hydrology models. To improve the reliability of ArcSWAT hydrology modeling in Beas River basin, advanced and easily available data collection is needed. According to Zhao et al. (2010) and Fuka et al. (2014), USA and Europe currently provide global reanalysis dataset for several applications for hydrology. Ward et al. (2011) and Fuka et al. (2014) described that for medium-sized watershed, there is a great advantage of reanalysis data with high spatial resolution for capturing high-frequency distribution. According to Kumar

T. Bhattacharya (✉) · D. Khare
Department of Water Resources Development and Management, Indian Institute of Technology Roorkee, Roorkee, Uttarakhand 247667, India
e-mail: bhattacharya.tanmoyee36@gmail.com

D. Khare
e-mail: kharefwt@gmail.com

M. Arora
Surface Water Hydrology Division, National Institute of Hydrology, Roorkee, Roorkee, Uttarakhand 247667, India
e-mail: manohararora19@gmail.com

A. Pandey et al. (eds.), *Water Management and Water Governance*, Water Science and Technology Library 96,
https://doi.org/10.1007/978-3-030-58051-3_15

et al. (2007) at the eastern part of Beas River basin, there is no rain gauge station where Parvati and Sainj rivers, two important tributaries, originate. ERA-INTERIM reanalysis data is used in this study because it is freely available dataset with all meteorological data needed for this research such as temperature, rainfall, wind speed, solar radiation and relative humidity. ERA-INTERIM is produced by ECMWF (Berrisford et al. 2009; Dee et al. 2011), and it is available for long period 1989 to recent date. We have investigated the adequacy of ERA-INTERIM data for hydrology modeling of Beas basin that has less rain gauge station through this study.

15.2 Study Area

The Beas River basin is the major part of Indus River basin which has a total length of 460 km and catchment area of 20,303 km^2. Beas River basin covers four districts of Himachal Pradesh, and they are Kangra, Hamirpur, Kullu and Mandi. It is situated at geographical co-ordinates 32°21′59″–31°16′09″ N latitude and 77°05′08″ E–74°58′31″ E longitude. Its altitude varies from 802 m near Pandoh dam up to 6600 m on the northeast borders (Lakhera et al. 2016). Some of the major tributaries are Tirthan and Sainj rivers near Larji, Parbati River near Bhuntar, Bakhli khad near Pandoh Dam and Sabari Nala near Kulu.

15.3 Data Collection and Methodology

15.3.1 Data Collection

The datasets required for this study area are:

1. Spatial data—Digital elevation model (DEM), slope map, land use land cover (LULC) and soil data.
2. Temporal data—Maximum temperature, minimum temperature, rainfall, wind speed, solar radiation and relative humidity.
3. ArcSWAT software is an interface for ArcGIS software and also an open version.

15.3.1.1 Digital Elevation Model

From digital elevation model, we obtain the elevation information for topography. The basin, drainage line and slope are also derived from DEM (Fig. 15.1a, b). The DEM is acquired from Cartosat-1 DEM version 3.1 obtained from geoportal of Bhuvan. The resolution of the DEM is 30 m.

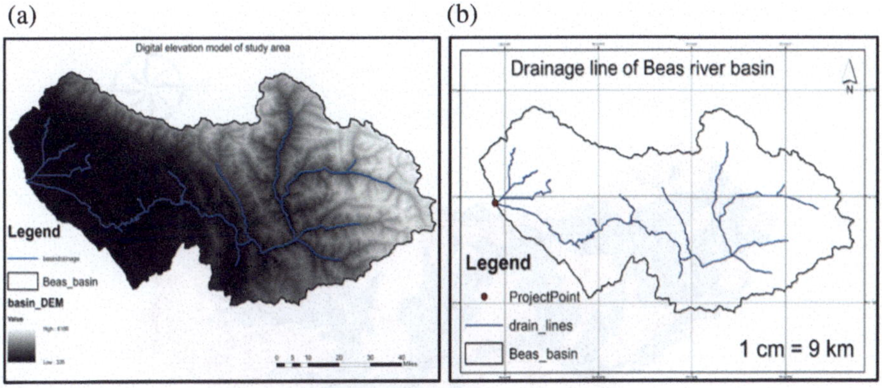

Fig. 15.1 **a** DEM and **b** drainage map of Beas River basin

15.3.1.2 Land Use/Land Cover

Land use/Land cover (LULC) change is of critical importance for land use planning and water resource management (Zhu and Li 2014). The LULC map obtained from MODIS (MCD12Q1) for a temporal coverage of 2001–2013 (Fig. 15.2). The

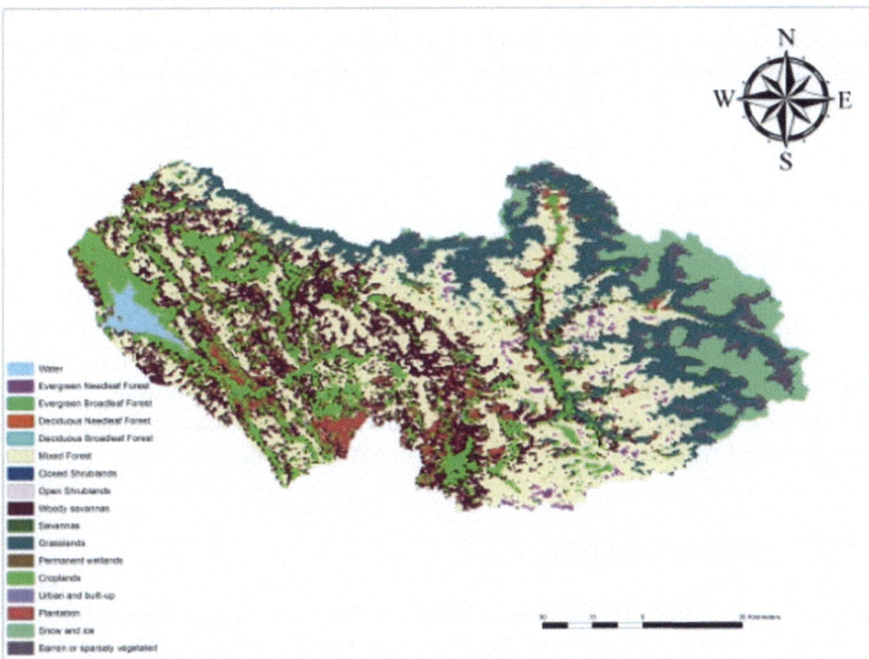

Fig. 15.2 Land use/Land cover map of Beas River basin

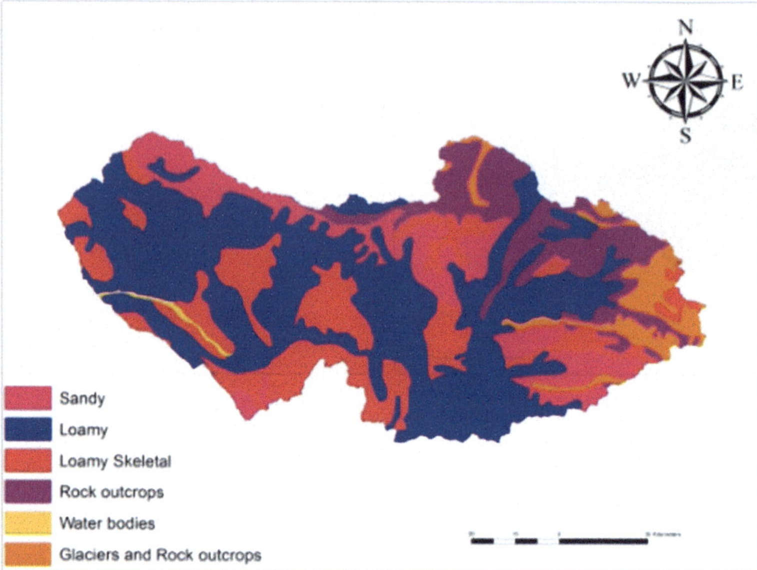

Fig. 15.3 Soil map of Beas River basin

resolution of the dataset is 500 m. A look-up table for land use is prepared based on SWAT LULC code, so that each LULC classes possess similar hydrological response. Cropland and plantation cover a major part of basin followed by mixed forest and grassland.

15.3.1.3 Soil Data

The soil map of the study area is obtained from National Bureau of Soil Survey and Land Use Planning, NBSS&LUP at 1:250,000 scale (Aggarwal et al. 2016). The major soils for Beas river basin are sandy, loamy, loamy skeletal, rock outcrops, water bodies and glaciers and rock outcrop (Fig. 15.3). The soil textural information of NBSSLUP is used for deriving soil properties such as available water content, hydraulic conductivity, bulk density, soil texture and organic carbon content. The appropriate soil type from user soil is linked to SWAT soil database code.

15.3.1.4 Meteorological Data

The rain gauge stations for Beas river basin are Pong dam, Pandoh dam, Bhuntar, Manali, Larji, Sainj, Banjar and Janjehali (Fig. 15.4). The weather inputs for ArcSWAT are maximum temperature, minimum temperature, rainfall, relative humidity, solar radiation and wind speed.

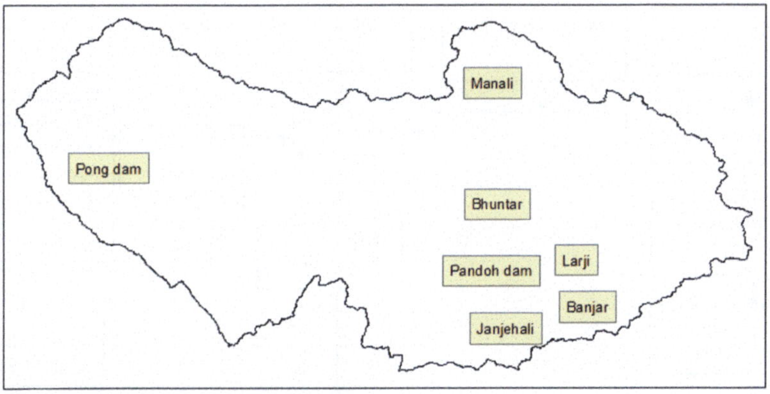

Fig. 15.4 Rain gauge stations of Beas River basin

Daily meteorological dataset for rain gauge stations is obtained from IMD. ERA-INTERIM dataset is available at https://apps.ecmwf.int/datasets/data/interim-full-daily. The dataset is of 3 hourly forecast dataset which is converted into daily data.

15.3.2 Methodology

ArcSWAT 2012 version model is integrated with ArcGIS software through an interface. Hydrology simulation has done by ArcSWAT hydrology model. It combines different spatial and temporal datasets and analyzes them to produce stream flow. The processing of dataset for application in ArcSWAT is done by ArcGIS software. The runoff is computed by SWAT using Soil Conservation Service (SCS) curve number technique (USDA 1972). The methodology for the runoff modeling using SWAT is depicted in flowchart as shown in Fig. 15.5. The model is evaluated by coefficient of determination (R^2), Nash–Sutcliffe efficiency (NSE) and root mean square error (RMSE). Hydrologic response units (HRUs) comprised of unique land cover, soil, slope and management combinations (Dile and Srinivasan 2014). The runoff is estimated individually for each HRU and routed to obtain the total runoff for the watershed (Thakur et al. 2016).

(a) **Coefficient of Determination (R^2)**

Generally, R^2 ranges from 0 to 1. A good correlation obtained for values between 0.6 and 1. R^2 is most often used in linear regression (Khare et al. 2014). A formula used in linear regression for a line where for a given set of data points the line most closely matches that data points. R^2 is calculated as:

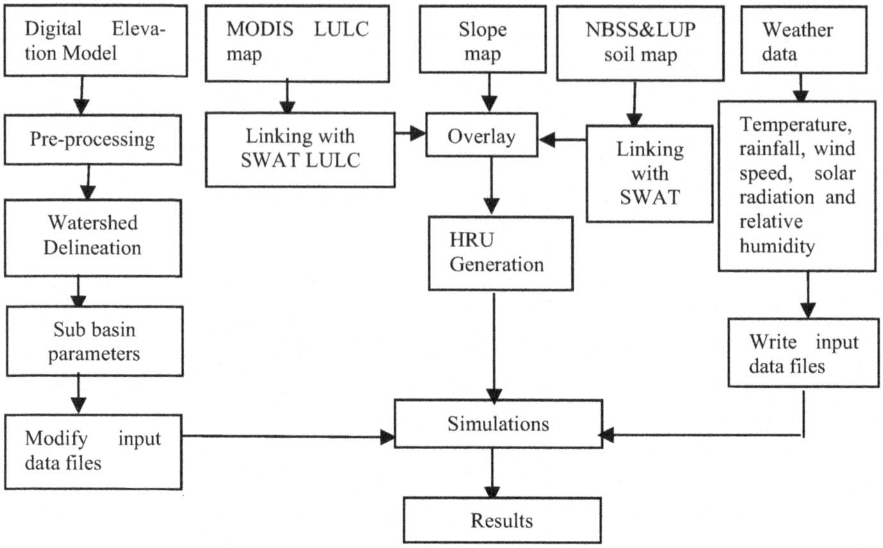

Fig. 15.5 Flowchart of rainfall–runoff analysis for Beas River basin using ArcSWAT

$$R^2 = \left[\frac{\sum_{i=1}^{n} \left(O_i - \overline{O}\right) - \left(P_i - \overline{P}\right)}{\sqrt{\sum_{i=1}^{n} \left(O_i - \overline{O}\right)\sqrt{\sum_{i=1}^{n} \left(P_i - \overline{P}\right)}}} \right] \tag{15.1}$$

where O is observed and P is predicted stream flow.

(b) **Nash–Sutcliffe efficiency (NSE)**

NSE indicates how well the plot of the observed value vs. the simulated value fits the 1:1 line. The value of NSE ranges between $-\infty$ and 1. There is a better model performance when NSE value is closer to 1 (Moriasi et al. 2007).

$$\mathrm{NSE} = 1 - \left[\sum_{(I=1)}^{N} \left(Q_{\mathrm{obs}} - Q_{\mathrm{sim}}\right)^2 \Big/ \sum_{(I=1)}^{N} \left(Q_{\mathrm{obs}} - Q_{\mathrm{obs}}^{\mathrm{mean}}\right)^2 \right] \tag{15.2}$$

Q_{obs} and Q_{sim} are the observed and simulated stream flows.

(c) **Root mean square error (RMSE)**

RMSE is measure of difference between simulated and observed discharge from model (Moriasi et al. 2007). It is expressed as:

$$\text{RMSE} = \sqrt{1/N \sum \overline{(Q_{\text{obs}} - Q_{\text{sim}})}^{2}} \tag{15.3}$$

Q_{obs} and Q_{sim} are the observed and simulated stream flows.

15.4 Results and Discussions

For Beas River basin, only three rain gauge stations are possible for ERA-INTERIM data, whereas totally eight rain gauge stations are available for IMD. The total amount of ERA-INTERIM rainfall is 60% more than rainfall obtained by rain gauge stations. Figure 15.6 shows the monthly average rainfall of rain gauge and ERA-INTERIM for years 2001–2004.

The proficiency of ERA-INTERIM data is tried to be proved by comparing the discharge using ERA-INTERIM data of three stations, eight rain gauge stations' data and three rain gauge stations' data. Observed discharge data is available at Thalot. Figure 15.7b shows the discharge comparison by ERA-INTERIM data for three stations, three rain gauge stations' data and observed data from 2001 to 2004.

From Fig. 15.7a, we conclude that the discharge obtained by ERA-INTERIM is better than discharge obtained by three rain gauge stations' data.

Moreover, in this research a small attempt has made to estimate the discharge using different number of rain gauge station data. It is clear from Fig. 15.7a that discharge obtained using eight rain gauge stations' data is more than three rain gauge stations' data. There is a decrease in discharge when the number of rain gauge station reduced. Moreover, the discharge of ERA-INTERIM using three rain gauge stations is close to the discharge of eight rain gauge stations. Thus, the suitability of ERA-INTERIM is proved for a data-scarce Beas River basin with less number and less scattered rain gauge stations.

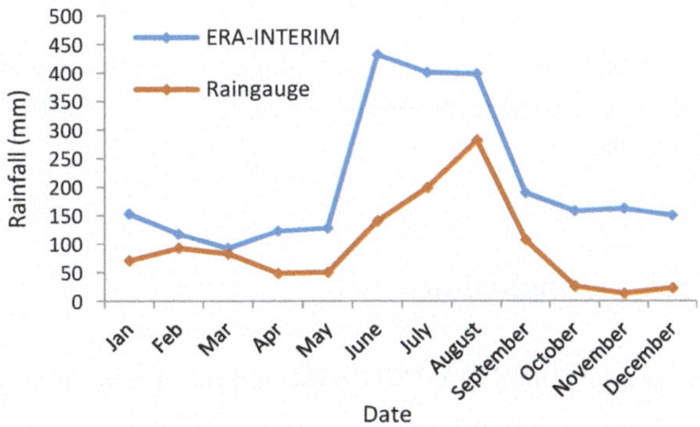

Fig. 15.6 Difference between monthly rainfall of ERA-INTERIM and rain gauge station data

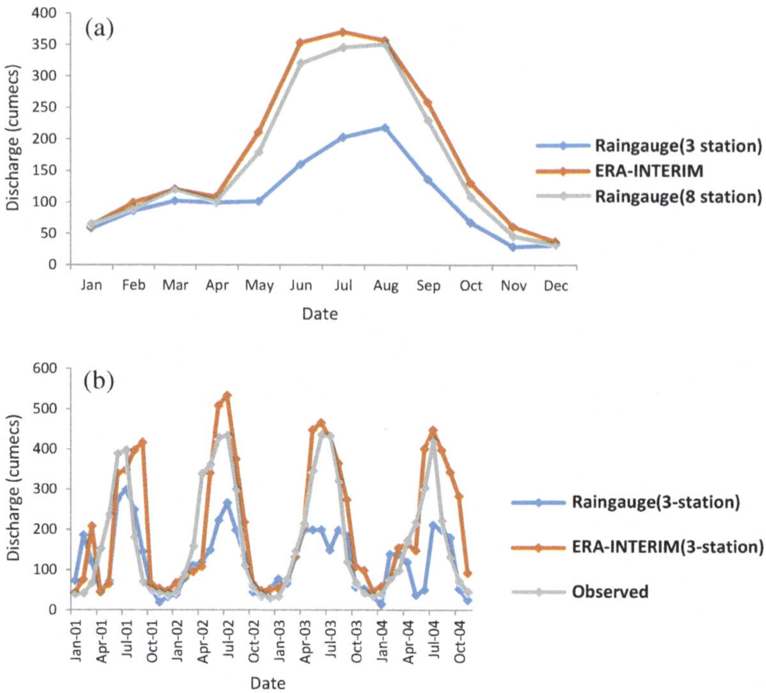

Fig. 15.7 Comparison of discharge using ERA-INTERIM and rain gauge data—**a** monthly average of 2001–2004 and **b** discharge for years 2001–2004

Table 15.1 RMSE, NSE and R^2 of discharge using ERA-INTERIM and rain gauge data

Data	RMSE	NSE	R^2
Eight rain gauge	103.34	0.64	0.68
Three rain gauge	160.25	0.55	0.53
ERA-INTERIM data	95.45	0.70	0.73

RMSE, NSE and R^2 are calculated for discharge data obtained by rain gauge and ERA-INTERIM with observed discharge to prove the accuracy of data is shown in Table 15.1.

15.5 Conclusions

After analyzing ERA-INTERIM and rain gauge data, we can conclude that

1. ERA-INTERIM data predicts better discharge data than IMD rain gauge data of three stations.
2. Increase of discharge data is due to increase of rain gauge stations.

3. ERA-INTERIM of three stations can be used as a substitute of IMD eight rain gauge stations' data.
4. ERA-INTERIM data is predictable for basin with less number of rain gauges. This is due to its better spatial distribution of rainfall and fine resolution.

References

Aggarwal SP, Thakur PK, Garg V, Nikam B, Chouksey A, Dhote P, Bhattacharya T (2016) Water resources status and availability assessment in current and future climate change scenarios for Beas river basin of North Western Himalaya. Int Arch Photogram Remote Sens Spat Inf Sci (XLI-B8):1389–1396

Berrisford P, Dee DP, Fielding K, Fuentes M, Kallberg P, Kobayashi S, Uppala SM (2009) The ERA-Interim Archive. ERA Report Series, No. 1. ECMWF: Reading, UK

Dee DP et al (2011) The ERA-Interim reanalysis: configuration and performance of the data assimilation system. Q J Royal Meteorol Soc 137:553–597

Dile Y, Srinivasan R (2014) Evaluation of CFSR climate data for hydrologic prediction in data scarce watersheds: an application in the Blue Nile River basin. J Am Water Resour Assoc 13(0074):1–16

Fuka DR, Walter MT, MacAlister C, Degaetano AT, Steenhuis TS, Easton ZM (2014) Using the climate forecast system reanalysis as weather input data for watershed models. Hydrol Process 28(22):5613–5623

Khare D, Singh R, Shukla R (2014) Rainfall-runoff modelling of Ajay river catchment using SWAT model. Int J Geol Earth Environ Sci 4:224–235

Kumar V, Singh P, Singh V (2007) Snow and glacier melt contribution in the Beas River at Pandoh Dam, Himachal Pradesh, India. Hydrol Sci 52(2):376–388

Lakhera N, Kumar N, Shankar AV (2016) Water balance study of Beas River, Himachal Pradesh, using ArcGIS technique. In: Proceedings of national conference: civil engineering conference—innovation for sustainability

Moriasi D, Arnold J, Van Liew M, Bingner R, Harmel R, Veith T (2007) Model evaluation guidelines for systematic quantification of accuracy in watershed simulations. Trans ASABE 50(3):885–900

Thakur S, Kant D, Hardaha MK, Sharma SK (2016) Land use/Land cover change in a hydrologic regime using SWAT model—a review. Int J Sci Eng Technol Res (IJSETR) 5(1):218–223

USDA (1972) National engineering handbook hydrology chapters. Soil Conservation Service. National engineering handbook, Section 15.4. Hydrology. Washington, D.C.

Ward E, Buytaert W, Peaver L, Wheater H (2011) Evaluation of precipitation products over complex mountains terrain: a water resources perspective. Adv Water Resour 34(10):1222–1231

Zhao TB, Fu CB, Zongjian K, Weidong G (2010) Global atmosphere reanalysis datasets: current status and recent advances. Adv Earth Sci 25(3):242–254 (in Chinese)

Zhu C, Li Y (2014) Long-term hydrological impacts of land use/land cover change from 1984 to 2010 in the little river watershed, Tennessee. Int Soil Water Conserv Res 2(2):11–22

Chapter 16
Impact of Land Use and Land Cover Change on Streamflow of Upper Baitarani River Basin Using SWAT

Raunak M. Prusty, Aparna Das, and Kanhu Charan Patra

Abstract Any changes in the land use and land cover pattern in a watershed can severely impact the hydrological properties of the watershed by changing its streamflow and groundwater characteristics which may lead to further changes in LULC pattern in the area. Hence, it is essential to assess the impact of LULC changes on various hydrological aspects of a watershed. Such an analysis can be carried out by using a simulated hydrological model of the watershed which integrates the GIS data of the area. This study deals with the impact of LULC change on the hydrology of the upper Baitarani river basin. The hydrological model of the study area was simulated in Soil and Water Assessment Tool (SWAT) on a daily time step concurrently from the year 1979 to 2013 using the LULC data of 1995 and 2013 as well as slope and soil data with a warm-up period of five years. The sensitivity analysis is carried out to find the most sensitive parameters which influence the streamflow of the basin. Calibration and validation are done to gauge the performance of the simulated model using global sensitivity approach of Sequential Uncertainty Fitting (SUFI2) algorithm in SWAT Calibration and Uncertainty Procedure (SWAT-CUP) software. Streamflow obtained from the model using 1995 LULC is calibrated from the year 1986 to 1992 and validated from the year 1993 to 1995 using observed streamflow data at Anandapur hydrological gauge station. Similarly, the model simulated using 2013 LULC is calibrated from the year 2004 to 2010 and validated from the year 2011 to 2013. The objective functions in terms of R^2, NSE and PBIAS in SWAT-CUP show that the SWAT model has worked satisfactorily. The analysis of LULC change in eighteen years from 1995 to 2013 shows that the area of forest cover, barren land, and agricultural land has decreased by 0.59%, 1.62% and 6.18%, respectively. In contrast, the built-up area has increased by 8.26% during the same duration. The

R. M. Prusty · A. Das (✉) · K. C. Patra
Department of Civil Engineering, National Institute of Technology, Rourkela, Odisha, India
e-mail: aparnadas.civil.engineer@gmail.com

R. M. Prusty
e-mail: 50raunak@gmail.com

K. C. Patra
e-mail: kcpatra@nitrkl.ac.in

A. Pandey et al. (eds.), *Water Management and Water Governance*, Water Science and Technology Library 96,
https://doi.org/10.1007/978-3-030-58051-3_16

comparison shows that due to land use and land cover change in the Upper Baitarani basin, the mean annual streamflow has increased by 3.78%.

Keywords Land use · Land cover · Hydrological modeling · SWAT · SWAT-CUP · SUFI2 algorithm

16.1 Introduction

Land use and land cover (LULC) change has an adverse impact on the environment (Ndulue et al. 2015) as well as on the hydrology of river basin, leading to changes in the river regime changes due to the temporal variation of runoff distribution (Kashaigili et al. 2008). In developing countries like India, agriculture is the primary sector for economic progress (Himani 2014). On the other hand, the rapid growth of population (Kulkarni et al. 2014) is a significant issue for India. To prepare the land for agriculture and housing facilities, the existing land cover gets continuously changed. As a result, the land changes its characteristics which influences the hydrological response of a watershed (Welde and Gebremariam 2017).

The Upper Baitarani basin, which is one of the most crucial water contributors in Odisha state, has undergone many transformations of LULC (Uniyal et al. 2015). SWAT is a river basin simulation model that works on a daily time interval. It is a useful tool for hydrologic modeling as long-term hydrological responses can be quantified through SWAT in very less time. SWAT model components include weather, hydrology, erosion/sedimentation, plant growth, nutrients, pesticides, agricultural management, stream routing and pond/reservoir routing for proper watershed modeling (Arnold and Fohrer 2005). The present study aims to assess the impact of the land use and land cover change on streamflow for the period 1995–2013 using SWAT.

16.2 Study Area

Baitarani River originates from the Guptaganga hill of Keonjhar District in Odisha, India, at an elevation of about 900 m above mean sea level and after travelling 360 km finally drains into the Bay of Bengal. About 93.7% of the total area of Baitarani river basin falls under the state of Odisha, and the remaining 6.7% falls under the state of Jharkhand. The basin drains an area of 10,982 km^2.

Anandapur gauge station (Fig. 16.1) divides the catchment area into upper and lower reaches. The study area can be located approximately between longitudes of 85°10′ E–86°23′ E and latitudes of 20°53′ N–22°15′ N and covers an area about 8619 km^2. Anandapur is situated at Keonjhar District of Odisha. The weather is tropical with average annual precipitation of 1442.53 mm. The average annual maximum temperature is 37.67 °C, and the annual minimum temperature is 20.32 °C. During

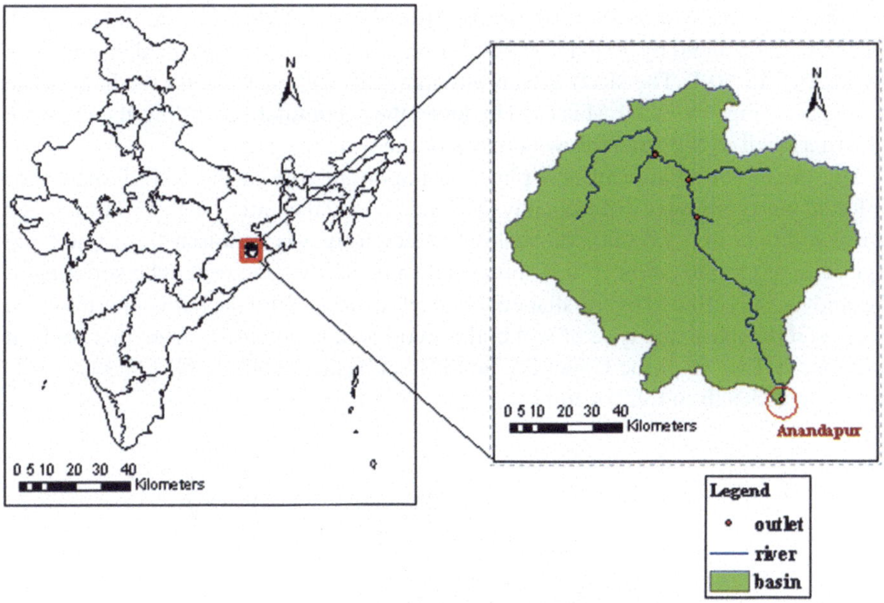

Fig. 16.1 Location of the study area

summer, the months of April and May are among the hottest months of the year when the temperature varies between 35 and 38 °C. The wind velocity is higher in April, May and June, whereas it is lesser in December and January (CWC 2014).

16.3 Input Data

16.3.1 Meteorological Data

The meteorological parameters such as temperature, precipitation and wind speed greatly influence the hydrologic cycle. That is why for SWAT model generation, meteorological data is essential. Meteorological data from 1979 to 2013 is acquired from the National Centers for Environmental Prediction (NCEP). The data is collected for six stations inside the study area.

16.3.2 GIS Data

Terrain data. As water flows from high elevation to low elevation, terrain data is necessary for hydrological modeling. The digital elevation model (DEM) of Baitarani river basin is obtained in the form of tiles from CIGAR-CSI Web site (https://srtm.csi.cgiar.org) with 90 by 90 m DEM resolution. These tiles are mosaiced into a single tile in ArcGIS 10.1. The study area is automatically delineated with the help of the outlet point (Fig. 16.2). DEM is used to depict the accumulation of flow, the drainage pattern as well as the streamflow network of the watershed area.

Soil data. The texture of soil plays an important role in the SWAT model, as different kind soils have different physical and chemical characteristics. Harmonized world soil data in a 1:5,000,000 scale in vector format is acquired from Food and Agricultural Organization (FAO) soil portal (https://www.fao.org). The soil map of the study area is clipped by the shapefile in ArcGIS 10.1. With the help of harmonized world soil database, the type of soil in the study area is obtained. From this study, it is observed (Fig. 16.3) the Upper Baitarani River basin contains I-Ne (28.38%), I-bc (6.65%), Nd50-2b (64.92%) and Lf95-1a (0.05%).

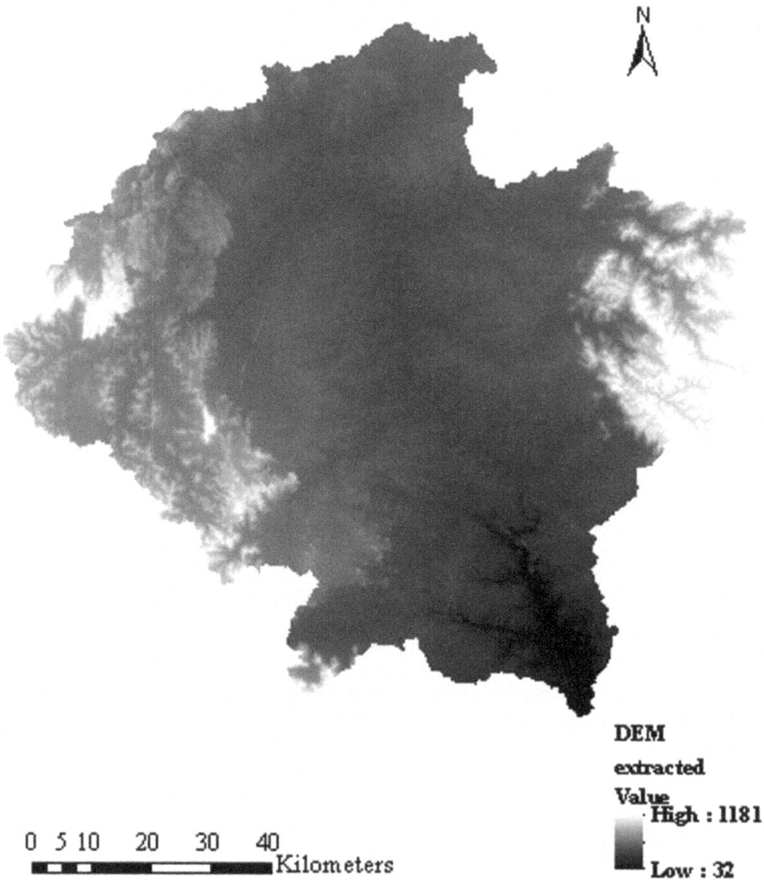

Fig. 16.2 Digital elevation model of Upper Baitarani basin

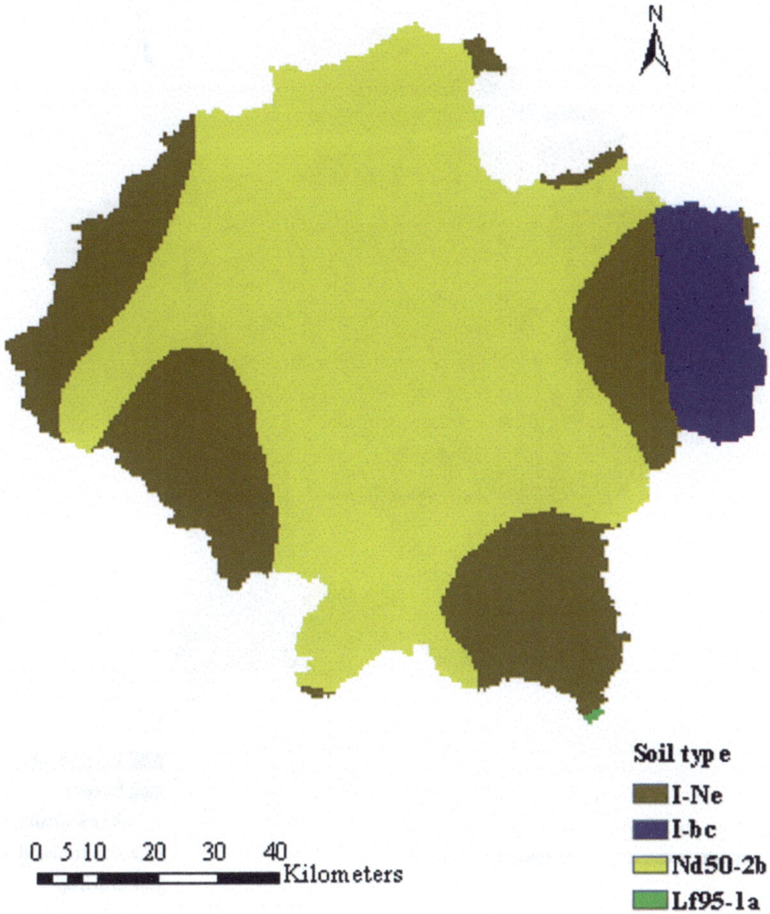

Fig. 16.3 Soil map of Upper Baitarani basin

Land use and land cover data. Land use is defined as the expenditure of land to produce goods and services. Hence, land use is dependent on the purpose for which the land is being used, whereas land cover means natural physical cover, as seen by eyes or remote sensing (Ndulue et al. 2015). LULC maps are prepared using USGS LANDSAT satellite imagery (https://earthexplorer.usgs.gov/). Using path 139 and 140 and row 45, LANDSAT 5 and 8 satellite images for 1995 and 2013 (spatial resolution 30 m) are acquired in the form of a tile. Then in ArcGIS 10.1, the tiles of both years are mosaiced individually. Using the shapefile of the study area, the mosaiced images are clipped separately. Each clipped image is classified into five categories of land use cover with the help of maximum likelihood algorithm in supervised classification tool of ArcGIS (Figs. 16.4 and 16.5). The five classes are water body, forest, agriculture, barren land and built-up area. Table 16.1 represents the quantified change of land use cover categories.

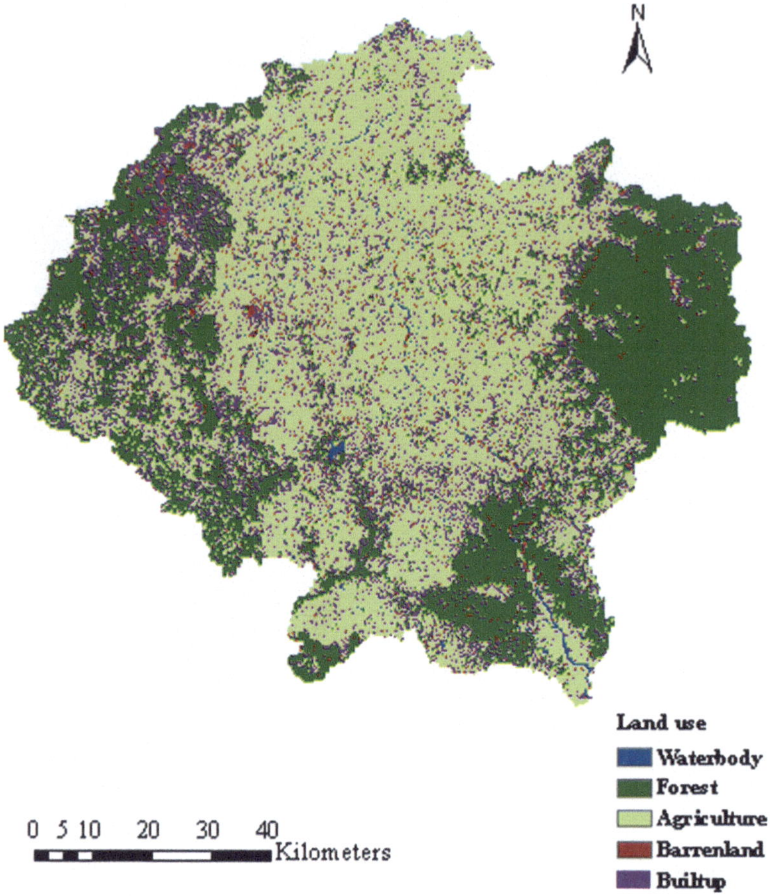

Fig. 16.4 Land use and land cover map for 1995

16.3.3 Hydrological Data

Daily discharge data (1979–2013) of Baitarani River at Anandapur stream gauge station (21°12′40″ N, 86°7′14″ E) is obtained from the Water Resource Information System of India (https://india-wris.nrsc.gov.in/wris.html). This data is further used for calibration and validation purposes of the SWAT model.

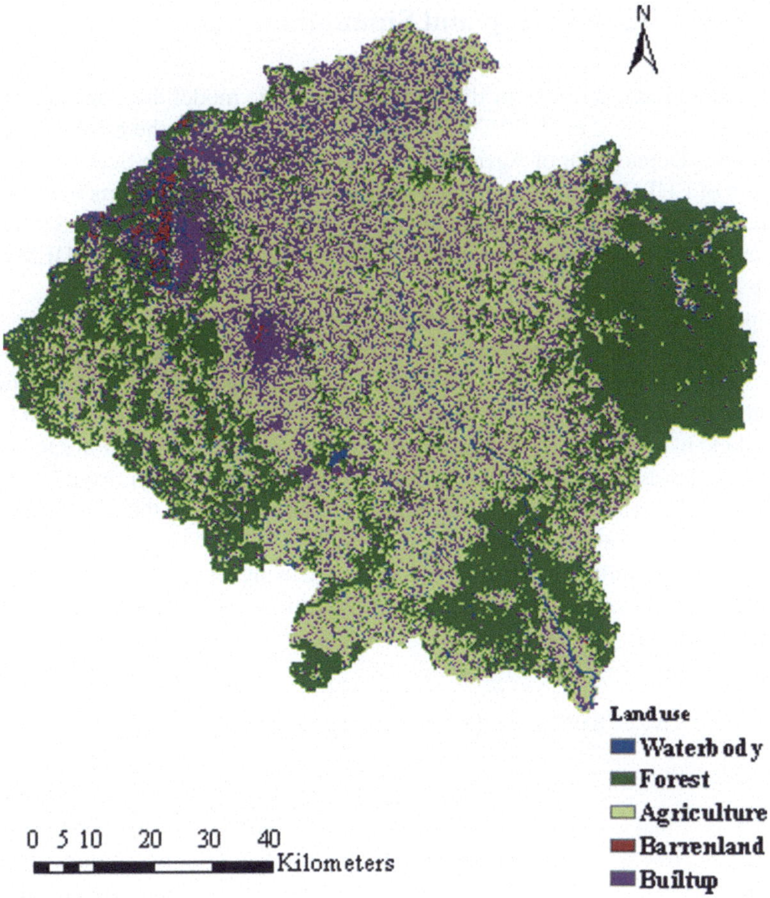

Fig. 16.5 Land use and land cover map for 2013

Table 16.1 Land use and land cover in Upper Baitarani basin

Id	Land cover specification	1995		2013		Change
		Area (km^2)	%	Area (km^2)	%	%
1	Water	82.74	0.96	93.95	1.09	+0.13
2	Forest	2773.59	32.18	2722.74	31.59	−0.59
3	Agriculture	4274.16	49.59	3741.51	43.41	−6.18
4	Barren land	178.42	2.07	38.79	0.45	−1.62
5	Built-up	1310.09	15.20	2022.01	23.46	+8.26
	Total area	8619	100	8619	100	–

16.4 SWAT Model Setup and Simulation

SWAT is a comprehensive, semi-distributed hydrologic model that runs at a continuous time step (Arnold et al. 2012). Dr. Jeff Arnold developed the SWAT model for United States Department of Agriculture—Agricultural Research Service to predict the impact of LULC changes and management on various components of a complex watershed under varying slope and soil conditions (Neitsch et al. 2009). SWAT uses spatial and temporal data as input that includes digital elevation model (DEM), land use map and soil map.

Temporal data includes hydrological data (streamflow) and weather data (precipitation, temperature, wind speed, relative humidity and solar radiation). SWAT divides a basin into multiple sub-basins depending on the topographic criteria and further subdivided into a number of hydrologic response units (HRUs) on the basis of homogeneous land use, soil and slope combination characteristics. Simulation can be done for different kinds of hydrological (runoff, water quality, evapotranspiration, etc.) and agricultural variables (crop yield, nutrients, etc.). Based on the requirement and available data, user can easily simulate the variables in SWAT.

Hydrological simulation of the basin is executed using ArcGIS extension of SWAT called ArcSWAT (Arnold et al. 2012). The SWAT model is developed and refined by the water balance Eq. (16.1) given as

$$\mathrm{SW}_t = \mathrm{SW}_o + \sum_{i=1}^{t}\left(R_{\mathrm{day}} - Q_{\mathrm{surf}} - E_a - W_{\mathrm{seep}} - Q_{gw}\right) \tag{16.1}$$

where SW_t = Final soil water content in mm, SW_o= Initial soil water content on day i (mm), R_{day}= Amount of precipitation on the day i (mm), Q_{surf} = Amount of surface runoff on day i (mm), E_a = Amount of evapotranspiration on day i (mm), W_{seep} = Amount of water entering the vadose zone from soil profile on day i (mm) and Q_{gw}= Amount of return flow on day i (mm). In this study, ArcSWAT v2012.10.1.18 (https://swat.tamu.edu/software/arcswat) is used. To set up the SWAT model, the study area is automatically delineated with respect to the outlet at Anandapur stream gauge station (21°12′40″ N, 86°7′14″ E) (CWC 2014). The Upper Baitarani basin has seven sub-basins and that were further subdivided into 270 HRUs based on input soil and land use/cover data. Then the model is simulated from 1979 to 2013 on a daily time step with a warm-up period of five years. The flowchart of complete methodology has been shown in Fig. 16.6.

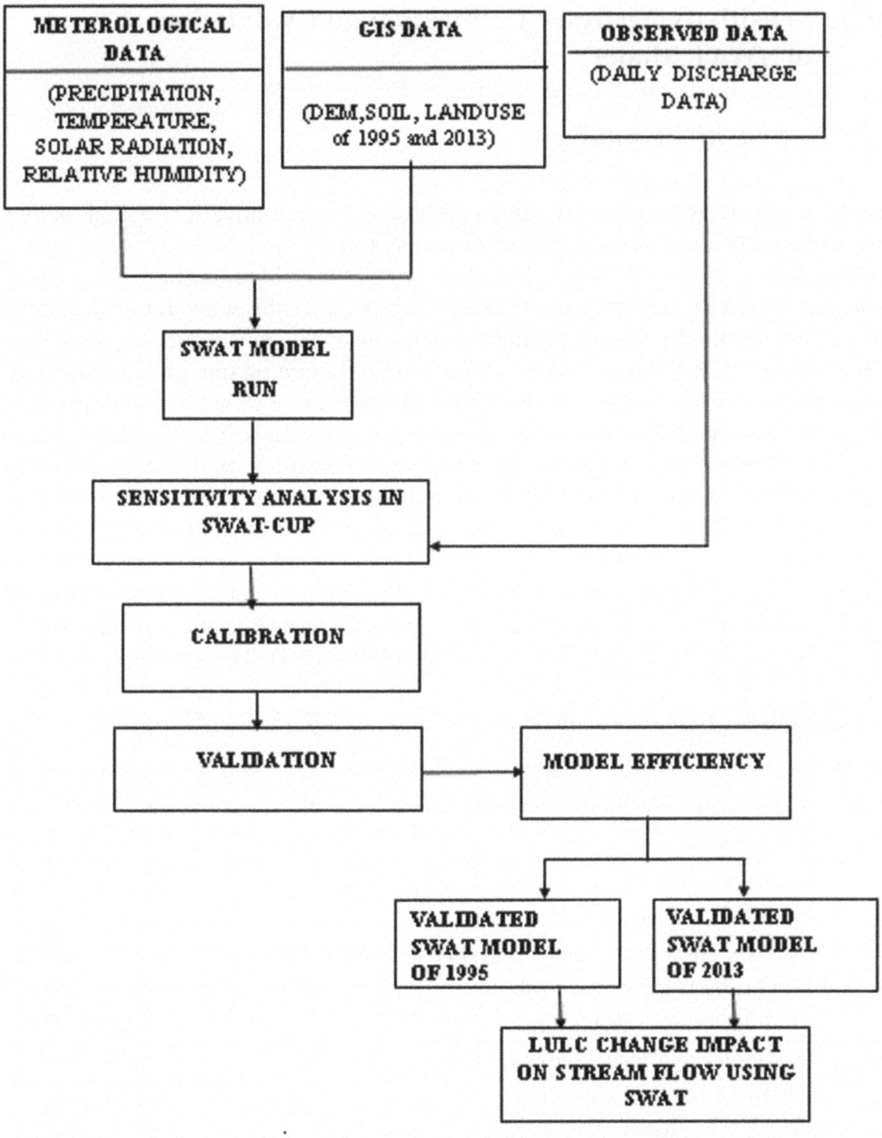

Fig. 16.6 Flowchart of methodology

16.5 Sensitivity Analysis, Calibration and Validation of SWAT Model

16.5.1 Sensitivity Analysis

Sensitivity analysis identifies the main parameters which affect the SWAT output flow using global sensitivity approach of Sequential.

Uncertainty Fitting (SUFI2) algorithm in SWAT Calibration and Uncertainty Procedure (SWAT-CUP) (Abbaspour et al. 2015). Here, in this work, fourteen parameters are taken initially to examine their sensitiveness. The range of these parameters is set from SWAT-CUP user manual. After an initial iteration run, global sensitivity approach checks the sensitivity of one parameter relative to another and arranges the parameters by ranks according to their *t*-stat and *p*-values. The allowable ranges and the best-fitted values are obtained for sensitivity analysis. In this study, fourteen streamflow influencing parameters are tested for sensitivity (Table 16.2); of these fourteen parameters, six parameters were found to be best fitted (Table 16.3).

The second case in which 2013 land use scenario model is used, the results shows that out of fourteen parameters (Table 16.4), only five parameters are very sensitive to streamflow. Those fourteen parameters are presented according to their sensitiveness measured by *p*-value and *t*-stat value (Table 16.4). The sensitivity analysis indicates, in this case, the most sensitive parameter for the streamflow is effective hydraulic conductivity in main channel alluvium (CH_K2). The other sensitive

Table 16.2 Sensitivity analysis of streamflow parameters for 1995 scenario model

S.no	Parameter name	*t*-stat	*p*-value
1	ALPHA_BNK (Base-flow alpha factor for bank storage)	14.535	0.000
2	CN2 (SCS runoff curve number)	10.966	0.000
3	SOL_AWC (Available water capacity of the soil layer)	−4.673	0.000
4	CH_K2 (Effective hydraulic conductivity in main channel alluvium)	− 3.086	0.002
5	SOL_K (Saturated hydraulic conductivity)	3.020	0.002
6	ESCO (Soil evaporation compensation factor)	−2.035	0.042
7	SLSUBBSN (Average slope length)	1.668	0.095
8	GW_REVAP (Groundwater "revap" coefficient)	−1.556	0.120
9	CH_N2 (Manning's "n" value for the main channel)	−1.301	0.193
10	REVAPMN (Threshold depth of water for revap or percolation to occur)	−1.129	0.259
11	GWQMN (Threshold depth of water in the shallow aquifer required for return flow to occur)	0.983	0.326
12	HRU_SLP (Average slope steepness)	0.816	0.414
13	ALPHA_BF (Base-flow alpha factor)	0.707	0.479
14	GW_DELAY (Groundwater delay)	0.397	0.691

Table 16.3 Ranges and best-fitted values of flow calibration parameters for 1995 model scenario

Rank	Parameter name	Fitted value	Minimum value	Maximum value
1	v_ALPHA_BNK.rte	0.949	0.000	1.000
2	r_CN2.mgt	0.150	−0.200	0.200
3	r_SOL_AWC.sol	0.063	−0.200	0.400
4	v_CH_K2.rte	117.625	5.000	130.000
5	r_SOL_K.sol	− 0.782	−0.800	0.800
6	v_ESCO.hru	0.875	0.800	1.000

Table 16.4 Sensitivity analysis of streamflow parameters for 2013 scenario model

S. no	Parameter name	*t*-stat	*p*-value
1	CH_K2 (Effective hydraulic conductivity in main channel alluvium)	9.319	0.000
2	CN2 (SCS runoff curve number)	6.321	0.000
3	ALPHA_BNK (Base-flow alpha factor for bank storage)	5.866	0.000
4	SOL_K (Saturated hydraulic conductivity)	3.320	0.000
5	GW_REVAP (Groundwater "revap" coefficient)	−2.022	0.044
6	REVAPMN (Threshold depth of water for revap or percolation to occur)	1.877	0.061
7	GWQMN (Threshold depth of water in the shallow aquifer required for return flow to occur)	−1.583	0.114
8	HRU_SLP (Average slope steepness)	1.347	0.179
9	CH_N2 (Manning's "n" value for the main channel)	1.273	0.204
10	SOL_AWC (Available water capacity of the soil layer)	0.963	0.334
11	ESCO (Soil evaporation compensation factor)	−0.912	0.362
12	ALPHA_BF (Base-flow alpha factor)	0.654	0.514
13	GW_DELAY (Groundwater delay)	0.508	0.612
14	SLSUBBSN (Average slope length)	−0.491	0.624

parameters are SCS runoff curve number (CN2), base-flow alpha factor for bank storage (ALPHA_BNK), saturated hydraulic conductivity (SOL_K) and groundwater evaporation coefficient (GW_REVAP). Rest eight parameters are not found to be sensitive to streamflow in the catchment as their p-values are greater than 5% (Anaba et al. 2017). The best-fitted values and ranges of these five most sensitive parameters are presented in (Table 16.5).

The extension (e.g., .mgt, .rte) refers to the SWAT input text file where the parameters are selected. The qualifier *r*_refers to the relative change in the parameter where the value from the SWAT database is multiplied by 1 plus a factor in the given range. Parameter *v*_refers to the substitution of a parameter by a value from the given range.

Table 16.5 Ranges and best-fitted values of flow calibration parameters

Rank	Parameter name	Fitted value	Minimum value	Maximum value
1	v_CH_K2.rte	117.625	5.000	130.000
2	r_CN2.mgt	0.070	−0.200	0.200
3	v_ALPHA_BNK.rte	0.097	−0.800	0.800
4	r_SOL_K.sol	0.463	0.000	1.000
5	v_GW_REVAP.gw	0.016	0.000	0.200

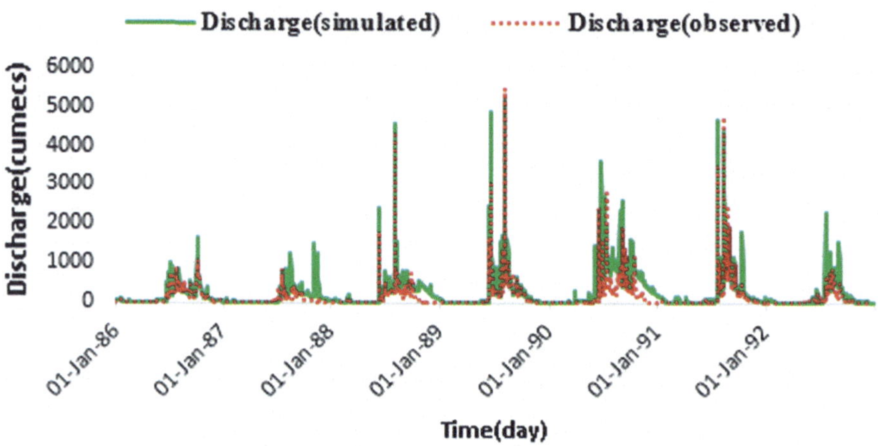

Fig. 16.7 Observed and simulated daily discharge during calibration period (1986–1992) for 1995 LU model

16.5.2 SWAT Model Calibration and Validation Result

Graphical representation of streamflow indicates the calibrated (Fig. 16.7) and validated (Fig. 16.8) results for 1995 LU model. It also depicts the calibration results showing a good match as compared to the validated streamflow results. Besides, the statistical objective functions in SWAT-CUP that evaluates the model performance are good (Table 16.6).

Similarly for 2013 LU model after the calibration (Fig. 16.9) and validation (Fig. 16.10), the graphical and objective functions indicate the 2013 LU model is equally good (Table 16.7).

16.5.3 Modeling Streamflow Response to Land Use Dynamics

Different scenarios are established to assess the impact of land use change on streamflow by SWAT 2012. The one factor at a time approach (Li et al. 2009) is considered.

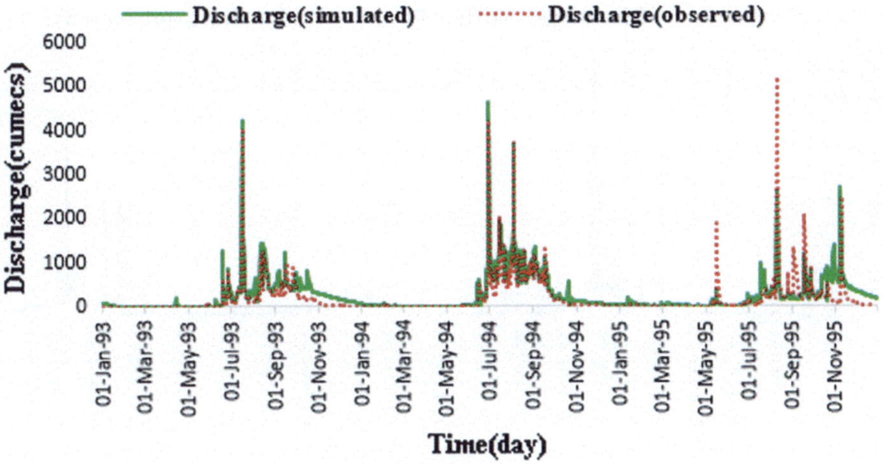

Fig. 16.8 Observed and simulated daily discharge during validation period (1993–1995) for 1995 LU model

Table 16.6 Calibration and validation of SWAT model using 1995 LU model

Stage of model	Statistical parameters		
	R^2	NSE	PBIAS
Calibration (1986–1992)	0.700	0.661	−15.0
Validation (1993–1995)	0.677	0.623	−17.0

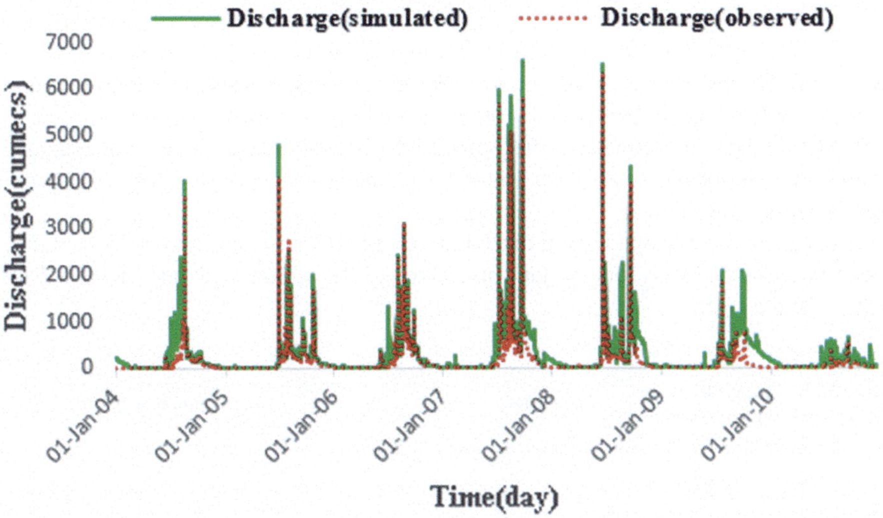

Fig. 16.9 Observed and simulated daily discharge during calibration period (2004–2010) for 2013 LU model

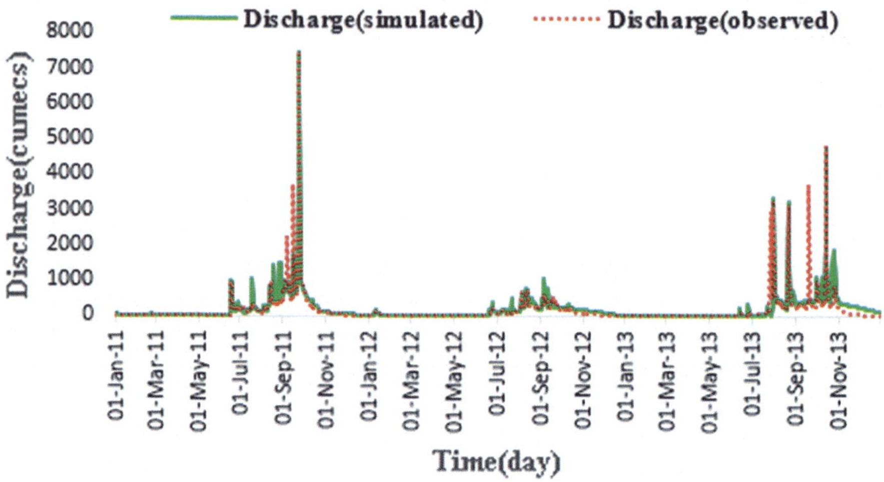

Fig. 16.10 Observed and simulated daily discharge during calibration period (2011–2013) for 2013 LU model

Table 16.7 Calibration and validation of SWAT model using 1995 LU model

Stage of model	Statistical parameters		
	R^2	NSE	PBIAS
Calibration (2004–2010)	0.685	0.621	−14.0
Validation (2011–2013)	0.637	0.611	−17.0

According to this approach, to analyze the effect of land use on streamflow, one factor at each time is changed while other factors are to be kept constant.

Land use data and climate of the years 2008–2013 were considered for scenario I; similarly, land use data of 1995 along with climate of the years 2008–2013 were considered for scenario II.

Comparing the average annual discharge due to 1995 land use and 2013 land use, it can be seen that only due to land use change, the streamflow has increased by 3.78% (Table 16.8).

Table 16.8 Streamflow reaction to land use change

Scenario	Mean annual streamflow (m^3/s)
I	229.69
II	221.33
Change (I–II) with respect to scenario II	+3.78%

16.6 Conclusion

The study shows that the Upper Baitarani River basin has undergone a significant change over eighteen years (1995–2013). Forest cover, agriculture and barren land have decreased significantly, whereas built-up area has increased due to urbanization.

In this study, the SWAT model is used to assess the impact of land use and land cover change on streamflow in Upper Baitarani basin. After calibration and validation, the objective function indicates in both the runs that the SWAT model has simulated streamflow results satisfactorily. This indicates the impact of LULC change is evaluated using SWAT simulated model of the Upper Baitarani River basin. Using two land use data for 1995 and 2013 under a constant climate, it is established that only due to land use change, annual average streamflow has increased by 3.78%. From this study, it can be concluded that there is a direct relationship between land use and land cover and streamflow as a change in land use can change the streamflow pattern. That is why for effective watershed management, the streamflow should be continuously assessed.

References

Abbaspour KC, Rouholahnejad Freund E, Ashraf Vaghefi S, Srinivasan R, Yang H, Klöve B (2015) A continental-scale hydrology and water quality model for Europe: calibration and uncertainty of a high-resolution large-scale SWAT model. J Hydrol, 524. https://doi.org/10.1016/j.jhydrol.2015.03.027

Anaba LA, Banadda N, Kiggundu N, Wanyama J, Engel B, Moriasi D (2017) Application of SWAT to assess the effects of land use change in the Murchison Bay catchment in Uganda. Comput Water Energy Environ Eng 20:24–40

Arnold JG, Fohrer N (2005) SWAT2000: current capabilities and research opportunities in applied watershed modeling. Hydrol Process 19:563–572

Arnold JG, Moriasi DN, Gassman PW, Abbaspour KC, White MJ, Srinivasan R, Santhi C, Harmel RD, Griensven AV, Van Liew MW, Kannan N, Jha MK (2012) SWAT: model use, calibration and validation. Am Soc Agric Biol Eng 55(4):1491–1508

Central Water Commission, Brahmaniand-Baitarini basin report, 2014 https://www.india-wris.nrsc.gov.in/Publications/BasinReports

Himani (2014) An analysis of agriculture sector in Indian Economy. IOSR J Humanit Soc Sci 19:47–54

Kashaigili JJ (2008) Impacts of land-use and land-cover changes on flow regimes of the Usangu Wetland and the Great Ruaha River, Tanzania. Phys Chem Earth 33:640–647

Kulkarni SS, Kulkarni SR, Patil SJ (2014) Analysis of population growth in India and estimation for future. Int J Innovative Res Sci Eng Technol 3(9):15843–15850

Li Z, Wen-Zhao L, Zang X, Zeng F (2009) Impacts of land-use change and climate variability on hydrology in an agricultural catchment on the Loess Plateau of China. J Hydrol, 272–288

Neitsch SL, Arnold JG, Kiniry JR, William JR (2009) Soil and water assessment tool theoretical documentation. AgriLife Research & Extension

Ndulue EL, Mbajiorgu CC, Ugwu SN, Ogwo V, Ogbu KN (2015) Assessment of land use/cover impacts on runoff and sediment yield using hydrologic models: a review. J Ecol Nat Environ (Academic Journals) 7:46–55

Uniyal B, Jha MK, Verma AK (2015) Assessing climate change impact on water balance components of a River Basin using SWAT model. Water Resour Manag 29

Welde K, Gebremariam B (2017) Effect of land use land cover dynamics on hydrological response of watershed: case study of Tekeze Dam Watershed, Northern Ethiopia. Int Soil Water Conserv Res 5:1–16

Williams JR, Nicks AD, Arnold JG (1985) Simulator for water resources in rural basins. J Hydraul Eng, 970–986

Chapter 17
Parameter Estimation of a Macroscale Hydrological Model Using an Adaptive Differential Evolution

Saswata Nandi and Manne Janga Reddy

Abstract To overcome the drawbacks faced by the traditional manual calibration of hydrological models, this study employs an adaptive differential evolution (DE) algorithm for automatic calibration of Variable Infiltration Capacity (VIC) hydrological model. In the DE algorithm, proper tuning of its control parameters is laborious and generally needs a great amount of time and resources. Therefore, a self-adaptive scheme is presented to enhance the efficacy of the basic DE. The proposed automatic parameter estimation scheme is applied for a case study and evaluated its performance using standard performance measures of coefficient of correlation (R^2), Nash–Sutcliffe coefficient (NSE), percent bias (PBIAS), and index of agreement (IoA). The findings from the study revealed that the adaptive DE was successful to optimize the unknown parameters of the VIC model accurately, which signified that the automatic calibration scheme is a credible alternative to the manual approach.

Keywords Hydrological modeling · Variable infiltration capacity model · Calibration · Differential evolution

17.1 Introduction

Several hydrological models are being formed and implemented for various hydrological studies across the globe over the last half-century. With the advancement of modern computational systems from the late 1960s, hydrologists focused on developing complex physically based hydrological models that are capable of simulating the movement of water using physical laws. Such complex hydrological models have been typically parameterized by extracting various catchment characteristics such as

S. Nandi · M. Janga Reddy (✉)
Department of Civil Engineering, Indian Institute of Technology Bombay, Mumbai 400076, India
e-mail: mjreddy@civil.iitb.ac.in

S. Nandi
e-mail: ce.saswata@gmail.com

A. Pandey et al. (eds.), *Water Management and Water Governance*, Water Science and Technology Library 96,
https://doi.org/10.1007/978-3-030-58051-3_17

land cover, soils, slope, and geomorphology. The model structure and its parameterization scheme determine the dependability of model predictions. But there are few parameters in every hydrological model that can never be explicitly determined from the field measurements, which introduces large uncertainties in determining appropriate parameter values. This requires hydrologist to perform the model calibration as calibration of a model helps to reduce parameter uncertainty and consequently the uncertainty in model simulation to improve the model performance. The most popular approach of hydrological model calibration is manual calibration, where unknown parameter values are estimated by adjusting the parameter values using trial and error method. Also, the performance of the calibrated model is assessed on the basis of visual findings by relating simulated and observed data, and some statistical performance measures for the purpose of supporting the result. Therefore, this calibration type is time-consuming and also subjective in nature. Because of the vast number of parameters in modern hydrological models and their inherent complexity in simulating hydrological processes, automated calibration methods are getting high popularity for parameter estimation, while reducing the overall calibration effort.

Automated calibration is especially useful when a model has multiple parameters that need to be estimated when there is no distinct connection between those parameters' values and the physical processes of the catchment and when different parameters are strongly inter-related (Ryu 2009). The practice of automated calibration method applied to physically based distributed hydrological modeling has become a growing research field (Madsen et al. 2002). Gradient-based local search methods (Johnston and Pilgrim 1976) and the population evolution-based global optimization techniques (Duan et al. 1992) have been extensively used for parameter estimation for hydrological models. Several past studies also have compared the local and global search techniques in rainfall-runoff modeling, from where a broad conclusion is achieved that the global optimization methods perform superior to local search techniques (Sorooshian et al. 1993; Madsen et al. 2002). Among different global optimization methods (Reddy and Kumar 2020), shuffled complex evolution method (SCE-UA), particle swarm optimization (PSO), and genetic algorithms (GA) are the frequently utilized optimization algorithms applied to hydrological model calibration (Duan et al. 1992; Jha et al. 2006; Jung and Karney 2006). The differential evaluation (DE) (Storn and Price 1997) is another kind of evolutionary algorithms which is simple yet effective and widely used in water resources systems for their operation, planning, and design. In spite of several merits, it is hardly used in the calibration of the hydrologic models (Zhang et al. 2009). Nevertheless, the skilful performance of DE relies on appropriate setting of its control parameters such as mutation factor (F) and crossover probability (CR), which are problem dependent and require a lot of computational effort in tuning those parameters. Since DE itself is a self-learning scheme, DE can evolve the control parameters (F and CR). Therefore, this study presents an innovative self-adaptation scheme for DE which will dynamically adjust the control parameters during its execution.

The main objective of the current study is to present self adaptive differential evolution (SaDE) algorithm for parameter estimation of the Variable Infiltration Capacity (VIC) model which is a macroscale hydrological model. The procedures of integrating the SaDE with VIC, the methods of dealing with the calibration of spatial model parameters, and specification of calibration and evaluation criteria are presented. The approach is illustrated for model calibration of the Sina river catchment, India, with 15 years observed flow hydrographs at the basin outlet as the calibration target. The remaining of this paper is structured in the following way: Sect. 17.2 presents the necessary steps in SaDE algorithm, a brief description of the VIC hydrological model and approach for combining SaDE with the VIC model. Section 17.3 introduces the study area and details regarding various data used in the study. Application and discussion of the results are presented in Sect. 17.4. Finally, the conclusions of the study are given in Sect. 17.5.

17.2 Methodology

17.2.1 SADE Method

Qin and Suganthan (2005) presented a self-adaptive approach to adopt the F and CR parameter values of the traditional DE algorithm. The present study adopted a modified version of Self-adaptive DE (SaDE) (Sirsant and Reddy 2018), whose details are summarized as follows:

(i) Generate initial population $\left(Z_{i,0} = \{z_{i,0}^1, z_{i,0}^2, \ldots, z_{i,0}^D\}\right)$ with random values within the upper and lower limits of the decision variables. Here, $z_{i,0}^j$ is the initial estimate of jth decision variable for ith individual.

(ii) Mutation factor F_i and crossover rate CR_i for each individual are set according to:

$$\begin{aligned} F_i &= N_i(0.5, 0.3) \\ \mathrm{CR}_i &= N_i(0.5, 0.1) \end{aligned} \tag{17.1}$$

where $N_i(0.5, 0.3)$ represents a random number from Gaussian distribution with a mean 0.5 and standard deviation 0.3.

(iii) For each individual, calculate the Nash–Sutcliffe efficiency (NSE) as a fitness score.

(iv) Now mutation is performed for each individual according to:

$$S_{i,G} = Z_{r_1,G} + F_i(Z_{r_2,G} - Z_{r_3,G}) \tag{17.2}$$

where $S_{i,G}$ is the mutant vector, and $Z_{r1,G}$, $Z_{r2,G}$ and $Z_{r3,G}$ are three arbitrarily selected individual from the current generation.

(v) Next, a trial vector $T_{i,G} = \left\{T_{i,G}^1, T_{i,G}^2, \ldots, T_{i,G}^D\right\}$ is produced by applying crossover as followed:

$$t_{i,G}^{j} = \begin{cases} s_{i,G}^{j} & \text{if rand}^{j}[0,1] \le CR_{i,G} \text{ or } (j = j^{\text{rand}}) \\ z_{i,G}^{j} & \text{otherwise} \end{cases} \tag{17.3}$$

where $t_{i,G}^{j}$, $s_{i,G}^{j}$ and $z_{i,G}^{j}$ are the jth parameter from generation G for ith trial vector, ith mutant vector, and ith population vectors, respectively; D is the dimension of each trial vector; rand^{j} is some random number from 0 to 1; and j^{rand} is a random value in the range of 1 to D.

(vi) Compute the fitness function as NSE values for each trial vector generated.

(vii) Compare the fitness values of the trial and respective target vector in the selection process to produces the next generation of population.

(viii) Repeat this process until the convergence conditions are fulfilled. Convergence will be attained only when there will be no increase in the fitness values for some selected number of iterations ($\approx$ 20 iterations) or when a maximum number of iterations are completed.

(ix) Finally, F and CR values are updated by mean of evolution. First, for a predefined number of iterations (e.g. 10 used in the current study), the values of F and CR providing trial vectors successfully are recorded. After completion of the specified iterations, the average values of the F and CR are upgraded as equivalent to the average of the successful F and CR values. A new set of mutation and crossover factors are then generated for each individual from Gaussian distributions with the updated mean values and previous standard deviation values.

17.2.2 VIC Model

The VIC model is a macroscale hydrologic model developed at Washington University by Liang et al. (1994). It has been widely used in various hydrological studies at global as well as regional scale. It is also a process-based model which can solve both the full water and energy balances to simulate surface runoff, baseflow, canopy interception, snowpack, evapotranspiration, and many other hydrological processes at each time step at each model grid. One distinctive feature of the model is that its consideration for multiple elevation bands, non-linear baseflow estimation, representation of sub-grid variability of vegetation classes, and incorporation of multiple soil layers with variable infiltration.

The VIC model works on a grid basis, and each model grid is partitioned into different land surface cover types. The Penman–Monteith equation is employed to calculate the corresponding evapotranspiration from each grid which is a summation of bare soil evaporation, vegetation transpiration, and canopy evaporation. Among the three soil layers in the VIC model, the upper two layers are intended to characterize the infiltration response of the catchment, and wherever the middle layer is wet, diffusion takes place from middle to top layer. The lowermost soil layer is responsible for baseflow, and it receives soil moisture from the middle layer by means of gravity drainage. The variable infiltration in the soil layer is expressed as:

$$i = i_m\left[1 - (1 - A)^{\frac{1}{b_i}}\right] \tag{17.4}$$

where b_i is a shape parameter; A is the saturated fraction of the basin; i_m is the maximum infiltration capacity, and i is the infiltration capacity.

Water balance for each grid in the model was estimated by the principle of conservation of mass (i.e. the water entering a grid essentially leave the grid or be stored within the grid) and the partitioning of precipitation into evapotranspiration, runoff, and water storage in the soil. Thus, the water balance can be expressed as:

$$\frac{\mathrm{d}S}{\mathrm{d}t} = P - E - R \tag{17.5}$$

where R, E, P, and $\mathrm{d}S/\mathrm{d}t$ are the runoff, evapotranspiration, precipitation, and change of soil water, respectively.

As the VIC model produces the runoff for individual grids, the model-simulated runoff must be routed to the outlet through the channel network using a separate routing scheme. The present study has adopted the routing scheme presented by Nijssen et al. (1997), which first brings the simulated runoff of grid cell into the cell outlet and then into the river network. A unit hydrograph theory is used for the routing within the grid cell, and for channel routing, the linearized Saint–Venant equations are utilized.

17.2.3 Coupling of SADE with VIC

The procedure of integrating SaDE with VIC is shown in Fig. 17.1. In the MATLAB environment, a linkage was built, so SaDE can communicate with VIC input and output files. During the optimization process, VIC model parameters are adapted automatically as specified by the objective functions of optimization. The procedure is continued until some stopping condition is fulfilled, e.g. a predetermined iteration count or the convergence of the fitness function is attained. The configuration of the model calibration involves the determination of the calibration parameters in addition to their possible ranges, as well as the specifying of the SaDE parameters.

The VIC hydrological model includes many parameters which need to be defined for the model setup. Although the majority of the model parameters can be estimated using the various ground observation data (e.g. soil map, land use, and topography), there are few parameters which cannot be measured directly. The six major parameters of the VIC model that need to be estimated are presented in Table 17.1. The choice of calibration parameters is made with reference to the previous studies of the model applications (Park and Markus 2014; Mishra et al. 2010). All other parameters

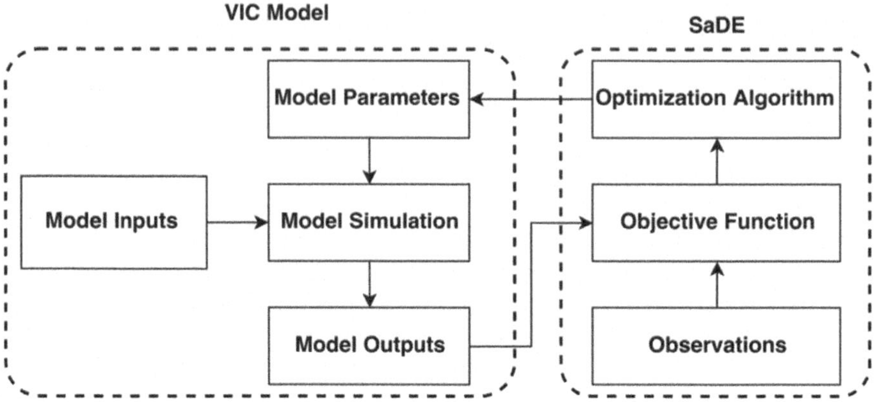

Fig. 17.1 Schematic of the VIC automatic calibration

Table 17.1 VIC model parameters selected for calibration, their description, units, and scales

Parameter	Description	Units	Range
b_i	Infiltration parameter	Dimensionless	0.0–0.4
D_s	Baseflow parameter[a]	Fraction	0.0–1.0
W_s	Baseflow parameter[b]	Fraction	0.0–1.0
d_1	First soil layer depth	Meters	0–0.5
d_2	Second soil layer depth	Meters	0.5–1.5
d_3	Third soil layer depth	Meters	0.5–2.0

[a]Fraction of maximum velocity of baseflow; and [b]Fraction of maximum soil moisture where non-linear baseflow occurs

of the VIC model are determined from the available observed data and not selected for calibration.

From the spatial viewpoint, the calibration parameters can be broadly classified into two groups: distributed parameters and lumped parameters. The distributed parameters are site-specific and have added physical meaning. The selected calibration parameters from the VIC model are also falling under the group of distributed parameter, and calibration for each of them will be difficult to realize. Therefore, in this study, a self-organizing map (SOM) based approach is adopted for fixing the spatial pattern (i.e. parameter regularization) of the model parameters. This simplification greatly decreases the effort in optimization and makes the calibration speedy and practical for models with a higher number of grids and long simulation period.

17.2.4 Performance Measures

To assess the VIC model simulations, standard performance measures such as coefficient of determination (R^2), percentage of bias (PBIAS), index of agreement (IoA), and Nash–Sutcliffe efficiency (NSE) are used in the study. The equations used for these performance measures are given below:

$$R^2 = \frac{\left(n\sum_{i=1}^{n} Y_i^{\text{obs}}.Y_i^{\text{sim}} - \left(\sum_{i=1}^{n} Y_i^{\text{obs}}\right)\left(\sum_{i=1}^{n} Y_i^{\text{sim}}\right)\right)^2}{\left(n\sum_{i=1}^{n}\left(Y_i^{\text{obs}}\right)^2 - \left(\sum_{i=1}^{n} Y_i^{\text{obs}}\right)^2\right).\left(n\sum_{i=1}^{n}\left(Y_i^{\text{sim}}\right)^2 - \left(\sum_{i=1}^{n} Y_i^{\text{sim}}\right)^2\right)} \tag{17.6}$$

$$\text{PBIAS} = \left[\frac{\sum_{i=1}^{n}\left(Y_i^{\text{obs}} - Y_i^{\text{sim}}\right) * 100}{\sum_{i=1}^{n} Y_i^{\text{obs}}}\right] \tag{17.7}$$

$$\text{IoA} = 1 - \left[\frac{\sum_{i=1}^{n}\left(Y_i^{\text{obs}} - Y_i^{\text{sim}}\right)^2}{\sum_{i=1}^{n}\left(\left|Y_i^{\text{sim}} - Y_{\text{mean}}^{\text{Obs}}\right| + \left|Y_i^{\text{obs}} - Y_{\text{mean}}^{\text{Obs}}\right|\right)^2}\right] \tag{17.8}$$

$$\text{NSE} = 1 - \left[\frac{\sum_{i=1}^{n}\left(Y_i^{\text{obs}} - Y_i^{\text{sim}}\right)^2}{\sum_{i=1}^{n}\left(Y_i^{\text{obs}} - Y_{\text{mean}}^{\text{obs}}\right)^2}\right] \tag{17.9}$$

where y_i^{obs} and y_i^{sim} are observed and simulated flow values for the ith time step; $y_{\text{mean}}^{\text{obs}}$ and $y_{\text{mean}}^{\text{sim}}$ are the mean values of observed and model-simulated flow values; n is the total number of time steps.

17.3 Case Study

17.3.1 Study Area

The Sina river basin, one of the major feeders to the Bhima River in India, is chosen as a case study. It starts in Ahmednagar district and runs southeast through Ahmednagar and Solapur districts to fall into the Bhima near Kudul in the south of Solapur, on the Maharashtra and Karnataka boundary. It is having a length of 180 km. The area of the basin is around 12,160 km^2. The Sina River is about 100–200 m broad and has steep banks. The bed of the river is commonly sandy but intermittently rocky. The climate is semi-arid in nature, and the annual rainfall is 852 mm in the basin, that is, much below to the country's average annual rainfall. Average daily temperature in the basin is around 27°C. The elevation varies from 561 m to 766 m in the basin. The location of the basin is shown in Fig. 17.2.

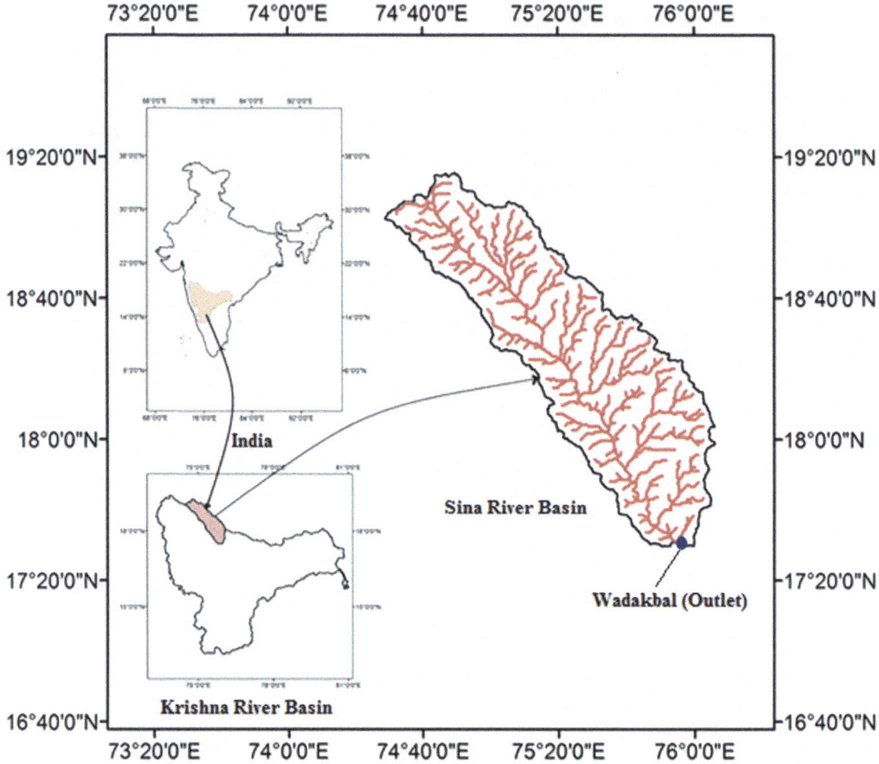

Fig. 17.2 Location map of Sina river basin

17.3.2 Data Used

Daily gridded temperature ($1^\circ \times 1^\circ$) and precipitation data ($0.25^\circ \times 0.25^\circ$) are obtained from the Indian Meteorological Department (IMD). Reanalysis wind speed data ($2.5^\circ \times 2.5^\circ$) are retrieved from the National Center for Environmental Prediction (NCEP). Advanced Very High-Resolution Radiometer (AVHRR) global land cover map was obtained from the University of Maryland (1 × 1 km). SRTM (90 m) DEM has been obtained from the National Remote Sensing Centre of India. Soil data were collected from harmonized world soil database by the Food and Agricultural Organization (FAO) with a 30 arc-second (~ 1 km) spatial resolution. The data of daily observed streamflows on the outlet of the study area were collected at Wadakbal Hydro-observation (HO) site from the WRIS-India.

17.4 Results and Discussion

Hydrological simulations are conducted both at a daily and monthly time step for the time period of 1980–1994. The automatic calibration of the VIC model is applied to nine out of fifteen years of simulation, and the remaining data are used for verification. The NSE indicator, i.e. maximization of the NSE, is considered as the objective function for the optimization of model parameters. The calibration is performed on two different time scales, i.e. on a daily basis and monthly basis. All computations were carried out on an Intel Core i7 −4790, 3.6 GHz processor using Linux distribution, and with the help of MATLAB and C ++ programming.

The relation between fitness function values of the best parameter set and a number of model evaluations is shown in Fig. 17.3. For daily model calibration, the initial best objective function value was 0.56. The objective function shows a clear convergence around the NSE value of 0.746. The number of model evaluation for achieving this convergence was 1471 with population size as 30. For monthly VIC calibration, initial best objective function value started at 0.93 and finally converged to a value

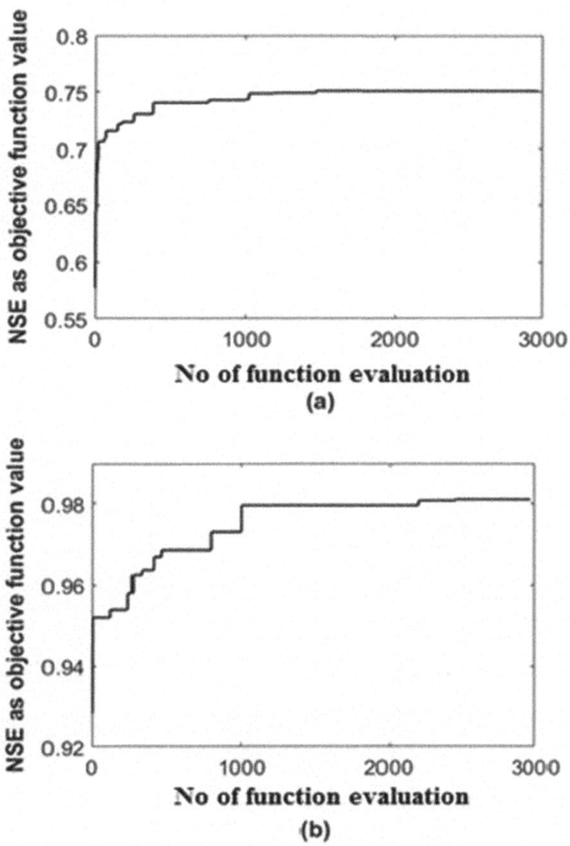

Fig. 17.3 Typical convergence characteristics of SaDE based on **a** daily calibration and **b** monthly calibration for VIC hydrological model

of 0.98. A total of 2481 model simulations were executed to attain this convergence. At the end of optimization, the standard deviation for objective function for all the population was found very close to zero (i.e. 0.1 and 0.08 for daily and monthly calibration, respectively) for both the daily and monthly calibration, which verified the fact that all the population initiated at the beginning of optimization algorithm and finally converged at the end of the optimization process.

The statistics for the optimized parameter set from the final generation of optimization is shown in Fig. 17.4. From this figure, it can be seen that for daily model calibration, parameters d_1, d_2, and d_3 showed significant variation among the different population. However, in the final iteration, these different parameters were giving a similar model performance, which signified the presence of parameter equifinality in the current calibration problem. Nevertheless, for monthly calibration, the variation for the d_1, d_2, and d_3 was smaller as compared to daily calibration problem. The box plots for the parameters b_i, D_s, and W_s were found to be very short for both daily and monthly calibration, which represents a high level of agreement between the

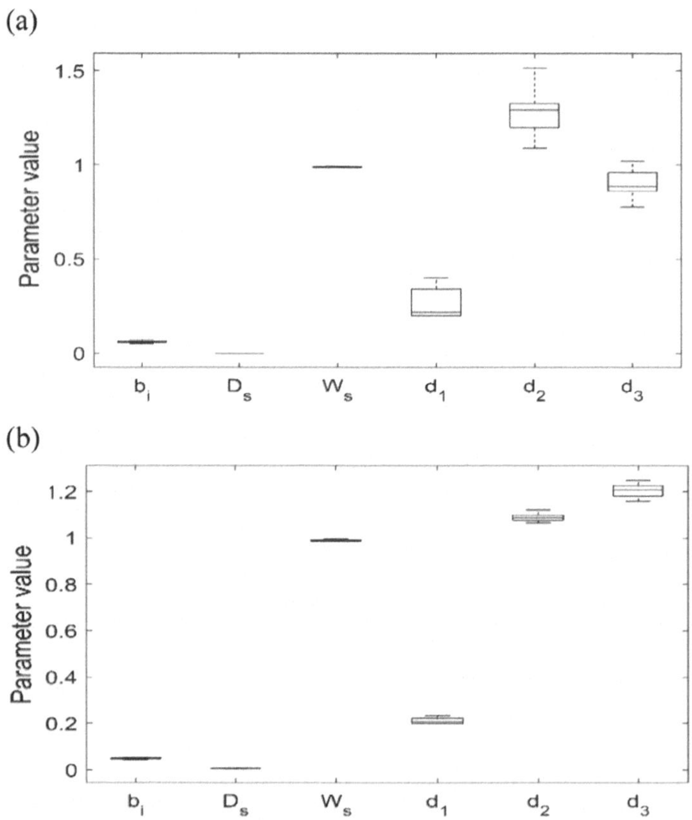

Fig. 17.4 Box plots of **a** daily and **b** monthly calibrated parameters from the last iteration of SaDE-based automated calibration

populations and suggests them as more identifiable parameters with less uncertainty in their estimation.

Daily streamflow data from 1980 to 1988 were used for the automatic calibration of the VIC model using SaDE algorithm, and the remaining data from 1st January 1989 to 31st December 1994 were used to validate the calibration process. Comparison of simulated daily and monthly average streamflows, using the optimized parameter set obtained from SaDE algorithm, with the observed flow is shown in Fig. 17.5. From this figure, it can be reckoned that in terms of observed discharge during calibration and validation periods, the models yielded comparable results for both daily and monthly scale of model operation. The peak values in the monthly simulation were perfectly captured. For daily simulation, similar kind of agreement can also be visible for most of the calibration and validation periods.

To further investigate the performance of the automatic calibration procedure, four efficiency measures for calibration and validation period are calculated on a daily and monthly basis and the values of performance statistics are presented in Table 17.2. Better model performance is realized if the values of PBIAS are closer to zero, and R^2, NSE, and IOA are closer to unity. According to Golmohammadi et al. (2014), a

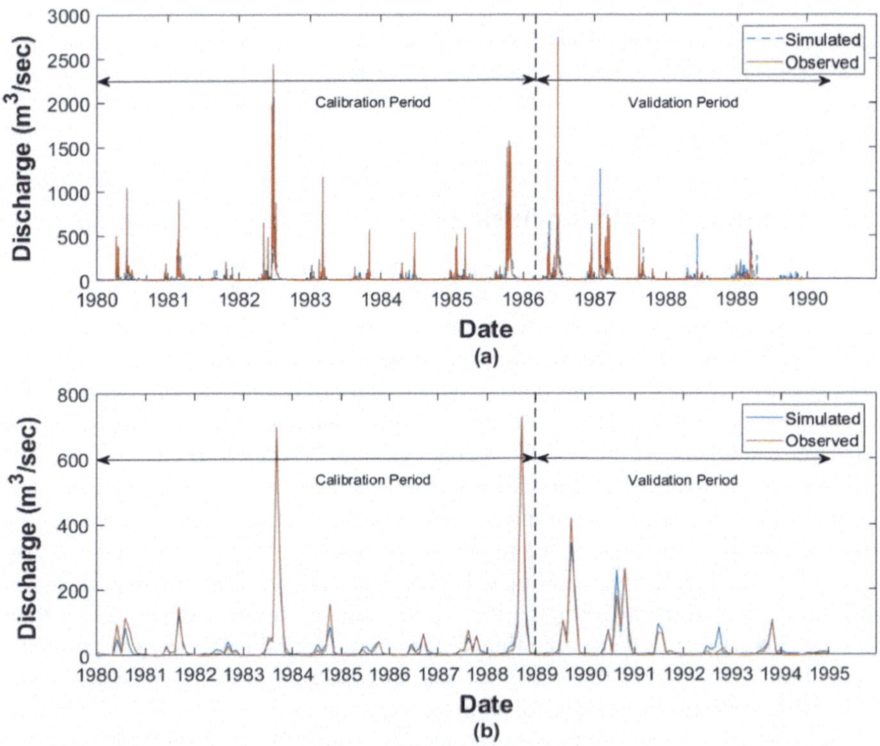

Fig. 17.5 Comparison of observed and VIC model-generated discharge at **a** daily and **b** monthly time scale during calibration and validation periods

Table 17.2 Performance measure values obtained for VIC model calibration using SaDE

	Performance measure values for daily model				Performance measure values for monthly model			
	NSE	R^2	PBIAS	IOA	NSE	R^2	PBIAS	IOA
Calibration	0.75	0.86	−0.25	0.91	0.98	0.99	−2.1	0.99
Validation	0.61	0.76	−19.3	0.86	0.92	0.96	−12.2	0.98

model can be considered to be calibrated well for streamflow simulation if monthly NSE $\geq$ 0.65, it will be considered "very good" if PBIAS is less than 10%, "good" if it is between 10 and 15%, and "fair" if the error is between 15 and 25 percentage (Van-Liew et al. 2007). Here, monthly PBIAS and NSE for calibration periods were obtained as 0.98 and −2.1, which are entirely satisfying such criteria. Even for daily model operation, during the calibration phase, all of the four-performance measures are having promising values. However, during model validation from 1989 to 1994, the PBIAS measures have higher negative values. PBIAS is representing estimates of bias percentage of simulated streamflows as compared to the observed flows. A negative value represents that model is overestimating the flow. This phenomenon can also be verified in Fig. 17.5. Looking back into the observed data, it was found that for the year 1992 and 1994, several data were found to be missing and many others were manually adjusted instead of recorded data. The presence of a large number of inconsistent data may be one reason for such model behavior.

17.5 Summary and Conclusions

In this study, an effective approach is presented for automated calibration of a physically based hydrological model by incorporating a model-independent parameter scheme based on a differential evolution algorithm. A self-adaptation strategy is adapted to evolve the mutations and crossover factor of traditional DE algorithm for improving its efficiency. The application of this approach is demonstrated for the Sina river catchment, in India with an application of VIC hydrological model. The VIC macroscale hydrological model is capable of simulating various hydrological processes in a catchment including surface runoff, baseflow, canopy interception, snowpack, evapotranspiration, and many others. Six distributed parameters are estimated using the SaDE-based automated calibration scheme. The distributed parameters are calibrated by categorizing them into a smaller number of groups by their physical characteristics and estimating all the parameter values for each such group. The results of this study illustrated the use of linking a macroscale hydrological model with SaDE for the successful parameter estimation of the hydrological model. This scheme used in the study is faster and less subjective than the trial and error method and may serve as an optimization tool to estimate the VIC model parameters. Without such techniques, calibration of the model will take much higher time and effort. Moreover, the calibration results from this study exhibit a great possibility of SaDE to be utilized in the calibration process of other complex hydrological models.

References

Duan Q, Sorooshian S, Gupta V (1992) Effective and efficient global optimization for conceptual rainfall-runoff models. Water Resour Res 28(4):1015–1031

Golmohammadi G, Prasher S, Madani A, Rudra R (2014) Evaluating three hydrological distributed watershed models: MIKE-SHE, APEX. SWAT Hydrol 1(1):20–39

Jha MK, Kumar A, Nanda G, Bhatt G (2006) Evaluation of traditional and nontraditional optimization techniques for determining well parameters from step-drawdown test data. J Hydrol Eng 11(6):617–630

Johnston PR, Pilgrim DH (1976) Parameter optimization for watershed models. Water Resour Res 12(3):477–486

Jung BS, Karney BW (2006) Hydraulic optimization of transient protection devices using GA and PSO approaches. J Water Resour Planning Manage 132(1):44–52

Liang X, Lettenmaier DP, Wood EF, Burges SJ (1994) A simple hydrologically based model of land surface water and energy fluxes for general circulation models. J Geophys Res: Atmos 99(D7):14415–14428

Nijssen B, Lettenmaier DP, Liang X, Wetzel SW, Wood EF (1997) Streamflow Simulation for Continental-Scale River Basins. Water Resour Res 33:711–724

Madsen H, Wilson G, Ammentorp HC (2002) Comparison of different automated strategies for calibration of rainfall-runoff models. J Hydrol 26:48–59

Mishra V, Cherkauer KA, Shukla S (2010) Assessment of drought due to historic climate variability and projected future climate change in the midwestern United States. J Hydrometeorol 11(1):46–68

Park D, Markus M (2014) Analysis of a changing hydrologic flood regime using the variable infiltration capacity model. J Hydrol 515:267–280

Qin AK, Suganthan PN (2005) Self-adaptive differential evolution algorithm for numerical optimization. Evol Comput 2:1785–1791

Reddy MJ, Kumar DN (2020) Evolutionary algorithms, swarm intelligence methods, and their applications in water resources engineering: a state-of-the-art review. H_2Open J 3(1):135–188. https://doi.org/10.2166/h2oj.2020.128

Ryu JH (2009) Application of HSPF to the distributed model intercomparison project: case study. J Hydrol Eng 14(8):847–857

Sirsant S, Reddy MJ (2018) Reliability-based design of water distribution networks using self-adaptive differential evolution algorithm. ISH J Hydraul Eng 24(2):198–212

Sorooshian S, Duan Q, Gupta VK (1993) Calibration of rainfall-runoff models: Application of global optimization to the sacramento soil moisture accounting model. Water Resour Res 29(4):1185–1194

Storn R, Price K (1997) Differential evolution—a simple and efficient heuristic for global optimization over continuous spaces. J Glob Optim 11:341–359

Van Liew MW, Veith TL, Bosch DD, Arnold JG (2007) Suitability of SWAT for the conservation effects assessment project: comparison on USDA agricultural research service watersheds. J Hydrol Eng 12(2):173–189

Zhang X, Srinivasan R, Zhao K, Liew MV (2009) Evaluation of global optimization algorithms for parameter calibration of a computationally intensive hydrologic model. Hydrol Processes 23(3):430–441

Chapter 18
Streamflow Response to Land Use–Land Cover Change Over the Subarnarekha River Basin, India

Pratik Deb, Ashok Mishra, and Soukhin Tarafdar

Abstract Land use is referred as one of the most important components of the global environment system. The present study inspects the impacts of ongoing land use change on the streamflow over the Subarnarekha River Basin, India, using a physically based, continuous, semi-distributed eco-hydrological long-term simulation model. Land use change dynamic in the studied river basin was evaluated through the analysis of six land use–land cover maps corresponding to the land use–land cover status for the periods of 1987, 1990, 1995, 2000, 2004 and 2008. An unsupervised classification approach followed by supervised classification was employed to develop land use–land cover maps. The results of the image classifications for a basin area of 19,121 km^2 showed the area under dense forest and open forest has decreased drastically by 52.06% and 43.44%, respectively, during 1987–2008 periods. The area of settlement and agricultural land has increased by 161.98% and 20.71%, respectively. This may be due to high population growth, urban expansion and increase of agricultural area. Soil–water balance was simulated by applying the SWAT at three stream gauging locations over the entire basin. The model-simulated streamflow was calibrated and validated for the periods of 1987–2001 and 2002–2011, respectively, by using observed streamflow. The model setup showed good agreement between observed and simulated streamflows both graphically and statistically with coefficient of determination (R^2) varying from minimum of 0.84 to maximum of 0.89 and Nash–Sutcliffe efficiency (NSE) from minimum of 0.85 to maximum of 0.88 during calibration and validation period for the three sub-basins of the Subarnarekha River

P. Deb (✉) · S. Tarafdar
Garhwal Regional Center, G. B. Pant. National Institute of Himalayan Environment (NIHE), Srinagar-Garhwal, Uttarakhand 246174, India
e-mail: deb.pratik26@gmail.com

S. Tarafdar
e-mail: taradar101s@gmail.com

A. Mishra
Agricultural and Food Engineering Department, Indian Institute of Technology Kharagpur, Kharagpur, West Bengal 721302, India
e-mail: amishra@agfe.iitkgp.ac.in

A. Pandey et al. (eds.), *Water Management and Water Governance*, Water Science and Technology Library 96,
https://doi.org/10.1007/978-3-030-58051-3_18

Basin. The calibrated–validated model was used to simulate the land use change impacts on streamflow by changing the land use in six different periods in which the most recent land use of the year 2008 showed an increasing effect on streamflow.

Keywords Land use–land cover · Satellite image processing · Hydrological modeling · SWAT · SWAT-CUP

18.1 Introduction

Land use–land cover change is one of the important issues all over the world, particularly in highly populated river basins. The land use–land cover pattern is the consequence of natural as well as rapid socioeconomic development of a region by human in space and time. Due to rapid urbanization and increase in agricultural practice, land is becoming a scarce resource. Therefore, information regarding land use pattern and their dynamics is crucial for the selection, planning and implementation of land use schemes to support the basic needs of increasing population. Inventory and monitoring the land use–land cover changes are necessary for modeling the change impact on environment at different scales and further understanding of climate change mechanism (William et al. 1994; Turner et al. 1995).

The increasingly reduced quantity and quality of utilizable water resources around the globe have become a challenging issue, which leads to threatening of food production, economic development, human health and natural ecosystems. With the increasing growth of global population, availability of water resources required for the sustenance of humans and the natural ecosystem are shrinking. Management of water resources is a growing concern worldwide and especially true for developing countries like India. Understanding of basic hydrological processes at the catchment scale is necessary to implement management programs for conservation of resources. The characteristics of catchments within a river basin are dynamic and vary both spatially and temporally (Blöschl and Sivapalan 1995; Strayer et al. 2003). Land use–land cover changes can significantly alter catchment water balance (Zhao et al. 2012) and may result in an increase of water scarcity conditions, thus contributing to a deterioration of living conditions (Wagner et al. 2013). In a system, accurate representation of the hydrologic processes plays an important role for the accurate prediction of streamflows. A fruitful way to improve this accuracy is by employing distributed hydrologic models. Distributed hydrological models can play as a useful and effective tool in the sustainable water resources management particularly on analyzing the effects of land use change (Fohrer et al. 2001; Eckhardt et al. 2003; Immerzeel et al. 2008; Ghaffari et al. 2010; Gosain et al. 2011; Wagner et al. 2013).

A large number of hydrological models have been developed over the last few decades such as tank model, UBC watershed model, HBV, TOPMODEL, MIKE SHE, EPIC (Singh 1995), Soil and Water Assessment Tool (SWAT) (Gassman et al. 2007), Pitman model (Hughes 2008), WEAP21 (Yates et al. 2005) and GR4J (Perrin et al. 2003). Among these, SWAT is most widely used by researchers in different parts of the world because of its ability to simulate hydrologic regimes at both daily and monthly scales (Bouraoui et al. 2005). SWAT can easily deal with different spatially distributed inputs, such as rainfall to match with different basin delineations. The advance model configuration enhances its acceptability for impressive function such as modeling in limited data regions and evaluating various scenarios (Engel et al. 1993; Spuill et al. 2000; Chanasyk et al. 2003; Bosch et al. 2004). In recent years, SWAT has been used by many researchers in different parts of the world to evaluate the impacts of land use–land cover (Fohrer et al. 2001; Eckhardt et al. 2003; Ghaffari et al. 2010; Wagner et al. 2013) on the catchment hydrology. In India, SWAT has been used for small-scale (Behera and Panda 2006; Pandey et al. 2009), mesoscale (Dhar and Mazumdar 2009) and large-scale (Immerzeel et al. 2008; Gosain et al. 2011) basin studies. These utilizations of SWAT model showed its power for modeling for catchment runoff and sediment yield, as well as to evaluate impacts of land use–land cover and climate change on the quantity and quality of land and water resources.

In this study, a hydrological modeling technique is conducted using SWAT model to assess the impacts of land use–land cover changes on the streamflow. Sensitivity analysis has been carried out to evaluate model calibration and validation (Sorooshian et al. 2008; Zheng and Keller 2008; Moradkhani and Sorooshian 2009; Deb 2017). The model calibration is a difficult process. It mainly depends upon the number of input parameters, complexity of the model and iterations. Several optimization techniques have been introduced in SWAT for calibration and validation: SUFI-2 (Abbaspour et al. 2004), GLUE (Beven and Binley 2014), MCMC (Vrugt et al. 2008; Vrugt and Braak 2011) and ParaSol (Yang et al. 2008). These techniques have been connected through SWAT-CUP (Abbaspour et al. 2007) and allow sensitivity analysis and uncertainty analysis of model parameters as well as structure. Abbaspour et al. (2004) applied the SUFI-2 algorithm to evaluate SWAT model performance. The main advantage of the SUFI-2 algorithm is that it attained a high-quality calibration result using a small number of model simulations (Yang et al. 2008). The main objectives of this study are to simulate the streamflow of a medium-size river basin using SWAT including model calibration and validation followed by the SUFI-2 algorithm and to analyze the streamflow response under different land use–land cover scenarios.

18.2 Study Area

To perform the designed research objectives, Subarnarekha River Basin (shown in Fig. 18.1) has been chosen as the study area. Subarnarekha, an interstate river originating from Chota Nagpur Plateau, flows through Singhbhum and Ranchi district of Jharkhand, West Midinapur district of West Bengal and Balasore district of Odisha. At the south of Dantan, it enters the Balasore district and finally falls into the Bay of Bengal near Kirtania in Balasore district of Odisha. The Subarnarekha River Basin has an area of 19,121 km^2 and is located between 21° 33′ 18″ N and 23° 18′ 00″ N latitude, 85° 11′ 00″ E and 87° 23′ 31″ E longitude.

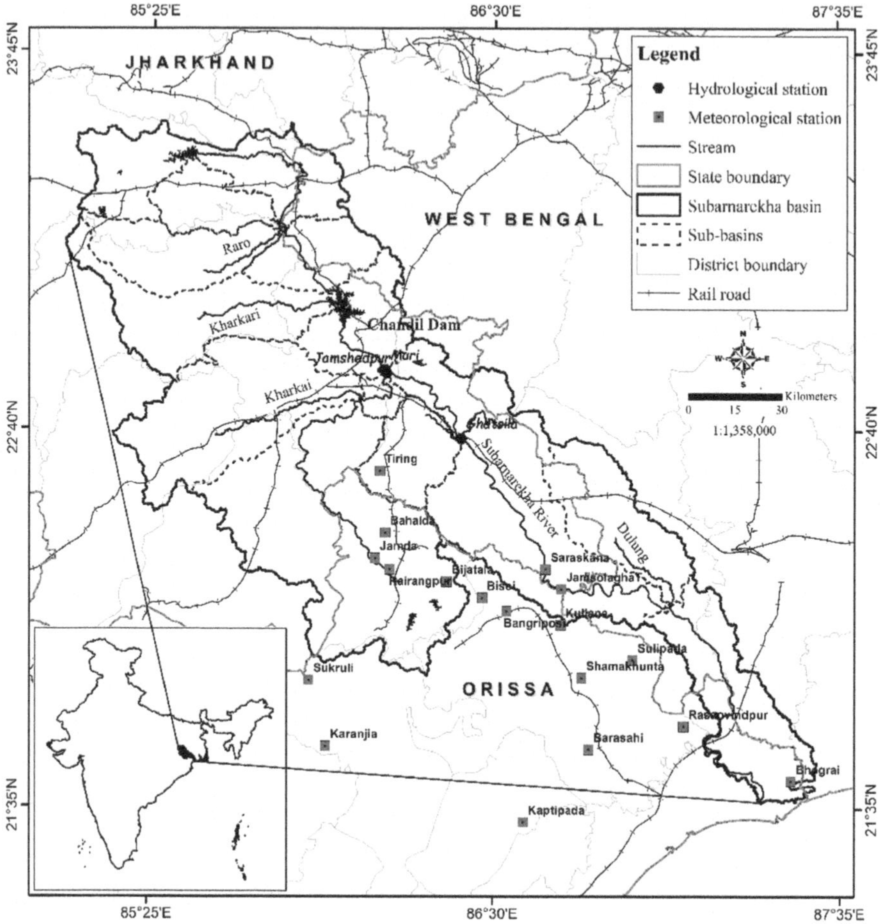

Fig. 18.1 Location of Subarnarekha River Basin including its sub-basins with distributed hydrological and meteorological stations

The total length of the river is around 400 km (Bhuyan et al. 2014). The basin is dominated by forested, mostly terrace land with slope in between one and twelve percent toward the eastern direction (Deb and Mishra 2016). The average annual rainfall over the entire basin has been found around 1498.2 mm (averaged from 1987 to 2011), wherein 80% of the normal rainfall occurs during southwest monsoon period (June–September). May is the hottest month of the year over the basin with the temperature rising to about 45 °C. The minimum temperature observed in the month of January was around 10 °C.

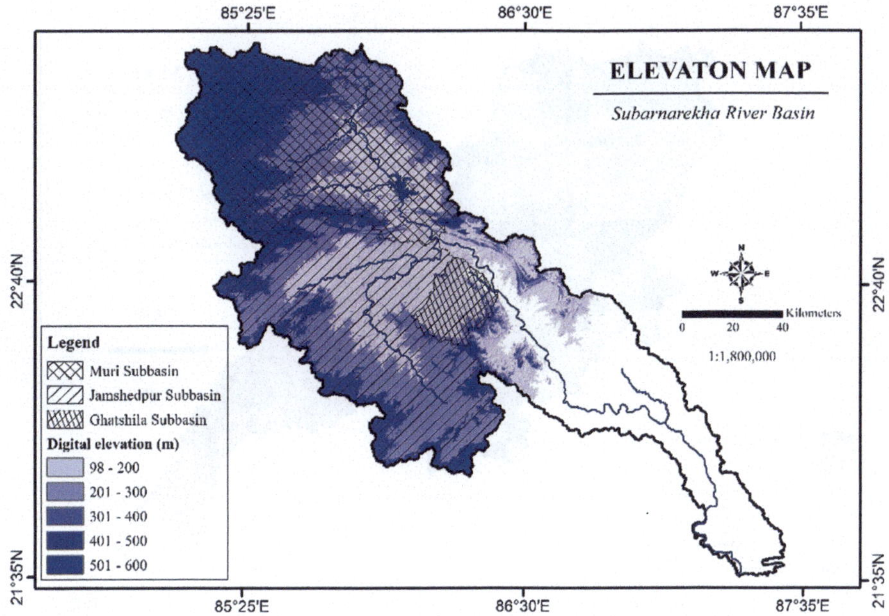

Fig. 18.2 Elevation variation over the Subarnarekha River Basin

18.3 Methodology

18.3.1 Digital Elevation Model (DEM)

Digital elevation model (DEM) is one of the major inputs for hydrological modeling studies. DEM helps to understand the flow pattern and flow behavior governing slow and fast runoff processes (Wagener and Wheater 2006; Patel et al. 2013; Yadav et al. 2014). In the present study, DEM based on ALOS phased array-type L-band synthetic aperture radar (ALOS PALSAR) data with a resolution of 12.5 m (shown in Fig. 18.2) is used. Fourteen tiles of readily processed ALOS PALSAR DEMs, covering entire study area, were acquired from the UAF Alaska Satellite Facility Web site (https://www.asf.alaska.edu/). All the DEM tiles were merged and projected to WGS_1984_UTM_Zone_45 N projection system to cover the entire study area.

18.3.2 Soil Map Preparation

The soil of the Subarnarekha River Basin is characterized as laterite soil mostly of residual and alluvial origin which is derived from old bedrock. River-borne alluvial deposit of unconsolidated materials is found over 11% of the basin area in relatively

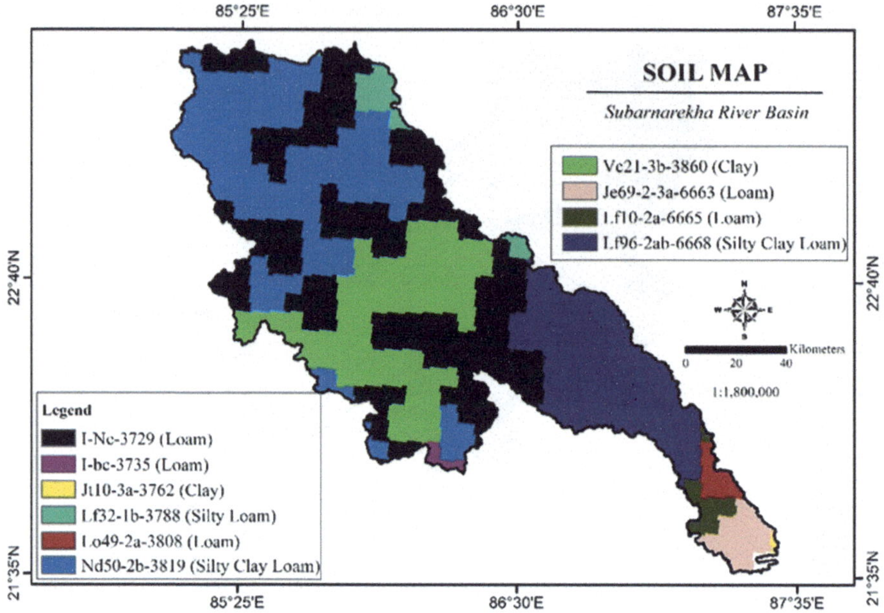

Fig. 18.3 Distribution of different soils over the Subarnarekha River Basin

lower valley portions and coastal plains. The remaining six percent of the basin mostly lies in the heavy rainfall zones, covered by laterites. The basin shows that severe soil erosion takes place in about 23.74% of the basin area, and very severe soil erosion takes place in 1.4% of the basin area. The spatial distribution of soils of the entire basin (Fig. 18.3) was downloaded from the Digital Soil Map of the World, which was developed by Food and Agriculture Organization (FAO) of United Nation. The major parts of the entire basin comprise sandy clay loam and loam soil, whereas the minor parts include sandy loam and clay soil.

18.3.3 Land Use–Land Cover Mapping

The land use–land cover in the study area is primarily agriculture, dense forest, open forest, mixed forest, settlements and waterbodies, which have undergone a substantial change during the last three decades. Figure 18.4 depicts the land use–land cover patterns of the study area during the periods of 1987, 1991, 1995, 2000, 2000 and 2008. Landsat data with a resolution of 30 m was employed in this study with the purpose to maintain the consistency in dataset. Based on the data availability, data quality and minimum percentage of cloud cover, four Landsat scenes related to each respective period were collected from the Earth Explorer Web site (https://earthexplorer.usgs.gov/). To avoid false changes in land use–land cover analysis,

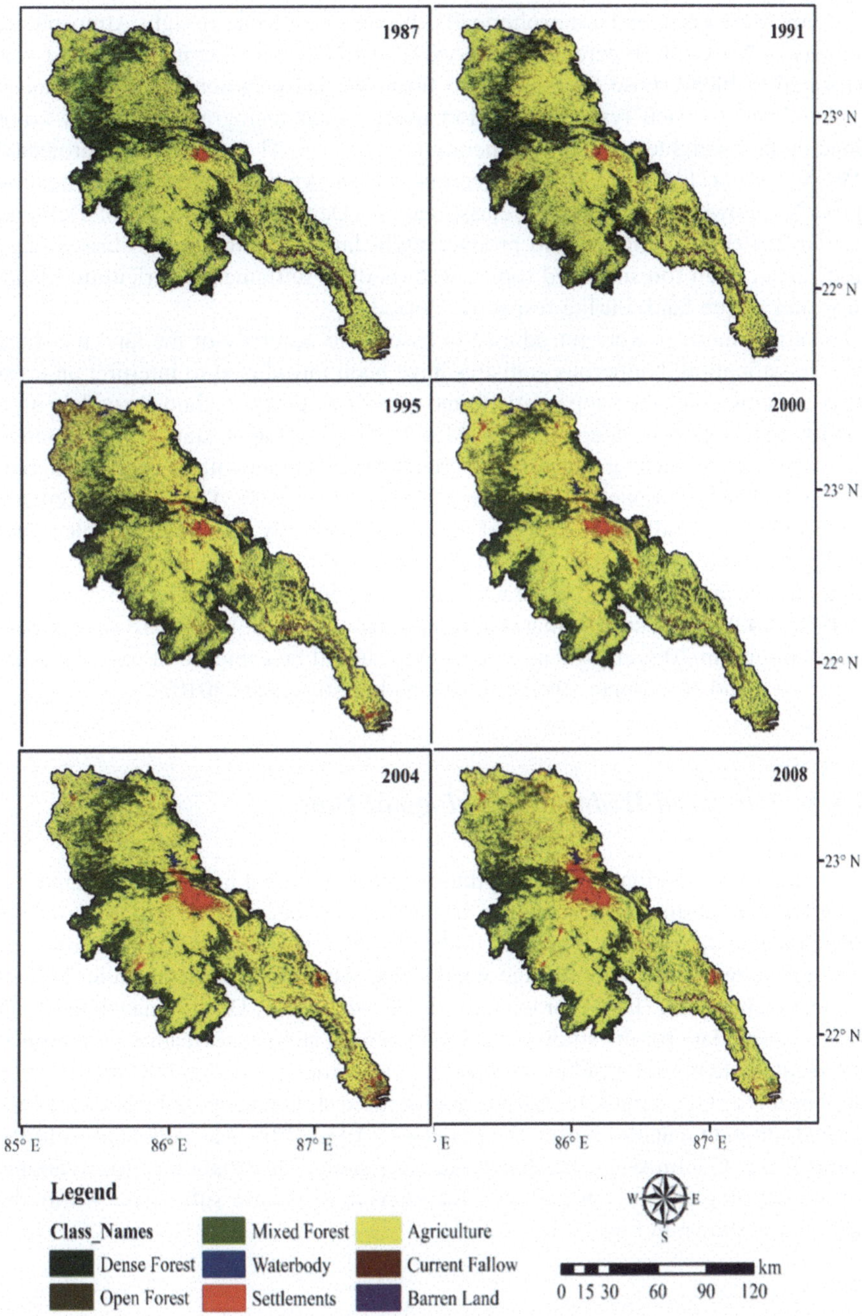

Fig. 18.4 Distribution of various land use–land cover categories over the Subarnarekha River Basin during 1987–2008

the scenes were corrected atmospherically through Fast Line-of-sight Atmospheric Analysis of Spectral Hypercube (FLAASH) in ENVI 5.1. Secondly, mosaic was performed to blend those four scenes to form one large radiometrically balanced image related to each respective period using equal histogram matching, color balancing and weighted seamline generation technique. The images were projected to WGS_1984_UTM_Zone_45N projection system. An unsupervised classification approach followed by supervised classification (Deb and Mishra 2016; Deb and Tarafdar 2019) was employed to produce eight land use–land cover classes (e.g., dense forest, open forest, mixed forest, waterbodies, settlements, agricultural land, fallow and barren land) during respective periods.

The error matrices were introduced to assess the accuracy of the land use–land cover classification. Numerous statistics have been introduced to interpret an error matrix, among which the kappa statistic and overall accuracy method are widely used by many researchers (Anderson 1976; Mas 1999). The kappa statistic is a discrete multivariate technique that compares an observed agreement with an expected agreement of a classified image (Congalton 1991; Banko 1998). The overall accuracy is determined by adding all diagonal elements (correctly identified samples) and dividing the sum of all errors (except diagonal numbers) in the error matrix. The kappa statistic and the overall accuracy of the classified land use maps were achieved as 90% and 0.93, respectively, that is acceptable for the hydrologic analysis, wherein at least minimum 70% classification accuracy should be achieved at regional scale (Thunnissen and Noordman 1997; Cheema and Bastiaansena 2010).

18.3.4 Observed Hydrometeorological Data

The weather information, viz., precipitation, maximum and minimum air temperature, relative humidity, solar radiation and wind speed, is required as inputs for the hydrological water balance. Observed daily rainfall, relative humidity, minimum and maximum temperature, and average wind speed were collected from India Meteorological Department (IMD) for the periods of 1987–2011. Due to unavailability of solar radiation data for the study area, SWAT inbuilt statistical weather generator is used to calculate the solar radiation based on the available weather information. The daily observed stage versus discharge data for three stream gauging locations (Muri, Jamshedpur and Ghatsila) during the periods of 1987–2011 was collected from the Central Water Commission, Bhubaneswar. The spatial distribution of the available hydrometeorological stations and their locations in respective sub-basins within the study area is shown in Fig. 18.1.

18.4 Hydrological Modeling

18.4.1 The Hydrological SWAT Model

Among all the hydrological models, Soil and Water Assessment Tool (SWAT) has been widely used by many researchers in different parts of the world to simulate the long-term impacts of land use–land cover and climate change on the quality and quantity of land and water resources (Gassman et al. 2007). SWAT is a process-based, comprehensive, long time simulation, deterministic and semi-distributed eco-hydrological model at river basin scale, developed by the United States Department of Agriculture-Agricultural Research Service (USDA-ARS) (Arnold et al. 1998; Arnold and Fohrer 2005; Srinivasan et al. 2010). It is operated through a GIS-based interface to allow the use of digital elevation map, land use–land cover maps and soil map as a basic input. The hydrological cycle, as simulated by the SWAT, is based on the soil–water balance equation (Neitsch et al. 2008):

$$\mathrm{SW}_t = \mathrm{SW}_0 + \sum_{t=1}^{n} \left(R_{\mathrm{day}} - Q_{\mathrm{surf}} - E_a - W_{\mathrm{seep}} - Q_{\mathrm{gw}}\right) \tag{18.1}$$

where

SW_t is the soil moisture content (mm of H_2O) at time t (day).

SW_0 is the initial moisture water content (mm of H_2O).

R_{day} is the amount of rainfall on day i (mm of H_2O).

Q_{surf} is the amount of surface runoff on day i (mm of H_2O).

E_a is the amount of evapotranspiration on day i (mm of H_2O).

W_{seep} is the amount of percolation and bypass flow exiting in the soil profile bottom on day i (mm of H_2O).

Q_{gw} is the amount of return flow on day i (mm of H_2O).

Conceptually, SWAT divides a watershed into number of sub-basins. In a simulation, the usage of sub-basins is beneficial particularly when different parts of a watershed are dominated by land use–land cover or variation of soil properties to influence hydrology. The user is able to reference different areas of the watershed to one another spatially by partitioning the watershed into sub-basins. Each stream channel is connected to each sub-basin and further divided into hydrologic response unit (HRU). HRU is defined by a unique combination of same land use, same slope and same soil type in a sub-basin, and SWAT simulates hydrology including vegetation growth and management practices at HRU level. From each HRU, water, sediment, nutrients and other pollutants are summarized in each sub-basin and later on routed through the stream network to the watershed outlet. A brief description of the various components and their interrelationships employed to simulate the processes and their interface in the SWAT can be found in Arnold et al. (1998), Neitsch et al. (2008), Arnold and Fohrer (2005) and Arc-SWAT documentation (Neitsch et al. 2011; Arnold et al. 2012).

18.4.2 Sequential Uncertainty Fitting Version 2 (SUFI-2)

SUFI-2 technique is a Bayesian framework. Based on the prior assessment, it incorporates new information to calculate posterior parameters (Haan 2002) for the events of interest. In SUFI-2, parameter uncertainty includes uncertainties from different sources such as input variables (e.g., precipitation), model parameterization, conceptualization and measured data. The 95% prediction uncertainty (95PPU) band computes the parameter uncertainty. The 95PPU is quantified at 97.5 and 2.5% levels of the cumulative distribution function of the output variables (Abbaspour et al. 2007). A Latin hypercube sampling (Abbaspour et al. 2007) is introduced to draw independent parameter sets. The systematic procedure for the application of SUFI-2 technique can be found in SWAT-CUP Manual (Abbaspour 2008).

18.4.3 SWAT Setup for Streamflow Simulation

Watershed delineation is the primary step in the parameterization of SWAT. Arc-SWAT automatically delineates a watershed based on the flow direction from each grid cell until either an outlet cell or the edge of DEM grid extent is completed. Watershed delineations were performed at those three available stream gauging locations. The parameterization of the stream reaches, and the sub-basin geomorphology (i.e., elevation distribution, area, slope and stream length) is done automatically through Arc-SWAT interface. The area delineated by the SWAT at three stream gauging locations, Muri, Jamshedpur and Ghatsila, was found to be 6353.493 km^2, 6262.828 km^2 and 723.3772 km^2, respectively.

The land use–land cover map and soil map, ArcGIS shape file format, were imported into the model, and their respective lookup tables were provided. Both the maps were made to overlay to result in the land use–land cover and soil distribution within the sub-basins. After characterizing land use–land cover and soil map, slope categories were selected. This leads to a creation of total 219 HRUs based on the uniform land use–land cover, soil type and slope using 10% as a threshold to avoid unnecessary large number of HRUs in the analysis.

One of the main sets of input to stimulate the hydrological processes (e.g., streamflow) in SWAT model is weather data. It comprises rainfall, minimum and maximum air temperature, relative humidity, average wind speed and solar radiation. Using the observed weather information, a statistical weather generator file WXGEN (.wgn) was prepared for the periods of 1987–2011 in order to generate future weather data as well as to fill the gap or missing record in the observed dataset, if any. The weather data for each sub-basin were then automatically assigned in the SWAT based on the location of closest weather station. The next step, after importing the weather data, was setting up some additional inputs: soil parameters, sub-basin, HRUs, main channel and groundwater to run SWAT model. Using the "write all inputs files" option, database files related to soil parameters, sub-basin, HRUs, main channel and groundwater were created. Finally, SWAT was setup and run for the streamflow simulations.

18.4.4 Model Performance Evaluation

Evaluation of the model performance is necessary in order to assess how measured values fitted with the model-simulated values. According to ASCE (1993) and Legates and McCabe (1999), both graphical comparison and quantitative statistics were followed for model evaluation. Two statistical indicators, coefficient of determination (R^2) and Nash and Sutcliffe efficiency (NSE), were used in this study, and those are expressed as:

$$R^2 = \left(\frac{\sum_{i=1}^{n} (O_i - O_{\text{avg}})(S_i - S_{\text{avg}})}{\sqrt{\sum_{i=1}^{n} (O_i - O_{\text{avg}})^2} \sqrt{\sum_{i=1}^{n} (S_i - S_{\text{avg}})^2}} \right)^2 \tag{18.2}$$

$$\text{NSE} = 1 - \left(\frac{\sum_{i=1}^{n} (O_i - S_i)^2}{\sum_{i=1}^{n} (O_i - O_{\text{avg}})^2} \right) \tag{18.3}$$

where

O_i is the ith measured value.

S_i is the ith simulated value.

O_{avg} is the average of the measured values.

S_{avg} is average of the simulated values and n is the total number of observations.

R^2 defines the amount of the total variance in the measured dataset that can be explained by the model simulation. It varies between 0 and 1, with higher values representing less error variance, and values >0.5 are considered satisfactory performance of the model (Santhi et al. 2001; Moriasi et al. 2007). Whereas NSE, a normalized statistic governs the relative magnitude of the residual variance compared to the observed data variance. It shows how well the plot of observed versus simulated data fits the 1:1 line. It ranges from $-\infty$ to 1.0, with NSE $= 1$ being the optimal value. Value in between 0 and 1 is treated as acceptable levels of performance, whereas

NSE < 0 represents the unacceptable performance of the model and in such case the mean observed information is considered better than model simulation.

18.5 Results and Discussion

18.5.1 Land Use–Land Cover Change

Figure 18.5 shows the land use–land cover distributions and its change over the Subarnarekha Basin and its sub-basins during different periods. Different types of forest and agricultural land mainly cover the entire basin (shown in Table 18.1). The remaining types include settlements, waterbodies, current fallow and barren land. However, the percentage of area under the agricultural land was higher than that of forest land over the entire study basin and in the sub-basins.

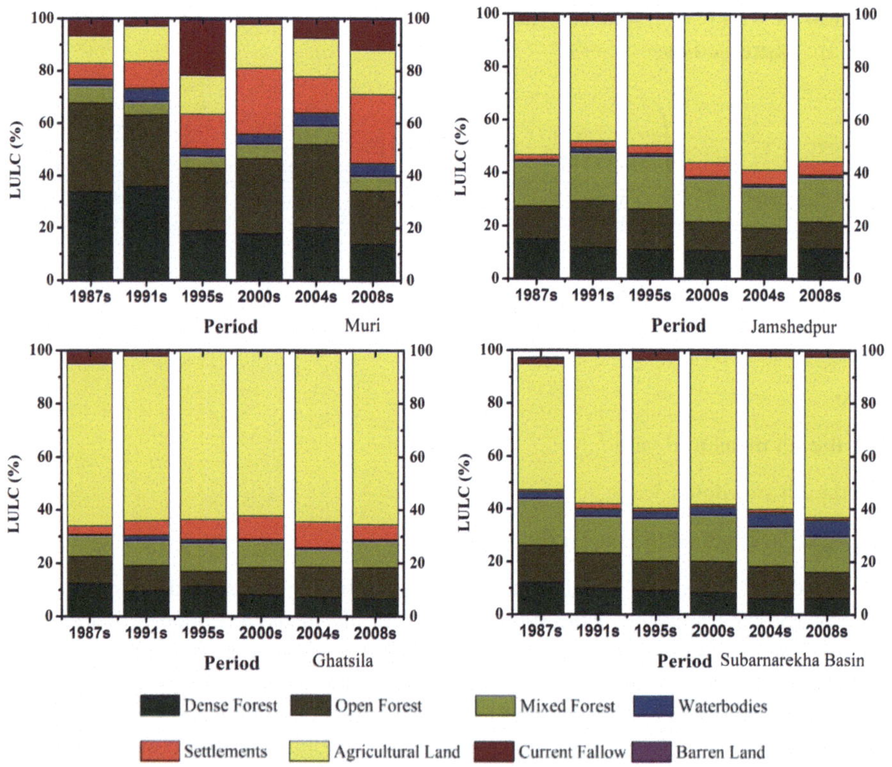

Fig. 18.5 Land use–land cover composition and its change in the Subarnarekha River Basin and its sub-basins

Table 18.1 Land use–land cover change during 1987–2008 over the Subarnarekha Basin and its sub-basins

Land use–land cover	Muri sub-basin		Jamshedpur sub-basin		Ghatsila sub-basin		Total basin	
	1987	2008	1987	2008	1987	2008	1987	2008
Dense forest	33.88	13.82	15.09	11.59	12.54	6.81	12.87	6.17
Open forest	33.65	20.30	12.28	9.95	9.98	11.43	13.95	7.89
Mixed forest	6.79	5.78	16.98	16.99	7.92	10.15	18.76	16.57
Waterbody	2.17	4.68	0.41	0.69	0.21	0.33	0.70	1.38
Settlements	6.25	26.39	1.94	5.05	3.29	5.63	1.66	6.89
Agriculture	10.62	17.04	50.67	54.92	61.15	65.42	47.8	57.7
Current fallow	6.32	11.59	2.57	0.78	4.91	0.22	2.56	2.85
Barren land	0.33	0.41	0.06	0.04	0.01	0	1.42	0.4

Figure 18.5 reveals the presence of semi-natural vegetation over the entire basin, with dense forest (canopy density > 40%) and open forest (10% < canopy density < 40%) covering the higher elevations in the northwest and southern part and mixed forest dominating the comparatively lower elevation. The middle and locale pockets of lower portions of the entire basin are mainly dominated by cities and surrounding settlements.

Figure 18.5 exhibits the area under the dense forest and open forest, which has drastically decreased from 12.87 to 6.17% and 13.95 to 7.89%. The area of settlement and agricultural land increased that support human population growth, city development, rural area expansion and agricultural development during 1987–2008 over the entire river basin. In the three sub-basins, the Muri sub-basin showed the maximum change in area under the dense forest and open forest which are decreased from 33.88 to 13.82% and 33.65 to 20.30% during 1987–2008 time periods, respectively.

Agricultural land occupies a sizable area. Rice, pulse, sugarcane, vegetables, oilseeds and other cereals are the major agricultural crops. Agricultural land, which covered 47.80% of the whole basin in 1987, has increased to 57.70% in the year of 2008 (shown in Table 18.1). Maximum change in the area under the agricultural land use was observed in the Muri sub-basin as compared to other two sub-basins, where the agricultural land has increased from 10.62 to 17.02% during 1987–2008 periods.

Settlements also include roads, urban area, industries and tourist spots. The industries like Heavy Engineering Corporation and Usha Martin Industries in Ranchi; Indian Aluminum and other industries at Muri; TELCO, Tata Steel, Indian Tube Company, etc., in Jamshedpur; and Uranium Corporation of India and Hindustan Copper Limited in Ghatsila are existing over the Subarnarekha River Basin. The increasing human populations resulting in high urbanization and agricultural development have resulted in increase of settlement by 161.98% during 1987–2008 periods. Here, some portions of agricultural land have converted into settlements. Few portions of forest also got transformed into settlements. Compared to the other two sub-basins, Muri shows the maximum change in area under the settlements which has increased from 6.25 to 26.39% during 1987–2008 due to higher industrial development in Ranchi and surrounding area. In Jamshedpur, the area under the settlements has increased from 1.94 to 5.05% during 1987–2008.

18.5.2 Parameter Sensitivity Analysis

Sensitivity analysis is the process of determining the significance of one or a combination of parameters with respect to the objective function or model output. Several methods have been developed to perform the sensitivity analysis (Veith and Ghebremichael 2009). Global sensitivity analysis was performed with the help of SWAT-CUP software. Seventeen physical and model process parameters (shown in Table 18.2) were used in the sensitivity analysis in order to capture the major processes represented by SWAT model. A t-test is employed to recognize the relative significance of each parameter. A value close to zero has more significance. Critical parameters were initial SCS runoff curve number for moisture condition II (CN2), base flow alpha factor (ALPHA_BF), available water capacity of soil layer (SOL_AWC) and depth from soil surface to bottom of layer (SOL_Z). Using the optimized value of different parameters, streamflow was estimated for those periods at three sub-basins: Muri, Jamshedpur and Ghatsila.

18.5.3 Calibration and Validation of SWAT

SWAT model has been calibrated on monthly scale for fifteen years (1987–2001) and validated (2002–2011) for streamflow at three stream gauging stations: Muri, Jamshedpur and Ghatsila within the Subarnarekha River Basin using land use–land cover for the year 1995. Calibration has been done through auto-calibration approach in SWAT-Calibration and Uncertainty Programs (SWAT-CUP) with Sequential Uncertainty Fitting Version 2 (SUFI-2). The graphical results of the SWAT model calibration for three gauging locations (Muri, Jamshedpur and Ghatsila) are shown in Fig. 18.6.

These results show that the model calibrated the streamflow at those stream gauging locations quite close to that of observed data. However, at Muri sub-basin, in the initial years, the peaks are not well matched as compared to other two sub-basins. Overall model captured the streamflow patterns quite correctly in all those three sub-basins. The validation of the model also shows quite good results as it follows the observed streamflow pattern quite well.

The statistical results of model calibration and validation are given in Table 18.3. These results show a quite high value of coefficient of determination (R^2) varying from a minimum 0.84 to a maximum 0.89 and Nash–Sutcliffe efficiency (NSE) from a minimum 0.87 to 0.88 during calibration and validation period among three

Table 18.2 Sensitivity analysis reports at three stations during calibration at monthly scale

Parameter name	Ghatshila			Jamshedpur			Muri		
	Rank	*t*-stat	*p*-value	Rank	*t*-stat	*p*-value	Rank	*t*-stat	*p*-value
v_esco.hru	16	1.66	0.10	16	1.58	0.11	16	1.46	0.14
a_gwqmn.gw	17	−2.12	0.03	17	−2.21	0.03	17	−2.03	0.04
r_cn2.mgt	1	14.08	0.00	1	2.41	0.02	1	−17.01	0.00
v_alpha_bf.gw	2	2.02	0.04	2	1.86	0.06	2	1.62	0.10
r_sol_awc(..).sol	3	0.00	1.00	3	−2.87	0.00	3	−4.60	0.00
r_sol_z(..).sol	4	0.02	0.99	4	0.02	0.98	4	0.02	0.98
v_revapmn.gw	5	0.33	0.74	5	0.50	0.62	5	0.44	0.66
v_ch_k2.rte	6	−0.11	0.91	6	0.28	0.78	6	0.36	0.72
v_gw_revap.gw	7	−0.34	0.74	7	−0.12	0.91	7	−0.10	0.92
v_epco.hru	8	0.68	0.50	8	−0.59	0.55	8	−0.47	0.64
v_canmx.hru	9	0.21	0.83	9	0.17	0.87	9	0.16	0.88
a_gw_delay.gw	10	−0.53	0.60	10	−0.40	0.69	10	−0.34	0.73
v_surlag.bsn	11	−0.72	0.47	11	−0.66	0.51	11	−0.77	0.44
r_sol_k(..).sol	12	−0.21	0.83	12	−0.25	0.80	12	−0.39	0.70
r_slsubbsn.hru	13	−1.29	0.20	13	−1.27	0.20	13	−1.18	0.24
r_sol.alb(..).sol	14	−0.66	0.51	14	−0.68	0.50	14	−0.41	0.68
v_ch_n2.rte	15	0.34	0.73	15	−0.01	0.99	15	0.18	0.86

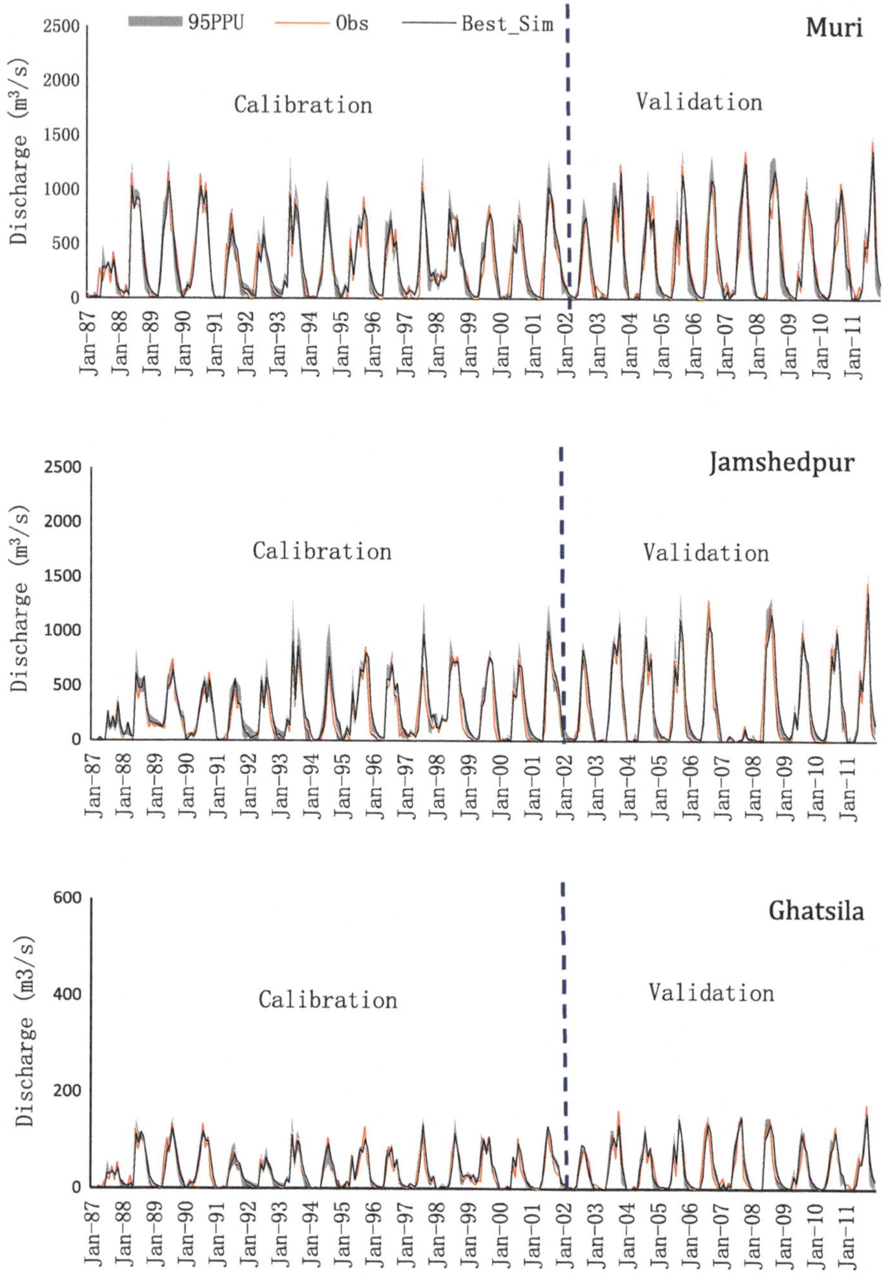

Fig. 18.6 Monthly calibration and validation of SWAT for streamflow at three stream gauging locations

Table 18.3 SWAT model performance in three calibrated sub-basins of the Subarnarekha Basin

Location	Statistical parameter	Streamflow (monthly)			
		Calibration		Validation	
		Obs	Sim	Obs	Sim
Muri	Mean	315.91	308.12	477.59	428.28
	SD	322.31	328.36	381.69	356.26
	NSE	0.86		0.85	
	R^2	0.84		0.88	
Jamshedpur	Mean	199.67	201.37	279.94	267.55
	SD	202.54	189.72	234.31	217.36
	NSE	0.88		0.88	
	R^2	0.86		0.85	
Ghatsila	Mean	38.75	34.99	44.69	48.2
	SD	40.22	38.14	47.08	41.08
	NSE	0.87		0.86	
	R^2	0.89		0.89	

gauging locations. These statistics are quite good along with quite near estimation of average streamflow by the model. The graphical as well as statistical results show quite satisfactory calibration and validation of the SWAT model for the Subarnarekha River Basin. Thus, the calibrated and validated model is used to assess the land use–land cover change impact on streamflow in the study river basin.

18.5.4 Impacts of Land Use–Land Cover Change on Surface Runoff

Three sub-basins (Muri, Jamshedpur and Ghatsila; Fig. 18.2) with observed land use–land cover changes were selected to assess the impacts of land use–land cover changes on streamflow at the sub-basin scale. After the calibration and validation by using baseline land use–land cover of 1995 (S3), SWAT model was simulated with five different scenarios: 1987 (S1), 1991 (S2), 2000 (S4), 2004 (S5) and 2008 (S6). The change in average monthly streamflow is used in order to simulate land use–land cover change impacts on streamflow dynamics. Change in monthly total streamflow for land use–land cover change scenarios S1, S2, S4, S5 and S6 was compared with the reference scenario S3 for those three sub-basins. An increase of total annual average streamflow between six scenarios is observed for all sub-basins followed by the order S1, S2, S3, S4, S5 and S6 (shown in Table 18.4). High rate of deforestation, urbanization and socioeconomic development rendered the classes under threatened and vulnerable and enhanced streamflow in all three sub-basins.

Table 18.4 Variations of annual average streamflow under different land use–land cover scenarios at three sub-basins

Sub-basin	1987s	1991s	1995s	2000s	2004s	2008s
Muri	368.26	378.03	387.37	399.58	415.32	440.18
Jamshedpur	313.13	329.75	334.49	344.43	364.74	380.94
Ghatsila	43.21	48.21	55.21	39.21	47.79	52.67

The conversion from dense forest to open forest has a higher impact toward the month of high precipitation (September). Hence, peak flows and high discharges increase with dense forest degradation. The results indicate that there is increasing trend of average annual streamflow in all three sub-basins due to degradation in forest cover and expansion in agricultural and settlement area.

The average monthly streamflow in all three sub-basins has been clearly depicted in Fig. 18.7, and average annual streamflow is given in Table 18.4. The data simulated from the present study show that annual average streamflow has been increased by 21.89% in Ghatsila sub-basin followed by Jamshedpur sub-basin (21.65%) while minimum (19.52%) was estimated in Muri sub-basin during 2002–2011.

18.6 Conclusion

The present study evaluated the impacts of ongoing and past land use–land cover changes on streamflow at sub-basin scale over the Subarnarekha River Basin using hydrological modeling and land use–land cover scenarios. The investigations showed that:

(i) The image classification study shows that the area under dense forest and open forest has drastically decreased in between 1987 and 2008 as changes were estimated to reduce from 12.87% to 6.17% and 13.95% to 7.89%, respectively, of the entire basin area (19,121 km^2). The area of settlement and agricultural land has increased (2.63% to 6.89% and 47.80% to 57.70%, respectively) which is due to urbanization, human population and increase of agricultural fields. Compared to other sub-basins, Jamshedpur sub-basin (6262.83 km^2) has undergone through the maximum change in area under the dense forest and open forest, where the area under those land covers has been decreased from 33.88% to 13.82% and 33.65% to 20.30% during 1987–2008 time periods, respectively. Jamshedpur also showed the maximum change in area under the settlements, where the area has been increased from 6.25% to 26.39% during 1987–2008 time periods.

(ii) The hydrological modeling study showed the calibration and validation of SWAT model brought to light a quite high value of coefficient of determination (R^2) varying from a minimum 0.84 to a maximum 0.89 and Nash–Sutcliffe efficiency (NSE) from a minimum 0.85 to 0.88 during calibration and validation

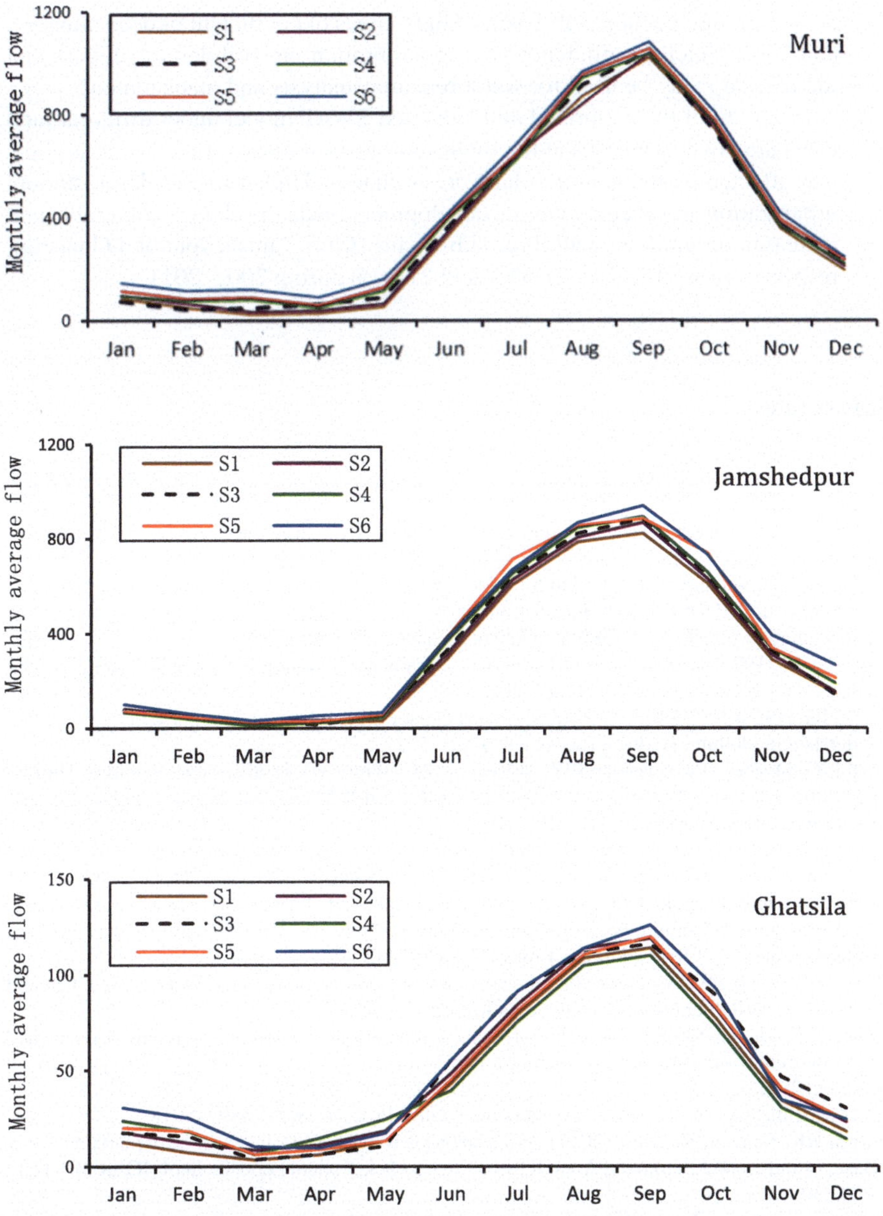

Fig. 18.7 Change in monthly streamflow for land use–land cover change scenarios S1, S2, S4, S5 and S6 compared to the baseline scenario S3 for the three sub-basins. Values shown are computed using the average monthly values of the evaluation period (2002–2011)

period among the three sub-basins. High values of coefficient of determination and Nash–Sutcliffe efficiency during calibration and validation suggest that model setup may be used for water resource analysis and management.

(iii) Application of the calibrated and validated SWAT model under different land use–land cover scenarios at the three sub-basins indicated that the streamflow was affected by the land use–land cover change. Higher rate of deforestation, urbanization and socioeconomic development made the classes vulnerable and increases streamflow in all three sub-basins (Muri, Jamshedpur and Ghatsila), respectively, by 19.52%, 21.65% and 21.89% during 2002–2011.

References

Abbaspour CK (2008) SWAT calibrating and uncertainty programs. A user manual. Eawag Zurich, Switzerland

Abbaspour KC, Johnson CA, Van Genuchten MT (2004) Estimating uncertain flow and transport parameters using a sequential uncertainty fitting procedure. Vadose Zone J 3(4):1340–1352

Abbaspour KC, Vejdani M, Haghighat S (2007) SWAT-CUP calibration and uncertainty programs for SWAT. In: MODSIM 2007 international congress on modelling and simulation, modelling and simulation society of Australia and New Zealand, 1596–1602

Anderson JR (1976) A land use and land cover classification system for use with remote sensor data (964). US Government Printing Office

Arnold JG, Fohrer N (2005) SWAT2000: current capabilities and research opportunities in applied watershed modelling. Hydrol Process 19(3):563–572

Arnold JG, Moriasi DN, Gassman PW, Abbaspour KC, White MJ, Srinivasan R, Santhi C, Harmel RD, Van Griensven A, Van Liew MW, Kannan N (2012) SWAT: Model use, calibration, and validation. Trans ASABE 55(4):1491–1508

Arnold JG, Srinivasan R, Muttiah RS, Williams JR (1998) Large area hydrologic modeling and assessment part I: model development. JAWRA J Am Water Resour Assoc 34(1):73–89

ASCE Task Committee on Definition of Criteria for Evaluation of Watershed Models of the Watershed Management Committee, Irrigation and Drainage Division (1993) Criteria for evaluation of watershed models. J Irrig Drainage Eng 119(3):429–442

Banko G (1998) A review of assessing the accuracy of classifications of remotely sensed data and of methods including remote sensing data in forest inventory

Behera S, Panda RK (2006) Evaluation of management alternatives for an agricultural watershed in a sub-humid subtropical region using a physical process based model. Agr Ecosyst Environ 113(1):62–72

Beven K, Binley A (2014) GLUE: 20 years on. Hydrol Process 28(24):5897–5918

Bhuyan NK, Sahu B, Rout SP (2014) Assessment of water quality index in Subarnarekha River Basin in and around Jharkhand Area. IOSR J Environ Sci, Toxicol Food Technol (IOSR-JESTFT) 8(11):39–45

Blöschl G, Sivapalan M (1995) Scale issues in hydrological modelling: a review. Hydrol Process 9(3–4):251–290

Bosch DD, Sheridan JM, Batten HL, Arnold JG (2004) Evaluation of the SWAT model on a coastal plain agricultural watershed. Trans ASAE 47(5):1493

Bouraoui F, Benabdallah S, Jrad A, Bidoglio G (2005) Application of the SWAT model on the Medjerda river basin (Tunisia). Phys Chem Earth, Parts A/B/C 30(8):497–507

Chanasyk DS, Mapfumo E, Willms W (2003) Quantification and simulation of surface runoff from fescue grassland watersheds. Agric Water Manag 59(2):137–153

Cheema MJM, Bastiaanssen WGM (2010) Land use and land cover classification in the irrigated Indus Basin using growth phenology information from satellite data to support water management analysis. Agric Water Manag 97(10):1541–1552

Congalton RG (1991) A review of assessing the accuracy of classifications of remotely sensed data. Remote Sens Environ 37(1):35–46

Deb P, Mishra A (2016) Forest cover change estimation using remote sensing and GIS—A study of the Subarnarekha River Basin, Eastern India. In: International conference on agriculture, food science, natural resource management and environmental dynamics: the technology, people and sustainable development, pp 165–171

Deb P (2017) A global sensitivity analysis tool for the parameters of a complex environmental model—SWAT. Rec Adv Environ Res 49

Deb P, Tarafdar S (2019) Land use land cover change and trend analysis of rainfall and temperature patterns in mid-Himalayan catchment using remote sensing data. Advancement in Basic and Applied Science. Ascent Publications, New Delhi, pp 243–263. ISBN: 978-93-84866-90-7

Dhar S, Mazumdar A (2009) Hydrological modelling of the Kangsabati river under changed climate scenario: case study in India. Hydrol Process 23(16):2394–2406

Eckhardt K, Breuer L, Frede HG (2003) Parameter uncertainty and the significance of simulated land use change effects. J Hydrol 273(1):164–176

Engel BA, Srinivasan R, Arnold J, Rewerts C, Brown SJ (1993) Nonpoint source (NPS) pollution modeling using models integrated with geographic information systems (GIS). Water Sci Technol 28(3–5):685–690

Fohrer N, Haverkamp S, Eckhardt K, Frede HG (2001) Hydrologic response to land use changes on the catchment scale. Phys Chem Earth Part B 26(7):577–582

Gassman PW, Reyes MR, Green CH, Arnold JG (2007) The soil and water assessment tool: historical development, applications, and future research directions. Trans ASABE 50(4):1211–1250

Ghaffari G, Keesstra S, Ghodousi J, Ahmadi H (2010) SWAT-simulated hydrological impact of land-use change in the Zanjanrood basin, Northwest Iran. Hydrol Process 24(7):892–903

Gosain AK, Rao S, Arora A (2011) Climate change impact assessment of water resources of India. Current Sci 356–371

Haan CT (2002) Statistical methods in hydrology. The Iowa State University Press.

Hughes DA (2008) Modelling semi-arid and arid hydrology and water resources: the southern Africa experience. In: Hydrological modelling in arid and semi-arid areas, pp 29–40

Immerzeel WW, Gaur A, Zwart SJ (2008) Integrating remote sensing and a process-based hydrological model to evaluate water use and productivity in a south Indian catchment. Agric Water Manag 95(1):11–24

Legates DR, McCabe GJ (1999) Evaluating the use of "goodness-of-fit" measures in hydrologic and hydroclimatic model validation. Water Resour Res 35(1):233–241

Mas JF (1999) Monitoring land-cover changes: a comparison of change detection techniques. Int J Remote Sens 20(1):139–152

Moradkhani H, Sorooshian S (2009) General review of rainfall-runoff modeling: model calibration, data assimilation, and uncertainty analysis. In: Hydrological modelling and the water cycle. Springer, Berlin, pp 1–24

Moriasi DN, Arnold JG, Van Liew MW, Bingner RL, Harmel RD, Veith TL (2007) Model evaluation guidelines for systematic quantification of accuracy in watershed simulations. Trans ASABE 50(3):885–900

Neitsch SL, Arnold JG, Kiniry JR, Williams JR (2008) Overview of soil and water assessment tool (SWAT) model. Soil and water assessment tool (SWAT) global applications. The World Association of Soil and Water Conservation, U.S.A., pp 03–23

Neitsch SL, Arnold JG, Kiniry JR, Williams JR (2011) Soil and water assessment tool theoretical documentation version 2009. Texas Water Resources Institute

Pandey A, Chowdary VM, Mal BC, Billib M (2009) Application of the WEPP model for prioritization and evaluation of best management practices in an Indian watershed. Hydrol Process 23(21):2997–3005

Patel DP, Gajjar CA, Srivastava PK (2013) Prioritization of Malesari mini-watersheds through morphometric analysis: a remote sensing and GIS perspective. Environ Earth Sci 69(8):2643–2656

Perrin C, Michel C, Andréassian V (2003) Improvement of a parsimonious model for streamflow simulation. J Hydrol 279(1):275–289

Santhi C, Arnold JG, Williams JR, Dugas WA, Srinivasan R, Hauck LM (2001) Validation of the swat model on a large rwer basin with point and nonpoint sources. JAWRA J Am Water Resour Assoc 37(5):1169–1188

Singh VP (1995) Computer models of watershed hydrology. Review

Sorooshian S, Hsu KL, Coppola E, Tomassetti B, Verdecchia M, Visconti G (eds) (2008) Hydrological modelling and the water cycle: coupling the atmospheric and hydrological models, vol 63. Springer Science & Business Media

Spruill CA, Workman SR, Taraba JL (2000) Simulation of daily and monthly stream discharge from small watersheds using the SWAT model. Trans ASAE 43(6):1431

Srinivasan R, Zhang X, Arnold J (2010) SWAT ungauged: hydrological budget and crop yield predictions in the Upper Mississippi River Basin. Trans ASABE 53(5):1533–1546

Strayer DL, Power ME, Fagan WF, Pickett ST, Belnap J (2003) A classification of ecological boundaries. Bioscience 53(8):723–729

Thunnissen HAM, Noordman EJM (1997) National land cover database of The Netherlands: classification methodology and operational implementation. Netherlands Remote Sensing Board (BCRS)

Turner II BL, Skole D, Sanderson S, Fischer G, Fresco L, Leemans R (1995) Land use land cover change science/research plan. IGBP Report no. 35

Veith TL, Ghebremichael LT (2009) How to: applying and interpreting the SWAT Auto-calibration tools. In: 2009 International SWAT conference proceedings, p 26

Vrugt JA, Ter Braak CJ (2011) DREAM (D): an adaptive Markov Chain Monte Carlo simulation algorithm to solve discrete, noncontinuous, and combinatorial posterior parameter estimation problems. Hydrol Earth Syst Sci 15(12)

Vrugt JA, Ter Braak CJ, Clark MP, Hyman JM, Robinson BA (2008) Treatment of input uncertainty in hydrologic modeling: doing hydrology backward with Markov chain Monte Carlo simulation. Water Resour Res 44(12)

Wagener T, Wheater HS (2006) Parameter estimation and regionalization for continuous rainfall-runoff models including uncertainty. J Hydrol 320(1):132–154

Wagner PD, Kumar S, Schneider K (2013) An assessment of land use change impacts on the water resources of the Mula and Mutha Rivers catchment upstream of Pune, India. Hydrol Earth Syst Sci 17(6):2233–2246

William ER, William BM, Turner BL (1994) Modeling land use and land cover as a part of global environmental change. Clim Change (28):45–64

Yadav SK, Singh SK, Gupta M, Srivastava PK (2014) Morphometric analysis of Upper Tons basin from Northern Foreland of Peninsular India using CARTOSAT satellite and GIS. Geocarto Int 29(8):895–914

Yang J, Reichert P, Abbaspour KC, Xia J, Yang H (2008) Comparing uncertainty analysis techniques for a SWAT application to the Chaohe Basin in China. J Hydrol 358(1):1–23

Yates D, Sieber J, Purkey D, Huber-Lee A (2005) WEAP21—a demand-, priority-, and preference-driven water planning model: part 1: model characteristics. Water Int 30(4):487–500

Zhao F, Xu Z, Zhang L (2012) Changes in streamflow regime following vegetation changes from paired catchments. Hydrol Process 26:1561–1573

Zheng Y, Keller AA (2008) Stochastic watershed water quality simulation for TMDL development–a case study in the Newport Bay watershed. JAWRA J Am Water Resour Assoc 44(6):1397–1410

Chapter 19
Hydrological Modeling of West Rapti River Basin of Nepal Using SWAT Model

Shekhar Nath Neupane and Ashish Pandey

Abstract In this study, hydrological modeling of the West Rapti River, Nepal was carried out for estimation of runoff and sediment yield using the Soil and Water Assessment Tool (SWAT) model. The water balance components of the SWAT model, viz. precipitation, surface runoff, lateral flow, groundwater recharge, actual evapotranspiration and potential evapotranspiration, were studied. The SWAT model setup was carried out for simulation of discharge and sediment yield on a monthly basis for the years 2000–2013 (14 years). Calibration and validation of the model were carried out using SWAT-CUP with Sequential Uncertainty Fitting (SUFI-2) technique. The model was calibrated and validated for the years 2003–2006 and 2007–2009, respectively, using measured streamflow and rating curve generated sediment data. The model performed well for both calibration and validation periods. The model showed reliable estimates of monthly runoff ($R_2 = 0.96$, NSE = 0.95, PBIAS = 4.7 and RSR = 0.22) and sediment yield ($R_2 = 0.71$, NSE = 0.68, PBIAS = 15.10 and RSR = 0.57) for the calibration period. During validation period, model results were lesser than the calibration period flow runoff ($R_2 = 0.78$, NSE = 0.78, PBIAS = 5.3 and RSR = 0.47) and sediment yield ($R_2 = 0.69$, NSE = 0.69, PBIAS = −9.70 and RSR = 0.56). The water balance study revealed that evapotranspiration is more predominant and accounting 48.60% of the average annual precipitation falling over the basin. The annual volume of water available at the basin outlet is 4.5 billion cubic meters (BCMs). The average annual sediment yield of the basin is 17.67 t/ha/year, and the study area lied under high erosion class. Further, the validated SWAT model was also employed for evaluation of the best management practices (BMPs) in the study area.

S. N. Neupane (✉)
Department of Water Resources and Irrigation, Jawalakhel, Lalitpur, Nepal
e-mail: neupaneshekhar@yahoo.com

A. Pandey
Department of Water Resources Development and Management, Indian Institute of Technology Roorkee, Roorkee 247667, Uttarakhand, India
e-mail: ashish.pandey@wr.iitr.ac.in

A. Pandey et al. (eds.), *Water Management and Water Governance*, Water Science and Technology Library 96,
https://doi.org/10.1007/978-3-030-58051-3_19

Keywords Hydrological modeling · Water balance · SWAT model · SWAT-CUP · SUFI-2 · Sediment yield

19.1 Introduction

Water is precious, and its proper use in integrated manner helps to bring prosperity to the society. The proper and accurate assessment of water availability from a basin has always been the concern of water resource planner, designer and developer. Hydrological models developed for basin scale give the total water availability of the basin, and the information can be used for formulation of different sectoral water resources project development plan. There is a long history of hydrological modeling, and it was first started by Irish engineer Thomas James Mulvaney who developed a single equation $Q_p = \text{CAR}$, relating peak flow not the whole hydrograph, with area (A), rainfall intensity (R) and the empirical constant (C) of the catchment commonly known as rational method (Beven 2012). After that in recent decades, many computer-based mathematical hydrological models have been developed by hydrologists, team of hydrologists and collaborating hydrological institutes (Dhami and Pandey 2013), and SWAT is one of them. SWAT has been extensively used by different researchers in different sectors. There are wide variation and diversification of SWAT application. Some past and present field application of SWAT model is hydrological simulation of gauged and un-gauged catchments, sediment yield modeling, hydropower potential assessment, effect of land use/land cover on runoff generation and sediment yield, soil water recharge, tile flow and related studies, snowmelt-related application, study of impact of climate change, drought analysis, pollutant loss studies, applications incorporating wetlands, reservoirs and other impoundment, land use impact on pollutant studies, sensitivity, calibration and uncertainty analysis, DEM resolution, soil and land use resolution effects and comparison of SWAT model with other models (Gassman et al. 2007). Bieger et al. (2013) used SWAT model to study the runoff generation, sediment yield and water balance in Xiangxi Catchment due to change in land use caused by construction of Three Gorges Dam in China. Uniyal et al. (2015) used SWAT model to study the impact of climate change on water balance of the Upper Baitarani River basin of India. Pandey et al. (2014) used SWAT model and spatial technologies to assess the water availability for hydropower potential of Mat River basin of Southern Mizoram, India. Dahal et al. (2016) used SWAT model to study the impact of climate change in hydrology of Bagmati River basin of Nepal. Likewise, Tripathi et al. (2005) used SWAT model to prepare effective management plan for critical subwatershed of Nagwan watershed, India. Though the SWAT model has not been used to large extent in Nepal, the review in worldwide application of SWAT model indicates that it is capable of and can be employed to simulate hydrological processes of West Rapti River basin with reasonable accuracy.

19.2 Materials and Methods

19.2.1 Study Area

The study area, West Rapti River basin (5281 km^2), lies in the mid-western region of Nepal, and it extends from 27° 56′ 50″ to 28° 02′ 30″ N latitude and from 81° 45′ 00″ to 81° 40′ 00″ E longitude as shown in Fig. 19.1. The West Rapti River originates from the middle hills of Nepal and then enters to the flat area and finally drains in India to join the Ganges River. The elevation of the study area ranges from 153 m amsl to 3626 m amsl, and the source of runoff is due to the monsoon rainfall and groundwater. The average annual rainfall of the study area is 1600 mm. The runoff generation is mainly due to monsoon rainfall extending from June to September, accounting about 80% of the total annual rainfall. Temperature in lower part of study area goes up to 46 °C in summer and falls below 2 °C in upper part in winter (Talchabhadel and Sharma 2014).

19.2.2 SWAT Model

The SWAT model developed by USDA is a physical-based continuous long-term yielding model which can be used to study the impact of different land management practices on water, sediment and agriculture chemical yield in large complex watershed under various condition of soil, land use and management over a long course of time (Neitsch et al. 2011). It is capable of simulating water balance taking into account of rainfall, evapotranspiration, surface and sub-surface runoff and deep aquifer recharge (Suryavanshi et al. 2017). It is a complete interdisciplinary watershed modeling tool. In SWAT model, watershed is divided into subwatershed and subwatershed further into hydrological response units (HRUs). HRUs are the areas of similar type of land use, slope and soil characteristics, and it represents the percentage of watershed area. Similarly, subwatershed is characterized by dominant land use, soil type and management practices (Gassman et al. 2007; Kalcic et al. 2015). In this study, ArcSWAT interface of the SWAT-2012 model was applied.

19.2.3 Hydrology

Whatever may be the study using SWAT model, water balance is the key force behind every process in a watershed. SWAT model simulates hydrological process based on following water balance equation (Neitsch et al. 2008).

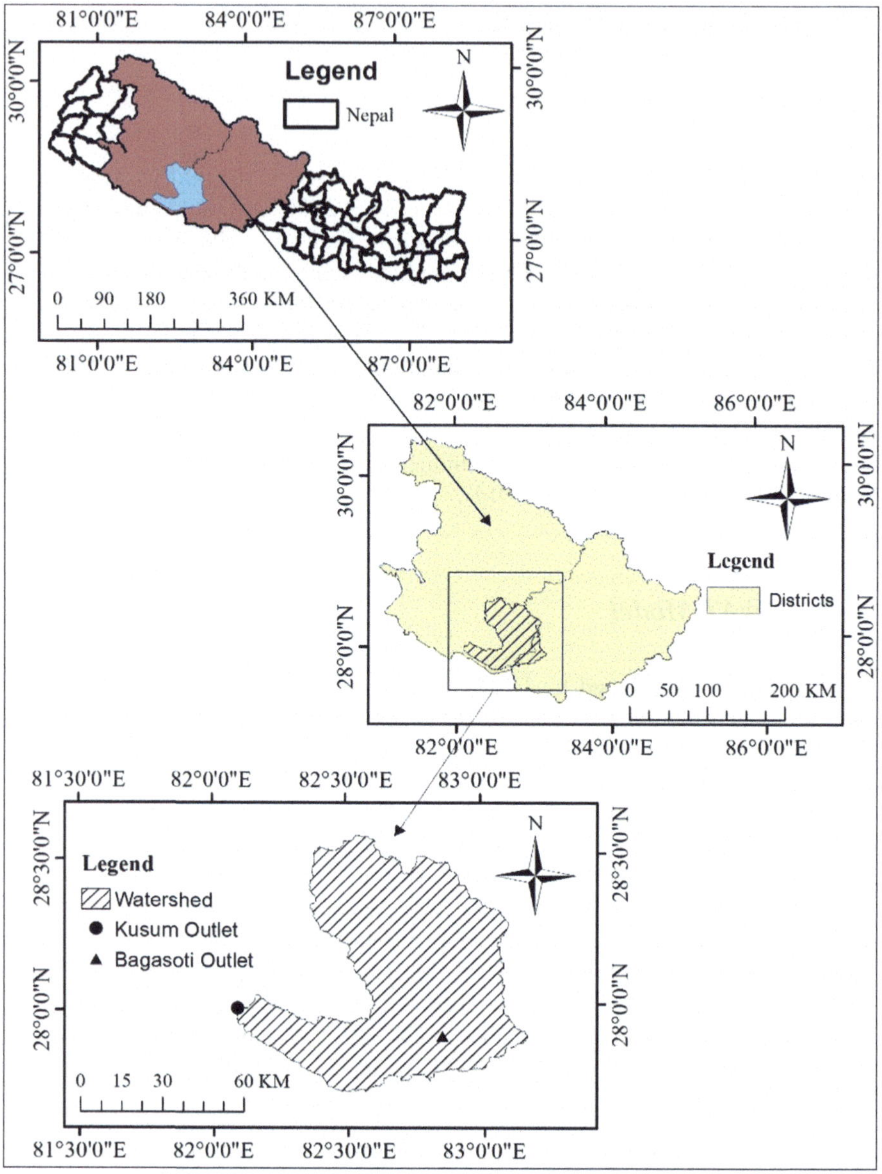

Fig. 19.1 Location map of the study area

$$\mathrm{SW}_t = \mathrm{SW}_0 + \sum_{i=1}^{1} \left(R_{\mathrm{day}} - Q_{\mathrm{surf}} - E_{\mathrm{a}} - W_{\mathrm{seep}} - Q_{\mathrm{gw}} \right) \tag{19.1}$$

where SW_t = final soil water content (mm H_2O), SW_0 = initial soil water content (mm H_2O), t = time in days, R day = amount of precipitation on day i (mm H_2O), Q_{surf} = amount of surface runoff on day i (mm H_2O), E_{a} = amount of evapotranspiration on day i (mm H_2O), W seep = amount of percolation and bypass exiting the soil profile bottom on day i (mm H_2O) and Q_{gw} = amount of return flow on day i (mm H_2O). To simulate runoff from daily rainfall in SWAT, SCS curve number method has been used in this study. The SCS curve number equation used in SWAT model is given as

$$Q = \frac{(P - I_{\mathrm{a}})^2}{(P - I_{\mathrm{a}}) + S} \tag{19.2}$$

where Q = depth of runoff in (mm), P = effective precipitation in (mm), I_{a} = initial abstraction of water in (mm) and S = maximum potential retention (mm). The initial abstraction of water (I_{a}) is the function of maximum potential retention S and can be expressed as $I = kS$, where $k = a$ constant value usually taken as 0.2 or 20%. So, $I_{\mathrm{a}} = 0.2\mathrm{S}$. Hence, Q and S can be obtained from Eqs. 19.3 and 19.4, respectively

$$Q = \frac{(P - 0.2S)^2}{(P - 0.8S)} \tag{19.3}$$

$$S = \left(\frac{25400}{\mathrm{CN}} \right) - 254 \tag{19.4}$$

The curve number drops as the soil approaches the wilting point and increases to near 100 as the soil approaches to saturation. Likewise, Modified Universal Soil Loss Equation (MUSLE) was used in the SWAT for sediment yield modeling.

19.2.4 Data Collection and Processing

Preliminary dataset required for the SWAT model as input is digital elevation model (DEM), climatological data (precipitation, maximum and minimum temperature, relative humidity, wind speed and solar radiation), and land use and soil data. Further, for validation and calibration of the model, river flow and sediment data are required and have been used in this study. The overall dataset used in this study is explained as below.

19.2.4.1 Digital Elevation Model (DEM)

Cartosat-1 DEM of 30 m resolution was downloaded from the Web site of National Remote Sensing Centre (NRSC) and Indian Space Research Organization (ISRO) open data and product archive portal. DEM gives the flow behavior and flow pattern within the watershed (Narsimlu et al. 2015). DEM has been used for delineation of watershed characteristics like boundary of watershed, drainage patterns and network, slope and length of channel. The watershed elevation varies from 153 to 3626 m above mean sea level.

19.2.4.2 Land Use/Land Cover (LU/LC)

In this study, "Land cover of Nepal 2010" at 30 m spatial resolution developed by the International Centre for Integrated Mountain Development (ICIMOD) was downloaded from Web site of ICIMOD. The LU/LC was then reclassified into seven land use/cover classes, and lookup table was generated to provide as an input in the SWAT model. The area under mixed forest was 59.03%, followed by agricultural area of 35.18% of the total area (Fig. 19.2).

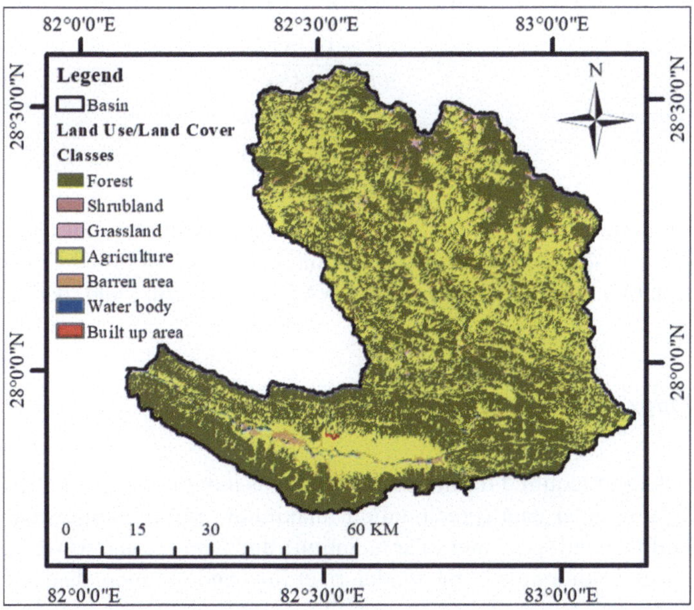

Fig. 19.2 LULC map of the study area

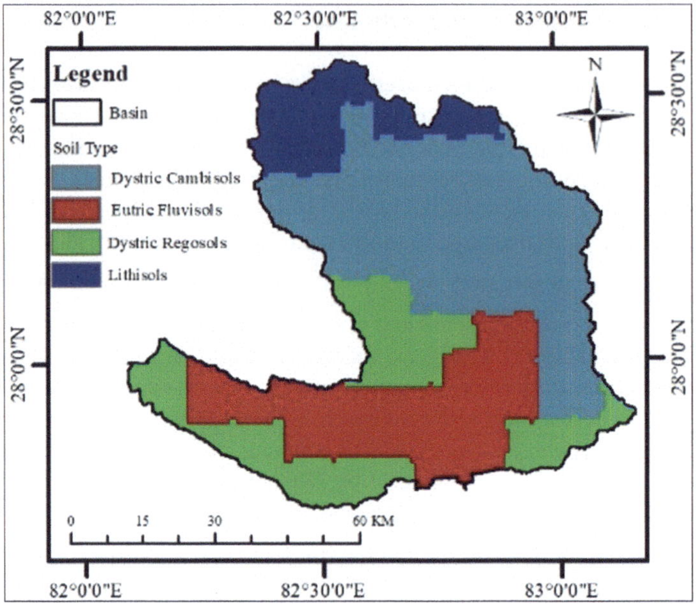

Fig. 19.3 Soil map of the study area

19.2.4.3 Soil Data

The soil map used for this study was prepared by analyzing and merging the soil and terrain (SOTER) database compiled by Food and Agriculture Organization (FAO) and Nepal's Survey Department that is freely available from the Web site of International Soil Reference and Information Centre (ISRIC)—World Soil Information (https://www.isric.org/) or can be obtained from Department of Survey, Nepal. Four types of soils, namely Dystric Cambisols, *Eutric* Fluvisols, Dystric Regosols and Lithosols, are identified, and physical properties of the available water content, soil texture, soil bulk density, hydraulic conductivity and organic matter were assessed for different layers. Out of four soil classes, first three soils are of loamy texture belonging to hydrological soil group (HYDGRP) C and remaining one is clay loam texture belonging to HYDGRP D (Fig. 19.3).

19.2.4.4 Meteorological Data

Daily rainfall data of different meteorological stations were collected from Department of Hydrology and Meteorology, Nepal (DHM, Nepal). Other climatological data like maximum and minimum temperature, solar radiation, relative humidity and wind speed were taken from global weather data for SWAT (https://globalweather.tamu.edu/).

19.2.4.5 Stream Discharge and Sediment

Daily discharge data in m^3/s, measured at Kusum and Bagasoti Gaun stations for the years 2003–2013, and the suspended sediment data in PPM measured at the Bagasoti Gaun station for the year 1978 and 1985–1988 were obtained from the Department of Hydrology and Meteorology, Nepal (DHM, Nepal). The sediment load in ton/day was calculated and sediment rating curve was developed using available sediment load and discharge data of Bagasoti Gaun station. The rating curve (power equation) developed was $S\left(\frac{\text{ton}}{\text{day}}\right) = 0.59Q^{1.97}$ with correlation coefficient, $R^2 = 0.75$ where S is sediment load and Q is the daily discharge in m^3/s. The correlation coefficient developed was weak because of scattered sediment data for same discharge. Therefore, to develop a suitable rating curve, the daily stream discharges were arranged in descending order and were regrouped into different classes. For lower discharge, narrower class interval was adopted, and for higher discharge, wider class interval was adopted as suggested by Khanchoul et al. (2007). The mean of streamflow and sediment load was calculated, converted to log values and plotted to obtain the rating equation. The rating equation obtained was $S\left(\frac{\text{ton}}{\text{day}}\right) = 1.64Q^{1.98}$ with $R^2 = 0.97$. Using this developed rating curve, sediment load was calculated and percentage error on sediment load estimation was calculated using relation, $\text{error}\% = \left(\frac{\text{rating curve estimate}}{\text{load from measurement}} - 1\right) \times 100$ and the percentage error was found to be -19% which is assumed to be satisfactory and the negative value indicated underestimation of sediment. Based on developed final sediment discharge rating equation, sediment load ton/day was calculated, and from the developed daily data series, monthly sediment data was developed which was used for calibration and validation of the SWAT model. The developed discharge sediment rating curve using all point data and the mean values of the class interval are given in Figs. 19.4 and 19.5.

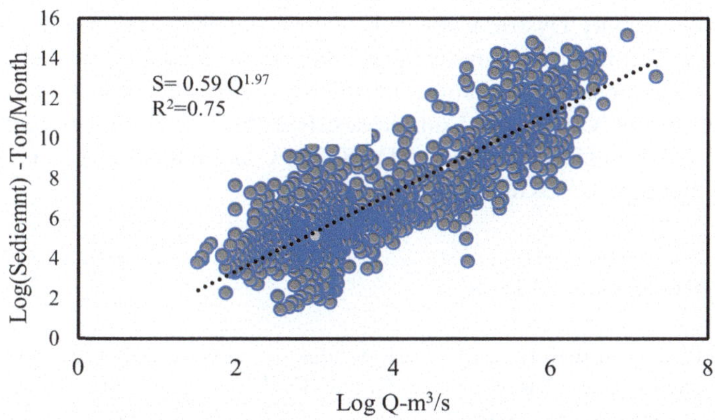

Fig. 19.4 Rating curve developed using measured suspended sediment concentrations and discharges

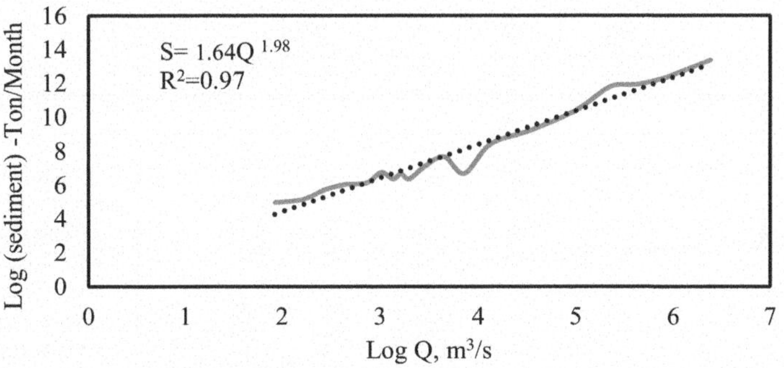

Fig. 19.5 Sediment rating curves developed using mean water discharges and mean suspended sediment concentrations of all data

19.2.5 SWAT Model Setup

19.2.5.1 Watershed Delineation

Watershed delineation was done within ArcSWAT interface. DEM projected to UTM 44 Zone was given as input and stream definition, outlet and inlet definition, selection of watershed outlets, and definition and calculation of subbasin parameters were the works carried under watershed delineation. In this study, watershed was divided into 18 subwatershed and same number of outlets including manually added two outlets. Stream networks and outlets were created considering flow accumulation from least area up to 155 km^2. The conceptual framework of SWAT model (Gull et al. 2017; Getache and Melesse 2012) is presented in Fig. 19.6.

19.2.5.2 Hydrological Response Units (HRUs)

Hydrological response units (HRUs) are the smallest spatial unit of the model consisting of similar land use, soil and slope within the subbasin (Kalcic et al. 2015). In ArcSWAT, LU/LC map, soil map and the delineated watershed were overlapped and multiple slope classes were given for HRUs definition (Worku et al. 2017). The LU/LC class and the soil type class were given using lookup table. Multiple HRUs definition was adopted, and a threshold level of 15, 10 and 15% for land use, soil and slope class was taken to eliminate minor land use, soil and slope class. As SWAT eliminates the area less than threshold value, it will recount the remaining land use, soil and slope as 100% (Gull et al. 2017). These threshold values were chosen as suggested on SWAT user's manual by Winchell et al. (2013). The process of HRUs definition using SWAT model is shown in Fig. 19.6.

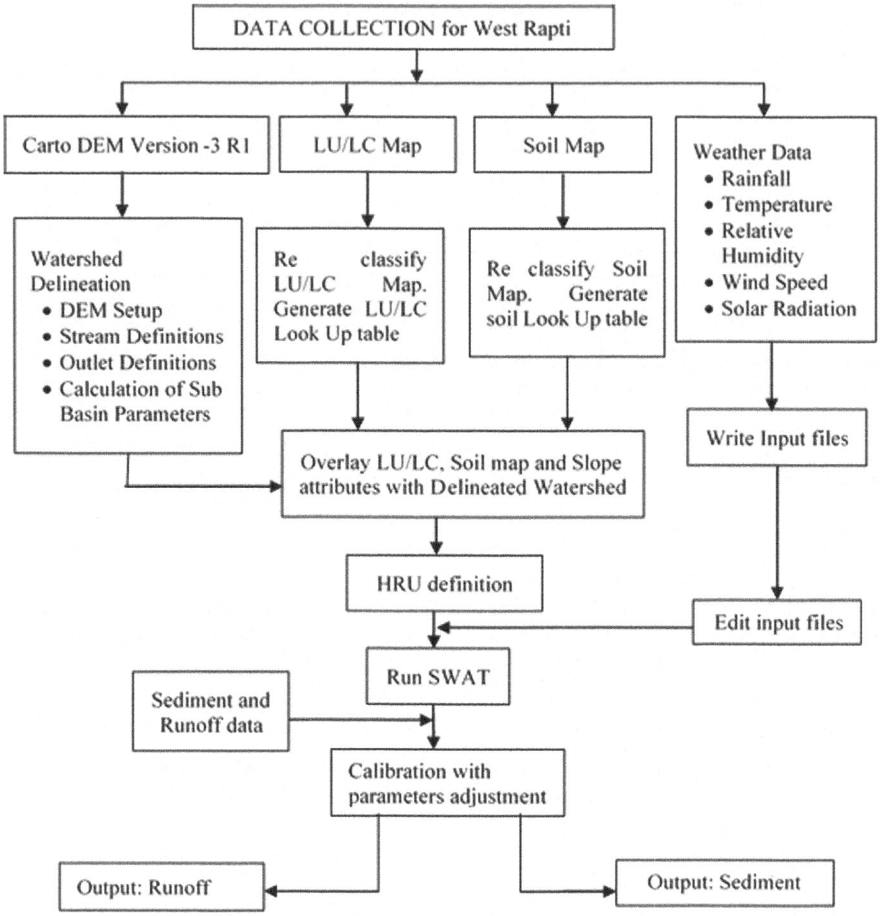

Fig. 19.6 Conceptual framework of SWAT model (Gull et al. 2017; Getache and Melesse 2012)

19.2.5.3 Uncertainty Analysis

SWAT-CUP: SWAT Calibration and Uncertainty Program was used for calibration, validation and sensitivity analysis. Within SWAT-CUP, Sequential Uncertainty Fitting (SUFI-2) program was used for verification of uncertainty arising from measured data, calibration parameters and model uncertainty. SUFI-2 accounts for all the sources of uncertainties and quantifies them in terms of *p*-factor and *r*-factor; i.e., the degree of uncertainty and goodness of fit was assessed by *p*-factors and *r*-factors. *P*-factor is the percentage of observations data captured within 95% PPU, and *r*-factor indicates the average thickness of the 95 PPU band divided by the standard deviation of the observed values showing the degree of uncertainty. The theoretical value of *p*-factor ranges from 0 to 100% that of *r*-factor ranges from 0 to ∞. The *p*-factor of 1 and *r*-factor of zero are the ideal conditions of simulation, i.e., exact matching of simulated data with observed ones. While calibration and validation of model, our concern is always getting reasonable values of these two factors. Most of the observations are captured in 95 PPU band (*p*-factor near to 1), and at the same time, smaller envelope (smaller *r*-factor) is desired. Therefore, a balance between *p*-factor and *r*-factor is required to judge the strength of calibration (Abbaspour 2015; Worku et al. 2017). For discharge, *p*-factor > 0.7 is recommended to be enough and *r*-factor around 1 depending upon situation would be desirable as per Abbaspour et al. (2015). For sediment, slight deviations (smaller *p*-factor and larger *r*-factor) on above values for discharge are acceptable as accurate estimation of sediment is quite tough job.

19.2.5.4 Performance Evaluation of Model

Performance evaluation of the model was carried out in SWAT-CUP using four objective functions, i.e., Nash–Sutcliffe efficiency, 1970 (NSE), coefficient of determination (R^2), ratio of root-mean-square error to the standard deviation of measured data (RSR) and the percentage bias (PBIAS). The R^2, NSE, PBIAS and RSR are given in Eqs. 19.5, 19.6, 19.7 and 19.8, respectively.

$$R^2 = \frac{\left[\sum_i (Q_{m,i} - \overline{Q}_m)(Q_{s,i} - \overline{Q}_s)\right]^2}{\sum_i \left(Q_{m,i} - \overline{Q}_m\right)^2 \sum_i \left(Q_{s,i} - \overline{Q}_s\right)^2} \quad (19.5)$$

$$\text{NS} = 1 - \frac{\sum_i (Q_m - Q_s)_i^2}{\sum_i \left(Q_{m,i} - \overline{Q}_m\right)^2} \quad (19.6)$$

$$\text{PBIAS} = 100 * \frac{\sum_{i=1}^{n} (Q_m - Q_s)_i}{\sum_{i=1}^{n} Q_{m,i}} \quad (19.7)$$

$$\text{RSR} = \frac{\text{RMSE}}{\text{Std Dev}_{\text{obs}}} = \frac{\sqrt{\sum_{i=1}^{n} (Q_m - Q_s)_i^2}}{\sqrt{\sum_{i=1}^{n} \left(Q_{m,i} - \overline{Q}_m\right)^2}} \quad (19.8)$$

where Q is the variable, '*m*' stands for measured, '*S*' stands for simulated values, bar stands for average, and *i* is the *i*th measured or simulated variable. The PBIAS value shows the deviation (in percentage) of simulated values from observed values (Van Liew et al. 2007). PBIAS is the measure of average tendency of the simulated values to be larger or smaller than their observed values. PBIAS with optimal value zero is assumed to be best, and deviation toward negative or positive from zero indicates model simulation is biased. The negative value of PBIAS is the condition of underestimation, and positive value is the condition of overestimation (Gupta et al. 1999). The ideal value for RSR is 0 and increases toward positive value. Zero value of RSR means zero root-mean-square error (RMSE) or residual variation indicating perfect simulation of model. So, lower the RSR, lower the RMSE and better the model simulation performance (Moriasi et al. 2007).

19.3 Results and Discussion

19.3.1 Model Sensitivity Analysis, Calibration and Validation

19.3.1.1 Sensitivity Analysis

Selection of sensitive parameters based on sensitivity analysis makes calibration and validation process easier and saves time. Therefore, based on SWAT, SWAT-CUP user manual and as suggested by Abbaspour et al. (2007), a list of parameters sensitive to flow and sediment was prepared. One at a time, sensitivity analysis was performed in SWAT-CUP and 22 parameters sensitive to flow and sediment were selected for calibration and validation. Further, for model calibration and validation in the beginning, as flow is the main controlling variable, parameters sensitive to flow only were selected (Abbaspour et al. 2007) and the calibration was done. After calibrating for flow, keeping flow parameter ranges as obtained from flow calibration, parameters sensitive to sediment were added again as suggested by Abbaspour et al. (2007). After that, global sensitivity analysis was carried out (Abbaspour 2015). List of parameters sensitive to flow and sediment with SWAT-CUP fitted value was presented as in Table 19.1 on the basis of *t*-stat and *p*-value. Higher the absolute *t*-stat value and smaller *p*-value, more sensitive the parameter is (Abbaspour 2015).

OAT sensitivity analysis is done for the identification of parameters that are sensitive to the model developed for the study area, and the global sensitivity analysis evaluates the effects of relative changes on a number of distributed parameters (selected from OAT sensitivity analysis) on the model output and ranks the parameters based on their final effects. OAT sensitivity analysis gives *t*-stat and *p*-value. Higher the absolute *t*-stat value and smaller *p*-value, the parameters are assumed to be more sensitive. Rank 1 is for the maximum effect, and lowest rank equal to the number of parameters chosen is for smallest effect.

The results of sensitivity analysis presented in Table 19.1 show that USLE support practice factor (USLE_P) is the most sensitive parameter followed by aquifer percolation coefficient and (RCHRG_DP) and so on. Other parameters like average channel slope along channel length (CH_S1), groundwater delays (GW_DELAY), Manning's n value for main channel (CH_N2), SCS runoff curve number (CN2), peak rate adjustment factor for sediment routing in the main channel (PRF_BSN), average slope of main channel (CH_S2), average slope steepness (HRU_SLP) and soil erodibility factor (USLE_K) are also ranked to top sensitive parameters, and average slope length (SLSUBBSN), soil bulk density (SOL_BD), effective hydraulic conductivity

Table 19.1 Final calibrated parameter with their global sensitivity rank and calibrated parameter with range

Rank	Parameter	Physical description	Range of calibrated parameter			*t*-stat	*P*-value	Sensitive to
			Min.	Max.	Fitted			
1	R_USLE_P.mgt	USLE support practice factor	−0.92	−0.66	−0.79	6.83	0.00	Sediment
2	V_RCHRG_DP.gw	Aquifer percolation coefficient	0.25	0.90	0.39	−6.78	0.00	Flow
3	V_CH_S1.sub	Average channel slope along channel length, mm/meter	−0.03	4.66	0.68	5.79	0.00	Sediment
4	V_GW_DELAY.gw	Groundwater delays, days	32.70	110.93	32.99	−5.61	0.00	Flow
5	V_CH_N2.rte	Manning's n value for the main channel	0.01	0.30	0.10	−3.79	0.00	Sediment
6	R_CN2.mgt	SCS runoff curve number for moisture condition II	−0.22	−0.11	−0.16	3.64	0.00	Flow, Sediment
7	V_PRF_BSN.bsn	Peak rate adjustment factor for sediment routing in the main channel	0.20	1.50	1.39	2.74	0.01	Sediment
8	V_CH_S2.rte	Average slope of main channel m/m	0.24	0.68	0.57	1.67	0.10	Sediment
9	R_HRU_SLP.hru	Average slope steepness, m/m	−0.15	0.02	−0.14	1.44	0.15	Flow, Sediment
10	R_USLE_K(..).sol	Soil erodibility factor	−0.19	0.27	0.17	1.43	0.15	Flow, Sediment
11	R_SOL_Z(..).sol	Depth from soil surface to bottom of layer, mm	−0.03	0.42	0.23	−1.40	0.16	Flow, Sediment
12	V_ADJ_PKR.bsn	Peak rate adjustment factor for sediment routing in the subbasin (tributary channels)	0.50	2.00	1.59	1.35	0.18	Sediment
13	V_CH_ERODMO(..). rte	Monthly channel erodibility factor	0.40	0.60	0.46	−1.30	0.19	Sediment

(continued)

Table 19.1 (continued)

Rank	Parameter	Physical description	Range of calibrated parameter			*t*-stat	*P*-value	Sensitive to
			Min.	Max.	Fitted			
14	V_CH_N1.sub	Manning's n value for tributary channel	−0.21	3.26	1.06	−1.17	0.24	Sediment
15	V_OV_N.hru	Manning's n value for overland flow	0.01	5.00	1.81	1.09	0.28	Sediment
16	V_ESCO.hru	Soil evaporation compensation factor	0.09	0.90	0.71	0.98	0.33	Flow
17	V_CANMX.hru	Maximum canopy storage	43.24	89.76	43.41	0.76	0.45	Flow
18	R_SLSUBBSN.hru	Average slope length	−0.35	0.05	−0.18	0.75	0.45	Sediment
19	R_SOL_BD(..).sol	Soil bulk density	−0.47	0.11	−0.02	0.73	0.46	Flow, Sediment
20	V_CH_K1.sub	Effective hydraulic conductivity in tributary channel alluvium	0.00	10.00	1.84	−0.41	0.68	Sediment
21	R_SOL_K(..).sol	Soil hydraulic conductivity	−0.94	0.22	−0.72	0.17	0.87	Flow, Sediment
22	V_CH_COV2.rte	Channel cover factor	0.36	1.00	0.58	0.07	0.94	Sediment

Note v_ means parameter value which is to be replaced by given value within the range, and r_ means the parameter value which is multiplied by (1+ the give value) in tributary channel alluvium (CH_K1), soil hydraulic conductivity (SOL_K1) and channel cover factor (CH_COV2) are ranked as least sensitive parameters and lie in the bottom part of the sensitivity table.

19.3.1.2 Model Calibration and Validation

The SWAT model was run for the period of 2000–2013, taking three years as warm-up period. Before calibration inconsistency in the hydrological data of the years 2003–2013 were checked by mass curve method and it was found that the data of the years 2010–2013 were inconsistent. The inconsistency in dataset may be mainly due to error in rating curve. Thus, model was calibrated and validated using shorter datasets, and in this context, Cui et al. (2015) studied the effect of duration of the observed dataset available to calibrate the distributed hydrological model SWAT in the Heihe Basin of China. They made comparison of results from calibration of single-year and three-year datasets of discharge, and the result obtained were same for both the cases of datasets. And it indicates that one can use discharge data of limited durations to calibrate the SWAT model effectively in poorly gauged basins or in case of basin with data availability for shorter period due to erroneous data recording. Thus, SWAT model was calibrated and validated using monthly datasets of shorter duration, i.e., from 2003 to 2006 (4 years-calibration) and from 2007 to 2009 (3 years-validation).

Moriasi et al. (2007) recommended statistics for SWAT model performance ratings as presented in Table 19.2 based on Saleh et al. (2000) and Van Liew et al. (2007). The calibration and validation results are presented in Table 19.3.

Based on the values of R^2, NSE, PBIAS and RSR and performance ratings provided in Table 19.3, it is confirmed that model performance is good for both calibration and validation.

19.3.2 Graphical Representation of Discharge and Sediment Yield

The SWAT model performance can be visualized by graphical way in the form of 95 PPU. The 95 PPU plots for sediment and discharge obtained from SWAT-CUP (calibration and validation) for different stations were as shown in Fig. 19.7.

Statistical as well as geographical evaluation showed that the simulation of streamflow and sediment yield during calibration is good and validation is satisfactory, but the 95 PPU plots above show that the SWAT model underestimates runoff most of the time during high flow (peak) period. This may be because of adoption of SCS curve number (CN) method for calculation of runoff. SCS CN method determines CN values considering moisture content of the soil of previous day and without considering the change in moisture content due to same day rainfall. The underestimation

Table 19.2 Performance ratings for recommended statistics for a monthly time step (Moriasi et al. 2007)

Rating	NSE	PBIAS		RSR	R^2
		Streamflow	Sediment		
Very good	0.75 < NSE ≤ 1.0	< ±10	< ±15	0 < RSR < 0.5	>0.5
Good	0.65 < NSE ≤ 0.75	±10 < PBIAS < ±15	±15 < PBIAS < ±30	0.5 < RSR < 0.6	
Satisfactory	0.5 < NSE ≤ 0.65	±15 < PBIAS < ±25	±30 < PBIAS < ±55	0.6 < RSR < 0.7	
Unsatisfactory	NSE ≤ 0.50	PBIAS < ±25	PBIAS < ±55	RSR > 0.70	<0.5

Table 19.3 Calibration (2003–06) and validation (2007–09) results

Parameters	Kusum (Flow)		Bagasoti (Flow)		Bagasoti (Sediment yield)	
	Calibration	Validation	Calibration	Validation	Calibration	Validation
p-factor	0.83	0.67	0.73	0.67	0.96	0.86
r-factor	0.40	0.42	0.41	0.46	0.64	1.12
R^2	0.96, Very Good	0.78, Very Good	0.90, Very Good	0.74, Very Good	0.71, Very Good	0.69, Very Good
NSE	0.95, Very Good	0.78, Very Good	0.90, Very Good	0.73, Good	0.68, Good	0.69, Good
PBIAS	4.70, Very Good	5.3, Very Good	4.70, Very Good	0.90, Very Good	15.1, Good	−9.70, Very Good
RSR	0.22, Very Good	0.47, Very Good	0.32, Very Good	0.52, Good	0.57, Good	0.56, Good

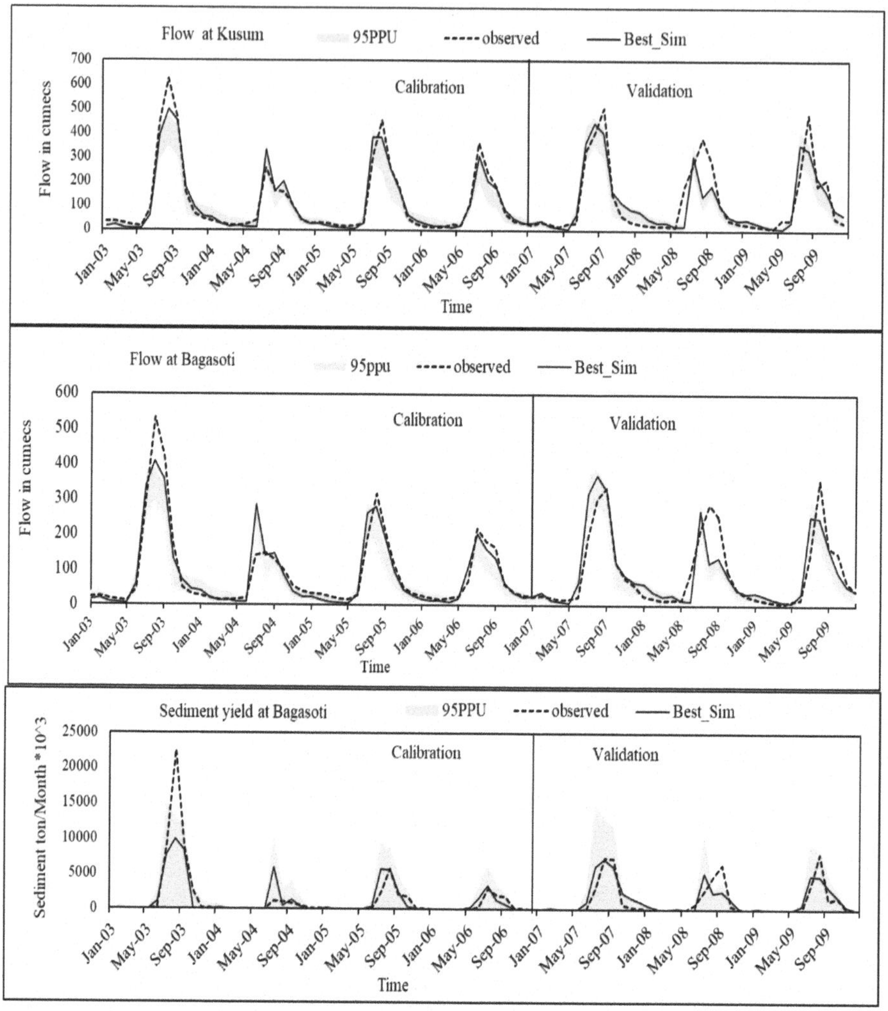

Fig. 19.7 95 PPU plots for streamflow and sediment yield

of peak runoff may be due to that SCS CN method defines a rainfall event as the sum of all rainfall that occurs during one day (Kim and Lee 2008; Gull et al. 2017).

Most of sediment transport occurs during high flow period. Therefore, it is difficult but necessary to capture these high flow events during model calibration and validation. Therefore, r-factor for sediment which indicates the thickness of 95 PPU envelope obtained was 0.64 (for calibration) and 1.12 (for validation) which were quite higher than that of flow but can be accepted. The *p*-factor, the percentage of observed data enveloped by modeling results obtained, for sediment was 0.96 for calibration and 0.86 for validation, which is also acceptable.

19.3.3 Water Balance and Sediment Yield

Water balance is a key concern and driving force irrespective of the problems that are studied in SWAT (Neitsch et al. 2011). The SWAT was calibrated and validated successfully, and re-run was carried out for the period of 2000–2013 considering the parameters' values of best simulation during the calibration. Finally, the SWAT outputs were analyzed to carry out the water balance study.

The average monthly distribution of water balance components and average annual yield are presented in Table 19.4. It is seen that average annual precipitation of the basin is 1677 mm and 48.60% (815 mm out off 1677 mm) of rainfall goes as annual evapotranspiration from the basin and this evapotranspiration value is quite high. The surface runoff from the basin is 273.00 mm, lateral sub-surface flow or interflow accounts 155.00 mm, base flow or return flow is 269.00 mm, and the total annual water yield is 854 mm. The total water volume available from the basin is 4.50 billion m^3. The amount of water that enters to deep aquifer is 169.00 mm, and this amount of water also contributes to streamflow somewhere out of the watershed.

Also, it is found that 80% of rainfall, 90% of runoff and 73% of water yield occur during four months of monsoon, i.e., from June to September.

The evapotranspiration was also found to be high during monsoon season with highest value of 119 mm in July and August. Sediment yield shows the proportional pattern with surface runoff. Surface runoff in July and August is 104 mm and 76 mm, respectively, accounting higher sediment yield of 6.04 and 4.73 ton/ha, respectively. The annual sediment yield of the study area is 17.67 ton/ha/year, and according to classification criteria mentioned by Singh et al. (1992) as presented in Table 19.5, the study area falls under high erosion class and needs attention on implementation

Table 19.4 Monthly distribution of yearly water balance components

Month	Rainfall (mm)	SURQ (mm)	GWQ (mm)	LATQ (mm)	ET (mm)	Water yield (mm)	Sediment (ton/ha)
Jan.	22	0.6	3.1	1.1	18.0	19.0	0.09
Feb.	39	3	1.4	2	31	16	0.31
Mar.	22	0	1.1	1	40	11	0.06
Apr.	44	0	0.5	1	58	8	0.05
May	120	2	0.3	2	105	9	0.19
Jun.	259	28	4.1	14	119	47	1.82
Jul.	483	104	36.7	49	119	194	6.04
Aug.	361	76	70.0	43	110	204	4.73
Sep.	250	47	73.8	33	97	177	3.45
Oct.	64	11	50.0	7	71	94	0.91
Nov.	5	1	20.9	0	29	46	0.01
Dec.	6	0	8.6	0	18	29	0.01
Annual	1677	273	271	155	815	854	17.67

Table 19.5 Area under different classes of soil erosion

SN	Sediment yield (ton/ha/year)	Area (km^2)	% area	Erosion class
1	0–5	984	19	Slight
2	5–10	588	11	Moderate
3	10–20	1987	38	High
4	20–40	1082	20	Very high
5	40–80	641	12	Severe
6	>80	5282	100	Very severe

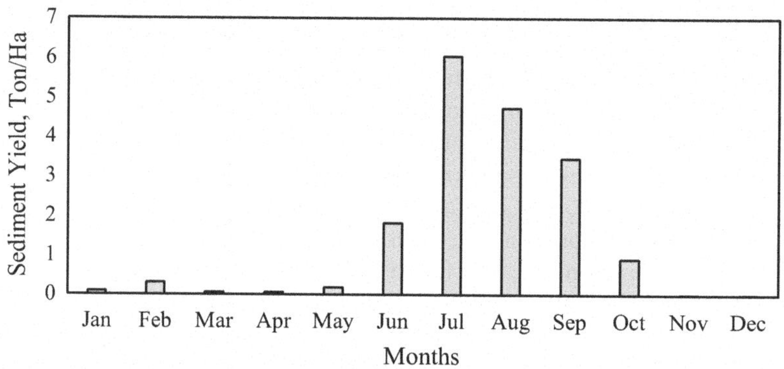

Fig. 19.8 Average monthly sediment yield

of BMPs for erosion control. The average monthly sediment yield was obtained as shown in Fig. 19.8. This shows that the average sediment yield is high during monsoon season, i.e., from June to September when streamflow and rainfall are also high. The maximum sediment yield was obtained in July which is 6.04 ton/ha and minimum at November and December with 0.01 ton/ha.

19.3.4 Spatial Distribution of Sediment Yield in West Rapti Watershed

SWAT is a powerful spatial analysis tool that helps in identification of sediment distribution from subbasin or even from HRU level giving the idea of erosion-prone areas in terms of average annual sediment yield, and this helps in implementation planning of some management option to control erosion from specific erosion-prone areas. The spatial distribution sediment yield in tons/ha is given in Fig. 19.9.

The average sediment yield (of the year from 2003 to 2013) for all 18 subbasins is presented in Fig. 19.11. The Bagasoti subwatershed (subbasin 13 in map) produces more sediment per hectare as compared to Kusum subwatershed (subbasin 15 in

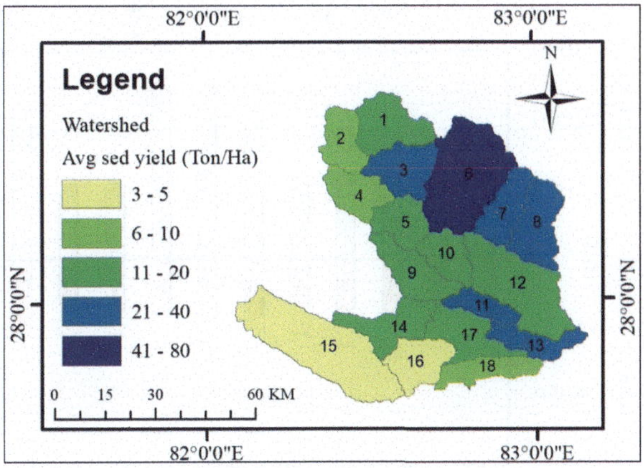

Fig. 19.9 Spatial distribution of sediment yield from SWAT model

map). Subbasin six produces high sediment compared to other subbasins. Also, from Table 19.6, it is clear that subbasinsix produces more runoff and hence the more sediment. Average annual water balance in the subbasins with balance closure in graphical form is shown in Fig. 19.10.

Table 19.6 Water balance components and sediment generated from subbasins

Subbasin	Area (km^2)	Rainfall (mm)	SURQ (mm)	GWQ (mm)	LATQ (mm)	ET (mm)	Water yield (mm)	Sediment (ton/ha)
1	290.81	1302.02	108.65	221.55	150.26	682.97	619.37	11.64
2	198.64	1302.02	114.82	237.34	132.37	668.96	633.37	9.24
3	257.29	1423.66	136.74	189.03	169.90	806.69	611.55	22.32
4	206.74	1260.49	111.05	172.10	126.26	739.85	506.94	7.49
5	193.05	1088.51	126.19	103.68	95.57	694.43	385.24	10.14
6	640.99	2513.25	432.15	513.33	386.78	854.12	1646.74	51.08
7	221.02	2513.25	421.08	516.97	381.84	863.78	1636.55	28.19
8	296.22	2513.25	454.58	533.11	366.98	819.24	1681.31	28.48
9	258.89	1088.51	152.38	76.06	44.24	763.60	316.12	17.70
10	188.39	1088.51	146.48	118.66	77.49	667.00	411.32	13.69
11	130.96	2251.65	404.70	365.64	212.28	1037.49	1202.89	30.79
12	530.69	1289.29	155.62	143.06	108.64	788.55	493.24	13.52
13	176.27	2251.65	465.21	382.60	142.52	1018.92	1220.50	25.29
14	260.79	1535.38	244.93	191.54	66.31	909.70	619.30	16.82
15	754.57	1221.96	229.17	135.69	17.86	755.21	464.68	2.82
16	228.93	1221.96	263.47	126.69	16.10	737.21	482.71	3.06
17	264.36	2251.65	406.41	414.38	125.61	1042.90	1196.15	16.20
18	182.47	2251.65	423.51	428.56	88.16	1040.19	1198.69	9.65

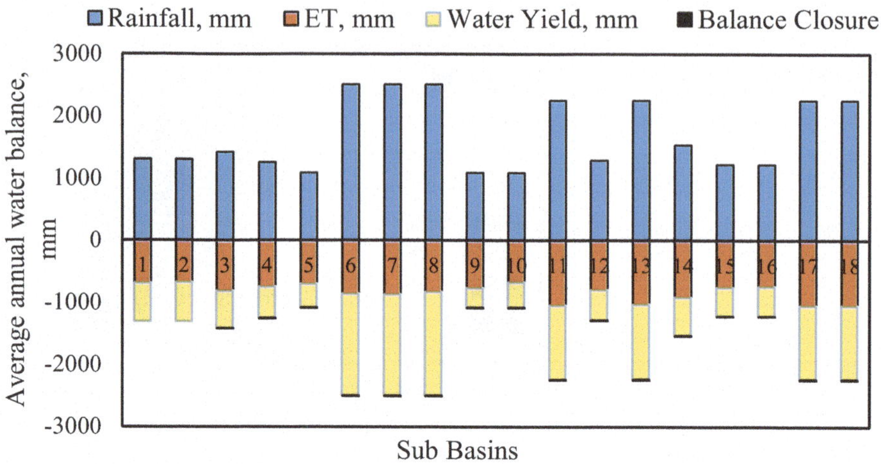

Fig. 19.10 Average annual water balance in the subbasins

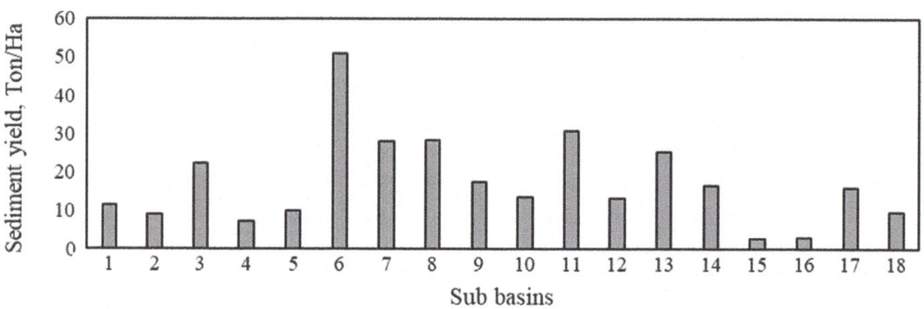

Fig. 19.11 Subbasin wise average annual sediment yield

19.4 Limitations of the Study

In this study, due to lack of the recent sediment data, rating curve was developed using past 30 years sediment data which was used for model calibration and validation. Therefore, the sediment yield should be rechecked using the model developed from neighboring catchment showing similar response. Also, single-site sediment calibration of the model has been used due to unavailability of data and response of different subbasin within watershed to sediment yield may be different which may limit the use of model for preparation of subbasin wise erosion or sediment yield control plan.

19.5 Conclusion

Followings conclusions are drawn from this study.

- Out of 1677 mm average annual rainfall in the river basin, about 48.6% (853 mm) accounts for evapotranspiration, 16.3% (273 mm) contributes to surface runoff, and 9.2% (155 mm) accounts as lateral flow, and
- 4.50 billion m^3 volume of water is available annually at the basin outlet (Kusum outlet), and its average annual discharge is Kusum 110 m^3/s.
- From monthly distribution of rainfall and water yield, 90% of runoff, 80% of rainfall and 73% of water yield occur during four monsoon months, i.e., June–September.
- The high value of evapotranspiration (119 mm) was found during monsoon season.
- 17.67 ton/ha/year sediment is produced annually from the basin, and the study area lies under high erosion class.
- The SWAT model can be effectively used for scientific planning, development and management of water resources in the data-scarce region like Nepal.

Acknowledgements The authors would like to acknowledge the officials from Department of Hydrology and Meteorology, Nepal, and Department of Survey, Nepal, for their kind cooperation during the collection of different data. Also, would like to express sincere thanks to ITEC, Ministry of External Affairs, India, for funding this study.

References

Abbaspour K (2015) SWAT-CUP: SWAT calibration and uncertainty programs—a user manual. Swiss Federal Institute of Aquatic Science and Technology, Eawag

Abbaspour KC, Rouholahnejad E, Vaghefi S, Srinivasan R, Yang H, Kløve B (2015) A continental-scale hydrology and water quality model for Europe: calibration and uncertainty of a high-resolution large-scale SWAT model. J Hydrol 524:733–752

Abbaspour KC, Yang J, Maximov I, Siber R, Bogner K, Mieleitner J, Zobrist J, Srinivasan R (2007) Modelling hydrology and water quality in the pre-alpine/alpine Thur watershed using SWAT. J Hydrol 333(2–4):413–430

Beven KJ (2012) Rainfall-runoff modelling: the primer. Rainfall-Runoff Modelling: The Primer: Second Edition

Bieger K, Hörmann G, Fohrer N (2013) The impact of land use change in the Xiangxi Catchment (China) on water balance and sediment transport. Reg Environ Change 15(3):485–498

Cui X, Sun W, Teng J, Song H, Yao X (2015) Effect of length of the observed dataset on the calibration of a distributed hydrological model. In: Proceedings of the international association of hydrological sciences, vol 368, pp 305–311

Dahal V, Shakya NM, Bhattarai R (2016) Estimating the impact of climate change on water availability in Bagmati Basin, Nepal. Environ Process

Dhami BS, Pandey A (2013) Comparative review of recently developed hydrologic models. J Indian Water Resour Soc 33:34–42

Gassman PPW, Reyes MMR, Green CCH, Arnold JJG (2007) The soil and water assessment tool: historical development, applications, and future research directions. Trans ASAE 50(4):1211–1250

Getache HE, Melesse AM (2012) The impact of land use change on the hydrology of the Angereb Watershed, Ethiopia. Int J Water Sci 1(4):1–7

Gull S, MA A, Dar AM (2017) Prediction of stream flow and sediment yield of Lolab Watershed using SWAT model. Hydrol Current Res 08(01):1–9

Gupta HV, Sorooshian S, Yapo PO (1999) Status of automatic calibration for hydrologic models: comparison with multilevel expert calibration. J Hydrol Eng 4(2):135–143

Kalcic MM, Chaubey I, Frankenberger J (2015) Defining soil and water assessment tool (SWAT) hydrologic response units (HRUs) by field boundaries. Int J Agric Bio Eng 8(3):1–12

Khanchoul K, Jansson MB, Lange J (2007) Comparison of suspended sediment yield in two catchments, northeast Algeria. Zeitschrift Für Geomorphologie 51(1):63–94

Kim NW, Lee J (2008) Temporally weighted average curve number method for daily runoff simulation. Hydrol Process 22(151):4936–4948

Moriasi DN, Arnold JG, Van Liew MW, Bingner RL, Harmel RD, Veith TL (2007) Model evaluation guidelines for systematic quantification of accuracy in watershed simulations. Trans ASABE 50(3):885–900

Narsimlu B, Gosain AK, Chahar BR, Singh SK, Srivastava PK (2015) SWAT model calibration and uncertainty analysis for streamflow prediction in the Kunwari River Basin, India, using sequential uncertainty fitting. Environ Process 2(1):79–95

Neitsch SL, Arnold JG, Kiniry JR, Williams JR (2008) Overview of soil and water assessment tool (SWAT) model. In: Soil and water assessment tool (SWAT) global applications. The World Association of Soil and Water Conservation, U.S.A., pp 03–23

Neitsch S, Arnold J, Kiniry J, Williams J (2011) Soil and water assessment tool theoretical documentation Version 2009. Texas Water Resources Institute

Pandey A, Lalrempuia D, Jain SK (2014) Assessment of hydropower potential using spatial technology and SWAT modeling in the Mat River of Southern Mizoram, India. Hydrol Sci J 6667:141217125340005

Saleh A, Arnold JG, Gassman PW, Hauck LM, Rosenthal WD, Williams JR, McFarland AMS (2000) Application of SWAT for the Upper North Bosque River Watershed. Trans ASAE 43(5):1077–1087

Singh G, Babu R, Narain P, Bhusan LS, Abrol IP (1992) Soil erosion rates in India. J Soil Water Conserv 47(1):97–99

Suryavanshi S, Pandey A, Chaube UC (2017) Hydrological simulation of the Betwa River basin (India) using the SWAT model. Hydrol Sci J 62(6):960–978

Talchabhadel R, Sharma R (2014) Real time data analysis of West Rapti River Basin of Nepal. J Geosci Environ Protect 1–7

Tripathi MP, Panda RK, Raghuwanshi NS (2005) Development of effective management plan for critical subwatersheds using SWAT model. Hydrol Process 19(3):809–826

Uniyal B, Jha MK, Verma AK (2015) Assessing climate change impact on water balance components of a river Basin using SWAT model. Water Resour Manage 29(13):4767–4785

Van Liew MW, Veith TL, Bosch DD, Arnold JG (2007) Suitability of SWAT for the conservation effects assessment project: comparison on USDA agricultural research service watersheds. J Hydrol Eng 12(2):173–189

Winchell M, Srinivasan R, Di Luzio M, Arnold JG (2013) Arcswat interface for SWAT2012: user, guide. Blackland Research Center, Texas AgriLife Research, p 464

Worku T, Khare D, Tripathi SK (2017) Modeling runoff–sediment response to land use/land cover changes using integrated GIS and SWAT model in the Beressa watershed. Environ Earth Sci 76(16):550

Chapter 20
Modelling of Groundwater Development Using Arc-SWAT and MODFLOW

Satavisha Ghosh, Sunny Gupta, and Susmita Ghosh

Abstract In the present study, surface water and ground water interaction in the canal command area was performed by coupled SWAT–MODFLOW model instead of MODFLOW model only. A method was proposed wherein the properties of hydrologic response units (HRUs) in the SWAT model were assigned to the cells in the MODFLOW model. The surface water flow model using Arc-SWAT was developed. The outputs from the Arc-SWAT such as evapotranspiration and groundwater recharge were used as inputs for the MODFLOW model. Finally, the groundwater table elevation was computed from developed MODFLOW model. The reliability study was also carried out on the study area for Arc-SWAT and MODFLOW model. The proposed methodology was applied in Yamuna-Hindon inter-basin. The water table fluctuation with respect to groundwater recharge/pumping can be visualized and facilitates the future sustainable management of groundwater for irrigation purpose in the study area.

Keywords Groundwater · Flow model · Visual MODFLOW · Arc-SWAT · Recharge estimation

S. Ghosh (✉)
Department of Civil Engineering, Mallabhum Institute of Technology, Bishnupur, Bankura 722122, West Bengal, India
e-mail: satavisha.dora.ghosh@gmail.com

S. Gupta
Department of Water Resources Engineering, WSP Consultants India Pvt. Ltd, Bangalore 560045, India
e-mail: guptasunny20@gmail.com

S. Ghosh
Department of Civil Engineering, National Institute of Technology Silchar, Silchar 788010, Assam, India
e-mail: susmita@civil.nits.ac.in

A. Pandey et al. (eds.), *Water Management and Water Governance*, Water Science and Technology Library 96,
https://doi.org/10.1007/978-3-030-58051-3_20

20.1 Introduction

Groundwater being one of the most important resources in this world is mainly used for domestic, industrial, and agricultural purposes. Due to overexploitation of groundwater, we face several technical/socio-economic problems like reduction of baseflow contribution to hydraulically connected stream, deterioration of groundwater quality, increase of pumping cost, drying up of wells, etc. A lowered water table may, in turn, cause other problems such as groundwater-related subsidence and saltwater intrusion (Kumar and Singh 2015). The planning of groundwater development is essential step to combat with aforesaid problems by keeping groundwater level within acceptable limit. Now, planning process comprises of feasibility study that are proposing the several alternative plans and screening out the best among, i.e. optimality study. For feasibility study, development of groundwater model is essential.

The groundwater planning/management is dependent upon the accurate identification of recharge/pumping rate in specific region where groundwater is significantly used with surface water. So, the surface water and groundwater have been studied conjunctively as both are strongly correlated in hydrologic cycle. Therefore, the relevant problems could be addressed by coupling the surface—groundwater model instead of groundwater model alone to understand the interrelation of ground and surface water for effective groundwater management.

Soil and Water Assessment Tool (SWAT) (Arnold et al. 1999) and MODFLOW (McDonald and Harbaugh 1988) are being widely used for groundwater management by addressing surface–groundwater model conjunctively. SWAT simulates surface water and shallow aquifer only. MODFLOW simulates flow processes occurring in the saturated zone. Therefore, there have some advantages and disadvantages both of two models. The combination of SWAT–MODFLOW model is firstly used by drawing the beneficial roles of two models by Sophocleous et al. (1999) to study the stream–aquifer interaction at Rattlesnake Creek basin, Kansas. Few studies related to integrating SWAT and MODFLOW model were available in literature those indicating the capability of this coupled model with some case studies (Perkins and Sophocleous 1999; Kim et al. 2008; Chu et al. 2010; Luo and Sophocleous 2011; Guzman et al. 2015; Singh and Shukla 2016; Semironi et al. 2017). The key feature of SWAT–MODFLOW is a hydrologic response unit (HRU) in SWAT can exchange with respective cells in MODFLOW and named as SWAT–MODFLOW and proposed method is illustrated in Musimcheon Basin, Korea (Kim et al. 2008). The pumping/recharge estimation for multi-aquifer is performed using integrating SWAT–MODFLOW model at Taiwan (Ke 2014).

The present study deals with the over exploitation of groundwater storage to meet the agricultural requirements in which the irrigational demands are fulfilled by groundwater withdrawal solely or by specified canal supplies added with the groundwater withdrawal. In this study, the recharge and evapotranspiration from surface water to groundwater is estimated using semi-distributed Arc-SWAT model. These recharge and evapotranspiration values are used as inputs to the fully distributed

Visual MODFLOW groundwater model. The spatial and temporal variation of state variables that is water table elevation are estimated from coupled SWAT–MODFLOW model. The above-mentioned methodology has been illustrated in Yamuna-Hindon Inter-basin, India.

20.2 Study Area

The chosen region for study is mainly agricultural area in Uttar Pradesh, India. The geographical area is having about 0.6 million hectares within the command area of Eastern Yamuna canal system (India) in between latitudes 29°18′ to 30°25′N and longitudes 77°1′30″ to 77°40′45″E (Fig. 20.1). Hindon, Yamuna River, and Sivalik hills are being bounded in the east, west, and north side, respectively. The two rivers combine in the southern part of that region. The aquifer 10 km apart from

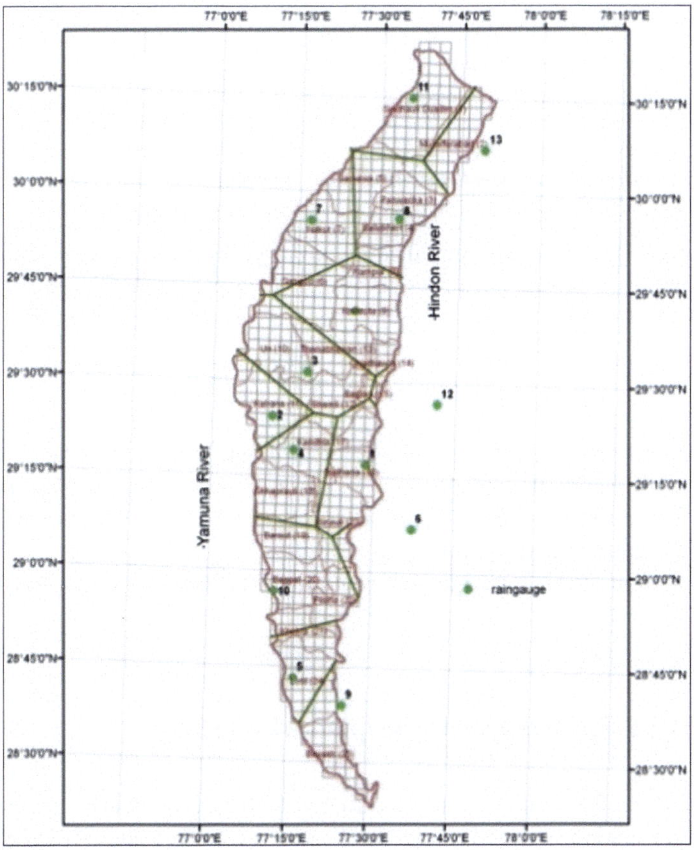

Fig. 20.1 Study area (Ghosh 2011)

these hills contains boulders, pebbles, and gravel with sand. The aquifer, after this section, mainly alluvial and unconfined, is spreading perpendicularly within a depth of 90 m below ground. The region has been experiencing by substantial rainfall in the rainy season (June to September). Moderate type of sub-tropical monsoon climate is observed in the area. In general, the temperature varies from about 40 °C in summer to below 2 °C in winter.

Eastern Yamuna canal originates from the left bank of Yamuna at Tejewala. In the study area, the existing canal supplies are inadequate to satisfy the irrigation water demands. As a result, there is an extensive ground water withdrawal in that region, and that withdrawal water is mostly utilized to fulfill crop water requirements. Accordingly, the steadily lowering trend of water table of the aquifer has shown. As such various works (Kashyap and Chandra 1982; Mishra 1987; CBIP 1987; Rathi 1997) was performed previously regarding the aquifer system of that region. The current study uses some relevant data and calibrated groundwater flow model from these earlier studies. Alam and Umar (2013) applied MODFLOW simulation model to simulate groundwater in Hindon-Yamuna region, western Uttar Pradesh. Ghosh and Kashyap (2012a; b) proposed two approximate simulations models like ANN and kernel model to simulate the groundwater flow. Ahmed et al. (2008) examined water balance in Krishni-Yamuna area, Uttar Pradesh.

20.2.1 Available Groundwater Flow Model

The governing 3D partial differential equation for groundwater system simulation is given below:

$$\frac{\partial}{\partial x}(K_{xx}\frac{\partial h}{\partial x}) + \frac{\partial}{\partial y}(K_{yy}\frac{\partial h}{\partial y}) + \frac{\partial}{\partial z}(K_{zz}\frac{\partial h}{\partial z}) + V = S_s\frac{\partial h}{\partial t} \quad (20.1)$$

where, x, y, and z are 3D axes in Cartesian coordinate system, h = head, K_{xx}, K_{yy}, K_{zz} = hydraulic conductivities along x-, y-, and z-axes, V = volumetric flux per unit volume and representing source and/or sink terms of water, S_S = Volumetric specific storage of the aquifer, and t = time.

For estimation of this recharge and evapotranspiration component Arc-SWAT is used.

20.3 SWAT–MODFLOW-Coupled Modelling

As the SWAT model contains semi-distributed features, the distributed parameters like hydraulic conductivity and storage coefficient are not considered as groundwater component. Therefore, in producing a meticulous to characterize the recharge from groundwater, it is very complicated to determine the water table/piezometric head distribution and pumping/recharge rate distribution. To facilitate the solution of this problem, the cells in the MODFLOW model are substituted with the properties of the hydrologic response units (HRUs) in the SWAT model. The distributed groundwater recharge rate and evapotranspiration could be simulated efficiently using aforesaid HRU–cell transformation interface. By taking into account the relation among the aquifer and the stream network to replicate boundary flow, the connection is accomplished.

20.3.1 ARC-SWAT

SWAT is a tool for soil and water assessment. Arc-SWAT is an ArcGIS-ArcView extension and graphical user input interface for SWAT. SWAT works on a daily time step and predicts the effect of land use and organization on surface and groundwater, sediment, and agrarian chemical produces in ungauged watersheds. The model is built on process, computationally efficient, and competent in uninterrupted simulation for extended occasions. Main model components comprise soil characteristics, weather, hydrology, and properties, vegetation development, nutrients, pesticides, bacteria and pathogens, and land organization.

The interflow and recharge evaluation are done with the help of Arc-SWAT. The inputs of SWAT are weather, soil, and temperature and land use land cover (LULC) data sets. A watershed separated into sub-watersheds, and sub-watersheds are further segmented into Hydrologic Response Units (HRUs). The HRU contains of homogeneous soil and land use management characteristics. Critical hydrologic processes in SWAT model can be written as follows in Eq. (20.2)

For determination of V,

$$V = GW + E - (R_i + R_r + R_s) \tag{20.2}$$

GW Groundwater withdrawal,
E Evapotranspiration,
R_r recharge due to precipitation,
R_i recharge due to applied irrigation,
R_s recharge from canal discharge.

The surface runoff calculation can be carried out by the model using the Soil Conservation Service (SCS) curve number method. The volume of runoff is anticipated from daily precipitation using the modified SCS-CN (Curve Number Method) and Green-Ampt methods. The model needs the input of (i) The DEM, (ii) land use, (iii) soil data, and (iv) weather data, i.e. daily rainfall and temperature. The hydrologic module of that model establishes a soil water balance at every time period based on daily data of rainfall, runoff, percolation, evapotranspiration, and base flow. Simulations have been executed at the HRU level and abridged in each sub-watershed.

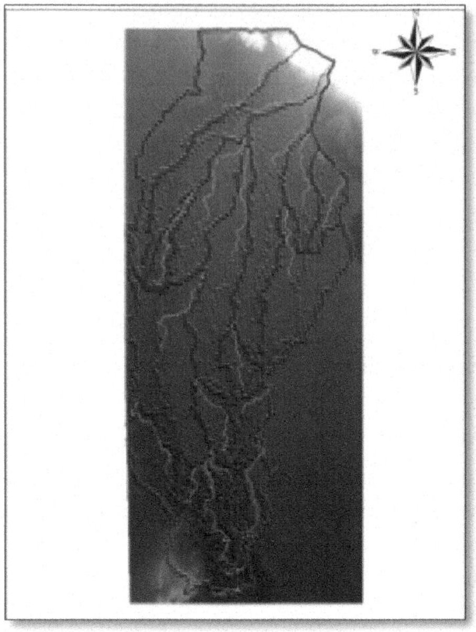

Fig. 20.2 DEM

20.3.2 Required Database of ARC-SWAT

Data required for the modeling using Arc-SWAT are mentioned below:

a. DigitalElevationModeldata- SRTM (Shuttle Radar Topography Mission) 1-Arc-Second Global from United States Geological Survey (USGS) Earth explorer (Fig. 20.2)
b. LandUseandSoildata-https://www.waterbase.org (Fig. 20.3)
c. Weatherdata- CFSR (Climate Forecast System Reanalysis) World Weather Dataset.

20.3.3 Modeling Using ARC-SWAT

For modeling using Arc-SWAT, DEM raster data is needed. DEM dataset was taken from USGS EarthExplorer. The development of SWAT model is described step-wise in below: Using the inputs given, the delineation of the Hindon-Yamuna basin was done and further subdivided in 22 sub-basins. After delineating watershed, the land use land cover map and soil map (Fig. 20.3) are inserted. Then the sub-basins are further divided to 757 Hydrologic Response Units (HRUs) (Fig. 20.4). After setting up the model, the simulation is done for 20 years taking 3 years of warm-up period into account.

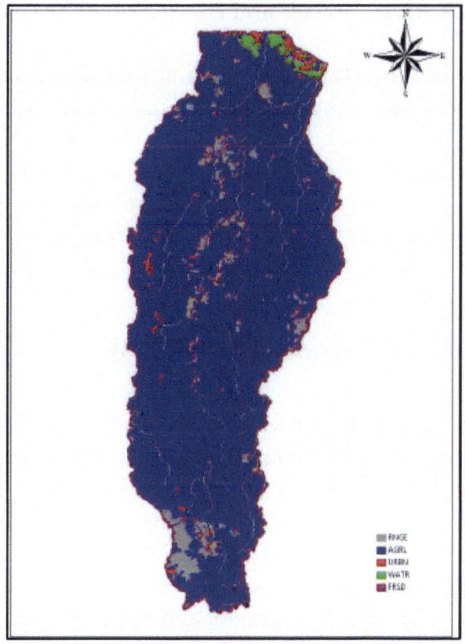

Fig. 20.3 LULC and soil map

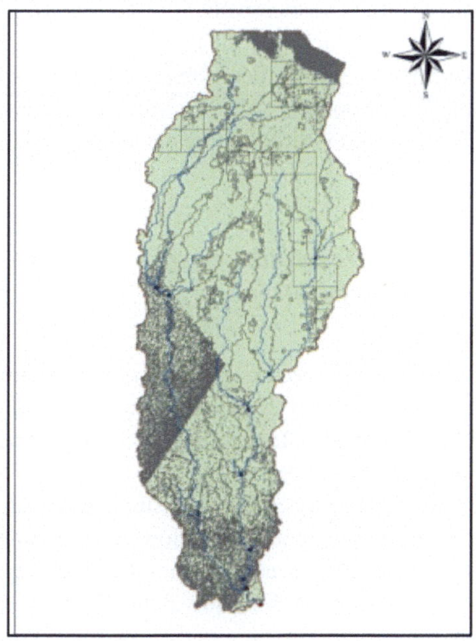

Fig. 20.4 HRU distributions

20.3.3.1 Result and Discussion of Developed ARC-SWAT Model:

The groundwater recharge, actual evapotranspiration, potential evapotranspiration, etc., sub-basins, HRUs wise daily basis are found by conducting simulation of Arc-SWAT model. For the present study, the average groundwater recharge and potential evapotranspiration are taken as input for the MODFLOW model.

The average actual evapotranspiration is calculated as 1.63 m. The average potential evapotranspiration is 6 mm/d.

20.3.4 Visual MODFLOW

Visual MODFLOW was the first commercially available graphical interface for the open-source groundwater modelling engine called MODFLOW. MODFLOW is the U.S.G.S. modular finite difference model which uses a computer code to resolve the groundwater flow equation. A 3D finite-difference groundwater flow model, MODFLOW is used to simulate the groundwater flow system. A single-layered groundwater flow model was created using the existing data associated with the aquifer features, rainfall, and other relevant data of the chosen region.

The region of this work is discretized by grids with 71 and 24 numbers of rows and columns, respectively, of 3.2 km uniform spacing (Ghosh and Kashyap 2012b). Aquifer of thickness 90 m is considered throughout the study area. The storage parameter (S_y) has been observed nearly space invariant and taken as 0.22 (CBIP 1987). The contour values of transmissivities are taken throughout the study area in x-direction and y-direction, respectively, and the value of the conductivity is 14.44 m/d.

20.3.5 Development of SWAT Visual MODFLOW Couple Model

The constant heads are given from the previous model mentioned above (Rathi 1997). The recharge and evapotranspiration component are taken from the output of Arc-SWAT analysis. Recharge of the study area is given along the sub-basins over the specified study area.

In the EVT package in Visual MODFLOW classic interface, the inputs are: the elevation of evapotranspiration extinction depth, the evapotranspiration surface and maximum evapotranspiration rate. The evapotranspiration surface elevation and the surface elevation are equivalent. Shallow depth of groundwater and plant growth are most important factors that influence the evapotranspiration extinction depth and the maximum evapotranspiration rate eventually. The evapotranspiration extinction depth is below where no evapotranspiration takes place. In this study, the maximum evapotranspiration rate and extinction depth have been considered equal to the

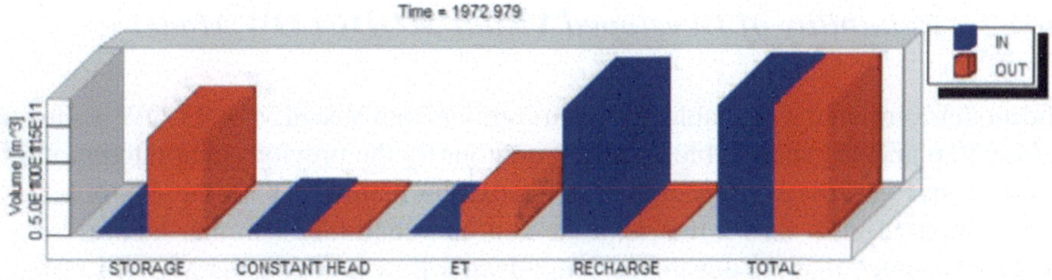

Fig. 20.5 Volume balance

average potential evapotranspiration and average evapotranspiration, respectively, calculated from simulated SWAT model (Ke 2014).

20.3.5.1 Result and Discussion of Developed Couple SWAT-Visual MODFLOW

The model is simulated along the stress period of 20 years with daily time step. The output from the Visual MODFLOW model is the water table elevation contour shown in Fig. 20.6. The volume balance check to the simulation results to validate the simulation model (Fig. 20.5). The volume balance equation can be written as follows:

$$I - O = \Delta S \tag{20.3}$$

where, I = Sum of all inflows; O = Sum of all outflows; ΔS = change in storage-all in stipulated time period.

20.4 Reliability Study

20.4.1 Reliability of Developed ARC-SWAT Model

The reliability of the developed SWAT model is verified by comparing with previously available calibrated model (Rathi 1997). The evapotranspiration rate and extinction depth computed from the SWAT model are compared with the above-mentioned pre-calibrated model. The value of evapotranspiration rate is 6 mm/day from Arc-SWAT model wherein the same parameter value is 5 mm/day in the pre-calibrated model. It is observed that the error is within 20% which is considerable. The extinction depth computed is equals to 1.63 m in Arc-SWAT wherein the value is 2 m for the above-mentioned parameter in the pre-calibrated model. So, it may be concluded that the model efficiency is in the acceptable accuracy range.

20.4.2 *Reliability of Developed Visual MODFLOW Model*

The model computed water table elevation contour from Visual MODFLOW is shown in Fig. 20.6 and the water table elevation contour by the previous calibrated models is shown in Fig. 20.7; for the reliability study, the nature of both the contours has been compared. It is shown that both the contours maintain a similar nature. So, it can be concluded that the developed groundwater flow model using MODFLOW in the present study is reliable with minimal amount of error.

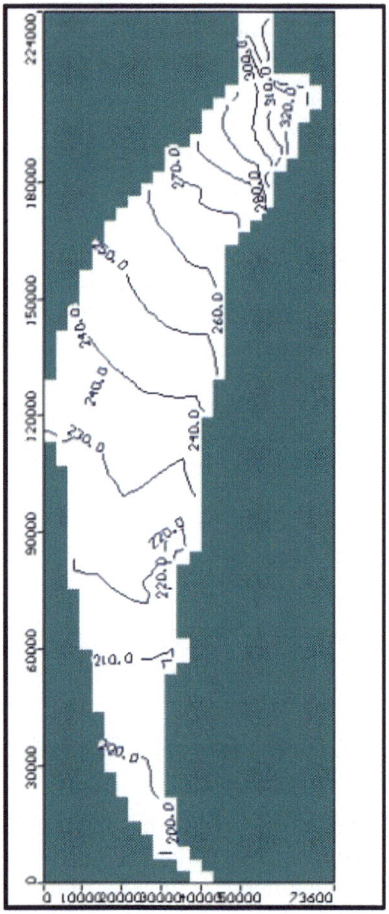

Fig. 20.6 Water table elevation contour of this model

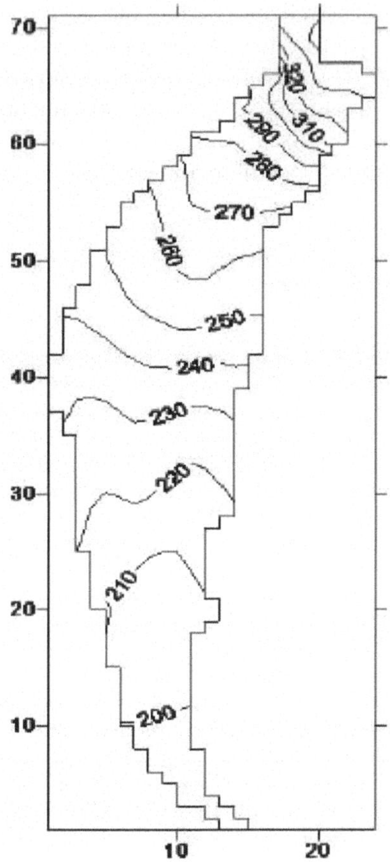

Fig. 20.7 Water table elevation contour of pre-calibrated model

20.5 Conclusion

In the present study, the conjunctive study of surface water and groundwater interaction is described by coupling of the respective models, i.e. Arc-SWAT and Visual MODFLOW model. Firstly, the surface water flow model using Arc-SWAT is made. The evapotranspiration and the groundwater recharge that have been computed from Arc-SWAT are taken as inputs for Visual MODFLOW model.

The groundwater flow model using Visual MODFLOW classic interface has been developed afterwards where recharge and evapotranspiration are taken from the above simulated Arc-SWAT model by HRUs–cells conversion. Finally, the spatial and temporal variation of the relevant state variable, i.e. groundwater table elevation is computed from combined SWAT–MODFLOW model that is illustrated in Yamuna-Hindon Inter-basin, India.

The reliability study has been carried out in the study area for Arc-SWAT with respect to evapotranspiration with the pre-calibrated model and the reliability of MODFLOW model had been carried out with respect to groundwater elevation.

In this study, the water table fluctuation regarding groundwater recharge/pumping can be visualized and this study facilitates the future planning/management of groundwater in regional basis.

References

Ahmed I, Umar R (2008) Hydrogeological framework and water balance studies in parts of Krishni–Yamuna interstream area, Western Uttar Pradesh, India. Environ Geol 53(8):1723–1730

Alam F, Umar R (2013) Groundwater flow modeling of Hindon Yamuna interfluve region. Western Uttar Pradesh, J Geol Soc India 82:80–90

Arnold JG, Allen PM (1999) Automated methods for estimating baseflow and groundwater recharge from streamflow records. J Am Water Resour Assoc 35(2):411–424

CBIP (Govt. of India) (1987) Conjunctive use of groundwater and surfacewater for irrigated agriculture: risk aversion. Project No.: Q(24)/83-P-III

Chu J, Zhang C, Zhou H (2010) Study on interface and frame structure of SWAT and MODFLOW-models coupling. Geophysical research abstracts vol. 12, EGU2010–4559, 2010, EGU General Assembly 2010

Ghosh S, Kashyap D (2012a) ANN-based model for planning of groundwater development for agricultural usage, Wiley, Ltd

Ghosh S (2011) Kernal function and ANN based planning of groundwater development for irrigation. PhD thesis. Indian Institute of Technology Roorkee

Ghosh S, Kashyap D (2012) Kernel function model for planning of agricultural groundwater development. J Water Resour Plann Manage 138(3):277–286

Guzman JA, Moriasi DN, Gowda PH, Steiner JL, Starks PJ, Arnold JG, Srinivasan R (2015) A model integrationframework for linking SWAT and MODFLOW. Environ Model Softw 73(2015):103–116

Kashyap D, Chandra S (1982) A distributed conjunctive use model for optimal cropping pattern. Proc Exeter Symp, July, IAHS Publ 135:377–384

Ke KY (2014) Application of integrated surface water-ground water model to multi aquifers modeling in Choushui river alluvial fan. Taiwan Hydrol Process 28:1409–1421

Kim NW, Chung IM, Won YS, Arnold JG (2008) (2008), Development and application of the integrated SWAT–MODFLOW model. J Hydrol 356:1–16

Kumar CP, Singh S (Feb 2015) Concepts and modelling of groundwater system, IJISET 2(2)

Luo Y, Sophocleous M (2011) Two-way coupling of unsaturated-saturatedflow by integrating the SWAT and MODFLOW models with application in an irrigation district in arid region of West China. J Arid Land 3(3):164–173. https://doi.org/10.3724/SP.J.1227.2011.00164

McDonald MG, Harbaugh AW (1988) A modular three-dimensional finite-difference ground-water flow model. US geological survey techniques of water resources investigations report book 6, Chapter A1, p 528

Mishra R (1987) Distributed aquifer response modeling in Yamuna-Hindon Doab. MTech dissertation, Indian Institute of Technology Roorkee, India

Perkins S, Sophocleous M (1999) Development of a comprehensive watershed model applied to study stream yield under drought conditions. Ground Water 37(3):418–426

Rathi S (1997) Numerical modeling of aquifer response in Yamuna-Hindon doab by. M.tech Dissertation, IIT Roorkee

Semiromi T, Majid KM (Sep 2017) A fully coupled SWAT-MODFLOW model to simulate surface-groundwater interactions: application to the gharehsoo river basin, Iran.44th annual congress of the international association of hydrogeologist. pp 23–27

Singh RM, Shukla P (2016) Groundwater system simulation and management using visual MODFLOW and Arc-SWAT by, Journal of Water Resource and Hydraulic Engineering. pp 29–35

Sophocleous MA, Koelliker JK, Govindaraju RS, Birdie T, Ramireddygari SR, Perkins SP (1999) Integrated numerical modeling for basin-wide water management: the case of the Rattlesnake Creek basin in Southcentral Kansas. J Hydrol 214(1–4):179–196

Chapter 21
Revisiting the Antecedent Moisture Content-Based Curve Number Formulae

Mohan Lal, S. K. Mishra, Ashish Pandey, and Dheeraj Kumar

Abstract The present study was carried out to explore the performance of the five existing and one proposed antecedent moisture condition (AMC)-based curve number (CN) conversion formulae. The comparison was made using the rainfall (P)–runoff (Q) data from 36 plots/watersheds of published literature and 27 plots of an agricultural field located at Roorkee, Uttarakhand, India. For developing of the proposed model, the CNs were derived from the 39 watersheds P–Q data using a standard initial abstraction ratio (λ) value as 0.20. As seen, the proposed formula out-performs all the existing formulae when tested for runoff estimation using a large set of field data. In the existing formulae, the Mishra et al. (Water Resour Manage 22:861–876, 2008b) perform superiorly followed by (Hawkins et al. Journal of Irrigation and Drainage Engineering 111:330–340, 1985). The multiple comparison results show that the runoff estimated by all existing formulae is not significantly different (at the 0.05 significance level) for the 24 tested watersheds. The proposed formula has also boosted the mean RMSE and R^2 values from 12.00 mm and 0.329 to 10.72 mm and 0.373, respectively, as compared to the best existing model (i.e. Mishra et al. Water Resour Manage 22:861–876, 2008b), which justify the use of former ones in the field.

M. Lal (✉) · D. Kumar
Department of Irrigation and Drainage Engineering, G. B. Pant,
University of Agriculture and Technology, Pantnagar, Uttarakhand 263145, India
e-mail: mohan841987@gmail.com

D. Kumar
e-mail: dheerajtejwan@gmail.com

S. K. Mishra · A. Pandey
Department of Water Resources Development and Management, Indian Institute of Technology Roorkee, Roorkee, Uttarakhand 247667, India
e-mail: skm61fwt@gmail.com

A. Pandey
e-mail: ashish.pandey@wr.iitr.ac.in

A. Pandey et al. (eds.), *Water Management and Water Governance*,
Water Science and Technology Library 96,
https://doi.org/10.1007/978-3-030-58051-3_21

Keywords Initial abstraction ratio · Curve number · Antecedent moisture content · Initial abstraction

21.1 Introduction

The Soil Conservation Service Curve Number (SCS-CN) currently known as the Natural Resources Conservation Service method was developed in 1954 followed by its documentation in Section 4 of the National Engineering Handbook (NEH-4). This handbook was published by the Soil Conservation Service, US Department of Agriculture in 1956. The SCS-CN method is the result of exhaustive field investigations carried out during 1930s and 1940s. Since its inception, this method has seen a number of development and applications. This method was originally developed for estimating runoff from small agricultural watersheds. But, nowadays, its usage has been extended from small agricultural watersheds to urban and forested watersheds. It is because most of the other methods are developed for gauged watersheds/catchments having complex type of approaches and many variables involved. But SCS-CN method is simple and applicable to ungauged watersheds as well for which minimum hydrologic information is available. It is simple, easy to understand and apply, stable, and useful for ungauged watersheds. Based on exhaustive field investigations carried out in the USA, curve numbers were derived for different land uses, soil types, hydrologic condition, and management practices and these are reported in NEH-4 (SCS 1972). This method is being used by many hydrologists and design engineers in their study because it is very simple and easy to apply for estimating the direct surface runoff. This method is very powerful and simple techniques for runoff estimation because it considered all the runoff producing characterises like soil type, land use, etc., and incorporating them into a single parameter known as curve number (CN). Presently, this method has also been coupled with many hydrologic, soil erosion, and water quality models due to requirement of its low data input. In contrast, this method has a well-documented weakness too. The one of the major weakness is its inability to consider rainfall intensity which makes it a 24-h model. Besides, slope is also another factor which was not included in original development.

In general, for any ungauged watershed, the CN is derived from commonly known national engineering handbook Chapter-4 (NEH-4) tables using the watershed characteristics such as soil type, land cover, land use, and previous five day rainfall (P_5) (Karn et al. 2016; Mishra et al. 2008a).The previous five-day rainfall plays an important role in runoff prediction as the CN varies with it; and therefore, to predict runoff precisely, this lumped parameter CN (Ponce and Hawkins 1996) must be determined accurately based on the P_5. In SCS method, the AMC is defined into three categories viz., dry moisture conditions (or AMC-1), average moisture conditions (or AMC-2), and wet moisture conditions (or AMC-3). It is worth to note that the AMC-2 status is considered as the reference condition and is the basis for CN values from NEH-4 tables (SCS 1971). Practically, the AMC-2 (or CN_2) CNs are first estimated, and then adjusted to AMC-3 (or CN_3) or AMC-1 (or CN_1) depending on the previous five-day rainfall amount (Mays 2005). In order to make runoff calculation simpler, these different AMC-dependent CN values are arranged and given in tabular form by NEH-4 (SCS 1956). Since the inception of these tables, only few studies have been conducted to check these AMC dependent CNs based on cumulative probability concept (Hjelmfelt et al. 1981). The Hjelmfelt et al. (1981) conducted an exhaustive study on AMC tables regarding conversion of one AMC into other. In their study, they found that the NEH-4 AMC tables (SCS 1971) described the AMC into three classes, AMC-3, AMC-2, and AMC-1 (or CN_3, CN_2 and CN_1), which account statistically for 90th, 50th, and 10th percentiles, respectively, of cumulative probability that a given rainfall will exceed the runoff depth.

In addition to above, a number of investigators have also made attempts to link the different AMC-dependent CNs to present them into mathematical form. In this regard, the work of Arnold et al. (1990), Sobhani (1975), Hawkins et al. (1985), Chow et al. (1988), and Mishra et al. (2008b) are notable one. All of these researchers have used the same NEH-4 AMC-dependent CN values and presented them into the forms as given in Table 21.1 through M1-M5. Keeping in the mind that most of the existing formulae were developed using same data source, a comprehensive study is conducted to compare and check the validity of these conversions using new and large dataset covering different characteristics such as slope, altitude, climate, area, etc.

21.2 SCS-CN Method

The general form of the SCS-CN equation is as given below:

$$Q = \begin{cases} \frac{(P-I_a)^2}{(P+S-I_a)}, & P > I_a \\ 0 & P \leq I_a \end{cases} \quad (21.1)$$

In Eq. 21.1, Q (mm) and P (mm), respectively, are the direct surface generated runoff and observed rainfall. I_a is known as the initial abstraction (mm). The I_a can be represented as a fraction of potential maximum retention (S), i.e., $I_a = \lambda S$. In this, λ is represented as initial abstraction ratio. A λ value of 0.2 is being used by current standard version of SCS-CN method (SCS 1972, 1985). However, many researchers demonstrated it as vague, and recommended a value of 0.05 or less for field applications (Lal et al. 2017; Zhou and Lei 2011).

Equation 21.1 (using $\lambda = 0.2$) can be solved for calculating the S value as (Hawkins 1993):

$$S = 5\left[(P + 2Q) - \left(4Q^2 + 5PQ\right)^{0.5}\right] \quad \text{for} \quad 0 < Q < P \quad (21.2)$$

Table 21.1 AMC-dependent formulae

Method ID	Method	AMC-1	AMC-3
M1	Sobhani (1975)	$CN_1 = \frac{CN_2}{2.334-0.01334\,CN_2}$	$CN_3 = \frac{CN_2}{0.4036+0.005964CN_2}$
M2	Hawkins et al. (1985)	$CN_1 = \frac{CN_2}{2.281-0.01281\,CN_2}$	$CN_3 = \frac{CN_2}{0.427+0.00573CN_2}$
M3	Chow et al. (1988)	$CN_1 = \frac{4.2CN_2}{10-0.058CN_2}$	$CN_3 = \frac{23CN_2}{10+0.13CN_2}$
M4	Arnold et al. (1990)	$CN_1 = CN_2 - \frac{20(100-CN_2)}{[100-CN_2+\exp\{2.533-0.0636(100-CN_2)\}]}$	$CN_3 = CN_2\exp[0.00673(100-CN_2)]$
M5	Mishra et al. (2008b)	$CN_1 = \frac{CN_2}{2.2754-0.012754CN_2}$	$CN_3 = \frac{CN_2}{0.430+0.0057CN_2}$
M6	Equation 21.4 fitted for CN_3 and CN_1 with 90% and 10% POE, respectively	$CN_1 = \frac{CN_2}{1.92192-0.00922CN_2}$ $R^2 = 0.472$	$CN_3 = \frac{CN_2}{0.50503+0.00495CN_2}$ $R^2 = 0.640$

After calculation of S value, the CN can be calculated by equation given below:

$$\mathrm{CN} = \frac{25400}{S + 254} \tag{21.3}$$

21.3 Materials and Methods

21.3.1 Study Area and Data Used

The present study uses the P-Q data from heterogeneous climate around the globe. A total P-Q data of 63 watersheds and plots collected from published literature and locally monitored were used. The details about used P-Q data are given in appendix "A" with brief characteristics of plots/watersheds such as climate type, number of P-Q events, area, rainfall, altitude, and location. As can be seen from appendix "A", out of 63, P-Q data of 27 study plots were monitored locally at Roorkee, Uttarakhand state (India). The remaining data covering the 36 small scale plots to large-scale watersheds were collected from published literature. This published literature data covers the different climatic conditions having different catchment characteristics.

As far as locally measured data concerns, natural P-Q data were monitored on experimental plots of size 22 m length and 5 m width. The slope of plots was varying as 5, 3, and 1%. The four different land uses: Maize, fallow land, black gram, and sugarcane were established for observing the P-Q data. The detailed information about plots construction and development can be found elsewhere (Lal et al. 2019).

In this study, the data from, respectively, 1–39 and 40–63 plots/watersheds were used in development and validation of proposed formulae.

21.3.2 Existing CN-Conversion Formulae

The formulae developed by Chow et al. (1988), Arnold et al. (1990), Hawkins et al. (1985), Sobhani (1975), and Mishra et al. (2008b) are shown in Table 21.1. As already mention in the introduction section, all these existing formulae were developed using the same source data, i.e. NEH-4 AMC table. Further, the details such as number of data points used, CN–range, data filtration, and smoothening procedure, etc., used in development of these formulae are not being discussed in this paper. Therefore, readers are advised to refer the Mishra et al. (2008b) and Lal et al. (2019) for exhaustive details about development of these existing formulae.

21.3.3 Proposed CN-Conversion Formulae

As can be seen from Table 21.1, all existing formulae can be represented in a general form given below:

$$\mathrm{CN_x} = \frac{\mathrm{CN_2}}{\mathrm{a + b\ CN_2}} \tag{21.4}$$

where $b = \frac{(1-a)}{100}$, and the values of $\mathrm{CN_x}$ will be as $\mathrm{CN_1}$ or $\mathrm{CN_3}$ based on values of a and b.

In this research paper, Eq. (21.4) was optimized for $\mathrm{CN_2}$ and $\mathrm{CN_1}$ or $\mathrm{CN_3}$ utilizing the large P–Q dataset from heterogenous climatic conditions. The proposed model M6 was derived optimizing Eq. (21.4) and presented in Table 21.1. In the development of proposed model M6, P-Q eventwise CN values were calculated using Eqs. 21.2 and 21.3. The different AMCs CN values were estimated using Hjelmfelt et al. (1981) concept and these values, i.e. $\mathrm{CN_3}$ and $\mathrm{CN_1}$ were further utilized in Eq. (21.4) for least square fitting.

Table 21.2 shows the derived $\mathrm{CN_1}$, $\mathrm{CN_2}$, and $\mathrm{CN_3}$ utilizing the 39-study plot/watershed dataset. Similarly, probability distributions of $\mathrm{CN_1}$ and $\mathrm{CN_3}$ with $\mathrm{CN_2}$ are shown in Fig. 21.1.

21.4 Performance Evaluation Criteria

The performance evaluation was carried out using coefficient of determination ($\mathrm{R^2}$), root mean square error (RMSE), Nash–Sutcliffe efficiency coefficient (E) (Nash and Sutcliffe 1970). The $\mathrm{R^2}$, E, and RMSE are expressed as:

$$R^2 = \left(\frac{\sum_{i=1}^{n} (Q_i - \overline{Q})\left(Q_{ci} - \overline{Q_c}\right)}{\left[\sum_{i=1}^{n} (Q_i - \overline{Q})^2 \sum_{i=1}^{n} \left(Q_{ci} - \overline{Q_c}\right)^2\right]^{0.5}} \right)^2 \tag{21.5}$$

$$\mathrm{E} = \left(1 - \frac{\sum_{i}^{n} (Q_i - Q_{ci})^2}{\sum_{i}^{n} (Q_i - \overline{Q})^2} \right) \tag{21.6}$$

$$\mathrm{RMSE} = \left(\frac{1}{n} \sum_{i}^{n} (Q_i - Q_{ci})^2 \right)^{1/2} \tag{21.7}$$

Table 21.2 Different AMC's CN values for the 39 plots datasets used in development of proposed formulae, i.e. M6

Watershed No.	M6		
	CN_2	CN_3	CN_1
1	81.24	88.83	70.01
2	79.88	91.35	65.29
3	81.09	92.12	67.41
4	78.08	88.28	66.03
5	79.21	90.08	63.15
6	77.75	81.78	63.67
7	80.88	90.36	74.68
8	79.78	85.58	73.36
9	81.49	92.08	77.67
10	79.79	90.76	72.98
11	79.17	89.92	67.13
12	80.30	91.60	62.30
13	92.35	97.73	81.47
14	88.10	95.35	80.89
15	86.29	94.35	76.82
16	73.90	78.60	70.30
17	81.94	94.13	70.61
18	90.00	96.08	83.86
19	87.71	94.04	67.39
20	78.71	88.21	60.11
21	82.65	89.31	63.28
22	83.65	88.40	71.88
23	69.66	87.10	58.86
24	73.71	86.52	52.91
25	86.52	92.08	78.63
26	87.92	92.08	86.11
27	86.63	91.50	59.25
28	82.60	89.02	78.41
29	80.17	88.06	73.83
30	87.52	97.60	68.52
31	89.52	93.42	77.69
32	90.19	93.14	77.17
33	87.29	91.04	76.36
34	87.97	91.59	73.89
35	83.92	91.86	75.50
36	73.09	82.75	61.84

(continued)

Table 21.2 (continued)

Watershed No.	M6		
	CN_2	CN_3	CN_1
37	74.77	79.89	68.96
38	76.91	86.08	56.60
39	81.81	90.42	67.48

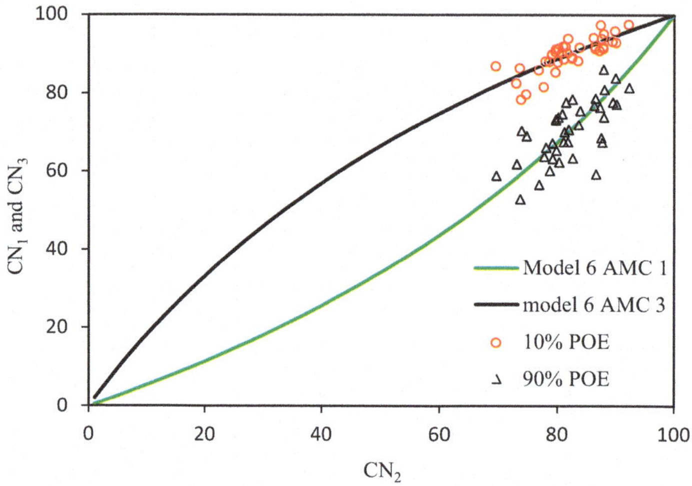

Fig. 21.1 Figure depicting the distribution of CN_1 and CN_3 with CN_2

where Q_{ci} (mm) and Q_i (mm) are the predicted and observed runoff, respectively, $\overline{Qc}$(mm) is the average of predicted runoff for all storm events, $\overline{Q}$(mm) is the average of observed runoff for all storm events, n is the total number of P-Q events.

Further, the improvement in performance of the proposed expressions over the existing ones was also evaluated using the r–statistic (Nash and Sutcliffe 1970; Lal et al. 2017; Senbeta et al. 1999) and it is expressed as:

$$r = \frac{(E_2 - E_1)}{1 - E_1} \times 100 \tag{21.8}$$

where E_1 and E_2 are the Nash and Sutcliffe efficiencies due to the existing model and proposed model, respectively. As per Senbeta et al. (1999) a value of r greater than 10% considered as significant improvement of the proposed expressions over existing one.

21.5 Results and Discussions

In this study, the performance evaluation of existing and proposed formulae was carried out using naturally observed P-Q data of 24 watershed/plots.

It is well-established phenomenon that the surface generated runoff is highly dependent on Antecedent moisture content (AMC) (Mishra et al. 2008b). Therefore, in estimation of runoff correctly, these derived CN_2, must be corrected for the desired AMC (i.e. CN_1 or CN_3). In this present study, first CN_2 were estimated using the data of 40–63 watersheds/plots. After estimation of CN_2, the CN_1 or CN_3 were calculated based on P_5 following Mays (2005).

In order to compare the existing (M1–M5) and proposed formulae (M6), the derived CN_2 of 40–63 watersheds/plots were first transformed into CN_1 or CN_3 using both existing (M1–M5) and proposed (M6) formulae. After estimation of AMC-based CNs, i.e. CN_1 or CN_3, the runoff was calculated using Eqs. 21.1, 21.2 and 21.3 employing $\lambda = 0.20$ into Eq. 21.1 (SCS, 1972, 1985); known as the standard procedure for runoff estimation. The runoff estimation performance of different AMC formulae was carried out through cumulative frequency distribution graph of different goodness of fit indices viz. R^2, E and RMSE. These graphs are shown in Figs. 21.2, 21.3, 21.4. As seen from these figures, the M6 found to show the relatively higher R^2, E and lower RMSE as compared to existing expressions. Among M1-M5 models, M5 found to show relatively higher R^2, E and lower RMSE as compared to other existing expressions.

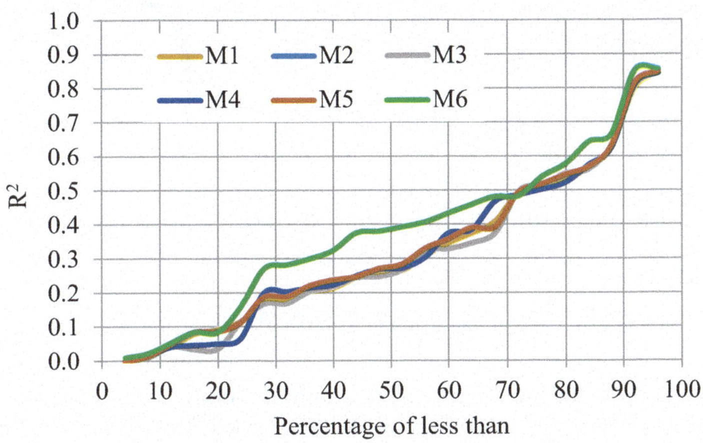

Fig. 21.2 Cumulative frequency distribution graph of R^2

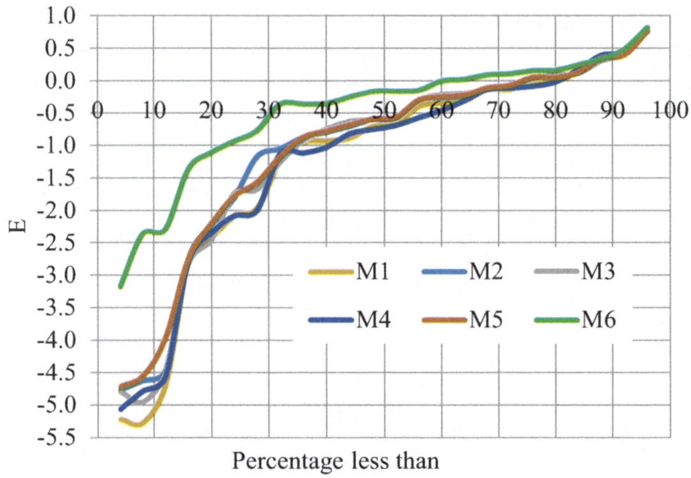

Fig. 21.3 Cumulative frequency distribution graph of E

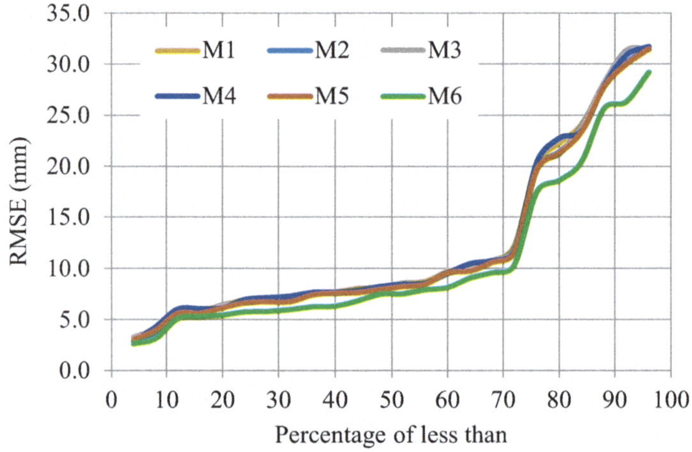

Fig. 21.4 Cumulative frequency distribution graph of RMSE

The details about the maximum (max), minimum (min), standard deviation (SD), and mean values of different indices viz., R^2, E, and RMSE yielded from the runoff estimation are shown in Table 21.3.

As seen from this table, R^2 values yielded from application of proposed model M6 varies between 0.00 and 0.858 with mean value of 0.373. Whereas among M1–M5 models, R^2 was varied from 0.00 to 0.851. Among existing, both M2 and M5 found to show relatively better performance as both were having same mean R^2 (=0.329) value. Based on R^2, M1 found to perform poorest among old set of formulae. Similarly, RMSE values yielded from application of proposed model M6 varies from 2.67 to

Table 21.3 Performance statistic of runoff estimation yielded from the analysis of 24 watersheds/plots P-Q data

Model ID	R^2				RMSE (mm)				E			
	Mean	SD	Max	Min	Mean	SD	Max	Min	Mean	SD	Max	Min
M1	0.322	0.237	0.847	0.00	12.344	8.713	31.44	3.08	−1.222	1.733	0.762	−5.27
M2	0.329	0.237	0.850	0.00	12.017	8.602	31.56	3.12	−1.062	1.598	0.775	−4.77
M3	0.325	0.234	0.851	0.00	12.166	8.721	31.52	3.17	−1.122	1.649	0.765	−4.96
M4	0.324	0.242	0.845	0.000	12.325	8.691	31.78	2.98	−1.205	1.651	0.775	−5.06
M5	0.329	0.238	0.850	0.00	12.001	8.569	31.60	3.13	−1.049	1.544	0.776	−4.71
M6	0.373	0.251	0.858	0.00	10.724	7.769	29.27	2.67	−0.466	0.972	0.829	−3.17

29.27 mm with mean value of 10.724 mm. Although this was varied between 2.98 and 31.78 mm for old set (i.e., M1-M5) of formulae. With a higher mean value of 12.00 mm, M5 followed by M2 performed best among old formulae. Based on RMSE, again M1 found to perform poorest. Further, for M6, the efficiency (E) was varied from −3.17 to 0.829 with mean value of −0.466. Once more, based on E too, M6 performed best of all among all tested models. Among existing, M5 followed by M2 performed best; M1 performed the poorest.

Lastly, by considering the mean values of RMSE, R^2 and E as overall yardstick criteria for evaluating the performance, these six models can rank as:

M6 > M5 > M2 > M3 > M4 > M1 (based on RMSE and R^2).
M6 > M5 > M2 > M4 > M1 > M3 (based on E).

The values of Table 21.3 and the depiction of Figs. 21.2, 21.3 and 21.4 lead to infer that M6 followed by M5 performed best of all. Similarly, among existing formulae, M5 followed by M2 performed best.

Further, r–statistic criteria (Eq. 21.8) was also employed to find out the runoff estimation improvement capability of proposed model M6 over best existing model M5 (i.e. M6 versus M5). The result of this analysis is shown in Fig. 21.5. As seen from this figure, M6 improved the runoff prediction efficiency (E) in 23 out of 24 tested watersheds as the r values varies from −2.35 to 72.74%. Further, the significant improvement was observed in 19 watersheds as the r–values was greater than 10%.

21.6 Conclusions

In Soil Conservation Service Curve Number (SCS–CN) method, accurate antecedent moisture condition (AMC)-based CN is necessary because antecedent soil moisture plays an important role in predicting the runoff from a given rainfall. The present work was carried out to evaluate the five existing and one proposed formula developed for changing the CNs from one AMC into another. The results analysis shows that the developed (M6) formulae perform best for conversion of CN_2 into CN_1 and CN_3

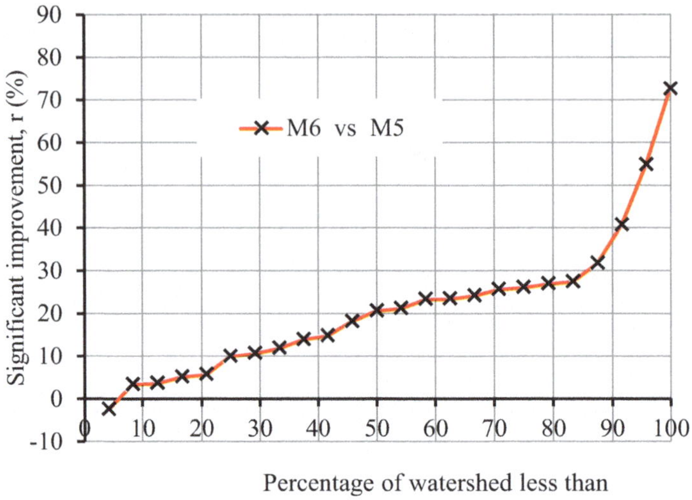

Fig. 21.5 Cumulative frequency distribution of improvement using the r criterion

in their application to field data. The proposed M6 performed best of all. Similarly, M5 performed best followed by M2 among existing formulae. The M6 has also significantly (based on r > 10%) improve the runoff prediction efficiency (E) in the 19 out of 24 tested watersheds when it compared with the best existing formula M5.

Acknowledgements Authors are thankful to Indian National Committee for Surface Water (INCSW) (Ministry of Water Resources, River Development and Ganga Rejuvenation (MoWRD&GJ), Government of India) Central Water Commission, RK Puram, New Delhi, for providing the fund under the R & D project "Experimental Verification of SCS Runoff Curve Numbers for Selected Soils and Land Uses" for monitoring P-Q data at Roorkee site. The authors also thankful to the researchers given in Appendix "A" whose published data have been used in our analysis.

Appendix A: Characteristics of Sixty-Three Watersheds/Plots Used in the Study

Watershed/Plot No	n	Area(km^2)	Rainfall (mm)	Altitude (m)	Climate type	Study location	References
1–15	Varies from 10–15	110×10^{-6}	1120–1500	266	Humid sub–tropical	Solani river catchment (India)	Data monitored at present site
16	40	765×10^{-6}	746	515	Semi-arid tropical	Experimental Plot in CRIDA, Hyderabad	Mandal et al. (2012)
17	10	18.7	1100	41.3	Humid subtropical	Canada de Los Chanchos basin (Uruguay)	Silveira et al. (2000)
18	21	448.2	296.4	1320–2960	Semi-arid	Kardeh Watershed (Northeast Iran)	Ebrahimian et al. (2012)
19	24	153	960	255–1330	Monsoon tropical climate with dry and wet seasons	Upper Lam Ta Kong watershed (Thailand)	Phetprayoon (2015)
20	25	713					
21	18	69.10	990	175–235	Temperate climate	Upper Little Vermilion River watershed (USA)	Walker et al. (2005)
22	18	12.90	940	312–798	Mediterranean semi-humid	Colorso stream catchment (Central Italy)	Brocca et al. (2008)
23	12	26.10	1087	–	Temperate	St. Esprit watershed (Canada)	Perrone and Madramooto (1997)
24	16	114	1420	3.6–72	Mediterranean climate	Simms Creek watershed (Florida, USA)	Melesse and Graham (2004)
25	40	1.90	1220	1835–3152	Alpine climate	Rio Vauz Basin catchment (Italian Dolomites)	Penna et al. (2011)

(continued)

(continued)

Watershed/Plot No	n	Area(km^2)	Rainfall (mm)	Altitude (m)	Climate type	Study location	References
26	27	237.80	850	859.50	Temperate	Catchment area of the Kamienica river (Poland)	Wałęga et al. (2015)
27	10	7.03	409	1080–1270	Semiarid continental monsoon climate	Liudaogou watershed (Northern China)	Xiao et al. (2011)
28	20	05	1286	1800	Humid and cold	Matash spring-fall mountainous rangeland (Iran)	Sadeghi et al. (2007)
29	20						
30	16	655	1100	1600	Mediterranean	Mdouar catchment in northern Morocco (Africa)	Tramblay et al. (2012)
31	15	137	930	249–887	Mediterranean	Niccone at Migianella (Italy)	Brocca et al. (2009)
32		104				Niccone at Reschio	
33		57				Vallaccia at Molino	
34		35				Vallaccia at P. Marte	
35		13				Colorso at P. Marte	
36	29	16.70	1016	184–1180	Subtropical	Wangjiaqiao watershed (China)	Shi et al. (2009)
37	14	1.14	542	1330–1707	Dry and continental	Qiaozi-West watershed (China)	Zhou and Lei (2011)
38	16	109	400	0–909	Subtropical Mediterranean	Rafina catchment in eastern Attica (Greece)	Massari et al. (2014)

(continued)

(continued)

Watershed/Plot No	n	Area(km^2)	Rainfall (mm)	Altitude (m)	Climate type	Study location	References
39	10	0.055	590	275–393	Humid-temperate continental	North Appalachian watershed (USA)	Wu et al. (1993)
40–51	Varies from 11–13	110×10^{-6}	1120–1500	266	Humid sub–tropical	Solani river catchment (India)	Data monitored at present site
52	31	9.624	1371	1057–1200	Tropical savanna	Capetinga catchment (Brazil)	Silva et al. (1999)
53	24	7.84	595	280–950	Mediterranean semi-arid	Upper Lykorrema, Penteli Mountain (Greece)	Soulis et al. (2009)
54	23	7.36		146–643		Entire Penteli Mountain (Greece)	
55	13	254	1500	575	Moist and temperate climate	Amicalola Creek watershed (USA)	Mishra and Singh (2004)
56	30	237.8	850	859.5	Temperate	Kamienica river catchment (Southern Poland)	Wałęga and Rutkowska (2015)
57	8	609.15	1200	26–911	Continental monsoon climate	Hoideok watershed (South Korea)	Moon et al. (2014)
58	12	208.4				Jungrangkyo watershed (South Korea)	
59	48	4661	1547	509–900	Sub-tropical and sub-humid	Mohegaon catchment (India)	Mishra et al. (2008a)
60	17	1.6	1370	3.7–10	Humid subtropical	Tributary of Huger Creek (South Carolina)	Epps et al. (2013)

(continued)

(continued)

Watershed/Plot No	n	Area(km^2)	Rainfall (mm)	Altitude (m)	Climate type	Study location	References
61	15	21	1000	2600–3000	semiarid	Godigne catchment, Tekeze river basin (Ethiopia)	Zelelew (2017)
62	42	0.02	867	217	Subtropical steppe as classified by Köppen classification;	Southcentral region of the state of Ceará (Brazil)	Andrade et al. (2017)
63	40	0.015					

(*n* number of rainfall events)

References

Andrade EM, Araújo Neto JR, Guerreiro MJS, Santos JCN, Palácio HAQ (2017) Land use effect on the CN model parameters in a tropical dry environment. Water Resour Manage. https://doi.org/10.1007/s11269-017-1732-4

Arnold JG, Williams JR, Nicks AD, Sammons NB (1990) SWRRB: A basin scale simulation model for soil and water resources management. Texas A&M University Press, College Station. p 142, 10 appendices

Brocca L, Melone F, Moramarco T (2008) On the estimation of antecedent wetness conditions in rainfall–runoff modeling. Hydrol Process 22:629–642

Brocca L, Melone F, Moramarco T, Singh VP (2009) Assimilation of observed soil moisture data in storm rainfall–runoff modelling. J Hydrol Eng 14(2):153–165

Chow VT, Maidment DR, Mays LW (1988) Applied hydrology. McGraw-Hill, New York

Ebrahimian M, Nuruddin AA, MohdSoom MAB, Sood AM (2012) Application of NRCS-curve number method for runoff estimation in a mountainous watershed. Caspian J Environ Sci 10(1):103–114

Epps T, Hitchcock DR, Jayakaran A, Loflin DR, Williams TM, Amatya DM (2013) Curve number method assessment for watersheds draining two headwater streams in lower coastal plain of South Carolina. J Am Water Resour Assoc 49(6):1284–1295

Hawkins RH, Hjelmfelt AT, Zevenbergen AW (1985) Runoff probability, storm depth and curve numbers. J Irrig Drainage Eng 111(4):330–340

Hawkins RH (1993) Asymptotic determination of runoff curve numbers from data. J Irrig Drainage Eng 119(2):334–345

Hjelmfelt AT, Kramer LA, Burwell RE (1981) Curve numbers as random variables. In: Rainfall-runoff relationship. Water Resources Publications, Littleton, CO, pp 365–370

Karn AL, Lal M, Mishra SK, Chaube UC, Pandey A (2016) Evaluation of SCS-CN inspired models and their comparison. J Indian Water Resour Soc 36(3):19–27

Lal M, Mishra SK, Pandey A, Pandey RP, Meena PK, Chaudhary A, Jha RK, Shreevastava AK, Kumar Y (2017) Evaluation of the Soil Conservation Service curve number methodology using data from agricultural plots. Hydrogeol J 25(1):151–167

Lal M, Mishra SK, Kumar M (2019) Reverification of antecedent moisture condition dependent runoff curve number formulae using experimental data of Indian watersheds. CATENA 173:48–58

Mandal UK, Sharma KL, Prasad JVNS, Reddy BS, Narsimlu B, Saikia US, Adake RV, Yadaiah P, Masane RN, Venkanna K, Venkatravamma K, Satyam B, Raju B, Srivastava NN (2012) Nutrient losses by Runoff and Sediment from an Agricultural Field in Semi-arid Tropical India. Indian J Dryland Agricu Res Dev 27(1):01–09

Massari C, Brocca L, Barbetta S, Papathanasiou C, Mimikou M, Moramarco T (2014) Using globally available soil moisture indicators for flood modelling in mediterranean catchments. Hydrol Earth Syst Sci 18:839–853

Mays LW (2005) Water resources engineering 2nd edn. Willey, Arizona, ISBN: 978-0-470-46064-1

Melesse AM, Graham WD (2004) Storm runoff prediction based on a spatially distributed travel time method utilizing remote sensing and GIS. J Am Water Resour Assoc 40(4):863–879

Mishra SK, Singh VP (2004) Validity and extension of the SCS-CN method for computing infiltration and rainfall-excess rates. Hydrol Process 18:3323–3345

Mishra SK, Jain MK, Suresh Babu P, Venugopal K, Kaliappan S (2008a) Comparison of AMC-dependent CN conversion formulae. Water Resour Manage 22:1409–1420

Mishra SK, Pandey RP, Jain MK, Singh VP (2008b) A rain duration and modified AMC-dependent SCS-CN procedure for long duration rainfall-runoff events. Water Resour Manage 22(7):861–876

Moon GW, Yoo JY, Ahn JH, Kim TW (2014) Comparative analysis of estimation methods for basin averaged effective rainfall using NRCS-CN method. J Korean Soc Civil Eng 34(2):493–503

Nash JE, Sutcliffe JV (1970) River flow forecasting through conceptual models, Part I—a discussion of principles. J Hydrol 10:282–290

Penna D, Tromp-van Meerveld HJ, Gobbi A, Borga M, Dalla Fontana G (2011) The influence of soil moisture on threshold runoff generation processes in an alpine headwater catchment. Hydrolo Earth Syst Sci 15:689–702

Perrone J, Madramootoo CA (1997) Use of AGNPS for watershed modeling in QUEBEC. Trans ASAE 40(5):1349–1354

Phetprayoon T (2015) Application of GIS-based curve number method for runoff estimation in agricultural-forest watershed, Thailand. KKU Res J 20(2):155–167

Ponce VM, Hawkins RH (1996) Runoff curve number: has it reached maturity? J Hydrol Eng 1(1):11–19

Sadeghi SHR, Mizuyama T, GhaderiVangah B (2007) Conformity of MUSLE estimates and erosion plot data for storm-wise sediment yield estimation. Terr Atmos Oceanic Sci J 18(1):117–128

SCS (1956, 1971, 1972, 1985) "Hydrology" national engineering handbook, supplement A, Section 4. Soil Conservation Service, USDA, Washington, DC

Senbeta DA, Shamseldin AY, O'Connor KM (1999) Modification of the probability-distributed interacting storage capacity model. J Hydrol 224:149–168

Shi ZH, Chen LD, Fang NF, Qin DF, Cai CF (2009) Research on the SCS-CN initial abstraction ratio using rainfall-runoff event analysis in the three gorges area, China. CATENA 77:1–7

Silva CL, Oliveira CAS (1999) Runoff measurement and prediction for a watershed under natural vegetation in central Brazil. Brazil J Soil Sci 23:695–701

Silveira L, Charbonnier F, Genta JL (2000) The antecedent soil moisture condition of the curve number procedure. Hydrol Sci J 45(1):3–11

Sobhani G (1975) A review of selected small watershed design methods for possible adoption to Iranian conditions. M.S. Thesis, Utah State University, Logan, UT

Soulis KX, Valiantzas JD, Dercas N, Londra PA (2009) Investigation of the direct runoff generation mechanism for the analysis of the SCS-CN method applicability to a partial area experimental watershed. Hydrol Earth Syst Sci 13:605–615

Tramblay Y, Bouaicha R, Brocca L, Dorigo W, Bouvier C, Camici S, Servat E (2012) Estimation of antecedent wetness conditions for flood modelling in northern Morocco. Hydrol Earth Syst Sci 16:4375–4386

Wałęga A, Michalec B, Cupak A, Grzebinoga M (2015) Comparison of SCS-CN determination methodologies in a heterogeneous catchment. J Mt Sci 12(5):1084–1109

Wałęga A, Rutkowska A (2015) Usefulness of the modified NRCS-CN method for the assessment of direct runoff in a mountain catchment. Acta Geophys 63(5):1423–1446

Walker SE, Banasik K, Northcott WJ, Jiang N, Yuan Y, Mitchell JK (2005) Application of SCS method on mild slope watershed. Southern cooperative series bulletin. Available at: https://s1004.okstate.edu/S1004/Regional-Bulletins/Modeling-Bulletin/paper98-draft1.html (Accessed 2016)

Wu TH, Hall JA, Bonta JV (1993) Evaluation of runoff and erosion models. J Irrig Drainage Eng 119(4):364–382

Xiao B, Wang QH, Fan J, Han FP, Dai QH (2011) Application of the SCS-CN model to runoff estimation in a small watershed with high spatial heterogeneity. Pedosphere 21(6):738–749

Zelelew DG (2017) Spatial mapping and testing the applicability of the curve number method for ungauged catchments in Northern Ethiopia. Int Soil and Water Conserv Res. https://doi.org/10.1016/j.iswcr.2017.06.003

Zhou SM, Lei TW (2011) Calibration of SCS-CN initial abstraction ratio of a typical small watershed in the Loess hilly-gully region. Scientia Agricultura Sinica 44(20):4240–4247 ((in Chinese))

Chapter 22
Effectiveness of Best Management Practices on Dependable Flows in a River Basin Using Hydrological SWAT Model

Santosh S. Palmate and Ashish Pandey

Abstract This study was carried out to evaluate the effectiveness of BMP implementation on annual dependable flows at different probabilities (40, 50, 60, 75 and 90%). Best management practices (BMPs) are utilized as the most viable solution for conservation and reduction in river flows. The Soil and Water Assessment Tool (SWAT), physically based semi-distributed hydrological model, was utilized to implement and evaluate the feasible BMPs for the Betwa river basin (BRB), a large agricultural river basin located in Central India. In this study, changes in dependable flows were analyzed using the model simulation outputs at pre-BMP and post-BMP conditions. Results show that strip cropping is the most effective agriculture land treatment reducing flows about 10–12%, while the grade stabilization structure is the most effective channel treatment reducing flows about 6–8% in all five dependability percentages. High sensitivity of runoff curve number (CN2) and main river channel slope (CH_S2) were mainly responsible for these reductions, and the significant changes in dependable flows. The study reveals effectiveness of BMP treatments (strip cropping and grade stabilization structure) on medium to high dependable flows can be recommended as the valuable assets for water resources planning and management in the Betwa river basin. In this context, the SWAT model is a useful tool assessing the flow characteristics of a river basin under various management treatments.

Keywords Best management practices · Betwa river basin · Dependable flow · SWAT model

S. S. Palmate (✉) · A. Pandey
Department of Water Resources Development and Management, Indian Institute of Technology Roorkee, Roorkee 247667, Uttarakhand, India
e-mail: psubhash@wr.iitr.ac.in

A. Pandey
e-mail: ashish.pandey@wr.iitr.ac.in

A. Pandey et al. (eds.), *Water Management and Water Governance*,
Water Science and Technology Library 96,
https://doi.org/10.1007/978-3-030-58051-3_22

22.1 Introduction

The water resources development and management in a region requires correct information on hydrological processes of a river basin. It is also essential to know water flow alteration regimes for determining contemporary water resources availability. Flow duration curve (FDC) technique generally used for analyzing water flow dependability at a site where no measure or very limited data record exists (Arora et al. 2005). Utilization of FDC is admitted in hydropower projects designing as well as water resources planning and management. The FDC application also includes rainfall–runoff modelling, reservoir sedimentation, water quality management, hydrologic alteration, and to check water flow data quality as well (Vogel and Fennessey 1995; Yu et al. 2002; Cole et al. 2003). Difficulty of water flow study arises due to complex hydrological processes, climatic and topographic variability with respect to space and time, and contemporary management, which controls this phenomenon. It is evident that climate change resulting unpredictable precipitation events, such as cloud bursting and heavy rainfall, can produce large variation in river flows, and hence affect the management of regional water resources. Thus, the variability of dependable flows under feasible management treatments should be incorporated into the analysis and design of a water resources project.

It is usually observed that in a large river basin, the average annual flows are about 10 to 20% more than the 75% dependable annual flows. However, in the small river basins about 100–1000 km^2, the average annual flows are observed to be much higher about 15–35% more than the 75% dependable flows. It means depending upon the size of a river basin and existing hydrological conditions, 10–35% excess water in comparison to the 75% dependable flow may be available for half of the period (50% dependable flow) (Nilsson et al. 2005; Smakhtin et al. 2007; Kusre et al. 2010). Flow varies annually as well as within a year, and the 70% to 90% of annual flows occur mostly during the monsoon months. Due to such flow variability both within a year and year to year, a considerable amount of water goes as waste, if, an adequate water management is not given, which can reduce the sufficient amount of utilizable water. The best management practices (BMPs) are being used as an adequate management treatment to sustain the available natural resources effectively (Tripathi et al. 2005; Srinivasan 2008; Pandey et al. 2009; Tuppad et al. 2010). Soil and water conservation, as well as river channel protection can be easily achieved by reducing the flows in a river basin after implementation of a feasible BMP (Arabi et al. 2008; Srinivasan 2008; Tuppad et al. 2010; Jang et al. 2017). Changes in water can affect the dependable flows, and hence the availability of utilizable water. Thus, it is crucial to study the effectiveness of BMP treatments on dependable flow in a river basin.

Looking to the aforementioned, an attempt has been made to examine the effectiveness of BMP treatments on dependable flows of the Betwa river basin (BRB), India. Streamflow values were simulated using a hydrologic model to analyze the effect of BMPs implemented in critical prone areas. The FDC has been used to estimate dependable flows at different probabilities, and their variations under different BMP treatments were applied for agriculture land and river channels of the BRB area.

22.2 Study Area and Data

22.2.1 Betwa River Basin (BRB)

The Betwa River, a tributary of Yamuna River, is located in the central part of India, geographically between 77° 05′ 38″ E and 80° 13′ 48″ E longitude and 22° 51′ 51″ N and 26° 3′ 5″ N latitude (Fig. 22.1). The BRB study area is an inter-state river basin located between Madhya Pradesh and Uttar Pradesh encompassing an area of about 43,936 km^2.

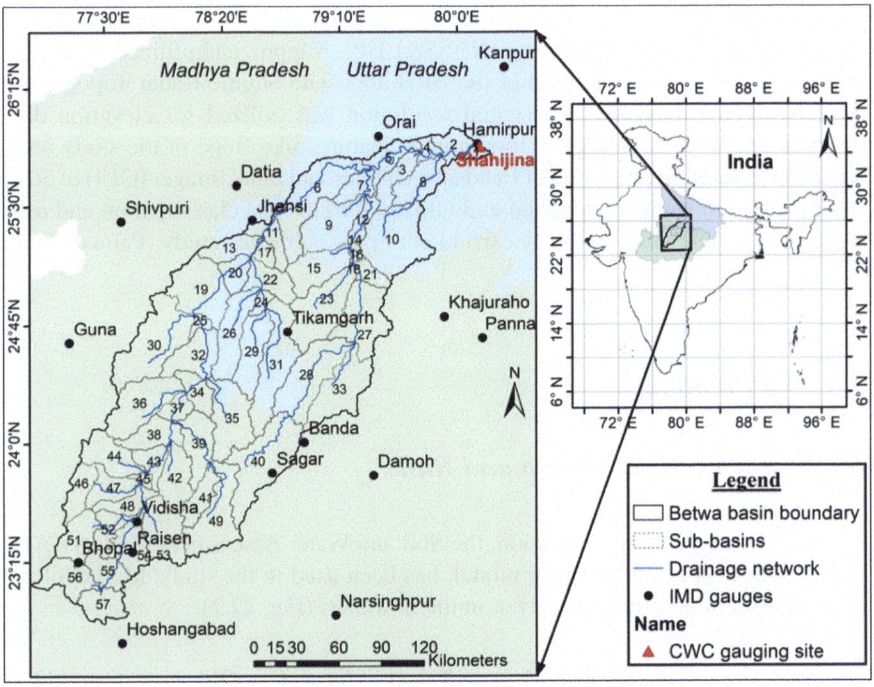

Fig. 22.1 Location map of the study area (Betwa river basin, India)

The study area is located in semi-humid and dry sub-humid climate region of Central India. Hydrological cycle is highly dependent on South-West monsoon circulation system (Chaube et al. 2011). Annual precipitation occurs in the form of rainfall ranges from 700 to 1200 mm in an average, and out of this, about 80% rainfall receives during monsoon months (Chaube 1988). About temperature, the minimum and maximum values are averaged to about 6.7 °C and 44.2 °C, respectively (Suryavanshi et al. 2014). The BRB has an undulating topography, elevation range from 61 to 715 m, and land slope variation up to 67%. Study area is dominated by existence of black cotton soil, agriculture and forest land, and surrounded by alluvial plains and Vindhyan plateau at northern and southern, respectively.

22.2.2 Data Acquisition

In this study, daily climate data records for rainfall and temperature were obtained from the India Meteorological Department (IMD), Pune, and river water flow measurements from Central Water Commission (CWC), New Delhi, for the years 2001–2013. Basic soil data information was procured from the National Bureau of Soil Survey and Land Use planning (NBSS&LUP), Nagpur, and utilized for preparation and generation of soil map of the BRB area. The Shuttle Radar Topography Mission (SRTM) data of 30 m spatial resolution was utilized for elevation data extraction and understanding the topographic features like slope of the study area. Furthermore, satellite imageries of Landsat 8 Operational Land Imager (OLI) of 30 m spatial resolution were downloaded and utilized for land use classification and map preparation, which was previously carried out in the published study (Palmate et al. 2017).

22.3 Methodology

22.3.1 SWAT Model Setup and Run

For river basin hydrology simulation, the Soil and Water Assessment Tool (SWAT), a water balance (Eq. 22.1)-driven model, has been used in the study. Methodology steps followed in this study are given in the flowchart (Fig. 22.2).

$$SW_t = SW + \sum\nolimits_{t-1}^{t} (R - Q - ET - P - QR) \tag{22.1}$$

where,

SW_t final soil water content (mm),
SW initial soil water content (mm),

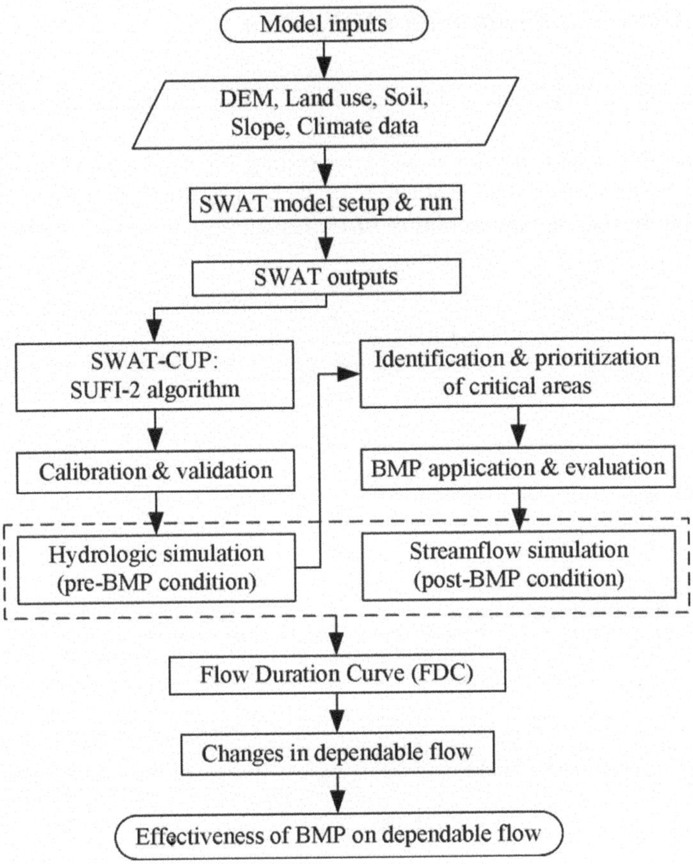

Fig. 22.2 Methodology flowchart adopted in the present study

t time (days),
R amount of precipitation (mm),
Q amount of surface runoff (mm),
ET amount of evapotranspiration (mm),
P percolation (mm), and.
QR the amount of return flow (mm).

The SWAT model setup was performed in ArcGIS environment, known by the name ArcSWAT, using several types of inputs and among them, the major inputs are elevation, land use, soil, slope, and climate data. Further, the model was used to run the complex hydrological processes of the BRB. The model simulation includes initial run for warm-up period of 2 years which allows the model fully functional as natural processes in the river basin.

22.3.2 Model Calibration and Validation

The model simulations were calibrated (2003–2009) and validated (2010–2013) on monthly time-scale by using the sequential uncertainty fitting version 2 (SUFI-2) algorithm in SWAT-Calibration and Uncertainty Program (SWAT-CUP). Model simulation performance was evaluated by the Nash–Sutcliffe Efficiency (*NSE*; Eq. 22.2) and the percentage of bias (*PBIAS*; Eq. 22.3).

$$NSE = 1 - \frac{\sum_{i=1}^{n} (x_i - y_i)^2}{\sum_{i=1}^{n} (x_i - \overline{x})^2} \tag{22.2}$$

where,

x_i *i*th observed value,
y_i *i*th simulated value,
$\overline{x}$ mean of observed data, and.
n total number of observations.

$$PBIAS = \left[\frac{\sum_{i=1}^{n} (x_i - y_i) \times 100}{\sum_{i=1}^{n} (x_i)} \right] \tag{22.3}$$

where,

PBIAS = deviation of data being evaluated, expressed as a percentage.

In this study, the model performance showed satisfactory to very good ratings for hydrologic simulation at the outlet (Shahijina gauging site) of the BRB area (Pandey and Palmate 2019).

22.3.3 BMP Simulation Using SWAT Model

After satisfactory evaluation, the model simulations were employed for identification and prioritization of critical areas in the BRB (Pandey and Palmate 2019). The SWAT model was further used to implement and evaluate the feasible BMPs in critical prone areas of the agriculture land and the river channels (Table 22.1). Considering the hydrological modelling simulation and sub-basin discretization performed using the SWAT, the selection of a BMP and representative parameters and their values was done based on the published literature studies, by communicating with

Table 22.1 List of best management practices (BMP) simulated using hydrological SWAT model

Best management practices (BMP)	
Agriculture land	1 Tillage management (Mold board plough, conservation tillage, Field cultivator, and zero tillage) 2 Contour farming 3 Residue management 4 Strip cropping
Main and tributary river channels	1 Grassed waterways 2 Streambank stabilization 3 Grade stabilization structures 4 Porous gully plugs 5 Recharge Structures

local water managers and farmers, agricultural development officers, and modelling expert scientists opinion (Chow 1959; Lane 1983; Tripathi et al. 2005; Bracmort et al. 2006; Narasimhan et al. 2017; Arabi et al. 2008; Srinivasan 2008; Pandey et al. 2009; Tuppad et al. 2010; Jang et al. 2017). These BMPs were implemented based on the functional capacity of a conservation practices suggested to represent a BMP in the SWAT model. Definition and purpose of BMPs were obtained from the Natural Resources Conservation Service (NRCS) standard practice code (USDA-NRCS 2008).

In order to evolve an appropriate soil and water conservation and main/tributary river channel protection strategy in the BRB, nine different BMPs have been implemented and evaluated in the present study as given in Table 22.1.

22.3.4 Flow Duration Curves (FDC)

The method used in finding the dependable flow is basically the FDC technique. The FDC helps to estimate and display the changes in water flow regimes with respect to percentages of time. It can be used to temporal characterization of hydrologic consequences in a river basin (Vogel and Fennessey 1994). The FDC has ability facilitating empirical distribution of flow characterization at various return periods on accounts of long-term data series. In this study, the FDCs for streamflow were developed using the Weibull distribution formula (Eq. 22.4). These curves were used to evaluate the severity of streamflow regimes in the BRB.

$$P_p = \frac{m}{N+1} \times 100 \tag{22.4}$$

where,

P_p percentage probability of the flow magnitude being equaled or exceeded,

m order number of the flow, and.
N number of data records.

FDC is constructed from streamflow values obtained from the SWAT model over a time interval of interest. It reflects the variability of streamflow during a typical period of BMP evaluation. In this study, the streamflow simulated before BMP implementation is considered as pre-BMP (or baseline) condition flow, and the streamflow simulated after implementation of each BMP is considered as post-BMP condition flow. Both the pre- and post-BMP condition flows have been further used in FDC plotting to study the changes in dependable flows of BRB area.

22.4 Results and Discussion

22.4.1 Effectiveness of Agricultural BMPs on Dependable Flow

In this study, baseline flow obtained before BMP implementation and the flow obtained after BMP implementation has been used to analyze the effect of a BMP on dependable flow. Results show that agricultural BMPs produce effect on dependable flows of the BRB (Fig. 22.3 and Table 22.2). Among all FDC plots showing changes in dependable flows, the FDC plot of strip cropping treatment indicated larger reduction (up to 12%) in dependable flow as compared to other agricultural BMPs (Fig. 22.3). This is due to high sensitivity of the curve number (CN2) parameter as well as sensitivity of the Universal Soil Loss Equation (USLE) crop management factor (USLE_C), USLE practice factor (USLE_P), and the Manning's roughness coefficient for overland flow (OV_N) parameters (Pandey and Palmate 2019). All these parameters were used to implement the strip cropping treatment in the SWAT model for reduction in the streamflow velocity. Other agricultural treatments, namely tillage management, contour cropping, and residue management, were also implemented using the aforesaid BMP parameters (Tripathi et al. 2005; Arabi et al. 2008; Pandey et al. 2009; Tuppad et al. 2010). However, an integrated effect of all the BMP parameters representing a strip cropping is usually high and thus, resulting more reductions in dependable flow of the BRB area.

Further, the changes in annual dependable flow have been studied using simulations obtained for baseline and post-BMP condition at 40, 50, 60, 75, and 90% probabilities as shown in Table 22.2. Among tillage management, the field cultivator is an effective treatment reducing water flow (up to 7%) compared to the other tillage treatments. Although, the depth of tillage for conservation tillage and field cultivator is same (about 100 mm), high tillage mixing efficiency (about 30%) of field cultivator causes more flow reductions. Thus, results show that strip cropping is the most effective agricultural treatment which induces reduction (up to 12%) in annual dependable flow. After strip cropping, contour farming can effectively reduce water flow (up to 10%) from the agriculture field. Flow reductions are mainly observed at low (40%)

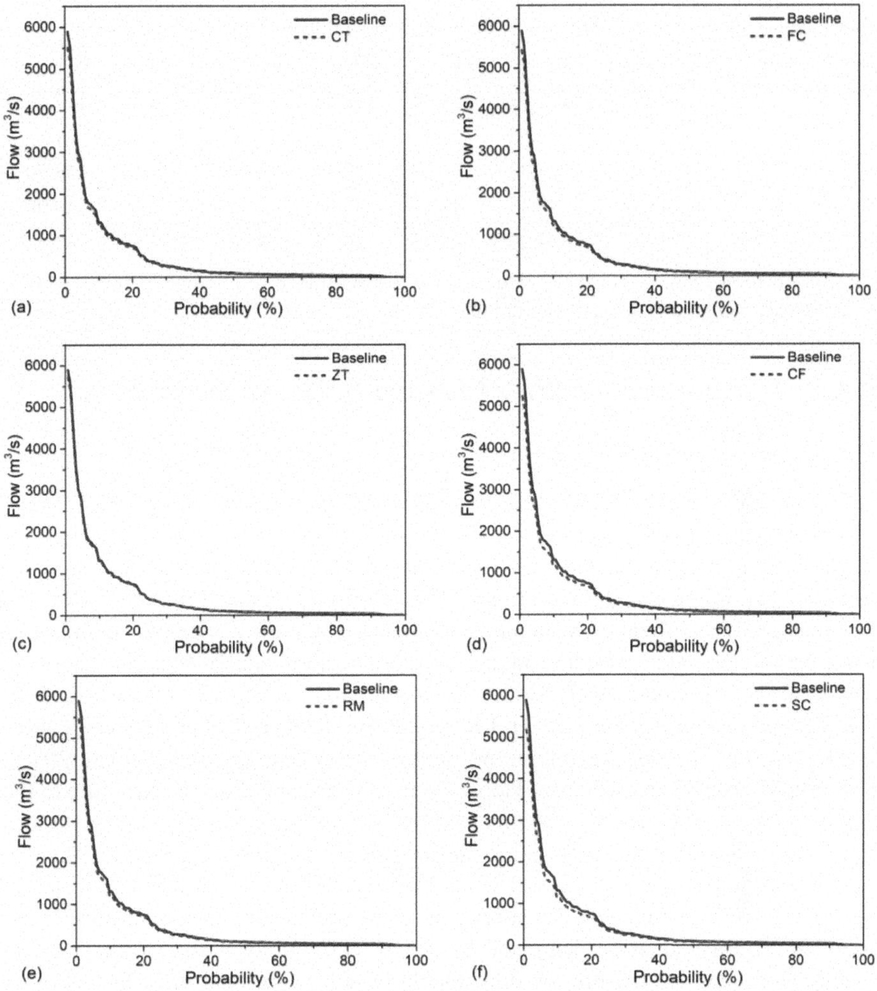

Fig. 22.3 Flow duration curves under different agricultural BMPs **a** conservation tillage (CT). **b** field cultivator (FC). **c** zero tillage (ZT). **d** contour farming (CF). **e** residue management (RM); and **f** strip cropping (SC)

probability and high (90%) probability. Very small increase in dependable flow has also been observed in the analysis which indicates that agriculture treatments can raise the surface water availability at 50%, 60%, and 75% probabilities.

Table 22.2 Changes in annual dependable flow (m^3/s) under different agricultural BMPs

Dependability (%)	Baseline	CT	FC	ZT	CF	RM	SC
40	144.70	137.02	134.94	142.54	130.24	135.51	127.48
50	90.97	93.53	92.11	97.30	88.90	92.50	81.42
60	65.66	67.33	66.31	70.05	64.00	66.59	58.50
75	45.48	44.15	43.48	45.93	41.96	43.66	40.90
90	37.12	34.96	34.43	36.37	33.23	34.58	32.52

Note Conservation tillage (CT); field cultivator (FC); zero tillage (ZT); contour farming (CF); residue management (RM); and strip cropping (SC)

22.4.2 Effectiveness of River Channel BMPs on Dependable Flow

Furthermore, effect of a BMP on dependable flow has also been analyzed for the main as well as tributary river channels of the BRB area. Similar to the previous analysis, this analysis indicated a small variation in FDC plotted using baseline simulation and a simulation under river channel BMP (Fig. 22.4 and Table 22.3). Results show that grade stabilization structure implemented in main channel is an effective river channel treatment which reduces up to 8% of dependable flow as compared to the other river channel BMPs. In this treatment, high sensitivity of average slope (CH_S2) and erodibility factor (CH_EROD) of the main river channel causes sufficient reductions in slope steepness and gully erosion (Pandey and Palmate 2019). These SWAT parameters were used to implement grade stabilization structure in the main river channel. Other river channel BMPs, namely grassed waterways, streambank stabilization, porous gully plugs and recharge structures, were implemented using the Manning's roughness coefficient for main channel (CH_N2) and tributary channel (CH_N1) having low sensitivity as compared to the sensitivity of CH_S2 parameter (Arabi et al. 2008; Tuppad et al. 2010). Thus, analysis shows that an effect of grade stabilization structure on dependable flow is more than the effect of other river channel BMPs.

Table 22.3 represents the changes in annual dependable flow between baseline and post-BMP condition at 40, 50, 60, 75, and 90% probabilities. Results show that grade stabilization structure is an effective treatment reducing dependable flow up to 8% compared to the other river channel BMP treatments. The grassed waterways and streambank stabilization structures implemented in main channel have shown a small amount of flow reduction (up to 2%) at 40–90% probabilities; however, dependable flow at 50, 60, and 75% probabilities increased up to 7%. Among tributary channel BMPs, both porous gully plugs and recharge structures induce nearly a same amount of reduction (up to 4%) in dependable flow of the BRB. In tributary channel, BMP also increases flow up to 5 at 50% and 60% probabilities. Hence, the analysis reveals that river channel BMPs have varying response to low, medium, and high dependable

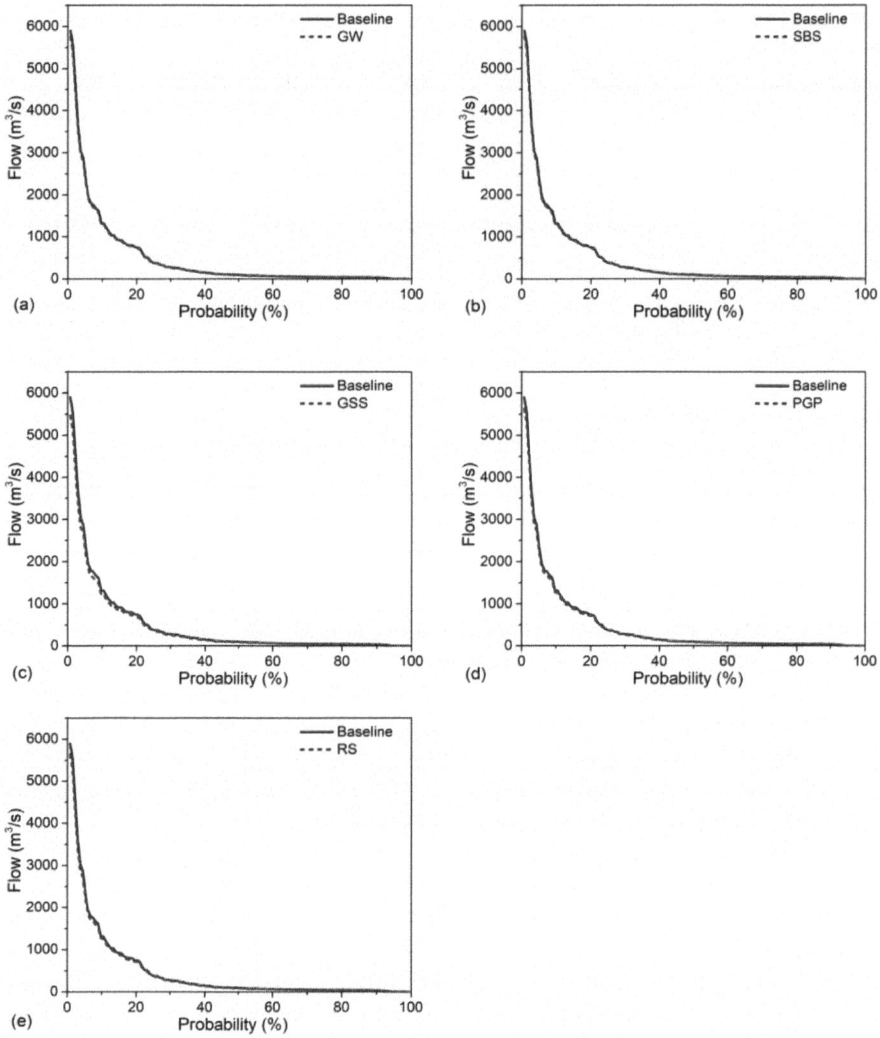

Fig. 22.4 Flow duration curves under different main and tributary river channel BMPs. **a** grassed waterways (GW). **b** streambank stabilization (SBS). **c** grade stabilization structure (GSS). **d** porous gully plugs (PGP). **e** recharge structures (RS)

flows. Furthermore, the effect of main channel BMPs on dependable flows is higher than the tributary channel BMPs in the BRB.

Table 22.3 Changes in annual dependable flow (m^3/s) under different main and tributary river channel BMPs:

Dependability (%)	Baseline	GW	SBS	GSS	PGP	RS
40	144.70	142.98	143.28	133.64	140.07	140.04
50	90.97	97.60	97.80	85.59	95.61	95.59
60	65.66	70.26	70.41	61.64	68.83	68.82
75	45.48	46.07	46.16	41.85	45.13	45.12
90	37.12	36.48	36.56	33.99	35.74	35.73

Note Grassed waterways (GW); streambank stabilization (SBS); grade stabilization structure (GSS); porous gully plugs (PGP); and recharge structures (RS)

22.5 Conclusions

The SWAT model outputs were used to study the effectiveness of BMP implementation on dependable flows at several % probabilities. The analysis was carried out for both agriculture land and river channels of the BRB area. From the study, following conclusions are drawn:

1. Field cultivator is an effective tillage treatment reducing water flow (up to 7%) due to high mixing efficiency (about 30%).
2. Strip cropping is the most effective agricultural BMP treatment reducing annual dependable flow 10–12%. High sensitivity of the curve number (CN2) as well as the USLE crop management factor (USLE_C), USLE practice factor (USLE_P), and the Manning's roughness coefficient for overland flow (OV_N) parameter causes combined impact on changes in water flow.
3. Dependable flow mainly reduces at medium (40%) and high (90%) probabilities. However, a small increase in 50, 60 and 75% dependable flow has also been observed which reveals variation in flow characteristics under different BMP treatments.
4. Further, the grade stabilization structure implemented in main river channel, consisting highly sensitive average main channel slope (CH_S2), is an effective river channel treatment reducing dependable flow 6–8% majorly at medium (60%) and high (90%) probabilities. Thus, effect of main channel BMP treatment on dependable flow variation is more as compared to the tributary channel BMP treatment.
5. Overall, an agricultural BMP can induce more effect on medium-to-high dependable flows; however, river channel BMP can be effective on predominantly up to medium dependable flow of a river basin.

Acknowledgements All authors sincerely acknowledge the Department of Water Resources Development and Management, Indian Institute of Technology (IIT) Roorkee, for providing workspace and other facilities to conduct this research work. Also, authors acknowledge the IMD, Pune; Yamuna Basin Organization, CWC, New Delhi, and NBSS&LUP, Nagpur for providing data required in the study. In addition, authors would like to acknowledge the Earth Explorer website (https://earthexplorer.usgs.gov/) for making availability of SRTM and Landsat 8 OLI data for downloading and their utilization in preparation of the model inputs. Furthermore, the first author sincerely acknowledges the Ministry of Human Resource Development (MHRD), Government of India (GoI), for providing financial assistance in the form of Ph.D. scholarship during his study at IIT Roorkee.

References

Arabi M, Frankenberger JR, Engel BA, Arnold JG (2008) Representation of agricultural conservation practices with SWAT. Hydrol Process 22(16):3042–3055

Arora M, Goel NK, Singh P, Singh RD (2005) Regional flow duration curve for a Himalayan river Chenab. Hydrol Res 36(2):193–206

Bracmort KS, Arabi M, Frankenberger JR, Engel BA, Arnold JG (2006) Modeling long-term water quality impact of structural BMPs. Trans ASABE 49(2):367–374

Chaube UC (1988) Model study of water use and water balance in Betwa Basin. J Ins Eng. India. Civil Eng Div 69(3):169–173

Chaube UC, Suryavanshi S, Nurzaman L, Pandey A (2011) Synthesis of flow series of tributaries in upper Betwa basin. Int J Environ Sci 1(7):1459

Chow VT (1959) Open-channel hydraulics. McGraw-Hill, New York, p 112

Cole RAJ, Johnston HT, Robinson DJ (2003) The use of flow duration curves as a data quality tool. Hydrol Sci J 48(6):939–951

Jang SS, Ahn SR, Kim SJ (2017) Evaluation of executable best management practices in Haean highland agricultural catchment of South Korea using SWAT. Agric Water Manag 180:224–234

Kusre BC, Baruah DC, Bordoloi PK, Patra SC (2010) Assessment of hydropower potential using GIS and hydrological modeling technique in Kopili river basin in Assam (India). Appl Energy 87(1):298–309

Lane LJ (1983) Chapter 19: transmission losses, soil conservation service (scs)–national engineering handbook, section 4. Hydrology. U.S, Government Printing Office, Washington, DC, p 19

Narasimhan B, Allen PM, Coffman SV, Arnold JG, Srinivasan R (2017) Development and testing of a physically based model of streambank erosion for coupling with a basin-scale hydrologic model SWAT. JAWRA J Am Water Resour Assoc 53(2):344–364

Nilsson C, Reidy CA, Dynesius M, Revenga C (2005) Fragmentation and flow regulation of the world's large river systems. Science 308(5720):405–408

Palmate SS, Pandey A, Mishra SK (2017) Modelling spatiotemporal land dynamics for a trans-boundary river basin using integrated Cellular Automata and Markov Chain approach. Appl Ggeography 82:11–23. https://doi.org/10.1016/j.apgeog.2017.03.001

Pandey A, Palmate SS (2019) Assessing future water–sediment interaction and critical area prioritization at sub-watershed level for sustainable management. Paddy Water Environ 17(3):373–382. https://doi.org/10.1007/s10333-019-00732-3

Pandey A, Chowdary VM, Mal BC, Billib M (2009) Application of the WEPP model for prioritization and evaluation of best management practices in an Indian watershed. Hydrol Process 23(21):2997–3005

Smakhtin V, Gamage N, Bharati L (2007) Hydrological and environmental issues of interbasin water transfers in India: a case of the Krishna River Basin. vol. 120. IWMI

Srinivasan R (2008) Bosque river environmental infrastructure improvement plan: phase I final report. TR-312, Texas Water Resources Institute. Texas A & M University, College Station

Suryavanshi S, Pandey A, Chaube UC, Joshi N (2014) Long-term historic changes in climatic variables of Betwa Basin. India. Theoret Appl Climatol 117(3–4):403–418

Tripathi MP, Panda RK, Raghuwanshi NS (2005) Development of effective management plan for critical subwatersheds using SWAT model. Hydrol Process 19(3):809–826

Tuppad P, Kannan N, Srinivasan R, Rossi CG, Arnold JG (2010) Simulation of agricultural management alternatives for watershed protection. Water Resour Manage 24(12):3115–3144

USDA-NRCS (2008) National handbook of conservation practices. USDA-NRCS, Washington, DC. Available at: https://www.nrcs.usda.gov/technical/standards/nhcp.html. Accessed 18 May 2009

Vogel RM, Fennessey NM (1994) Flow-duration curves. I: new interpretation and confidence intervals. J Water Resour Plann Manage 120(4):485–504

Vogel RM, Fennessey NM (1995) Flow duration curves II: A review of applications in water resources planning 1. JAWRA J Am Water Resour Assoc 31(6):1029–1039

Yu PS, Yang TC, Wang YC (2002) Uncertainty analysis of regional flow duration curves. J Water Resour Plann Manage 128(6):424–430

Chapter 23
An Analytical S-Curve Approach for SUH Derivation

Pravin. R. Patil, S. K. Mishra, Sharad K. Jain, and P. K. Singh

Abstract Traditional S-curve/S-hydrograph method is used for deriving the desired duration (τ-*hr*) unit hydrographs (UHs) from known *D-hr* parent UH. Such *D-hr* UHs are seldom available for the ungauged basins. Hence, analytical S-curve approach/inflection S-shaped (IS) model (Patil and Mishra, J Hydrol Eng 21:06016010, 2016) has been proposed for smooth-shaped *D-hr* UH derivation over ungauged basins, where the optimal parameters of the S-curve were estimated using Central Water Commission (CWC)-based *D-hr* UHs peak discharge (Q_p) and time to peak (t_p) as constraints. The proposed analytical S-curve-based *D-hr* synthetic UH (SUH) proves its supremacy over CWC-based *D-hr* SUH by skipping the erratic manual fitting of SUH to preserve the unit runoff volume, unproductive time as well as efforts and unnecessary calculations involved. The sum of analytical S-curve-based *D-hr* SUH ordinates equals the volume of 1 cm direct runoff depth over the entire catchment area or equilibrium discharge of the S-curve justifies the SUH derived through proposed approach. The analytical SUHs are of comparable accuracy as CWC SUHs, as justified by the exact peaks (relative error = 0) and runoff volumes preserved with the SUH shapes generated.

Keywords UH · SUH · Analytical S-curve approach · Peak flow rate · Time to peak · Equilibrium discharge

---

Pravin. R. Patil (✉) · S. K. Jain · P. K. Singh
National Institute of Hydrology Roorkee, Roorkee 247667, Uttarakhand, India
e-mail: prpatil25@gmail.com

S. K. Jain
e-mail: s_k_jain@yahoo.com

P. K. Singh
e-mail: pushpendras123@gmail.com

S. K. Mishra
Department of Water Resources Development and Management, Indian Institute of Technology Roorkee, Roorkee 247667, Uttarakhand, India
e-mail: skm61fwt@iitr.ac.in

A. Pandey et al. (eds.), *Water Management and Water Governance*, Water Science and Technology Library 96,
https://doi.org/10.1007/978-3-030-58051-3_23

23.1 Introduction

Hydrologists experienced that, rainfall–runoff transformation is complex to figure out due to unforeseen spatial and temporal instability of basin characteristics and rainfall patterns (Tokar and Markus 2000; Geetha et al. 2008; Yoshitani et al. 2009; Bhadra et al. 2010). It comprises movement of rainfall through different phases and its transformation to runoff in the natural or manmade channels. Its accurate quantitative estimation is essential to the hydrologists for the purpose of flood control/defenses, irrigation, drainage, water supply, hydropower, navigation, recreation, aquatic–wildlife propagation, wetlands modelling, lake water budget, and climate impact analysis. These are vital elements of sustainable water resource management and watershed development (Brooks et al. 2003; Ito et al. 2006, 2007; Bhar and Dwivedi 2005). Design engineers particularly need the design flood of specific return period for fixing the waterway capacity and foundations of a bridge, culvert, and cross drainage works depending on their life and importance to ensure safety as well as economy. An under/over-estimation of design flood results in the failed/uneconomic structure. The UH derived from the rainfall–runoff records (Sherman 1932) depicts time distribution of runoff at the outlet resulting from excess rainfall. Most of the basins are characterized by inadequate rainfall–runoff records by means of which design predictions can be made. It is due to inaccessibility; high costs of measuring structures/instruments and their maintenance; consequences of severe natural disasters or of human-endorsed high-risk activities—riots and wars; and mishandling or unexpected loss of data.

For such ungauged basins, the surface-runoff hydrograph for a given storm may be approximated by two techniques, i.e., use of a recorded hydrograph resulting from identical storm distributed over a physically similar area or use of a SUH. The degree of similarity between the prominent runoff producing characteristics of the selected basins might limit the success of the former method. An erroneous approximation of the true hydrograph may possible if characteristics are not closely alike. The success of the SUH depends on its reliability. The SUHs obtained without rainfall–runoff data and dependent on empirical relationships relating hydrograph shape (UH parameters) to basin characteristics, attempts to extend the UH applicability to ungauged basins.

A joint effort by Hydrology Directorate (Small Catchments/Regional Studies) of CWC, Research Designs and Standards Organization (RDSO) of the Ministry of Transport (Railways), Road Wing of the Ministry of Transport (Surface), and India Meteorological Department (IMD), in pursuance of the recommendations of the Khosla Committee of Engineers delivered flood estimation procedures (e.g. CWC 1984, 1987, 1995, etc.). For this purpose, the country (India) has been divided into 26 hydrometeorologically homogeneous subzones and 21 flood estimation reports covering studies for 24 subzones have been published from time to time. Such reports involve regional method based on hydrometeorological approach employing SUH and design storm of a specific return period. The physiographic characteristics of the sub-catchments and their respective representative *D-hr* UHs (obtained from rainfall-runoff data) have been correlated by regression analysis and the equations for *D-hr* SUH of the subzone were derived. The CWC approaches suggest manual fitting of SUH through Q_p, t_p, W_{50}, W_{75}, WR_{50}, WR_{75} and t_b derived using catchment characteristics, which is quite cumbersome and time consuming. Hence, an alternate SUH approach using few salient points/characteristics of the UH for its fitting need to be developed.

The law of growth states that, system grows exponentially until its maximum capacity/asymptotic or upper limit (NakiCenovic 1988), where the growth rate slows and eventually saturates or levels off producing characteristic S-shaped curve (Stone 1980). Hence, the S-shaped curve has applicability to several natural mechanisms. The S-hydrograph/S-curve/summation-curve is the hydrograph of direct runoff due to continuous uniform effective rainfall intensity of infinite duration, and is hypothetically smooth. The sigmoid curve resembles symmetrical S-curve only, but natural S-curves emerged from skewed UHs are unsymmetrical in nature. The hunting-free shape of unsymmetrical S-curve can be well reproduced analytically by using mean value function of the inflection S-shaped (IS) software reliability growth model (Ohba 1984), developed to analyze the software failure detection process by modifying popularly used logistic population growth curve (Verhulst 1838, 1845, 1847) as verified by Patil and Mishra (2016). It assumes S-shaped software reliability growth as the faults in a programme are mutually dependent, i.e., some faults are not detectable before some others are eliminated. Patil and Mishra (2016) concluded that analytical S-curve approach is accurate as well as expeditious than conventional S-curve approach. No alternate references of this model as an S-hydrograph have been found.

23.2 Theoretical Considerations

23.2.1 CWC-Based SUH

The equations developed by CWC for deriving *D-hr* SUHs of the different hydrometeorological subzones are as follows:

Lower Godavari subzone-3(f) (CWC 1995)	Upper Indo-Ganga plains subzone-1(*e*) (CWC 1984)	Mahi and Sabarmati subzone-3(*a*) (CWC 1987)	
Peak per unit area $(m^3/s/km^2)$			
$q_p = \frac{1.842}{(t_l)^{0.804}}$	$= \frac{2.030}{(L/\sqrt{S})^{0.649}}$	$= \frac{1.161}{(t_l)^{0.635}}$	(23.1)
Peak discharge $Q_P = q_p \times A \quad (m^3/s)$			(23.2)
Lag time (hr)			
$t_l = 0.348 \times \left(\frac{LL_c}{\sqrt{S}}\right)^{0.454}$	$= \frac{1.858}{(q_{pc})^{1.038}}$	$= 0.433 \times \left(\frac{L}{\sqrt{S}}\right)^{0.704}$	(23.3)
Time to peak $t_p = t_l + (D/2) \quad (hr)$			(23.4)

(continued)

(continued)

Lower Godavari subzone-3(f) (CWC 1995)	Upper Indo-Ganga plains subzone-1(*e*) (CWC 1984)	Mahi and Sabarmati subzone-3(*a*) (CWC 1987)	
Time base (*hr*)			
$t_b = 4.589 \times t_l^{0.894}$	$= 7.744 \times t_l^{0.779}$	$= 8.375 \times t_l^{0.512}$	(23.5)
Width of SUH at 50 and 75% of the Q_P (*hr*)			
$W_{50} = 2.353/(q_p)^{1.005}$	$= 2.217/(q_p)^{0.990}$	$= 2.284/(q_p)^{1.00}$	(23.6)
$W_{75} = 1.351/(q_p)^{0.992}$	$= 1.477/(q_p)^{0.876}$	$= 1.331/(q_p)^{0.991}$	(23.7)
Width of the rising side of SUH at 50 and 75% of the Q_P (*hr*)			
$WR_{50} = 0.936/(q_p)^{1.047}$	$= 0.812/(q_p)^{0.907}$	$= 0.827/(q_p)^{1.023}$	(23.8)
$WR_{75} = 0.579/(q_p)^{1.004}$	$= 0.606/(q_p)^{0.791}$	$= 0.561/(q_p)^{1.037}$	(23.9)
Volumetric equality condition Volume of SUH (volume of 1 cm excess rainfall depth over the entire catchment) = Equilibrium discharge, $Q_{eq} = \frac{AI}{0.36} = \frac{Ad}{0.36D}$ m^3/s = Sum of SUH ordinates (ΣQ_i) at a time step *D-hr*			(23.10)

A = catchment area (km^2), L = length of the main stream (*km*), L_c = length to watershed centroid from its outlet (*km*), S = equivalent stream slope (*m/km*), $I = d/D$ = uniform rainfall intensity (*cm/hr*), d = 1 cm depth, and D = unit duration of the excess rainfall pulse/parent UH (*hr*).

23.2.2 Analytical Inflection S-Shaped (IS) Model-Based SUH

Mean value function of the IS model (Ohba 1984) transformed into equivalent hydrological form (Patil and Mishra 2016) could be utilized to analytically represent time-varying *D-hr* parent S-curve (Eq. 23.11) or its lagged version (Eq. 23.12).

$$S_{(t)} = \frac{Q_{eq}}{1 + ce^{-bt}}\left(1 - e^{-bt}\right) \quad \text{for } t \geq 0;\; Q_{eq}, b, c > 0$$
$$\text{when } t = 0, \quad S_{(t)} = 0; \;\&\text{ if } t = t_b, \quad S_{(t)} = Q_{eq} \tag{23.11}$$

$$S_{(t-\tau)} = \frac{Q_{eq}}{1 + ce^{-b(t-\tau)}}\left(1 - e^{-b(t-\tau)}\right) \quad \text{for } t \geq 0;\; Q_{eq}, b, c, \tau > 0 \tag{23.12}$$

where e = base of the natural logarithm (2.71828); b = growth rate parameter which specifies width/steepness of S-curve, i.e., as $b\uparrow, f(t) \rightarrow Q_{eq}$ more rapidly; c = point of inflection, if $c > 0, f(t) \uparrow$ when $b > 0$ and $f(t) \downarrow$ when $b < 0$. Here, $f(t) \rightarrow Q_{eq}$ as $t \rightarrow t_b$ and when $t \rightarrow 0, f(t) \rightarrow 0$. τ = location parameter that shifts S-curve function in time keeping its shape constant (Meyer 1994). The upper/asymptotic limit (Q_{eq}) bounds the function, past which the output cannot grow. Q_{eq} occurs at $t = t_b$ = time base of *D-hr* UH.

The 'difference (Z) of *D-hr* S-curves [$S_{(t)}$- $S_{(t-\tau)}$] displaced by desired UH duration (τ-*hr*)' denotes the characteristics of desired duration UH (i.e. τ-*hr* UH) as derived by Patil and Mishra (2016).

$$\text{Time to peak of } \tau - \text{hr UH} = t = t_p = \frac{\tau}{2} + \frac{1}{b}\log c \tag{23.13}$$

$$\text{Peak of } \tau - hr \;\; UH = Q_p = \frac{Z_{\max}}{(\tau/D)} = \frac{Q_{eq}\left(1+\frac{1}{c}\right)\left[\frac{e^{b\tau/2}-1}{e^{b\tau/2}+1}\right]}{(\tau/D)} \tag{23.14}$$

23.3 Performance Evaluation

Following statistical indices were used for evaluating the analytical IS model SUHs against CWC SUHs.

23.3.1 Nash–Sutcliffe Efficiency (η_{NS})

The η_{NS} (Nash and Sutcliffe 1970) is a normalized statistic that determines the relative magnitude of the residual variance (noise) compared to the measured data variance (information). It takes scale of 0 to 100% and is expressed as:

$$\eta_{NS}\,(\%) = \left(1 - \frac{\sum_{i=1}^{N}(U_{oi} - U_{ci})^2}{\sum_{i=1}^{N}(U_{oi} - U_{av})^2}\right) \times 100 \tag{23.15}$$

where U_{oi} is the ith ordinate observed, U_{ci} the ith ordinate computed, U_{av} is the average of observed ordinates, and N the total number of ordinates in the series. Values between 0 and 1 are generally viewed as acceptable levels of performance, whereas $\eta_{NS} < 0$ indicates that the mean observed value is a better predictor than the simulated value, which indicates unacceptable performance.

23.3.2 Relative Error in Peak (RE-Q_p)

$$RE - Q_p = \left(\frac{Q_{pobs} - Q_{pcomp}}{Q_{pobs}}\right) \times 100 \tag{23.16}$$

where Q_{pobs} and Q_{pcomp} represent the observed and computed peak discharge, respectively.

23.4 Procedure in Steps for SUH Derivation

1. Sketch the *D-hr* SUH through the points (Q_p, t_p, W_{50}, W_{75}, WR_{50}, $WR_{75,}$ and t_b) estimated by using CWC approach (Eqs. 23.1–23.9).
2. Estimate the Q_{eq} assuming 1 cm excess rainfall depth over the known catchment area within *D-hr*. Sum the derived SUH ordinates (ΣQ_i) at a time step *D-hr* and check for volumetric equality condition (Eq. 23.10).
3. In case the volumetric equality condition is unsatisfied, the falling limb and/or rising limb be suitably modified to get the correct volume under the *D-hr* SUH taking care to get its smooth shape too. This is quite cumbersome and erratic.
4. In order to overcome the above-mentioned difficulties, analytical IS model has been applied as follows.
5. For known Q_{eq} and assumed *b* & *c,* fit the *D-hr* parent S-curve analytically (Eq. 23.11).
6. Lag the *D-hr* parent S-curve analytically by τ*-hr* (Eq. 23.12). Here, $\tau = D$ as we are deriving *D-hr* SUH.
7. Derive *D-hr* SUH by taking difference of both the S-curve ordinates and dividing it by the ratio ($\tau/D = D/D$) at each time step.
8. By assuming $\tau = D$ in the analytical IS model derived Eqs. (23.13 & 23.14) and equating them to CWC derived t_p(Eq. 23.4) and Q_p (Eq. 23.2), respectively, which are used as constraints in order to find the optimal parameters (*b* and *c*) of analytical *D-hr* parent S-curve using SOLVER routine of Excel.
9. The SOLVER maximizes Nash–Sutcliffe efficiency-η_{NS} (objective function, Eq. 23.15) which minimizes the overall deviation between derived *D-hr* SUHs especially at peaks (as verified by Eq. 23.16). The optimal *b* and *c* thus obtained results in the *D-hr* SUH having identical Q_p and t_p as of CWC-SUH.

23.5 Results and Discussions

The SUHs (Fig. 23.1) derived for the sub-catchment defined by Railway Bridge No. 269 located in Lower Godavari Subzone-3(f) has been used to exemplify the adequacy of the proposed analytical IS model over conventional CWC (1995) approach. The empirical relationships proposed by CWC (1995) are to derive salient points of 1-*h* SUH; hence, $D = 1$-*h*. For known $L = 27.70$ km, $L_c = 11.20$ km, and $S = 3.87$ *m/km* (Table 23.1), the $t_l = 3.46$ h is estimated using Eqs. (23.3), which is further used to derive $q_p = 0.68$ $(m^3/s)/km^2$and $t_b = 13.93$ h ≈ 14 h from Eq. (23.1) and (23.5), respectively. The t_p derived using Eq. (23.4) with known t_l and D is 3.96 $\approx$ 4 h. The $q_p$is converted to $Q_p = 164.22$ m^3/s (Eq. 23.2) using known $A = 242$ km^2(Table

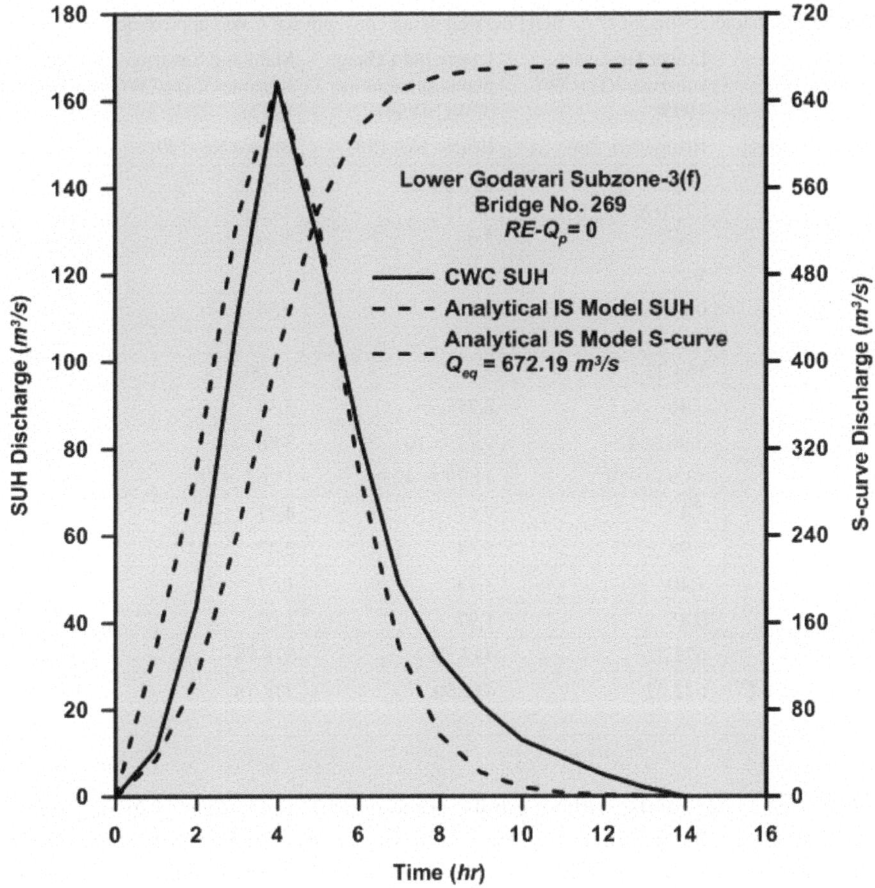

Fig. 23.1 1-*h* S-curve and SUHs derived for the sub-catchment defined by railway bridge no. 269

23.1). The widths of the SUH at 50 and 75% of the Q_p derived using q_p are W_{50} = 3.47 h (Eq. 23.6) and W_{75} = 1.98 h (Eq. 23.7), respectively. Similarly, widths of the rising side of the SUH derived are W_{R50} = 1.40 h (Eq. 23.8) and W_{R75} = 0.85 h (Eq. 23.9). Estimated parameters (Q_p, t_p, W_{50}, W_{75}, WR_{50}, $WR_{75,}$ and t_b) were plotted and joined carefully to get 1-*h* SUH. The ordinates of the 1-*h* SUH were summed up to get its volume ($\Sigma Q_i = 672.22\ m^3/s$) and compared with the volume of 1 cm direct runoff depth (*d*) over the entire catchment area [$Q_{eq} = (A \times d)/(0.36 \times D)$ = (242 km^2 × 1 cm)/(0.36 × 1-*h*) = 672.22 m^3/s]. If the estimated SUH volume $\neq Q_{eq}$, then the rising and falling limbs of the 1-*h* SUH were adjusted manually to achieve volumetric equality which is quite lengthy. The volumetric equality attained justifies the efforts made while manually modifying an SUH.

The proposed analytical IS model generates smooth 1-*h* SUH for the Railway Bridge Sub-catchment No. 269 using the CWC derived Q_p = 164.22 m^3/s and t_p =

Table 23.1 Characteristics of *D-hr* SUH derived using conventional CWC approaches

Subzones →	Lower Godavari subzone-3(f) (CWC 1995)	Upper Indo-Ganga plains subzone-(*e*) (CWC 1984)	Mahi and Sabarmati subzone-3(*a*) (CWC 1987)	Unit
Characteristics↓	Bridge No. 269	Bridge No. 1307	Bridge No. 129	
A L & L_c S	242 27.70 & 11.20 3.87	322.20 56.32 3.65	136.36 33.50 3.26	km^2 km m/km
D	1	2	1	hr
q_p	0.68	0.23	0.54	$\frac{m^3/s}{km^2}$
Q_P	164.22	72.76	73.00	m^3/s
t_l	3.46	8.71	3.38	hr
t_p	$3.96 \approx 4$	$9.71 \approx 10$	$3.88 \approx 4$	hr
t_b	$13.93 \approx 14$	$41.79 \approx 42$	$15.63 \approx 16$	hr
W_{50}	3.47	9.67	4.27	hr
W_{75}	1.98	5.44	2.47	hr
WR_{50}	1.40	3.13	1.57	hr
WR_{75}	0.85	1.97	1.07	hr
Q_{eq}	672.22	447.50	378.78	m^3/s
Volume of CWC SUH	672.22	447.50	378.78	m^3/s

$3.96 \approx 4$ h as constraints, without cumbersome manual adjustment of its ordinates. In order to do so, the parameters of analytical S-curve assumed initially were varied optimally ($b = 0.96$ and $c = 29.08$, Table 23.2) using SOLVER routine of Excel to get closer match ($\eta_{NS} = 91.89\%$) of both the SUHs (especially at their peaks). As a result, the relative error in SUHs peak is zero (Fig. 23.1). The analytical S-curve attains $Q_{eq} = 672.19\ m^3/s$ at $t = t_b$ of *D-hr* SUH (Fig. 23.1). The analytical 1-*h* SUH ($\Sigma Q_i = 672.19\ m^3/s$) too satisfies the volumetric equality with $Q_{eq} = 672.22\ m^3/s$

Table 23.2 Parameters of *D-hr* parent S-curve (Analytical IS model) calibrated using SOLVER

Subzones→	Lower Godavari subzone-3(f) (CWC 1995)	Upper Indo-Ganga plains subzone-(*e*) (CWC 1984)	Mahi and Sabarmati subzone-3(*a*) (CWC 1987)	Unit
Parameters↓	Bridge No. 269	Bridge No. 1307	Bridge No. 129	
$\tau = D$	1	2	1	hr
b	0.96	0.31	0.72	
c	29.08	16.07	12.49	
Volume of Analytical IS model SUH	672.19	447.48	378.68	m^3/s

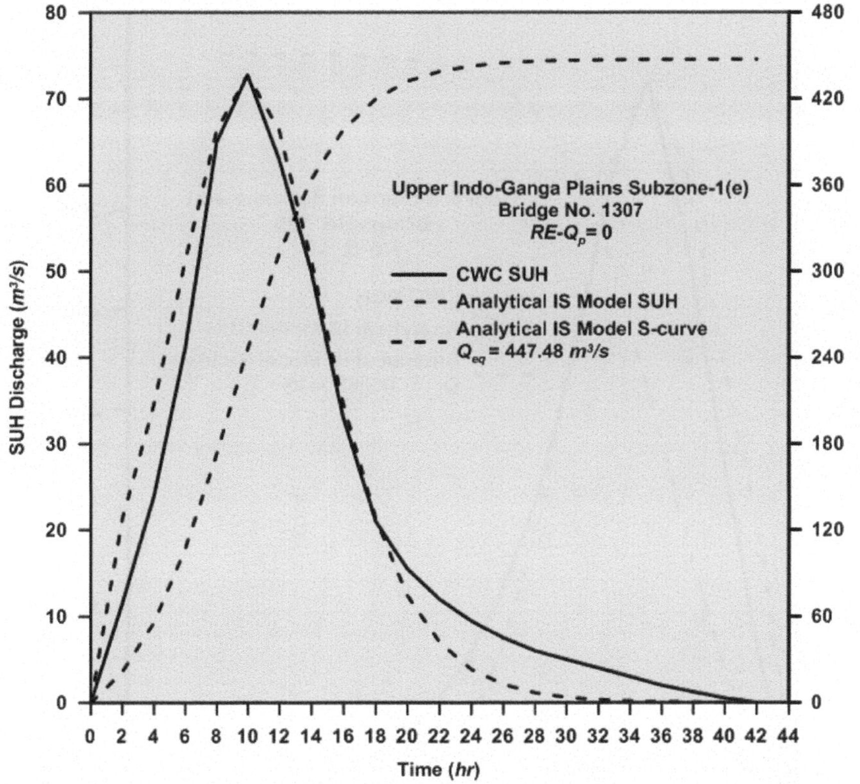

Fig. 23.2 2-*h* S-curve and SUHs derived for the sub-catchment defined by railway bridge no. 1307

(Eq. 23.10). The proposed analytical SUH approach skips the unnecessary parameters (W_{50}, W_{75}, WR_{50}, and WR_{75}) as required for SUH fitting through CWC approach. Similarly, *D-hr* SUHs (Figs. 23.2 and 23.3) were derived for the Railway Bridge Sub-catchment No. 1307 and 129 located in Upper Indo-Ganga Plains Subzone-1(*e*) and Mahi and Sabarmati Subzone-3(*a*), respectively, employing the CWC approach and analytical IS model indicates better correlation as verified by respective η_{NS} (95.68 and 95.39%). The smooth *D-hr* SUHs reproduced analytically indicate the effort-free/error-free reshaping of the CWC based *D-hr* SUHs. The analytical *D-hr* SUHs attain the requisite volumetric accuracy without erratic adjustments.

The peak, time to peak, time base, and volume of the derived SUHs are justifiable w.r.t. the catchment area, mainstream lengths, stream slope, and duration of the SUH used (Table 23.1 and 23.2, Figs. 23.1, 23.2 and 23.3). Highest peak discharge in the Railway Bridge Sub-catchment No. 269 is attributed to steep slope and shortest stream length. The SUH peaks obtained for Railway Bridge Sub-catchment No. 1307 and 129 were justified as the SUH peak has an inverse relation with the duration of SUH. The longer time base in case of Railway Bridge Sub-catchment No. 1307 is

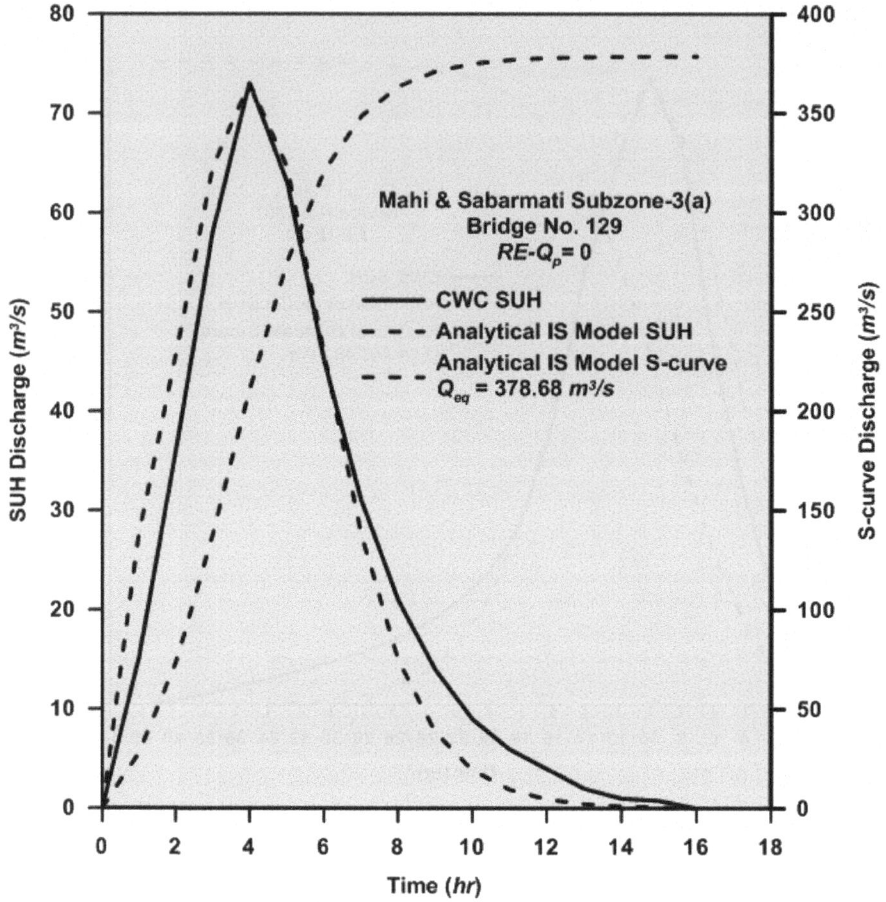

Fig. 23.3 1-*h* S-curve and SUHs derived for the sub-catchment defined by railway bridge no. 129

attributed to longer stream length which extends the time of concentration. Whereas the shorter time base in case of Railway Bridge Sub-catchment No. 129 is attributed to shorter stream length which contracts the time of concentration.

23.6 Conclusions

The following conclusions can be derived from the study:

1. The basic utility of the conventional S-curve method, i.e. derivation of the desired duration (τ-*hr*) UHs from known *D-hr* parent UH has been altered precisely in order to generate its input, i.e. *D-hr* parent SUHs over ungauged catchments.

2. By reformulating the conventional S-curve method analytically with IS model and assuming desired SUH duration identical to that of parent SUH duration ($\tau = D$), the derivation of *D-hr* parent SUH can be possible using easily measurable physiographic characteristics.
3. The analytical relationships used as constraints relating *D-hr* SUH characteristics (Q_p and t_p) to analytical S-curve parameters (b, c and τ) prove their adequacy over different hydrometeorological zones.
4. The derived *D-hr* SUHs meets the criteria for UH derivation.
5. The characteristics of the analytical *D-hr* SUH (Q_p, t_p and t_b) are valid w.r.t. the employed physiographic (A, L, L_c and S) characteristics and SUH duration (D).
6. The proposed analytical IS method can be combined with any synthetic approach deriving the *D-hr* SUHs Q_p and t_p precisely.
7. Erratic manual sketching of the *D-hr* SUHs has been skipped unknowingly using Q_p, t_p and t_b for analytically fitting an SUH.

Acknowledgements The authors wish to thank NIH Roorkee, India, for providing all the necessary research facilities.

References

Bhadra A, Bandyopadhyay A, Singh R, Raghuwanshi NS (2010) Rainfall-runoff modeling: comparison of two approaches with different data requirements. Water Resour Manage 24(1):37–62

Bhar AK, Dwivedi VK (2005) A brief report on survey of lakes of India. national seminar on survey, conservation, and utilization of water resources, Kolkata

Brooks KN, Ffolliott PF, Gregersen HM, DeBano LF (2003) Hydrology and the management of watersheds. Iowa State Press, Ames

CWC (1984) Flood estimation report for upper Indo-Ganga plains subzone-1(e)—A method based unit hydrograph principle. hydrology directorate (Small Catchments), central water commission, New Delhi

CWC (1987) Flood estimation report for mahi and sabarmati subzone-3(a). directorate of hydrology (small catchments), central water commission, New Delhi

CWC (1995) Flood estimation report for lower godavari subzone-3(f)-revised. directorate of hydrology (regional studies), central water commission, New Delhi

Geetha K, Mishra SK, Eldho TI, Rastogi AK, Pandey RP (2008) SCS-CN-based continuous simulation model for hydrologic forecasting. Water Resour Manage 22(2):165–190

Ito Y, Momii K, Nakagawa K, Takagi A (2006) Estimation model of lake level in lake ikeda-hydrologic budget of a lake as a water resource. Trans Jpn Soc Irrig, Drainage Reclam Eng 74(3):65–72

Ito Y, Momii K, Nakagawa K (2007) Impact of agricultural water management on lake water budget: a case study of lake ikeda, Japan. IAHS Publ 315:179–185

Meyer PS (1994) Bi-logistic growth. Technol Forecast Soc Change 47(1):89–102

Nash JE, Sutcliffe JV (1970) River flow forecasting through conceptual models, part-i: A discussion of principles. J Hydrol 10(3):282–290

NakiCenovic N (1988) U.S. transport infrastructures in cities and their vital systems. National Academy Press, Washington, D.C

Ohba M (1984) Software reliability analysis models. IBM J Res Dev 28(4):428–443

Patil PR, Mishra SK (2016) Analytical approach for derivation of oscillation-free altered duration unit hydrographs. J Hydrol Eng 21(11):06016010

Sherman LK (1932) Streamflow from rainfall by the unit graph method. Eng News Record 108:501–505

Stone R (1980) Sigmoids. bulletin in applied. Statistics 7(1):59–119

Tokar A, Markus M (2000) Precipitation-runoff modeling using artificial neural networks and conceptual models. J Hydrol Eng 5(2):156–161

Verhulst PF (1838) Notice sur la loi que la population poursuit dans son accroissement. Corresp Math Phys 10:113–121

Verhulst PF (1845) Recherches mathématiques sur la loi d'accroissement de la population. Nouv. M´em. Acad R Sci B-lett Brux 18:1–45

Verhulst PF (1847) Deuxième mémoire sur la loi d'accroissement de la population. M´em Acad R Sci Lett B-Arts Belg 20:142–173

Yoshitani J, Chen ZQ, Kavvas ML, Fukami K (2009) Atmospheric model-based streamflow forecasting at small, mountainous watersheds by a distributed hydrologic model: application to a watershed in Japan. J Hydrol Eng 14(10):1107–1118

Chapter 24
Effect of Land Use on Curve Number in Steep Watersheds

C. B. Singh, S. K. Kumre, S. K. Mishra, and P. K. Singh

Abstract Runoff is an important and valuable variable used in planning of water resources and design of hydraulic structures. A number of models have been developed to calculate runoff from a rainfall event. The Soil Conservation Service Curve Number (SCS-CN) methodology is one of the most widely accepted event-based methods and is extensively used for estimation of surface runoff for a known precipitation event from small un-gauged agricultural watersheds. The model is satisfactorily established in hydrologic engineering. The main cause for its wide applicability lies in the fact that it is easy to use, and it incorporates major runoff generating watershed characteristics: soil type, land use, surface condition, and AMC. The only parameter curve number CN is critical for exact runoff prediction. According to the SCS-CN concept, the runoff quantity agricultural watershed depends on the above four major watershed characteristics. The CN values resulting from exhaustive investigations in the United States for all soil and land uses have been investigated and reported in National Engineering HandbookChapter-4 (NEH-4). Since the inception of SCS-CN method, only a few or no efforts appear to have been made to justify curve number rationality to watersheds in other countries. The slope was excluded in its original development, and it is included as a factor in the recently developed new models. Investigations were carried out on agricultural plot of size (12.0 × 3.0 m^2) located Toda Kalyanpur, Uttarakhand, India to calculate the effect of slope,

C. B. Singh · S. K. Kumre (✉) · S. K. Mishra
Department of Water Resources Development and Management, Indian Institute of Technology Roorkee, Roorkee 247667, Uttarakhand, India
e-mail: skumre159@gmail.com

C. B. Singh
e-mail: cbsingh200@gmail.com

S. K. Mishra
e-mail: skm61fwt01@gmail.com

P. K. Singh
Water Resources Systems Division, National Institute of Hydrology, Jal VigyanBhawan, Roorkee 247667, Uttarakhand, India
e-mail: pushpendras123@gmail.com

A. Pandey et al. (eds.), *Water Management and Water Governance*, Water Science and Technology Library 96,
https://doi.org/10.1007/978-3-030-58051-3_24

soil type, AMC, and land use on the runoff and runoff curve number in selected three grades of 8, 12, and 16% with same Hydrologic Soil Group (HSG) "A." There were nine0 plots of three land uses, maize, finger millet, and fallow lands for investigation. As expected, the conclusion of land use on runoff curve number was such that the fallow lands showed the largest runoff and the CN values. With the increase of slope, the CN and runoff quantity increased which we got in 16% slope, fallow land. The effect of soil was more prominent than the slope. The soil was, however, the same for all experimental plots, i.e. HSG-A. The SCS-CN parameter potential maximum retention (S) showed an inverse relation with the measured AMC value.

Keywords Curve number · CN-P relationship · NEH-4 · Rainfall-runoff relation · Soil conservation service curve number method · Hydrologic soil groups (HSG)

24.1 Introduction

Hydrology is the science that deals with the occurrence, distribution, and movement of water on the earth surface in space and time. Hydrologists and engineers are much concerned with the calculation of runoff coming out from rainfall event. It requires in all respect whether it is for flood prediction or flood control or water resources assessment or design of hydraulic structures. The surface runoff is an essential parameter for the assessment of water yield potential of a watershed (Swain et al. 2018). The surface runoff is the function of many variables, e.g. rainfall duration and intensity, soil moisture, land use, land cover, soil type, and slope of watershed; i.e. it depends on climatic as well as physiographic condition of watershed (Swain et al. 2017a, b). For runoff computation from a given rainfall, there are so many models available, but probably one of the most widely used methods is SCS-CN method. It is because most of the methods are for gauged watershed, having complex type of approaches and many variables involved in. But SCS-CN method is simple and applicable to un-gauged watershed for which minimum hydrologic information is available. The SCS-CN method was originally developed by Soil Conservation Service (now known as Natural Resources Conservation Service, NRCS) of United States Department of Agriculture (USDA) in 1954, and results were documented in National Engineering Handbook, Hydrology Section-4 in 1956, popularly known as SCS-CN, NEH-4.The current version of NEH-4 is NEH 630 (USDA, NRCS 2003). SCS-CN method is widely accepted for predicting surface runoff in small agricultural watershed because of its simplicity and limited number of parameters required for runoff prediction (Ponce and Hawkins 1996).The SCS-CN method converts the given rainfall into surface runoff by the use of a single parameter CN which represents the runoff potential of watershed characterized by hydrologic soil type, land use and treatment, ground surface condition, and antecedent moisture condition (AMC). The SCS-CN method is based on the water balance equation and two fundamental hypotheses. The first proportionality concept of the method relates the two orthogonal hydrological processes of surface water and ground water; the second hypothesis relates to the atmospheric process. Qualitatively, the method broadly integrates all three major processes of the hydrologic cycle, and it can thus form one of the fundamental concepts of hydrology (Mishra et al. 2012). SCS-CN methodology was developed in United States by carrying out exhaustive field investigation for the main purpose of soil conservation which was mandatory to do under the Public law 83–566.CNs were derived which are given in SCS-CN, NEH-4, for different land use, soil type, hydrologic condition in spite of its simplicity; the application of CN procedure leads to a diversity of interpretations and confusions due to ignorance of its limitations (Hawkins 1979b; Hjelmfelt 1991; McCuen 1982). The main difficulties in application are to classify the hydrologic soil group and determination of AMC. Basically, SCS-CN was generated for US condition but is being used worldwide along with many researches for sorting out the limitations. Besides the calculation of runoff, it is used in assessment of soil erosion and sediment yield, environment and water quality, etc. Besides others, CN depends on land use, Hydrologic Soil Group (HSG), and AMC. On the basis of several infiltration tests conducted in USA, soils were divided into four HSGs (A to D) according to their minimum infiltration rates (SCS 1986). Soil type A indicates the highest rate of infiltration and D the lowest. Soil depth and percolation and drainage conditions in the subsoil also influence runoff response. Despite its simplicity, the application of the CN procedure leads to a diversity of interpretations and confusion due to ignorance about its limitations (Hawkins et al. 1985; Hjelmfelt 1980). Difficulties are mainly related with the classification of soil group derived for USA and determination of AMC, which is the index of watershed wetness (Chaudhary et al. 2013). The method has undergone rigorous review and has been recognized to perform well without impairing its simplicity. The works of Ponce and Hawkins (1996), Mishra and Singh (2002b, 2004b, 2013), Mishra et al. (2006c, 2005, 2012), Huang et al. (2006), and Jain et al. (2006) are noteworthy.

24.2 Study Area and Data Collection

The experimental site is located in village Toda Kalyanpur (Latitude: 290 50′9″ N, Longitude: 770 55′ 21″ E) near Roorkee, district Haridwar of Uttarakhand state in India. It is situated about 6 km southeast of IIT Roorkee. Its elevation/altitude is 266 m above mean sea level. The area experiences the sub-tropical climate. Most of the rainfall occurs during the period of June to September. Precipitation varies from 1200 to 1500 mm. Average maximum temperature varies in between 20 and 40 °C, and relative humidity is 35 to 97%. All the data using herein are directly measured in field, and no secondary data were used from outside as it was the experimental work. The required equipment's were established in the farm itself. Rainfall was measured using non-recording type rain gauge; soil moisture was measured by soil moisture meter "Field scout TDR 300″ with probe length of 20 cm. Double-ring infiltrometer was used to compute the infiltration rate of the soil. The runoff generated from each plot can be measured through the collection chamber made at the end of each plot

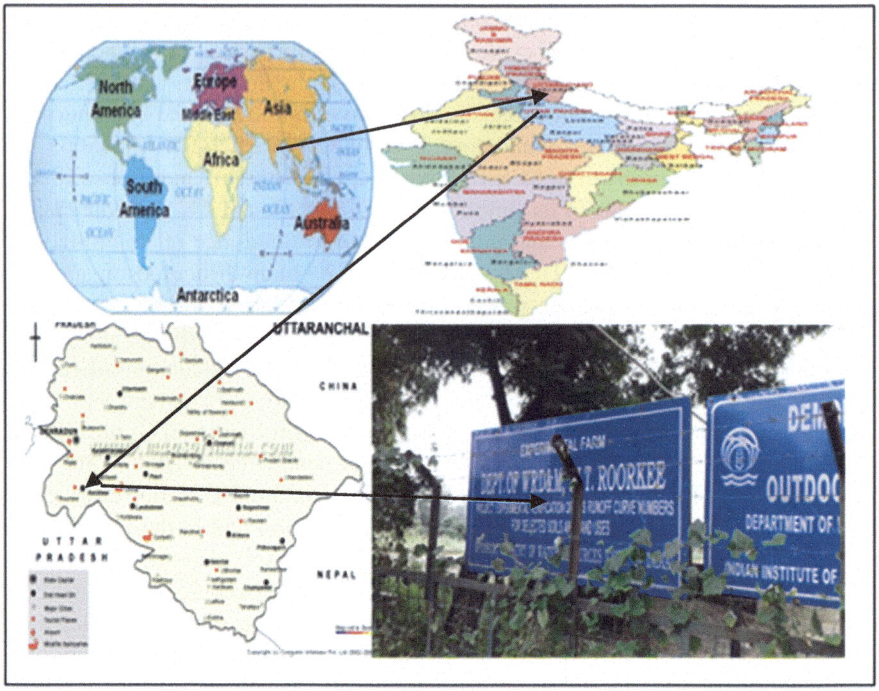

Fig. 24.1 Locations of sampling sites in the study area. (*Source* Google earth)

connected with an approach channel. Soil sample from each plot was collected and analyzed in the laboratory to know the soil type. (Fig. 24.1).

24.3 Methodology

The SCS-CN method is mostly used for estimation of runoff from given precipitation event especially in small agricultural watersheds. Its experimental verification for Indian conditions has not been attempted and published in the hydrologic literature, particularly at plot scale. To analyze the effect of soil, land use, AMC, and slope in Indian conditions, an experimental farm has been established by the Department of Water Resources Development and Management, IIT Roorkee in the village Toda Kalyanpur, near Roorkee, district Haridwar, India. The field has been divided into three different slopes at 8, 12, and 16%, and again each slope was arranged into three plots having different land uses. During a precipitation event, the resulting runoff from every plot was passed through a multi-slot divisor connected through a converging channel to tank/chamber of size 1mx1mx1m at the outlet. The runoff was measured in terms of depth of storage water, and multi-slot divisor was used for reducing the frequency of filling up of the chamber. The precipitation was measured by non-recording rain gauge installed at the site. For AMC of the soil, soil moisture was measured daily using TDR 300 during monsoon. In situ double-ring infiltrometer test was carried out for classifying HSG according to minimum infiltration rate.

24.4 Application

This section deals with step-by-step procedure associated with computation of CN from observed precipitation and runoff using the slope-adjusted CN from all the nine experimental plots for assessing the impacts of land use on runoff from steep watersheds.

24.4.1 Preparation of Experimental Plots

Three plots of three different grades were prepared viz. slopes of 8, 12, and 16. Each grade of plots was sub-divided into three small plots of size 12.0 × 3.0 m^2, leading to nine plots.

24.4.2 Measurement of Rainfall and Runoff

24.4.2.1 Rainfall

Precipitation data were measured from non-recording type rain gauge installed at the site. It was measured at every 8.30 to 9.00 AM from non-recording rain gauge. The rainfall measurement started from June 19, 2017 and it continued till recent rainfall of September 24, 2017. Total 19 rainfall events was observed. It was also found that runoff was generated by rain events of more than 9.6 mm.

24.4.2.2 Runoff

Runoff accumulated for every rainfall event from every plot was obtained in a chamber of size 1mx1mx1 m^3. Multi-slot divisor was established in the approach channel leading from plot to the chamber. Multi-slot divisor had five numbers of slots, and runoff was collected in the chamber coming only from the middle slot and that of other slots was transferred outside the chamber so that dimension of the chamber and the cost of project reduced. The depth of runoff quantity in the tank was measured using a steel tape. Since collecting tank and conveyance passage were open to atmosphere, quantity due to direct precipitation was also deducted to estimate the actual runoff. The water volume thus measured was multiplied five times to compute actual runoff volume, and it was divided by plot area to get the runoff depth.

24.4.3 Measurement of Infiltration Rate and Soil Moisture

24.4.3.1 Infiltration Test

Double-ring infiltrometer tests were carried out in all nine plots to determine the minimum infiltration rate of each of the plots. According to the minimum infiltration rate of the soil, HSG of the plot was decided. Two concentric rings of 30–45 cm^2 Dia. and 30 cm height was inserted into the soil so that about 10 cm was above the ground. Water was poured in both the rings up to a fixed level and drawdown in a certain interval of time was noted for inner ring only. The experiment was continued for at least 3–4 h until two consecutive readings of the same order were obtained, and it took around 9 days to finish these tests in all nine plots.

24.4.3.2 Antecedent Moisture Condition (θ_o)

Daily soil moisture of all plots was collected using field scout TDR 300 with a probe length of 20 cm. Three spot point values were measured in the field, and average of three values was calculated for soil moisture of the plot. It yielded directly the Volumetric Water Content (VWC) in percentage. The soil holding water was preceded for the whole monsoon and remained extended for as and when necessary after the monsoon. It was definitely observed the soil moisture prior to every rain event.

24.4.3.3 Soil Grain Size Analysis

Soil collected from every plot was carried to the laboratory in plastic bags. The soil was dried in an oven and mixed clearly such that structure of the soil was become disturbed. Then, it was passed through the sequence of sieves for 20 min. The retained soil on various sieves was weighed. It was compared with the previously weighed soil sample whether the loss of weight was more than 2%. The output was tabulated and analyzed to get the part of sand, silt, and clay which was the prime concern of the experiment. It was then used to differentiate HSG obtained from the infiltrometer test.

24.4.4 *Estimation of Curve Number for Experimental Plots*

In this research work, CNs were estimated using (i) NEH-4 table based on HSG and land use for all the three land use and 5% land slope, (ii) observed rainfall and runoff data from all the nine experimental plots, and finally (iii) the slope-adjusted CNs based on steps (i) and (ii). The estimated CNs were finally used for runoff estimation from all nine experimental plots.

24.5 Results and Discussion

The SCS-CN method is one of the simplest and most popular methods to compute runoff from given rainfall events. It was developed for the un-gauged small agricultural watershed. As there were twenty-one rainfall events that were able to generate runoff from all nine plots of different three grades, the effect of land use and slope of watershed was analyzed. From the observed rainfall and runoff pair, the potential maximum retention (S) and corresponding runoff curve number value were derived for all slopes and crops. CN values were derived for each storm event. In this study, all nine plots were agricultural lands. It was cultivated with row crops like maize, finger millet, and one plot in each slope was kept uncultivated as a fallow. Fallow lands were believed to generate more runoff than others, and to some extent, it was true. In this study, for slope of 12%, fallow land had the highest runoff and largest CN (Figs. 24.2–24.4). The effect of slope on runoff and CN is analyzed between 8 and 16% plot. It is seen that the runoff and CN values are lower for 8% plot than for 16% plot. Larger the slope and, in turn, the velocity generated, the shorter will be the opportunity time for infiltration to occur, and therefore, higher will be the runoff (or

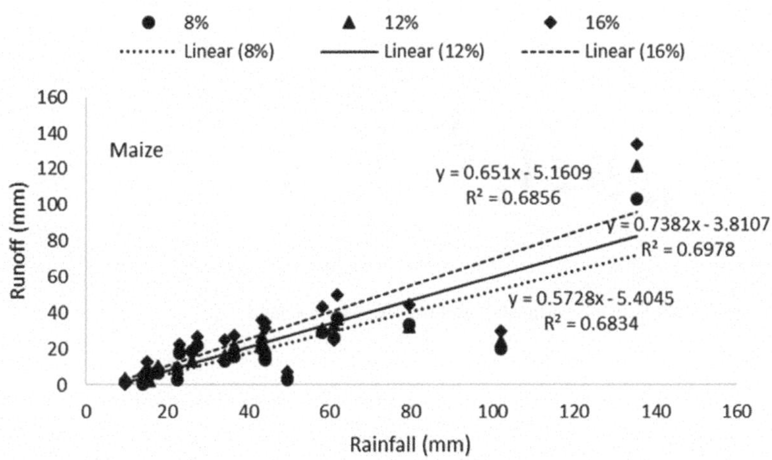

Fig. 24.2 Relationship between rainfall and runoff in maize crop

CN) generated. Thus, the larger slope is to produce larger runoff (or CN), and vice versa, which is in contrast with the observations for highest rain event. Table 24.1 shows the observed runoff and curve number corresponding to the observed rainfall. (Figs. 24.5–24.7).

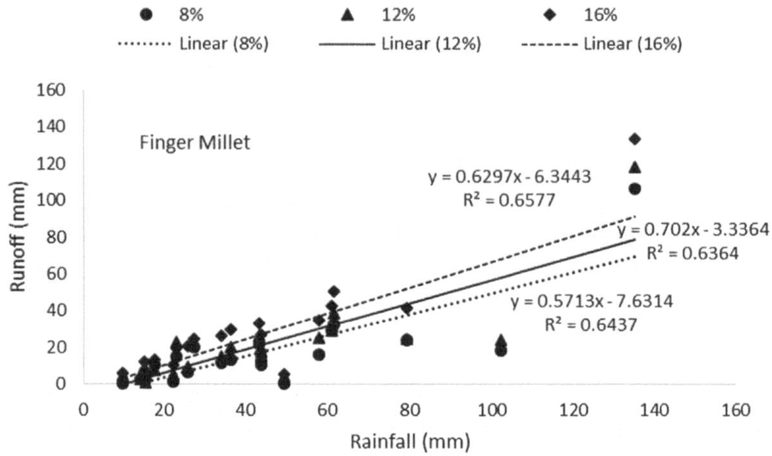

Fig. 24.3 Relationship between rainfall and runoff in finger millet

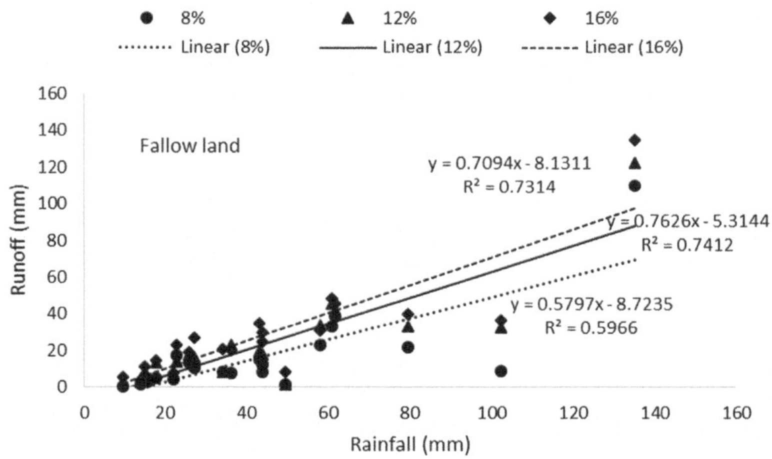

Fig. 24.4 Relationship between rainfall and runoff in fallow land

Table 24.1 Direct runoff and runoff curve numbers for different watersheds

Event no	Date	Rainfall (P) mm	Runoff (Q) (mm) for different watershed slopes								
			8%			12%			16%		
			Maize	Finger millet	Fallow	Maize	Finger millet	Fallow	Maize	Finger millet	Fallow
1	6/19/17	44.0	14.2	13.0	8.3	20.1	17.8	15.4	31.9	27.2	29.6
2	6/26/17	34.2	13.4	11.6	8.1	16.5	15.1	8.1	24.8	26.2	20.6
3	6/28/17	135.2	103.6	106.8	109.9	121.8	118.5	122.5	134.0	133.5	134.80
4	6/29/17	17.7	6.4	10.6	5.0	10.6	7.8	13.3	9.2	13.3	14.7
5	6/30/17	15.0	7.1	6.8	5.4	9.8	9.1	8.2	13.0	12.3	10.9
6	7/6/17	36.4	16.1	13.3	7.8	21.7	20.3	23.1	27.2	30.0	20.3
7	7/24/17	14.0	0.7	2.8	1.4	4.1	2.8	2.8	5.5	2.8	4.1
for8	8/2/17	79.5	33.2	24.2	22.1	32.5	24.9	33.2	44.3	41.5	40.1
9	8/3/17	9.6	2.0	0.6	0.6	3.4	4.7	3.4	0.6	6.1	5.7
10	8/7/17	27.4	21.6	20.2	10.5	23.7	23.0	16.1	26.5	24.8	27.00
11	8/10/17	43.4	19.9	22.0	15.1	26.9	26.2	19.9	35.9	33.1	34.7
12	8/19/17	22.3	2.9	1.6	4.6	9.2	5.7	6.4	9.9	10.4	7.9
13	8/22/17	58.1	29.0	16.3	23.3	30.9	25.3	33.7	43.4	35.1	30.9
14	8/23/17	15.5	5.3	2.5	2.5	2.5	1.2	5.3	6.7	3.9	2.5
15	8/25/17	61.8	37.3	32.4	38.7	36.6	38.7	41.4	49.8	50.5	45.6
16	9/1/17	44.0	15.0	10.8	12.2	20.5	15.0	17.8	34.4	17.8	24.7
17	9/1/17	23.0	17.4	15.3	17.4	21.6	23.0	13.8	22.5	19.5	23.0
18	9/2/17	61.1	26.3	29.1	33.2	33.2	29.1	45.7	24.9	42.9	48.5
19	9/3/17	26.0	19.1	6.6	13.5	13.5	9.3	17.7	17.7	20.5	19.1

(continued)

Table 24.1 (continued)

Event no	Date	Rainfall (P) mm	Runoff (Q) (mm) for different watershed slopes								
			8%			12%			16%		
			Maize	Finger millet	Fallow	Maize	Finger millet	Fallow	Maize	Finger millet	Fallow
Event no	Date	Rainfall (P) mm	Runoff curve numbers (CN) for different watershed slopes								
			8%			12%			16%		
			Maize	Finger millet	Fallow	Maize	Finger millet	Fallow	Maize	Finger millet	Fallow
1	6/19/17	44.0	84.0	82.9	77.5	88.7	86.9	85.0	95.3	93.0	94.2
2	6/26/17	34.2	89.2	87.6	83.8	91.6	90.5	83.8	96.3	96.9	94.2
3	6/28/17	135.2	89.0	90.1	91.3	95.5	94.3	95.7	99.6	99.4	99.9
4	6/29/17	17.7	93.6	96.8	92.0	96.8	94.9	98.3	95.9	98.3	98.9
5	6/30/17	15.0	96.0	95.8	94.5	97.8	97.4	96.8	99.3	99.0	98.4
6	7/6/17	36.4	90.1	87.8	81.9	93.7	92.9	94.4	96.5	97.7	92.9
7	7/24/17	14.0	85.9	91.7	88.5	93.8	91.7	91.7	95.3	91.7	93.8
8	8/2/17	79.5	79.4	73.3	71.7	78.9	73.8	79.4	85.6	84.2	83.4
9	8/3/17	9.6	94.3	90.4	90.4	96.3	97.6	96.3	90.4	98.5	98.2
10	8/7/17	27.4	97.8	97.2	91.0	98.7	98.4	95.0	99.7	99.1	99.9
11	8/10/17	43.4	88.9	90.3	85.1	93.1	92.7	88.9	97.3	96.1	96.8
12	8/19/17	22.3	84.8	81.0	87.8	93.1	89.4	90.3	93.7	94.1	91.9
13	8/22/17	58.1	87.1	77.7	83.3	88.2	84.8	89.8	94.5	90.5	88.2
14	8/23/17	15.5	94.0	90.0	90.0	90.0	86.3	94.0	95.4	92.3	90.0

(continued)

Table 24.1 (continued)

Event no	Date	Rainfall (P) mm	Runoff (Q) (mm) for different watershed slopes								
			8%			12%			16%		
			Maize	Finger millet	Fallow	Maize	Finger millet	Fallow	Maize	Finger millet	Fallow
15	8/25/17	61.8	90.0	87.3	90.7	89.6	90.7	92.0	95.6	95.9	93.9
16	9/1/17	44.0	84.7	80.5	82.0	89.0	84.7	86.9	96.4	86.9	91.6
17	9/1/17	23.0	97.8	96.8	97.8	99.5	100.0	96.0	99.8	98.7	100.0
18	9/2/17	61.1	83.9	85.6	88.1	88.1	85.6	94.2	82.9	93.0	95.4
19	9/3/17	26.0	97.3	87.7	94.1	94.1	90.8	96.6	96.6	97.9	97.3

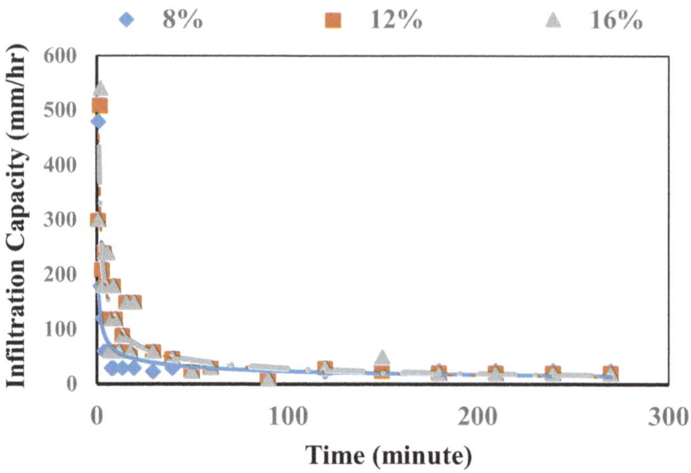

Fig. 24.5 Infiltration curve (Maize)

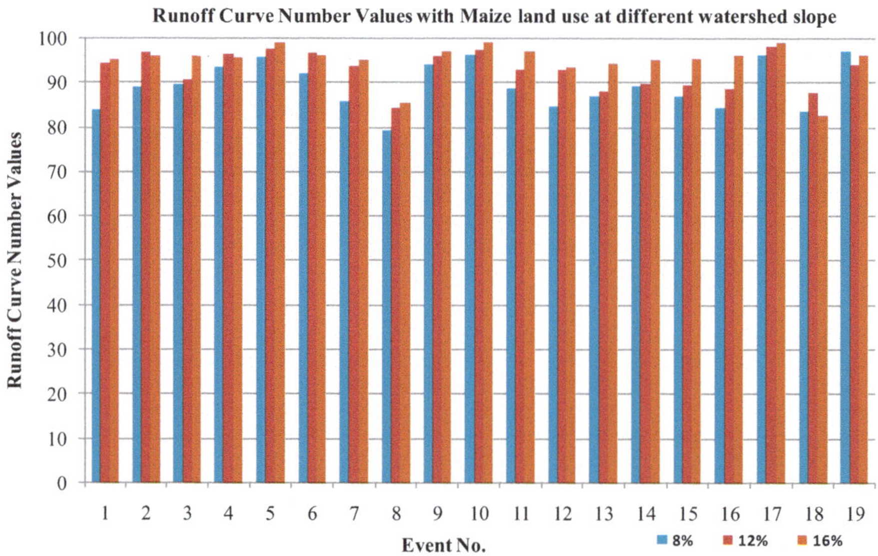

Fig. 24.6 Effect of maize on CN in different slope

24.6 Conclusions

The plot experiments were conducted to investigate the effects of land use on watershed runoff for steep slopes, effects of watershed slope on CN and runoff and to test the performance of the slope-adjusted SCS-CN method in predicting runoff from

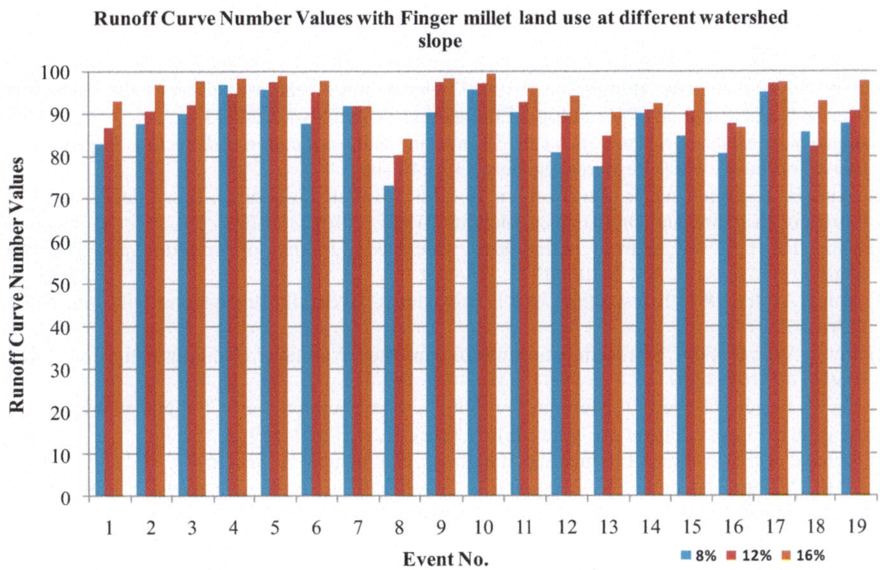

Fig. 24.7 Effect of finger millet on CN in different slope

different land uses and slopes using SCS-CN model from natural rainfall, consequent observed runoff, and other experimental works. The field had three slopes of 8, 12, and 16%, and each slope had three plots of size 12 m x 3 m. HSG of all the fields was the same as "A," and initial hydrologic condition of all the fields was poor (with no grass/vegetal cover). The land use of the field was agricultural and cultivated with maize, finger millet, and one plot from each slope was left fallow. The following conclusions are derived from the study:

The outputs of the study of the effects of soil, land use, slope, and AMC on runoff and CN are homogenous with the usual expectations. The SCS-CN parameter S shows a well-known inverse relation with measured AMC.

The effect of slope is not as important as that of soil on both runoff and CN; and thus, it is viable that a plot of higher grade may give lesser runoff depends on soil type. But in our case, the plot with the highest slope generated the highest runoff as well as CN because of same soil group in all experimental plots. Therefore, the fallow land produces the highest runoff and CN from all the slopes.

For more accurate runoff prediction, the effect of slope adjustment was investigated significant for slopes more than5%. In calculating the computed potential maximum retention and computed runoff, the slope adjustment formula was employed.

References

Bhuniya PK, Mishra SK, Berndtsson R (2003) Discussion on the estimation of the confidence interval for CN. Journal of Hydrologic Engineering, ASCE 8(4):232–233

Chaudhary A, Mishra SK, Pandey A (2013) Experimental Verification of Effect of Slope on CN. Journal of Indian Water Resources Society 33(1):40–46

Hawkins H, Hjelmfelt AT Jr, Zevenbergen W (1985) Runoff Probability, storm depth, and CN. Journal of Irrigation and Drainage Engineering, ASCE 111(4):330–339

Hawkins RH (1979) CN from partial area watersheds. Journal of Irrigation and Drainage Engineering Division, ASCE 105(HY4):375–389

Hjelmfelt AT Jr (1991) Investigation of CN procedures. Hydraulic Engineering Division, ASCE 117(6):725–737

Hjelmfelt AT Jr (1980) CN procedure as infiltration method. Journal of Hydraulics Division, ASCE 106(HY6):1107–1110

Huang M, Jacgues G, Wang Z, Monique G (2006) A modification to the SCS-CN method for steep slopes in the Loess Plateau of China. Hydrol Process 20(3):579–589

Hydrology, (1985) National Engineering Handbook. Supplement A, Sect, p 4

Jain MK, Mishra SK, Babu PS, Venugopal K, Singh VP (2006) An enhanced CN model incorporating storm duration and non-linear Ia-S relation. Journal of Hydrologic Engineering, ASCE 11(6):631–635

McCuen RH (1982) A guide to hydrologic analysis using SCS Methods. Prentice-Hall Inc., Englewood Cliffs, New Jersey

Mishra SK, Singh VP (2002) “SCS-CN based hydrologic simulation package. In: Singh VP, Frevert DK (eds) Mathematical models in small watershed hydrology”. Water Resources Publication, Littleton, Co, pp 391–464

Mishra SK, Singh VP (2004) Validity and extension of the SCS-CN method for computing infiltration and rainfall-excess rates. Hydrol Process 18:3323–3345

Mishra, S.K., and Singh, V.P (2013), “Soil conservation SCS-CN methodology (Vol.42)”, .Springer Science & Business Media

Mishra, S.K., A. Pandey, & V.P. Singh, (2012) “Special Issue on SCS-CN Methodology”. Journal of Hydrologic Engineering, ASCE, Nov. 2012.

Mishra SK, Jain MK, Pandey RP, Singh VP (2005) watershed area- based evaluation of the AMC-dependent SCS-CN-based rainfall-runoff models. Hydrol Process 19:2701–2718

Mishra SK, Tyagi JV, Singh VP, Singh R (2006) SCS-CN-based modelling of sediment yield. J Hydrol 324:301–322

Mishra, S.K., Rawat, S.S., Pandey, R.P., Chakraborty, S., Jain, M.K., and Chaube, U.C. (2012)."The relation between CN and PET”. Journal of Hydrologic Engineering, Manuscript HEENG-1496.

Ponce VM, Hawkins RH (1996) “CN: Has It Reached Maturity?”Journal of Hydrologic Engineering. ASCE 1:11–19

SCS (1986) 2003)’,Hydrology’, National Engineering Handbook, Supplement A, Section 4, Chapter 10. Soil Conservation Service, USDA, Washington, D.C.

Swain S, Verma MK, Verma MK (2018) Streamflow estimation using SWAT model over Seonath River Basin, Chhattisgarh, India. In: Hydrologic modelling, water science and technology library, vol 81. Springer Singapore, pp 659–665

Swain S, Patel P, Nandi S (2017a) Application of SPI, EDI and PNPI using MSWEP precipitation data over Marathwada, India. In: Geoscience and remote sensing symposium (IGARSS) 2017, pp 5505–5507

Swain S, Patel P, Nandi S (2017b) A multiple linear regression model for precipitation forecasting over Cuttack district, Odisha, India. In: 2nd International conference for Convergence in Technology (I2CT) 2017. IEEE, pp 355–357

Chapter 25
Performance Evaluation of a Rainfall Simulator in Laboratory

V. G. Jadhao, Rupesh Bhattarai, Ashish Pandey, and S. K. Mishra

Abstract Performance evaluation of a rainfall simulator installed in the laboratory of the Department of Water Resources Development and Management, IIT Roorkee, India was carried to characterize the simulated rainfall. About 360 simulations were made using six different header heights (2.0, 2.5, 3.0, 3.5, 4.0, and 4.5 m) and ten different pressure heads (0.2, 0.4, 0.6, 0.8, 1.0, 1.2, 1.4, 1.6, 1.8, and 2.0 kg/cm^2) on six different sets (1/8G-SS4.3 W, 1/4G-SS10W, 4G-SS14W, 3/8G-SS17W, 1/2G-SS30W, and 1/2GG-SS40W) of hydraulic nozzles having different orifice diameters and three sets of nozzles. The intensity and uniformity coefficients of the rainfall were assessed for all the nozzles at different header heights and operating pressures ranges. Results obtained from the simulations revealed that the measured rainfall intensities ranged from 17.38 to 231.58 mm/hr for operating pressure heads of 0.2 to 2.0 kg/cm^2. Post-hoc comparisons were made by one-way analysis of variance followed by Tukey –Kramer HSD test to determine if significant ($p \leq 0.05$) differences existed between nozzle intensities at different heights and pressures within groups. The uniformity coefficients (C_u) of nozzles with orifice diameter of 2.0, 2.80, 3.60, 4.0, 5.60, and 6.40 mm were 92.20, 91.24, 97.25, 92.91, 91.38, and 94.52%, respectively. The optimum values of C_u were found at header heights between 3.5 to 4.5 m and in the operating pressure heads of 0.6 to 1.0 kg/cm^2. The rainfall simulator experiment suggested optimal values of pressure and header heights for further analysis of runoff and sediment yield.

V. G. Jadhao (✉) · R. Bhattarai · A. Pandey · S. K. Mishra
Department of Water Resources Development and Management, Indian Institute of Technology, Roorkee 247667, Uttarakhand, India
e-mail: vikasjadhao@gmail.com

R. Bhattarai
e-mail: rupesbhattarai@gmail.com

A. Pandey
e-mail: ashish.pandey@wr.iitr.ac.in

S. K. Mishra
e-mail: skm61fwt01@gmail.com

A. Pandey et al. (eds.), *Water Management and Water Governance*, Water Science and Technology Library 96,
https://doi.org/10.1007/978-3-030-58051-3_25

Keywords Rainfall simulator · Laboratory simulation · Header height · Uniformity coefficient · Optimization

25.1 Introduction

In general, rainfall, runoff, and erosion are initiated by the impact of rain drops on bare or nearly bare soils, which detaches and splashes soil particles and subsequently transports them down slope as part of overland flow. Net erosion rate (sediment mass/unit area) is a function of both rain splash and overland flow transport. Runoff, as overland flow, carries the most erodible silt and very fine sand particles from the soil surface as the water flows downhill. It is difficult to study the soil infiltration characteristics, relative runoff, and soil erodibility in the absence of natural rain. However, rainfall simulator experiment may provide an opportunity to control number of variables that are govern by natural precipitation, and it is desirable to extend field observations made under natural rain in a controlled way using simulated rain (Sousa and Siqueira 2011; Sousa et al. 2017).

Nowadays rainfall simulator has become an ideal tool for investigating the process dynamics of soil erosion and surface hydrology and to create the database of various rainfall parameters for hydrological studies (Iserloh et al. 2013). The rainfall simulator can create various rain events of required characteristics (Dunkeley 2008; Lascelles et al. 2000). It is a time- and cost-effective method to overcome the dependency on natural rainfall and to understand the rainfall characteristics (Herngren et al. 2005).

The practical measures for reduction in soil loss require the accurate understanding of various factors which contribute to erosion process (Bisaro et al. 2014). The variables such as intensity of precipitation, its duration, soil moisture and soil type, land use and land cover are responsible for surface runoff by virtue of which erosion causes. In the overall phenomenon of soil erosion, the intensity and drop size of raindrop play very vital role, while in behavior, natural rainfall showed a loose correlation between the drop size, energy of raindrop, and their intensity (Kinnell 1981, 1987; Rosewell 1986).

In the study of different hydrological processes, one has to select specific type of rainfall simulator (Mutchler and Hermsmeier 1965). To create and apply uniform rate of rainfall, simulators are used to evaluate the runoff under controlled conditions (Kibet et al. 2014). Various studies can be found in specialized literature where a rainfall simulator has been used to analyze the different processes involved in erosion (Sangüesa et al. 2010; Arnaez et al. 2007; Fister et al. 2012). Rainfall simulators are capable to eliminate the unpredictable and erratic variability of natural storm by producing them on demand, quick, and as per necessity. The process-based mathematical models such as rainfall-runoff-sediment transport models which are conceptual, empirical, or process based should be calibrated and validated for use. The unique data produced by rainfall simulators are having vital importance for this purpose (Aksoy and Kavvas 2005).

Though many rainfall simulators have been developed by various researchers with different utility, but limited studies are available on to create lateral uniformity spray overlap with the help of an array of nozzles. The simulated rainfall characteristics depend upon the nozzle type used, pressure applied, and arrangement of the nozzles in the array (Aksoy et al. 2012). To study the range of intensities of rainfall generated and drop distribution over the laboratory scale plot by six different sizes of nozzles at different header heights and pressure levels, present study was undertaken in IWM Laboratory of the Department of Water Resources and Management Department, IIT Roorkee, India.

25.2 Methodology

25.2.1 Rainfall Simulator

The rainfall simulator used in this study was designed in the IWM Laboratory of the Department of Water Resources and Management Department, IIT Roorkee, India. The rainfall simulator consists of iron frame for nozzle mounted on four-wheels, pumping unit with flow meter and nozzles. The size of frame used is 3.0×1.5 m^2 which can support 11 nozzles (Fig. 25.1). The height of frame (Header height) can be adjusted up to 6.0 m from the ground with the help of chain pulley mechanism.

Fig. 25.1 Rainfall simulator setup used in the experiment

Table 25.1 Nozzle scheme with their specified characteristics (From a header height of 2 m)

Full-Jet nozzle 1	Nozzle equivalent orifice diameter (mm) 2	Capacity at 10 psi pressure (gpm) 3	Pressure applied in range of (kg/cm^2) 4	Average flow rate* (lpm) 5	Rainfall intensity (mm h^{-1}) 6
N4.3	2.00	0.43	0.2–2.0	0.69–2.77	17.38–51.94
N10	2.80	1.00	0.2–2.0	1.92–6.45	47.97–136.21
N14	3.60	1.40	0.2–2.0	2.80–10.08	42.27–132.37
N17	4.00	1.70	0.2–2.0	3.01–11.44	65.76–208.38
N30	5.60	3.00	0.2–1.0	4.51–12.89	126.67–230.30
N40	6.40	4.00	0.2–0.6	6.28–13.08	130.24–231.58

*Flow rate at pressure range given in column4 is observed at a header height range from 2.0 m to 4.5 m in the IWM laboratory of the WRDM, IIT Roorkee

A tray of size 2.50 m × 1.25 (3.125 sq. m) and depth of 0.56 m (Volume 1.75 m^3) is an integral part of the rainfall simulator for conducting rainfall-runoff-sediment studies. The tray slope can be adjusted from 0 to 20% with the help of slope adjustment device. The six different types of wide angle FullJet, G-style nozzles feature a solid cone-shaped spray pattern with a round impact area, and spray angles of 102° to 120° at 10 psi (0.7 bar) nozzles viz. B1/8GG-SS4.3 W(N4.3), B1/4GG-SS10W (N10), B1/4GG-SS14W (N14), B3/8GG-SS17W (N17), B1/2GG-SS30W (N30), and B1/2GG-SS40W (N40)of Spraying System Co. are used for laboratory performance evaluation. The details of the nozzle scheme used on rainfall simulator frame are provided in Table 25.1. They produce a uniform spray of medium to coarse drops across their entire spray area. This yields excellent results in spraying applications for complete coverage of an area or zone. This uniform spray distribution results from a unique vane design and exacting internal proportions. The nozzle scheme with their specified characteristics of average flow rate and rainfall intensity was observed in the laboratory, and the mean result from triplicate is given in Table 25.1.

25.2.2 Experimental Design

The experimental setup used in this study consist of a tank, pump, flowmeter, hose pipe with two pressure gauges, frame with provision of 11 numbers of nozzles supported in a framework resting on four columnar pipes with chain pulley mechanism. The nozzles were supplied with clean and constant flow of water from water source for simulation of rainfall. The frame with nozzles were connected to a centrifugal pump by means of hose pipe and stable and assured electric supply to pump (Fig. 25.1). The setup is having two pressure gauges in between, one near to pump (in between pump and flowmeter) and second near to the frame to provide

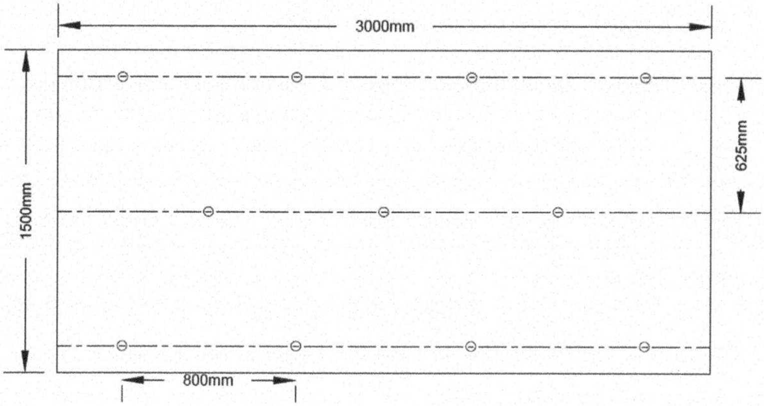

Fig. 25.2 Nozzle spacing on frame

appropriate control on the water pressure for nozzle with increment or decrement in header height. A flowmeter for flow measurement is also provided near to the pump. The tray filled with soil mass is placed beneath the rainfall simulator and gridded (in a size of 0.5 × 0.5 m) to measure the uniformity of rain distribution. At each grid corner, fifteen beaker of 500 ml capacity each with a diameter of 8.5 cm were placed on the tray ground and uniformly distributed with a distance 0.5 × 0.5 m to measure the distribution and intensity of rainfall.

A set of centrally aligned three nozzles of the frame (Fig. 25.2) spaced at 80 cm are sufficient to cover the given size tray. Other accessories such as split-eyelet connector, adjustable ball fittings, strainers, pressure regulators, pressure relief valves, control valves, solenoid valves, and check valves are also used in fittings of rainfall simulator.

25.2.3 *Test Procedure*

The observations were taken for hydraulic nozzles having different orifice diameters and three sets of nozzles at different height ranging from 2.0 to 4.5 m at an interval of 0.5 m of header height. The simulator was run for 10 min at each header height with pressure range from 0.2 to 2.0 kg/cm^2 observations. About 360 simulations were performed for all the nozzles. Six different header heights, i.e. 2.0, 2.5, 3.0, 3.5, 4.0, and 4.5 m (H2, H2.5, H3, H3.5, H4, andH4.5) and ten different pressure heads viz. 0.2, 0.4, 0.6, 0.8, 1.0, 1.2, 1.4, 1.6, 1.8, and 2.0 kg/cm^2 were designated in design of experimentation with a view of significant variations in intensity and uniformity coefficient. The uniformity coefficient was determined for six different sets (N4.3, N10, N14, N17, N30, and N40) of the nozzle at each height and pressure.

25.2.3.1 Discharge of Nozzle and Rain Intensity

The average discharge from all three nozzles was determined by diverting their flows with the help of elbow and 5 m long PVC pipe, and discharge was collected for the specific time in different buckets (Kibet et al. 2014). Simultaneously, the observations from flow meter were noted for each header height and the specified pressures.

The volume of water collected in each beaker was converted into rainfall intensity (mm/h), and mean of all was estimated at different pressures and different header heights. The analysis of variance (ANOVA) was also performed to check the hypothetical performance of the variables and to analyze their degree of freedom.

25.2.3.2 Uniformity Coefficient

Fifteen beakers of 500 ml capacity and 8.5 cm diameter were distributed uniformly over the entire area of the tray at $0.5 \times 0.5\ \mathrm{m}^2$ grid. The individual beakers yielding a volume at different locations on tray surface were weighed in terms of volume and average depth of rainfall estimated. The uniformity coefficient was then estimated by using Christiansen's (1942) method (Eq. 25.1) as it is most commonly used statistical method for evaluating uniformity (Warrick, 1983).

$$C_u = 100\left(1 - \frac{\sum X}{mn}\right) \tag{25.1}$$

In which x is the deviation of individual observations from the mean value of m and n is the number of observations. A simulator with uniformity coefficient greater than 80% is considered as satisfactory for rainfall simulation (Loch et al. 2001).

25.2.3.3 Evaluation of Rainfall Simulator

The intensity and uniformity coefficients of the rainfall were assessed for all the nozzles at different header heights and operating pressures ranges and used for calibration of the rainfall simulator (Pall et al. 1983). The results obtained from the simulations were analyzed for requirement of pressure and header height for various nozzles to produce the rainfall intensity range to be used in future work to match with storm intensities of rain received in selected basin (Singh et al. 1999).

The linear coefficient of determination (R^2) for various header heights and pressures for each nozzle was estimated to check the significance or non-significance of difference obtained between intensity and pressure of individual nozzles. One-way analysis of variances (ANOVA) was also performed to compare the means of observations of intensity and pressure hypothetically at 5% significance level. The $P < 0.05$ was considered statistically significant for all the tests.

Furthermore, multiple comparisons were performed by using Tukey – Kramer test honestly significant difference (HSD) procedure (Tukey 1949; Saunders and Blume 1981) employing following equation.

$$\text{Critical range} = \frac{M_1 - M_2}{\sqrt{MS_w\left(\frac{1}{n}\right)}} \tag{25.2}$$

where M is mean of observations within group, MS_w is mean of sum of squares between the groups of observations to compare, and n indicates the total number of observations in each group.

25.3 Results and Discussion

25.3.1 Performance Evaluation of Rainfall Simulator

The performance of the simulator was evaluated on basis of the intensity of rainfall and uniformity coefficient of the spray nozzle. The intensity of the nozzles on various pressures and at different header heights with average nozzle discharge (lpm) at specified pressure was determined to find out the maximum value of average precipitation of the study area with a return period of at least 10 years with a level of probability of occurrence of 90% and lasting for 15 min. (Navas et al. 1990).

25.3.2 Rainfall Intensity at Different Pressure

The average flow at various operating pressure heads of 0.2 to 2.0 kg/cm^{2}is shown in Table 25.1. For calibration of rainfall simulator, rainfall intensity was measured and analyzed. The values of measured rainfall intensities ranged from 17.38 to 231.58 mm/hr at operating pressure heads of 0.2 to 2.0 kg/cm^2. The relationship between rainfall intensity and pressure at nozzle at different header height is presented in Fig. 25.3. The coefficient of determination for the relationship obtained for various header heights is shown in Table 25.2. All the relationships between rainfall intensity and pressure at different header height exhibited positive correlation between the rainfall intensity and pressure heads.

The nozzle sizes below 4.0 mm remains operative and produces rainfall at1.0 kg/cm^2 pressure; however, the nozzle sizes of 5.60 and 6.40 mm were not operative at this pressure. Further, simulated rainfall intensity ranges between 126.67 and 231.58 mm/hr at pressure below 0.6 kg/cm^2.

The one-way analysis of variance for intensity is shown in Table 25.3. This indicates that the null hypothesis cannot be rejected ($F < F_{crit}$) for N14 as there is non-significant difference between the means of the observed intensity of rainfall simulated by the various nozzles at different header heights. The null hypothesis is rejected ($F > F_{crit}$) for all other five nozzles.

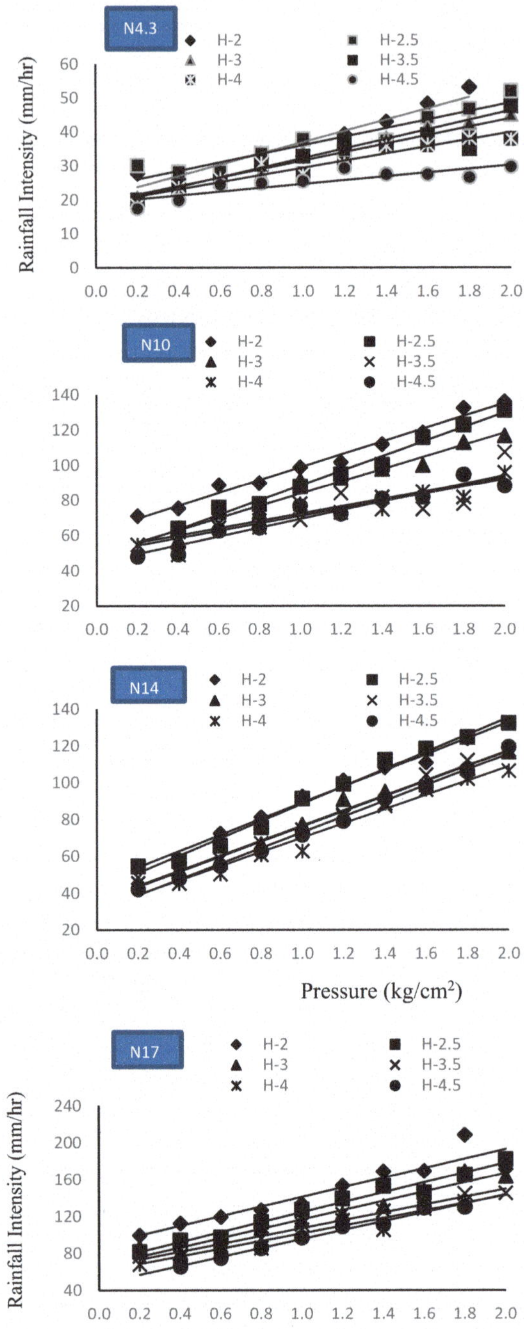

Fig. 25.3 Relationship between rainfall intensity and pressure at different header height

Fig. 25.3 (continued)

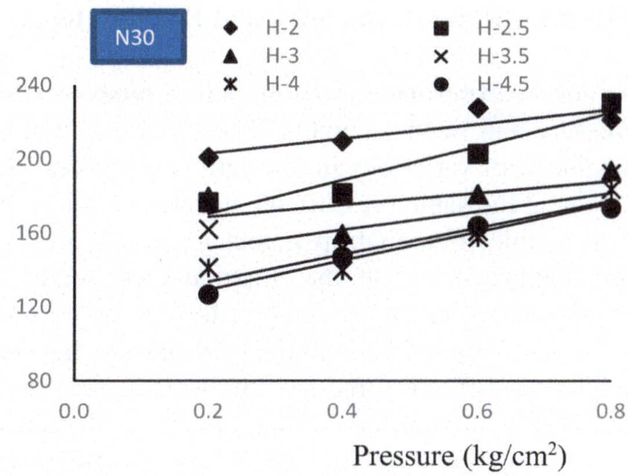

Table 25.2 Coefficient of determinations (R^2) for intensity at various pressures

Pressure (kg/cm^2)	N4.3	N10	N14	N17	N30	N40
0.2	0.7816	0.9895	0.9996	0.8337	0.9514	0.8898
0.4	0.8671	0.8378	0.8777	0.8869	0.8271	0.8210
0.6	0.8593	0.8943	0.7935	0.8696	0.8303	0.9514
0.8	0.4723	0.6500	0.8600	0.9637	0.8498	Non-operative
1.0	0.6684	0.6288	0.7913	0.8931	Non-operative	
1.2	0.7013	0.9604	0.9357	0.9404		
1.4	0.7956	0.8587	0.6814	0.9143		
1.6	0.9398	0.7716	0.6128	0.8253		
1.8	0.9120	0.6886	0.7448	0.8733		
2.0	0.8716	0.9893	0.5694	0.6495		

Table 25.3 One-way ANOVA for rainfall intensity at different pressure

Statistical parameters	N4.3	N10	N14	N17	N30	N40
F	3.38002	4.93314	1.31982	3.49396	6.917	14.8175
P	0.01007	0.00093	0.27026	0.00861	0.00092	0.00132
F_{crit}	2.38944	2.3966	2.39295	2.3966	2.77285	3.97152

As the null hypothesis was rejected for nozzles other than N14, further test was carried out to find the inequality critical range whether actual difference between every single group of the observations is showing significant difference. As per the Tukey–Kramer test, a higher difference in simulation of rainfall intensity was observed at higher header heights, i.e., in between range 3.5 to 4.5 m as in the multiple comparison significant difference observed at these header heights.

25.3.2.1 Rainfall Intensity and Header Height

As per the experimental design, all six types of nozzles were evaluated for different pressure and header heights. The observed rainfall intensity from various header heights with variations in pressure is shown in Fig. 25.4. With increase in header height, at the same pressure the nozzle exhibits simulated rainfall intensity.

A visible effect in drop diameter was also observed such as with increase in pressure and header height, the drop diameters decrease. The linear range of coefficient of determination for the relationship between rainfall intensity and header heights is shown in Table 25.4. All the relationships between rainfall intensities and header heights at different pressure implies inverse relationship between the variables I and H. It is also observed that, nozzles with orifice size of 2 mm is unaffected by header height and pressure variations. No correlation was found between intensity and header height in case of N40.

As $p < 0.05$ in case of one-way analysis of variances performed for intensity at various header heights (Table 25.5), the null hypothesis cannot be rejected. But as $F > F_{crit}$ indicates that there is a significant difference between the overall means of the simulated rainfall by individual observations at similar pressure values flashed inverse relationship between the header height and intensity.

Significant difference between the means of observations of intensities within the group of observations of nozzle of different sizes was observed in Tukey –Kramer test. No correlation was observed for the individual nozzle on intensity of rainfall at different operating pressure.

25.3.2.2 Average Nozzle Discharge and Pressure

As explained in the experimental design, the average discharge of the nozzles operated in specified header height and pressure was estimated (lpm). The coefficients of multiple determination (R^2) values obtained for the said linear relationship are shown in Table 25.6. This indicates a very strong correlation between pressure and average discharge.

The hypothetical results of the one-way analysis of variances performed for the average nozzle discharge showed non-significance difference (Table 25.7).

25.3.2.3 Evaluation of Uniformity Coefficient

In order to accurately replicate the design, the uniformity coefficient was estimated. To assess the accuracy and reproducibility of produced rainfall and to generate quantitative information about the homogeneity of simulated rainfall, the uniformity of rainfall distribution was carried out. The uniformity coefficient was estimated for each prescribed pressure and header heights on a grid size of 0.5×0.5 m. The optimized results of various nozzles are shown in Table 25.8.

Fig. 25.4 Relationship between rainfall intensity and header heights at different pressure

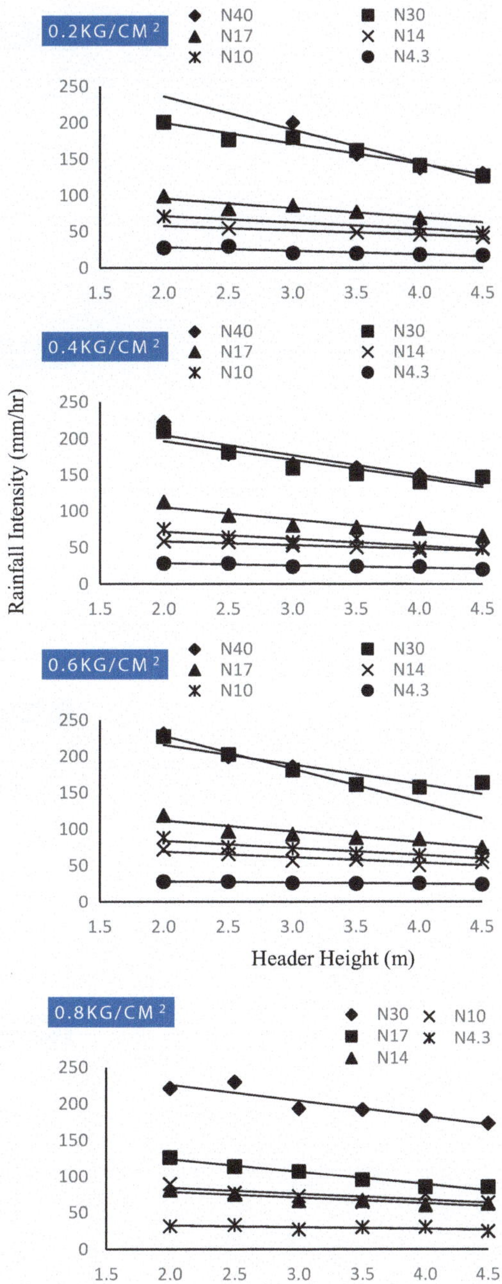

Fig. 25.4 (continued)

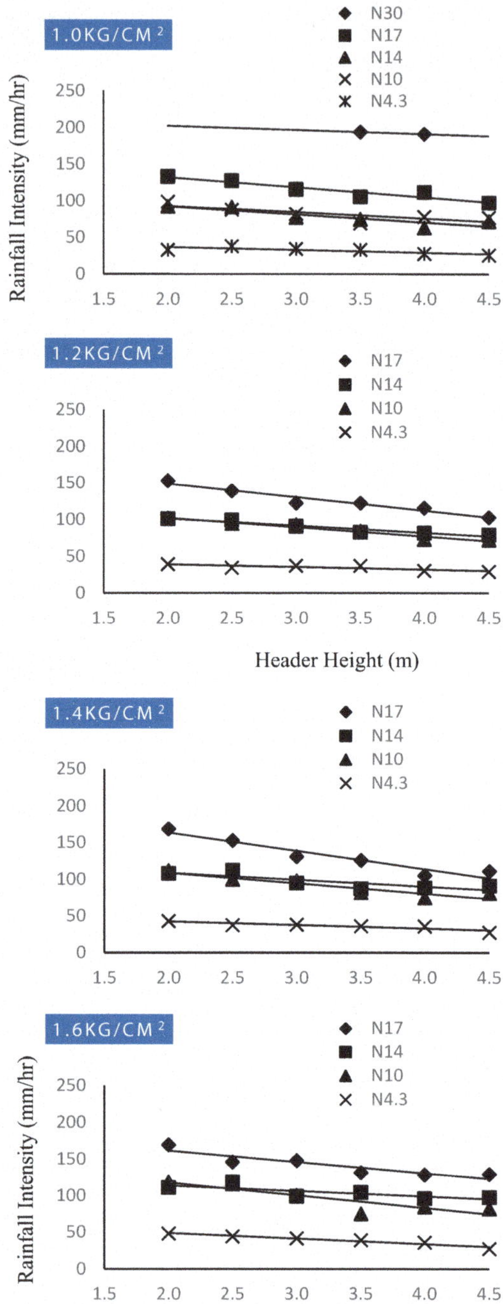

Fig. 25.4 (continued)

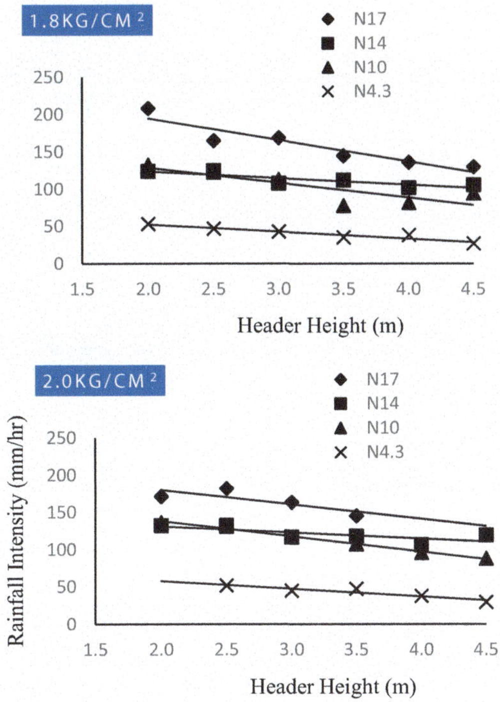

Table 25.4 Coefficient of determinations (R^2) for intensity at various header heights

Nozzle	Header heights					
	H2	H2.5	H3	H3.5	H4	H4.5
N4.3	0.9251	0.8825	0.9782	0.8784	0.9168	0.7403
N10	0.9835	0.9859	0.9517	0.6635	0.8316	0.9223
N14	0.9850	0.9860	0.9830	0.9840	0.9690	0.9931
N17	0.8741	0.9749	0.9595	0.9793	0.9200	0.9820
N30	0.7330	0.9254	0.3419	0.5399	0.8450	0.9770

The rainfall simulated by the medium size nozzle showed uniform distribution as compared to lower and higher orifice sizes nozzles. However, the performance of the nozzle N14 was excellent at every header heights and pressure levels.

25.4 Conclusion

The rainfall simulator was designed, modified, and calibrated on the basis of simulated rainfall intensity and its distribution at various header heights and pressure

Table 25.5 One-way ANOVA for rainfall intensity at different header heights

Statistical parameter	Pressure at nozzle (Kg/cm2)									
	0.2	0.4	0.6	0.8	1.0	1.2	1.4	1.6	1.8	2.0
F	50.91	71.01	95.51	133.32	5.76	59.59	40.56	59.78	37.76	57.51
P	0.0	0.0	0.0	0.0	0.0	0.0	0.0	0.0	0.0	0.0
F_{crit}	2.66	2.53	2.57	2.76	2.82	3.1	3.1	3.1	3.1	3.34

Table 25.6 Coefficient of determinations (R^2) for average nozzle discharge

Nozzle	Header heights					
	H2	H2.5	H3	H3.5	H4	H4.5
N4.3	0.9273	0.9485	0.9656	0.9842	0.9644	0.9731
N10	0.9672	0.9821	0.9844	0.9685	0.9708	0.964
N14	0.9748	0.9746	0.9656	0.9794	0.9665	0.9469
N17	0.9043	0.9749	0.9663	0.9364	0.9813	0.9531
N30	0.9395	0.9831	0.9669	0.9773	0.9412	0.9893

Table 25.7 One-way ANOVA for average nozzle discharge

Statistical parameter	Nozzle					
	N4.3	N10	N14	N17	N30	N40
F	0.110	0.131	0.050	0.102	0.287	0.221
P	0.990	0.984	0.998	0.991	0.914	0.942
F_{crit}	2.393	2.397	2.393	2.397	2.711	3.972

Table 25.8 Nozzle-wise optimized results

Properties	N4.3	N10	N14	N17	N30	N40
Pressure (Kg/cm^2)	1.00	1.60	1.00	0.60	0.60	0.40
Header height (m)	4.00	3.50	3.50	3.50	3.50	4.00
C_u(%)	92.20	91.24	97.25	92.91	91.38	94.52
Intensity (mm/hr)	27.61	75.00	74.45	88.39	161.13	149.73

heads. The rainfall intensity simulated by all six types of nozzles ranges between 17.38 and 231.58 mm/hr for operating pressure heads of 0.2 to 2.0 kg/cm^2. The correlation between the rainfall intensity, header heights, and pressure heads was also evaluated and revealed that a positive correlation is implied between the intensity of rainfall and pressure heads. Significant difference was observed in intensity of a nozzle at different pressure and header heights. The rainfall intensity decreases with increase in header heights for all the nozzles. The higher size nozzles with orifice diameter of 5.60 and 6.40 mm were operative at a pressure range of 0.2 to 0.6Kg/cm. Significant difference was observed between the average intensity of rainfall simulated by each type of nozzle at higher header height, i.e., > 3.5 m. Out of six nozzles, four nozzles, i.e., N10, N14, N17, and N30 showed maximum C_u at 3.5 m header heights. The average discharge rate measured at nozzle was found in direct proportion with the pressure head. The uniformity coefficient was in greater range in case of small size nozzles at higher header height. The uniformity coefficient was found to increase with increase in pressure head. The suggested optimal values of pressure and header heights obtained and ranges of rain intensities produced by various nozzles will be used for further study of erosivity by natural rain in a basin.

References

Aksoy H, Kavvas ML (2005) A review of hillslope and watershed scale erosion and sediment transport models. CATENA 64(2):247–271

Aksoy H, Unal NE, Cokgor S, Gedikli A, Yoon J, Koca K, Eris E (2012) A rainfall simulator for laboratory-scale assessment of rainfall-runoff-sediment transport processes over a two-dimensional flume. CATENA 98:63–72

Arnaez J, Lasanta T, Ruiz-Flaño P, Ortigosa L (2007) Factors affecting runoff and erosion under simulated rainfall in Mediterranean vineyards. Soil Tillage Res 93(2):324–334

Bisaro A, Kirk M, Zdruli P, Zimmermann W (2014) Global drivers setting desertification research priorities: insights from a stakeholder consultation forum. Land Degrad Dev 25(1):5–16

Christiansen JE (1942) Irrigation by Sprinkling. California Agriculture Experiment Station Bulletin, No, p 670

Dunkerley D (2008) Rain event properties in nature and in rainfall simulation experiments: a comparative review with recommendations for increasingly systematic study and reporting. Hydrol Process: An Int J 22(22):4415–4435

Fister W, Iserloh T, Ries JB, Schmidt RG (2012) A portable wind and rainfall simulator for in situ soil erosion measurements. CATENA 91:72–84

Herngren L, Goonetilleke A, Sukpum R, de Silva DY (2005) Rainfall simulation as a tool for urban water quality research. Environ Eng Sci 22(3):378–383

Iserloh T, Ries JB, Arnáez J, Boix-Fayos C, Butzen V, Cerdà A, Gómez JA (2013) European small portable rainfall simulators: a comparison of rainfall characteristics. CATENA 110:100–112

Kibet LC, Saporito LS, Allen AL, May EB, Kleinman PJ, Hashem FM, Bryant RB (2014) A protocol for conducting rainfall simulation to study soil runoff. J Visualized Exp (86):e51664

Kinnell P (1981) Rainfall intensity-kinetic energy relationships for soil loss prediction. Soil Sci Sot Am J 45:153–155

Kinnell P (1987) Rainfall energy in eastern Australia: intensity-kinetic energy relationships for Canberra, A.C.T. Aust J Soil Res 25:547–553

Lascelles B, Favis-Mortlock DT, Parsons AJ, Guerra AJT (2000) Spatial and temporal variation in two rainfall simulators: implications for spatially explicit rainfall simulation experiments. Earth Surf Process Landforms 25:709–721

Loch RJ, Robotham BG, Zeller L, Masterman N, Orange DN, Bridge BJ, Bourke JJ (2001) A multi-purpose rainfall simulator for field infiltrationand erosion studies. Soil Res 39(3):599–610

Mutchler CK, Hermsmeier LF (1965) A review of rainfall simulators Trans. ASAE (Am. Soc Agric. Eng.) 8:67–68

Navas A, Alberto F, Machín J, Galán A (1990) Design and operation of a rainfall simulator for field studies of runoff and soil erosion. Soil Technol 3(4):385–397

Pall R, Dickinson WT, Beals D, McGirr R (1983) Development and calibration of a rainfall simulator. Canadian Agricu Eng 25(2):181–187

Rosewell CJ (1986) Rainfall kinetic energy in eastern Australia. J Climate Appl Meteorol 25(11):1695–1701

Sangüesa C, Arumí J, Pizarro R, Link O (2010) A rainfall simulator for the in-situ study of superficial runoff and soil erosion. Chilean J Agric Res 70(1):178–182

Saunders JA, Blume DE (1981) Quantitation of major tobacco alkaloids by high-performance liquid chromatography. J Chromatogr A 205(1):147–154

Singh R, Panigrahy N, Philip G (1999) Modified rainfall simulator infiltrometer for infiltration, runoff and erosion studies. Agric Water Manag 41(3):167–175

Sousa Júnior SFD, Siqueira EQ (2011) Development and calibration of a rainfall simulator for urban hydrology research. In: Proceedings of 12th international conference on urban drainage, porto Alegre, Brazil (11–16)

Sousa Júnior SFD, Mendes TA, Siqueira EQ (2017) Development and calibration of a rainfall simulator for hydrological studies. Braz J Water Resour 22(0)
Tukey JW (1949) Comparing individual means in the analysis of variance. Biometrics 5(2):99–114
Warrick AW (1983) Interrelationships of irrigation uniformity terms. J Irrig Drainage Eng 109(3):317–332

Chapter 26
Algorithms of Minimal Number of Sensors Placement Using Pressure Sensitivity Analysis for Leak Detection in Pipe Network

Manish Kumar Mishra and Kailash Jha

Abstract In the proposed work, algorithms have been developed to minimise the placement of number of sensors to detect leaks in the looped pipe network using pressure sensitivity analysis. The algorithm of sensor placement uses the pressure sensitivity matrix which is the differences between the pressures calculated without leak and with leak values at nodes by EPANET simulation. Leakages are simulated as a constant demand that is assumed at junction. Binarised matrix is obtained by normalised pressure sensitivity matrix using threshold. The algorithm starts with the assumption that every node has a sensor. The minimum number of sensor is obtained using binarised matrix in such a way that every leak present in the network should be detected and isolated. The implementation of present algorithms shows the placement of sensors is minimised up to 15–20% of total number sensor placed at nodes. The problem identifies the best location of minimum number of sensors in the pipe network.

Keywords Pressure sensitivity analysis · Isolability · Detectability · EPANET · Threshold

26.1 Introduction

Leak is a type of pandemic in water pipe distribution. Due to financial cost and potential risk of public health and environment, leaks generate significant interest on water pipe distribution network. Such type of problem holds significant meaning to the society, struggling to supply the water supply of increasing demands. Thus, the

M. K. Mishra (✉) · K. Jha
Department of Mechanical Engineering, Dayananda Sagar University, Bengaluru 560068, Karnataka, India
e-mail: manish0546@gmail.com

K. Jha
e-mail: kailash0403@gmail.com

A. Pandey et al. (eds.), *Water Management and Water Governance*, Water Science and Technology Library 96,
https://doi.org/10.1007/978-3-030-58051-3_26

management of such type of problem becomes an important aspect for the managers of water supply networks (Nicolini and Patriarca 2011; Shu and Zhang 2010). Ground penetrating radar and acoustic listening device are the physical inspection of the leak in pipeline (Farley and Trow 2003; Colombo et al. 2009). These techniques require to shut down and isolate the effected part or whole system. The complete process may take few days to months with plenty of significant water wastage. The inspection of leaks has done on the observation on routine basis. When the fluctuation in demand is increased at night to day consumption abnormally or major losses are suspected, leak detection techniques are applied. A simpler leak detection and a location method have great economic concern.

A number of attempts have been made for the detection and isolation of leak in the water supply system (Wu et al. 2011; Colombo et al. 2009), summarised the recent and previous work on transient-based detection of leak and describe state of the art on the area. Mashford et al. (2009) have proposed a technique to obtain the information of leak size and location with desirable accuracy by performance assessment of support vector machine. Verde (2001) has tested and designed a multi-leak detection system and solves the problem using flow and pressure sensors at extreme duct by analytical redundancy. In inverse transient analysis, pressure data has been analysed during the incidence of transitory event through the minimisation of the difference of measured and estimated parameter (Covas and Ramos 2001; Kepler et al. 2011). Leak can be easily detected where flow measurements are available since the mass balance equation easily establish in the pipe network. Ragot and Maquin (2006) have used a model-based approach to detect fault. Fuzzy residual analysis for diagnosis strategy has used to detect and isolate the fault on sensors. However, the measurement of the flow in large pipe distribution network is very practical, but in it is not satisfactorily work in the case where there is dense network of pipe and only flow can measured at entrance is known. To deal with this situation, water companies give the attention on a viable solution for pressure sensor instalment in the district metered areas. The pressure sensors are cheap in cost and easily available to install and maintain. Model-based leak detection and isolation techniques have begun with the influential research work of Pudar and Liggett (1992) that expressed it as a least squares estimation problem. The estimation of parameter of water distribution network model is difficult task (Savic et al. 2009). Nonlinear equation caused the difficulties in water distribution network system and for estimating the parameters very few measurement are available which causes the undetermined problem. Pérez et al. (2011) developed a model-based method for detecting and localising the leaks, using pressure sensitivity analysis of junctions in water supply system. This method used the pressure residuals (difference between measured value and their estimated value) analysis and compared with a given threshold. Threshold has been used in keeping the model's uncertainty and noise in mind. The comparison of the residual against the threshold shows the possible leak present at nodes. Due to demand uncertainty at node and noise in measurement, the performance of this approach decreases, while in ideal condition it shows good efficiency. Casillas et al. (2012) improved this methodology by taking the analysis in a time horizon and show the comparison a number of leak detection and isolation method available. Blesa et al. (2010) developed a technique for leak detection based on pressure residual evaluation by using linear parameter varying (LPV) model. The discrepancies between LPV model and measurement obtained by sensors show the fault in the system. The zonotopes are used to avoid false alarms. Goulet et al. (2013) developed a strategy for placement of sensors and leak detection, based on model falsification of error domain. The approach falsifies the parameter for which the discrepancies between measurements and estimations are greater than the maximum reasonable error. Maximum reasonable faults are obtained through linking estimated and measurement uncertainty and noise. Casillas et al. (2015) improved the sensor placement approach for leak localisation using integer optimisation approach problem. The formulation minimises the overlapping leak signature from the leak signature space. Ferrandez-Gamot et al. (2015) have proposed the new approach to combine the use of modal and classifier. Residuals are generated by models and analyse by classifier taking the residual sensitivity in consideration. The model calibration is an impartment problem of water distribution network because the leakages are not all real (Wu et al. 2011). Shu and Zhang (2010) & Nicolini and Patriarca (2011) have used genetic algorithms to calibrate the model. Cheng and He (2010) have proposed a methodology of optimisation to calibrate the demands at junction of the network. Singular value decomposition is used to know and recognise the model parameter by solving sensitive coefficient matrix.

This work presents a model-based strategy to detect and localise the leak. The aim of this research is to obtain the minimal pressure sensor placement in pipe network for detection and localisation of leak. The method used the pressure variation at node which is produced due to uncertain change in demand (Pudar and Liggett 1992). The detection of leak is obtained by comparing pressure data of a network without leak and with a leak at time at all nodes of the network. The difference between measured and estimated pressure called residual which is evaluated to get signature matrix. The simulation model on EPANET is used to generate pressure data with a leak at time and without leak at all node. The present work proposed a minimal sensor placement strategy by analysing threshold independent of nodal pressure. The analysis of threshold is done in such a way that comparison of threshold to the sensitivity matrix gives the same leak signature at each time and count the sensor available at the same time. Different algorithms are developed and integrated to detect and localise the leak as well as sensor placement in this work.

Section 26.2 reviews the simulation modelling of water distribution network on EPANET used in the proposed methodology. Recent advancement in leak detection and isolation strategy is described in Sect. 26.3 describe the foundation of model-based leak detection and isolation technique. Signature matrixes for leak are obtained through pressure sensitivity matrix which describe in Sect. 26.4. Describes the leak signature matrix evaluation and comparison with threshold to binary signature matrix. Section 26.5 describes the optimal sensor placement strategy. Finally, Sect. 26.6 gives the summary and conclusions.

26.2 Water Distribution Systems Modelling

A water supply network consists of pumps, tank/reservoir, and pipes. Pumps are usually used for overcome the difference of elevation energy losses by friction and roughness of pipes. Pipe network may contain different component like flow-control device and pressure regulating valve. The only aim of water supply system is to supply and satisfy consumers demand.

Two basic governing equations for steady-state conditions are mass balance equation and energy equation. The inflows and outflows through the system are equal according to the law of conservation of mass. Therefore, the inflow and outflow should be balanced in the pipe and node of network.

$$\sum Q_{\text{in}} - \sum Q_{\text{out}} = Q_{\text{dmd}} \tag{26.1}$$

where Q_{in} and Q_{out} are the inflow and outflow at junction and Q_{dem} is the junction demands. Law of energy conservation states that difference between two end points of pipe is equal to difference of addition of energy and frictional losses. The equation for energy conservation can be written for a path between two fixed grade junctions of pipe.

$$\sum_{i \in J_p} H_{P,j} - \sum_{i \in I_p} H_{L,i} = \Delta E \tag{26.2}$$

where $h_{P,j}$ head increased by pump j, $h_{L,I}$ is the head loss across, $h_{P,j}$ pump head at j, and ΔE head loss in the path. The head loss for the path is represented as follows:

$$h_L = K Q^r \tag{26.3}$$

where H_{L} is the head loss, Q^r is the pipe flow, and K is pipe constant.

Water distribution system is modelled using EPANET software. Networks are simulated and pressures are calculated at each node with a leak at each node and without leak at node. In this paper, leak assume at node. In such case, leak can be seen as an additional demand at nodes. Simulated leaks presented are the 3% of total demand of water pipe network system.

26.3 Methodology of Leak Detection and Isolation

The present methodology is based on model-based diagnosis that has already been used for the leak detection and localisation (Ragot and Maquin 2006; Pérez et al. 2011). Two basic tasks are leak detection and leak isolation. Leak detection has been checked the consistency of observed behaviour and isolation of faulty part has done by fault isolation. The check of consistency has done by evaluating the residual $r(t)$ and obtained through measured input $i(t)$ and output $o(t)$ signals (pressure) using installed sensors.

$$r(t) = \Psi(i(t), o(t)) \tag{26.4}$$

where Ψ residual generator function depended on of type detection technique (Gertler 1998) and "t" is the time instance. At each time "t", instant residual is compared with statically obtained threshold value taking noise and uncertainty in consideration. When the value of the residual is higher than the threshold, then the fault is in the system otherwise system assume working fine.

In this paper, leak detection has been done by a passive method using threshold value. Residual evaluation provides observed fault signatures.

$\Phi(t) = [\Phi_1(t), \Phi_2(t)\ldots\ldots\ldots \Phi_n(t)]$ where each elements are given as follows:

$$\Phi_i(t) = \begin{cases} 0 & \text{if } r_i(t)| \le \tau_i(t) \\ 1 & \text{if } r_i(t)| \ge \tau_i(t) \end{cases} \tag{26.5}$$

where τ_i is threshold associated with the residual $r_i(t)$.

26.4 Leak Sensitivity Analysis

The effect of leak on pressure at node is evaluated in this section. If the process is repeated to each node with leak and compared without leak at is each time of a leak imposed in the node, the sensitivity matrix (Pudar and Liggett 1992) is obtained as follows:

$$S = \begin{pmatrix} \frac{\partial p_1}{\partial f_1} & \cdots & \frac{\partial p_1}{\partial f_n} \\ \vdots \cdots & & \vdots \\ \frac{\partial p_n}{\partial f_1} & \cdots & \frac{\partial p_n}{\partial f_n} \end{pmatrix} \tag{26.6}$$

where s_{ij} measured the effect of leak f_j on pressure at junction p_i. Sensitivity matrix S cannot easily calculated by analytically because the water pipe network is a complex problem and has nonlinear and non-explicit equation. The proposed work generates the sensitivity matrix by introducing the same leak in each node and measure the increment in the pressure. It implies n (number of node in network) number of node simulation for n number of pressure at node. Some of the sensor shows more sensitivity for any of the leak present in network. Thus, the normalisation of sensitivity matrix has done for the comparable information about the junction. The normalised sensitivity matrix is obtained by dividing the each element of each row of the sensitivity matrix to the corresponding maximum value of that row. This generates the normalise sensitivity matrix:

$$\overline{S} = \begin{pmatrix} \frac{s_{11}}{\sigma_1} \cdots \frac{s_{1n}}{\sigma_1} \\ \vdots \cdots \quad \vdots \\ \frac{s_{n1}}{\sigma_n} \cdots \frac{s_{nn}}{\sigma_n} \end{pmatrix} \tag{26.7}$$

where $\sigma_i = \max(s_{i1} \ldots\ldots, s_{\text{in}})$. It shows how a leak is most relevant to itself and maximum normalised sensitivity is shown in diagonal. The column of this matrix corresponds to nodes with leak and row corresponds to nodes with sensors.

The normalised sensitivity matrix used to formulate fault signature matrix (FSM). The element of the FSM equal to zero there is no fault or fault affected the pressure at node *i* and equal to 1 when fault effected pressure at node *i*.

Algorithm for binarised sensitivity matrix

Input ....
for each node i
    calculate the simulated pressure without leak P_{wtl}
    for each leak j
        calculate the simulated pressure with same leak at each node for a leak P_{wl}
        calculate the residual at each node of network for a leak at all node
        $s_{ij}=P_{wtl}-P_{wl}$
    end
end

for each row of matrix (sensor)
find the maximum value of this row (σ_i)

calculate $\bar{s}_{ij} = \dfrac{s_{ij}}{\sigma_i}$

if $\bar{s}_{ij} < s_{th}$

$\bar{s}_{ij}^{\,b} = 0$

else

$\bar{s}_{ij}^{\,b} = 1$

end

end
end

This algorithm shows that how a sensor linked to leaks. The matrix obtained by this algorithm is binary matrix in which at leak is present at any node the value is 1 otherwise 0.

Minimal sensor placement algorithm

A minimal sensor placement strategy is the configuration of sensors that minimise the total economical cost and considering to identify all fault in the system at the same time. A pipe network system considered as a graph network, where each edge represents the pipe of the network, and vertices are represented junctions such as sources and demands of the pipe network.

The algorithm starts with the binarisation of normalised sensitivity matrix as describe in Sect. 26.4. Each row of binarised sensitivity matrix corresponds to a location sensor at a junction and each column corresponds to leak at a junction. Thus, if an element of binarised sensitivity matrix comprises "1", it means that sensor is install at the corresponding row would able to detect the single leak associated to column of this element. For any particular distribution, a set of groups of indiscernible leaks appear, each group with n_i leaks. The aim of the minimal sensor placement algorithm is to minimise the sensors by minimising the set of leak having same signature. In the present work, sensors are minimised with the consideration that every column should present at one non-zero element.

Optimal sensor placement algorithm

Input: s_{th} is the binarisation threshold

$\overline{S}_{i,j}$ is the normalized matrix

$\overline{S}^{b}{}_{i,j}$ *binarised matrix*

$\overline{S}^{b}{}_{r}$ *binary matrix after row zero*

S_n *Numbar of signature*

S_{n1} *numbar of signature of* $\overline{S}^{b}{}_{r}$

s_{th}*=0.1: 0.01: 0.90*

for each s_{th}

if $\overline{S}_{i,j} > s_{th}$

$\overline{S}^{b}{}_{i,j} = 1$

If $\overline{S}_{i,j} < s_{th}$

$\overline{S}^{b}{}_{i,j} = 0$

Compute the number of signature for each binary matrix and save in an array

for each column of $\overline{S}^{b}{}_{i,j}$

decimal value(j)= bi2de($\overline{S}^{b}{}_{j}$*)*

if decimal value(j)= decimal value(j+1)

decimal value(j+1)=0

end

end

S_n *=non-zero of decimal value array*

For each $\overline{S}^{b}{}_{i,j}$

For each row

$\overline{S}^{b}{}_{r}$ *=make row zero of* $\overline{S}^{b}{}_{i,j}$

$S_{n1} = S_n$ *of* $\overline{S}^{b}{}_{r}$

If S_n *of* $\overline{S}^{b}{}_{r} < S_n$ *of* $\overline{S}^{b}{}_{i,j}$

Do not make zero that row
If any column of $\overline{S}^{b}{}_{r}$ *become zero decimal*
value
Do not make zero that row
End
End
End
Obtain sensor= Compute the non-zero row
Minimum number of sensors=min (obtain sensor)
end

26.5 Result and Discussion

A number of pipe networks are simulated for optimal sensor placement. Models are simulated using EPANET software. Model_1 as shown in Fig. 26.1 has 35 nodes and 68 pipes are simulated with a leak at each node and without leak at each node to calculate pressure discrepancies at each node. The total demand of model_1 is 746 LPS. The leak impose on each node is 3–4% of total demand of the network. The simulated leak for model_1 is 10 LPS.

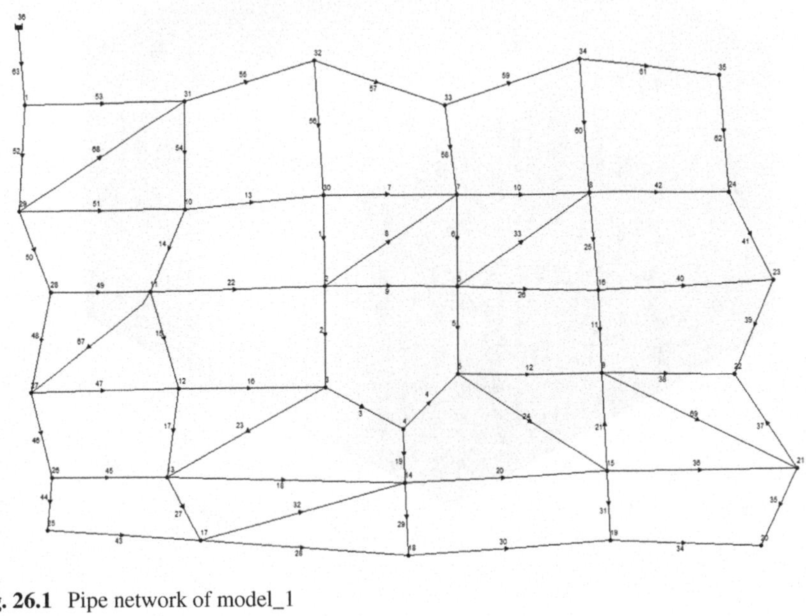

Fig. 26.1 Pipe network of model_1

The difference between the pressure with leak and without leak at each node produces the pressure sensitivity matrix. The sensors are placed at each node to obtain the pressure fluctuation. Some leaks show more sensitivity to the pressure fluctuation. That is why the pressure sensitivity matrix is normalised row by row to the maximum value of the same row. Figure 26.2 shows the normalised sensitivity matrix of model_1.

The normalised sensitivity matrix is binarised using different threshold (0.1:0.01:0.99) and saved. Each element of the binarised matrix is equal to "1" when the threshold is less or equal to the element, otherwise, its value will be zero. Element "1" of binary matrix shows the effect of leak at the respective nodes. Element of binary matrix "0" shows that there is no effect of leak at the respective position nodes.

Figures 26.3, 26.4, 26.5, 26.6 and 26.7 show the binarised matrix at 0.5, 0.75, 0.84, 0.96, and 0.99 threshold, respectively, for the model. Each element of normalised sensitivity matrix compared with these thresholds. Elements of the binarised sensitivity matrix are equal to "1" when the threshold is less or equal to the element and it is equal to 0 when the threshold is greater than the threshold. The vertical axis shows the node with sensors and the horizontal axis shows the nodes with leaks. Every

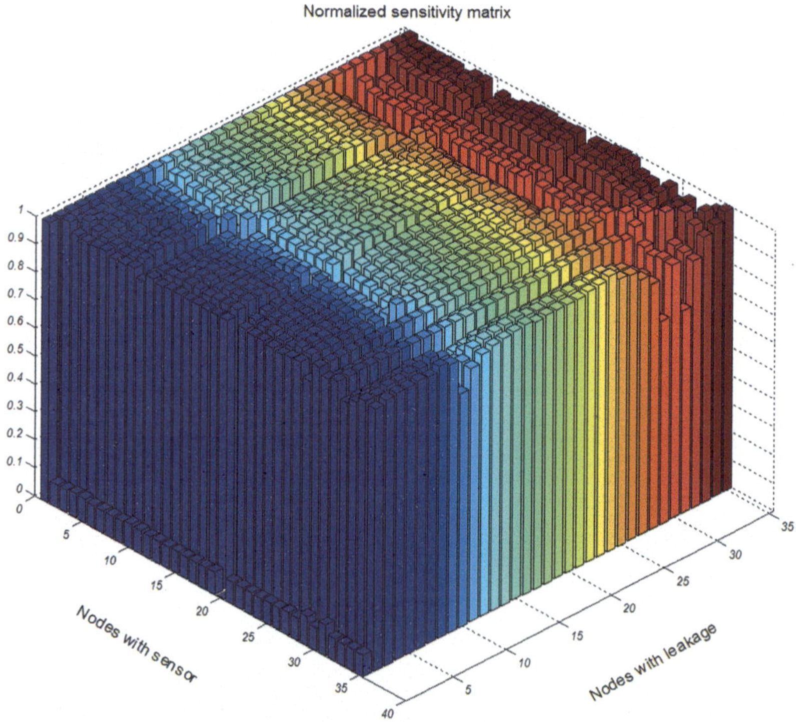

Fig. 26.2 Normalised sensitivity matrix for model_1

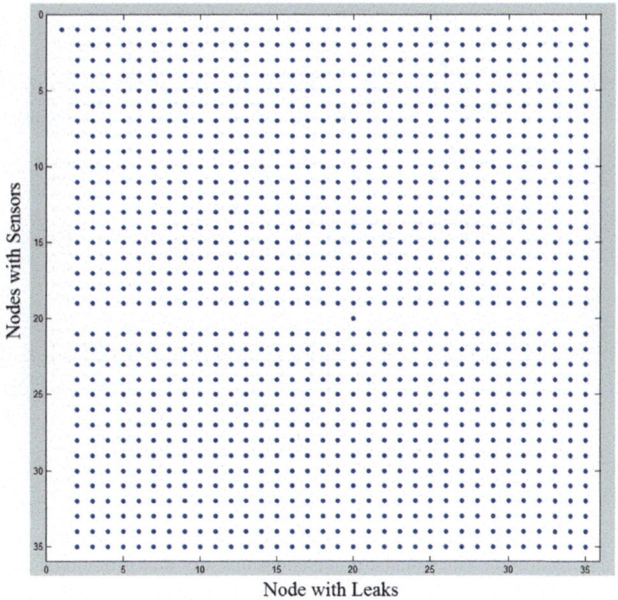

Fig. 26.3 Binarised with 0.5 threshold

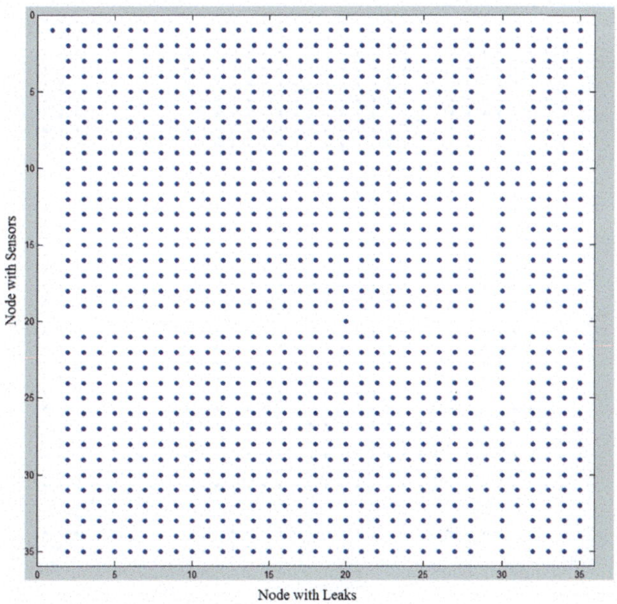

Fig. 26.4 Binarised with 0.75 threshold

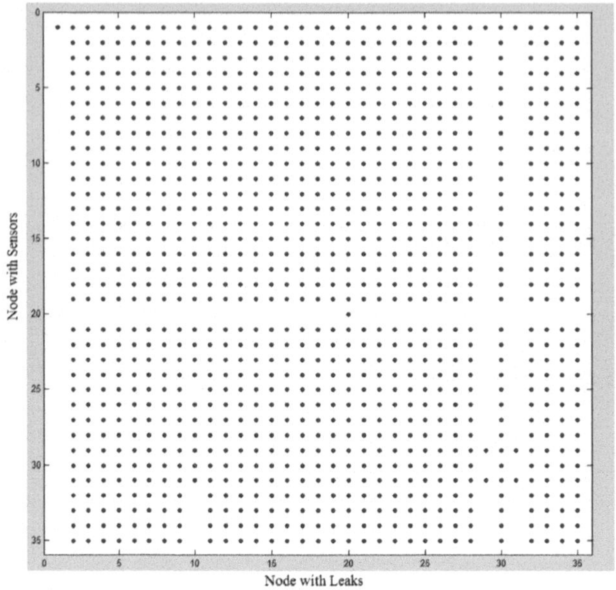

Fig. 26.5 Binarised with 0.84 threshold

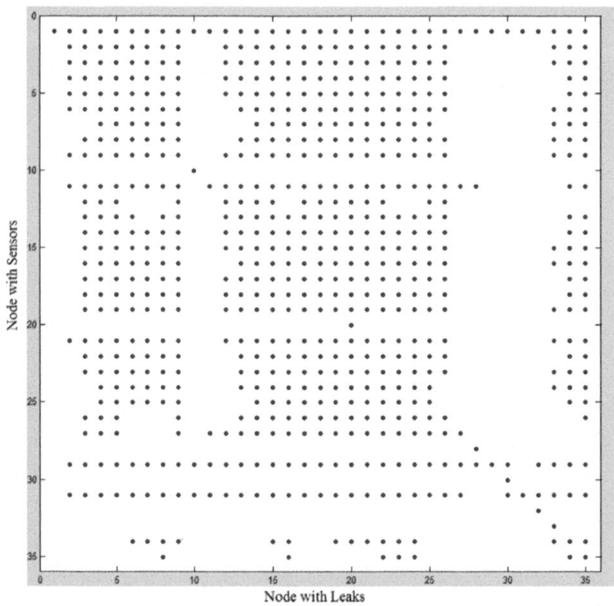

Fig. 26.6 Binarised with 0.96 threshold

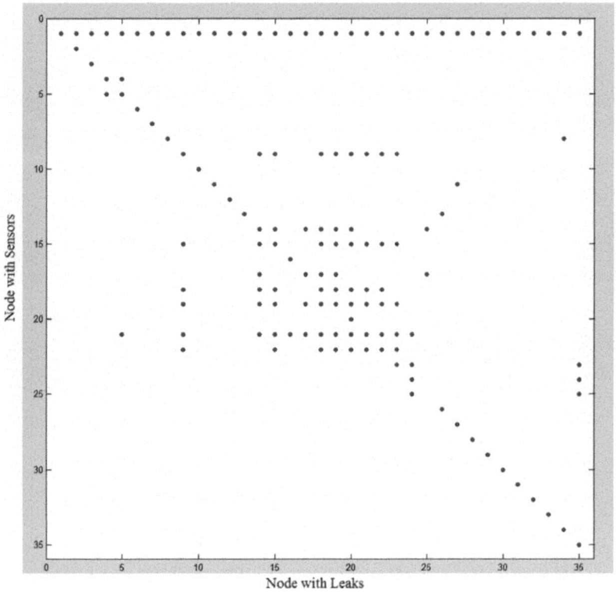

Fig. 26.7 Binarised with 0.99 threshold

column in the binarised matrix shows leak with node and its leak signature at respective node. When the value of threshold is less, then the frequency of number of 1's in the binarised matrix is more than the 0's and vice versa. Leak signature is count after binarisation of matrix of each leak. Leak signatures are the different combination of 0 and 1in each column for each node with leak. If two or more columns show the same kind of combinations, then they count the same leak signature.

The minimisation of sensors at nodes is based on the row deletion. The deletion of row is depends on the number of signature of leaks. Row deletion starts with first row with the assumption that the number of the signature of leaks after deletion are same and every column has at least 1 non-zero value. It is also required that the detectability and isolability are checked at the same time.

Figures 26.8, 26.9, 26.10, 26.11 and 26.12 show the minimum sensor required at different thresholds. Figure 26.13 shows the minimum sensor required at threshold 0.5. This configuration is obtained from binarised matrix by deleting the row. While deleting the rows, keep in mind that the leak signature remains the same as before deleting the rows. While deleting the row, every column should at least one non-zero value.

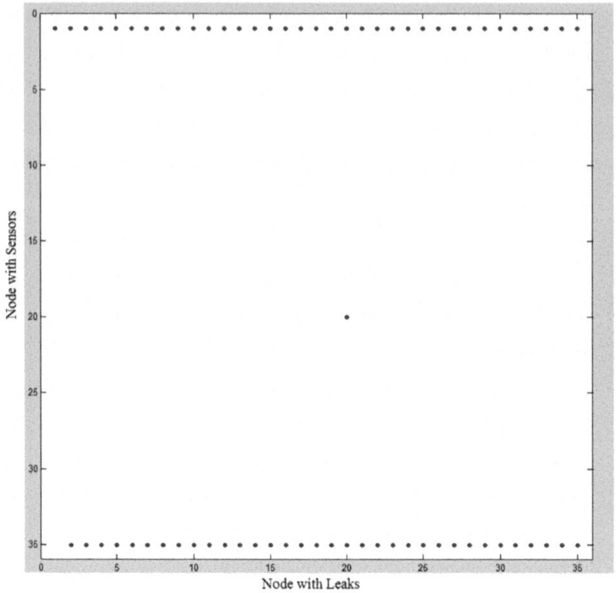

Fig. 26.8 Three sensors at 0.5 threshold

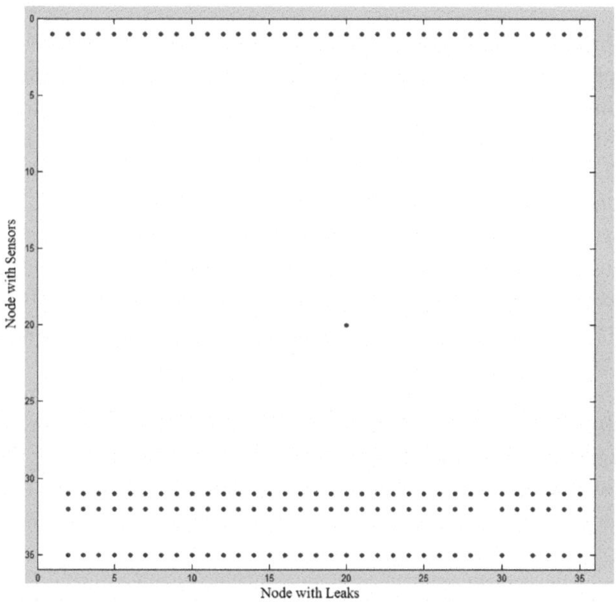

Fig. 26.9 Five sensors at 0.75 threshold

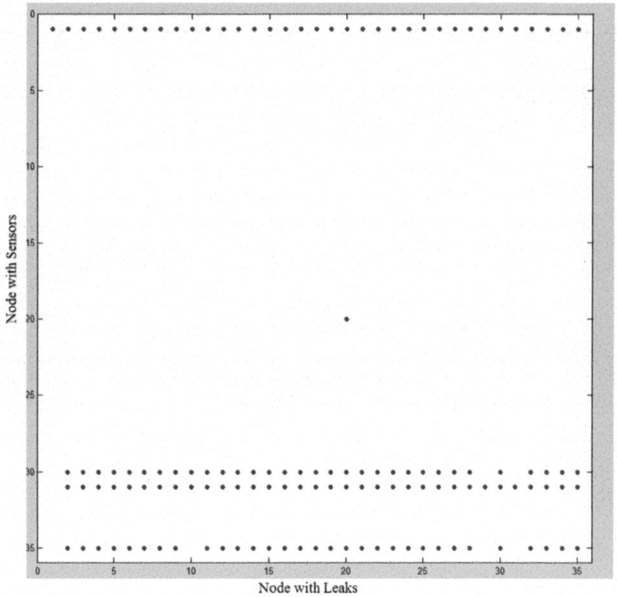

Fig. 26.10 Sensors at 0.84 threshold

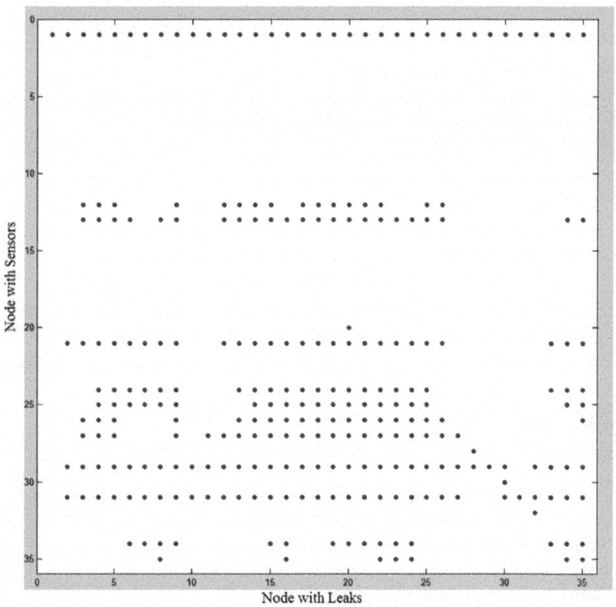

Fig. 26.11 Sensors at 0.96 threshold

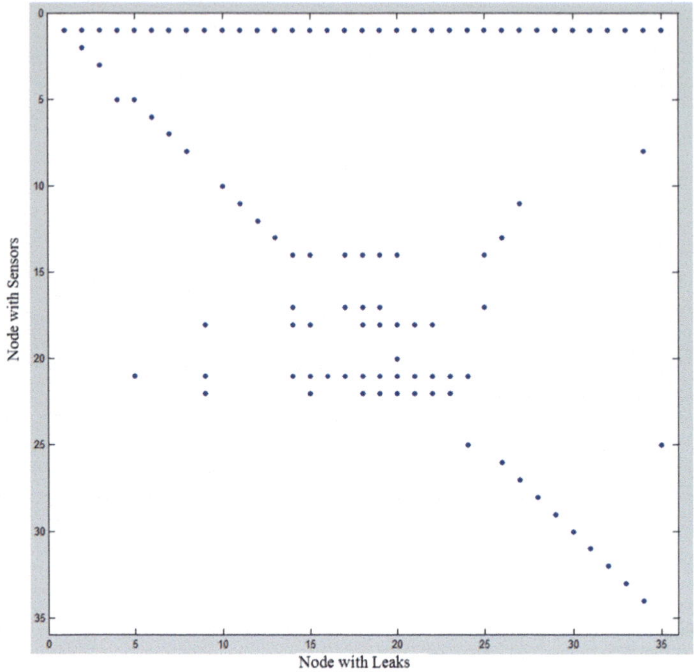

Fig. 26.12 Sensors at 0.99 threshold

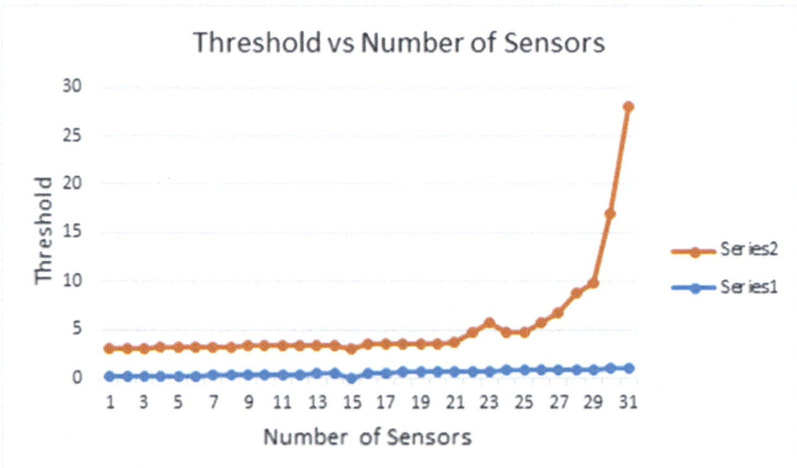

Fig. 26.13 Threshold versus number of sensors

Figures 26.8, 26.9, 26.10, 26.11 and 26.12 show how many sensors are required and sufficient for detecting all the leaks at different 0.5, 0.75, 0.84, 0.96, and 0.99 thresholds. Sensors required to detect the leaks are 3, 5, 5, 16, and 27 at different 0.5, 0.75, 0.84, 0.96, and 0.99 thresholds, respectively. It has been observed that at threshold 0.5 only 3 sensors are required for detecting all the faults in the network.

Figure 26.13 shows an increasing train of thresholds with respect to number of sensors. An example of same network is tested with six leaks at different positions of the node of the network. The positions of leaks are assumed at node numbers 1, 4, 9, 17, 28, and 33 in the network. The test result in Fig. 26.14 shows that only two sensors are required to detect the same leaks at threshold 0.5. The whole procedure repeated for the 6 leak positions in the network. Figure 26.14, 26.15, and 26.16 show the test result of example. It shows different locations of leak and can detect only by two sensors. From Fig. 14, one can conclude that two sensors which are already obtained by proposed methodology can detect all leaks in the present example.

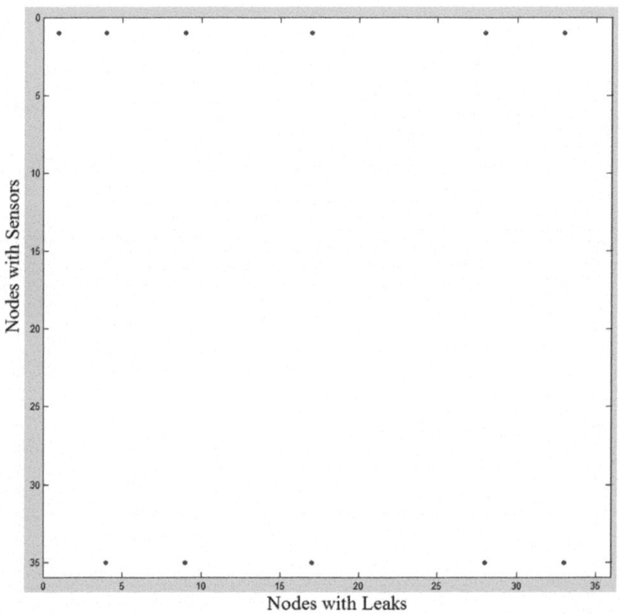

Fig. 26.14 Sensors at 0.5 threshold of test result

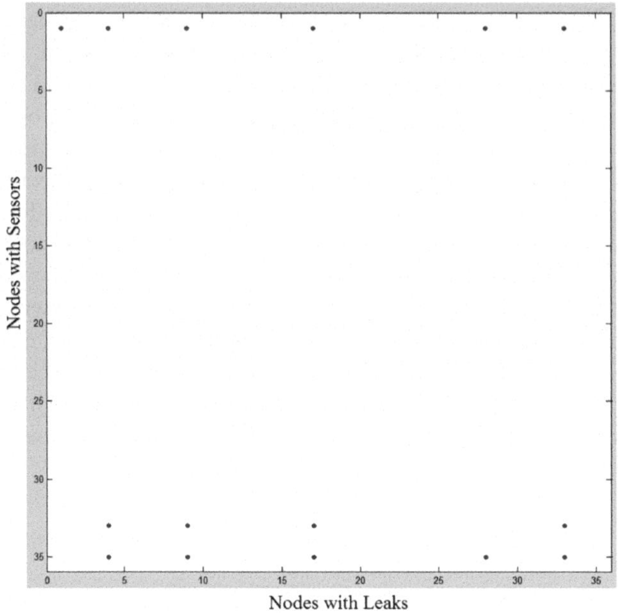

Fig. 26.15 Sensors at 0.9 threshold of test result

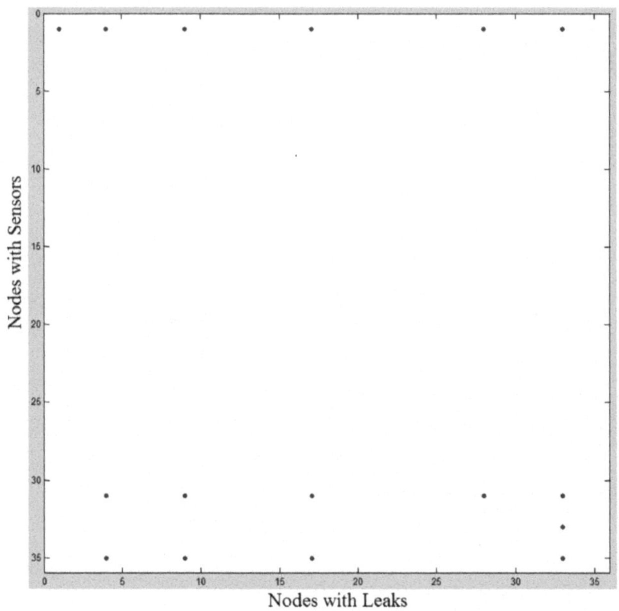

Fig. 26.16 Sensors at 0.96 threshold of test result

26.6 Conclusion

A minimal sensor placement method based on the pressure sensitivity analysis of nodes in a looped pipe network has been implemented. The minimal sensor placement methodology is developed using model-based diagnosis.

In order to obtain maximum isolability with the reasonable number of sensors, a minimum number of sensor placement strategies have been proposed. The objective was to minimise the number of node with sensors, detecting the same leak with each sensor. The information about leakage has been obtained by pressure sensitivity matrix. Pressure sensitivity matrix has been obtained using EPANET software. Leaks have been detected successfully by different number of sensors depending on the thresholds.

References

Blesa J, Puig V, Saludes J, Vento J (2010) Leak detection, isolation and estimation in pressurized water pipe networks using LPV models and zonotopes. IFAC Proc Vol 43(14):36–41

Casillas MV, Garza-Castañón LE, Puig V (2012) Extended-horizon analysis of pressure "sensitivities for leak detection in water distribution networks". In: 8th IFAC symposium on fault detection, super vision and safety of technical processes. Elsevier, pp 570–575

Casillas MV, Garza-Castañón LE, Puig V (2015) Sensor placement for leak location in water distribution networks using the leak signature space. IFAC-PapersOnLine 48(21):214–219

Cheng W, He Z (2010) Calibration of nodal demand in water distribution systems. J Water Resour Plann Manag 137(1):31–40

Colombo A, Lee P, Karney B (2009) A selective literature review of transient-based leak detection method. J Hydro-Environ Res 2:212–227

Covas D, Ramos H (2001) Hydraulic transients used for leak detection in water distribution systems. In: 4th international conference on water pipeline systems, pp 227–241, BHR Group

Farley M, Trow S (2003) Losses in water distribution networks: a practitioner's guide to assessment, monitoring and control, IWA, London, UK

Ferrandez-Gamot L, Busson P, Blesa J, Tornil-Sin S, Puig V, Duviella E, Soldevila A (2015) Leak localization in water distribution networks using pressure residuals and classifiers. IFAC-PapersOnLine 48(21):220–225

Gertler JJ (1998) Fault detection and diagnosis in engineering systems, Marcel Dekker

Goulet J-A, Coutu S, Smith IFC (2013) Model falsification diagnosis and sensor placement for leak detection in pressurized pipe networks. J Adv Eng Inf 27:261–269

Kepler A, Covas D, Reis L (2011) Leak detection by inverse transient analysis in an experimental PVC pipe system. J Hydro-Inf 13:153–166

Mashford J, De Silva D, Marney D, Burn S (2009) An approach to leak detection in pipe networks using analysis of monitored pressure values by support vector machine. In: Third international conference on network and system security, pp 534–539

Nicolini M, Patriarca A (2011) Model calibration and system simulation from real time monitoring of water distribution networks. In: 3rd international conference on computer research and development (ICCRD), vol 1, pp 51–55

Pudar RS, Liggett JA (1992) Leaks in pipe networks. J Hydraulic Eng ASCE 118(7):1031–1046

Pérez R, Puig V, Pascual J, Quevedo J, Landeros E, Peralta A (2011) Methodology for leakage isolation using pressure sensitivity analysis in water distribution networks. Control Eng Pract 19(10):1157–1167

Ragot J, Maquin D (2006) Fault measurement detection in an urban water supply network. J Process Control 16:887–902

Savic DA, Kapelan Z, Jonkergouw P (2009) Quo vadis water distribution model calibration? Urban Water J 6(1):3–22

Shu S, Zhang D (2010) Calibrating water distribution system model automatically by genetic algorithms. In: International conference on intelligent computing and integrated systems (ICISS), pp 16–19

Verde C (2001) Multi-leak detection and isolation in fluid pipelines. Control Eng Pract 9(6):673–682

Wu ZY, Farley M, Turtle DDS, Mulay M, Boxall J, Mounce S, Kleiner Y, Kapelan Z (2011) Water loss reduction. Bentley Systems Publications, Exton, Pennsylvania, USA

Chapter 27
Rainwater Harvesting System Planning for Tanzania

Msafiri Mussa Mtanda, Sakshi Gupta, and Deepak Khare

Abstract Water scarcity is a challenge in many parts of the world including Tanzania. Collecting rainwater wherever it falls from the sky should be considered as valuable technique in areas struggling to cope with potable water needs. The current water shortage in Arusha, Tanzania, is badly affecting the living standards due to the increasing water demands of a growing population. In this study, information based on the context of average number of household, land use, water use and mean monthly rainfall is used in planning and design of the system. The present paper demonstrates analysis for designing a domestic rainwater harvesting system, which identifies the relationship between rainwater potential, demand and storage capacity by bridging demand–supply gap. Rainwater harvesting systems can be adopted where municipal water supply systems have failed to meet community's requirements.

Keywords Rainwater harvesting system · Rainwater potential · Storage capacity · Arusha · Tanzania

27.1 Introduction

Rainwater harvesting is a technique of collecting rainwater that runs off from rooftops, parks, roads, open grounds, etc., and stored into natural reservoirs or tanks, or the infiltration of surface water into subsurface aquifers before it is lost as surface runoff

M. M. Mtanda (✉) · D. Khare
Department of Water Resource Development and Management, Indian Institute of Technology Roorkee, Roorkee 247667, Uttarakhand, India
e-mail: mtanda18@yahoo.com

D. Khare
e-mail: deepak.khare@wr.iitr.ac.in

S. Gupta
Graphic Era (Deemed to be) University, Department of Civil Engineering, Dehradun 248002, Uttarakhand, India
e-mail: sakshilibran@gmail.com

A. Pandey et al. (eds.), *Water Management and Water Governance*, Water Science and Technology Library 96,
https://doi.org/10.1007/978-3-030-58051-3_27

for future use to meet the demands of human consumption or activities. According to John Mbugua 1999, "*rainwater harvesting*" can be undertaken through several ways by capturing the rainfall where it falls either in villages, farms and local catchments or towns from watershed management, building roof-tops and taking measures to keep it clean and store that water to be consumed when required. Rainwater harvesting is a good fit option which is technically feasible, easy to operate and affordable even to poor communities (Hatibu and Mahoo 2000; Manyama 2006).

27.1.1 Background

Rainwater harvesting almost practised worldwide with different purposes. According to Baguma, rainwater in Malaysia is used for commercial domestic purposes including car washing in parking lots and also in South Australia used to supplement drinking water supply for nearly 40% of households (Baguma 2012).

27.1.2 The Case of Arusha District

The site visit around the town (Arusha Urban District, Tanzania) has revealed that the current water shortage situation is seriously affecting the residents and the urban centre is experiencing a noteworthy blast in land improvement, new enterprises, neighbourhood and worldwide associations and higher learning establishments. The Arusha Urban Water Supply Authority is hardly working to identify new water sources including recovering nine old wells that were dug in 1980 and however have never been utilized since then. The Authority relies upon filling its water reserves by tapping from natural springs and 16 wells of which are now drying up. The Authority can produce approximately 45,000,000 L of water per day which can simply solve only 50%, whereby total population needs 93,270,000 L per day (AUWSA 2012). According to Water Authority Director, there is a need of adjusting the city's system, including repairing the matured pipelines and tapping alternative sources in order to fully eradicate water issues in Arusha District (NEPAD Water Centres 2014).

27.2 Materials and Methods

27.2.1 Study Area

Arusha Urban District (or Arusha City Council) is one of the seven districts of the Arusha Region Northeast Tanzania, within East Africa.

Geographically, it lies on Latitude 3°22′21″ S, and Longitude 36°41′40″ E. It is bordered to the east by Meru District and by Arusha Rural District surrounding the south, west and north as shown in Fig. 27.1. Despite Arusha being near to the equator, annual temperature differs from 13 and 30 °C with an average around 25° of which cool dry air is prevailing for much of the year which keeps temperatures comparatively low and controls humidity (Deluxe and Mkomwa 2012). Arusha has an elevation of 1387 m (approximately equal to 4600 feet) on the southern slopes of the fifth-highest mountain in Africa, Mount Meru (height of 4562.13 m that is 14,968 feet). According to the 2012 Tanzania National Census, Arusha Urban District has a population of 416,442 with an approximation of 4% growth per year (National Bureau of Statistics 2013; NBS 2014). Master plan covers 267 km^2 with density

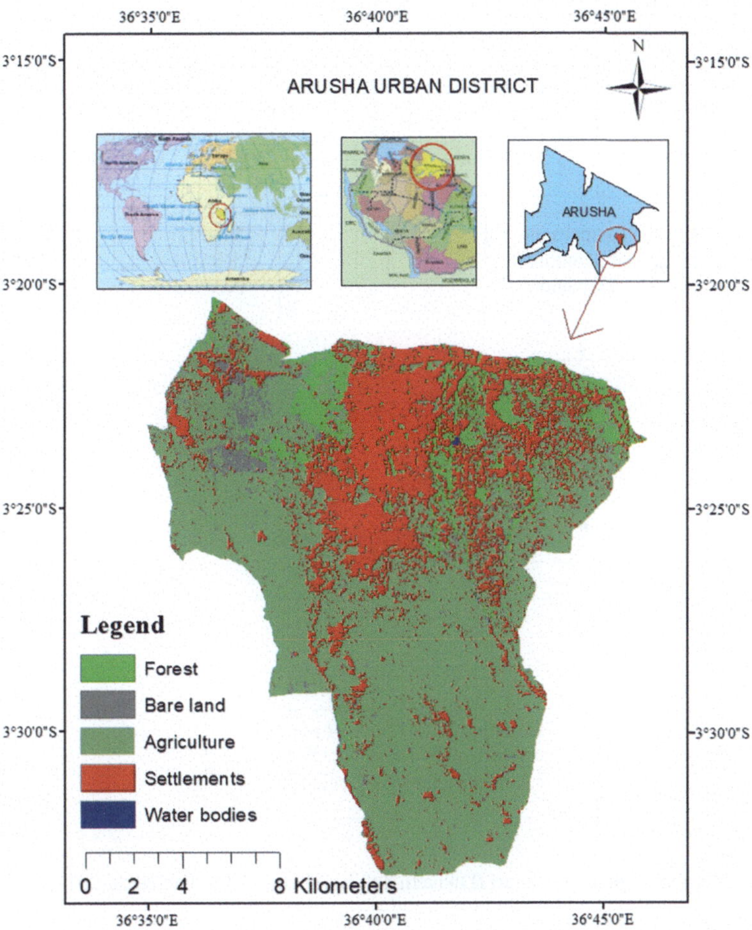

Fig. 27.1 Location and land use map of study area

of 1560 per square kilometre, while more than 70% of the city area falls under unplanned settlements (National Bureau of Statistics 2013).

27.2.2 Methodology

The present study is based on field visit done during May 2017, to acquire primary data of water scarcity status of the study region. The secondary data regarding the availability of rainfall, size and type of roof dominated on the area, population, water demand, etc., were collected from authorized respective departments and analysed with regard to methodological flow chart (Fig. 27.2) (Dwivedi and Bhadauria 2009; Panhalkar 2011).

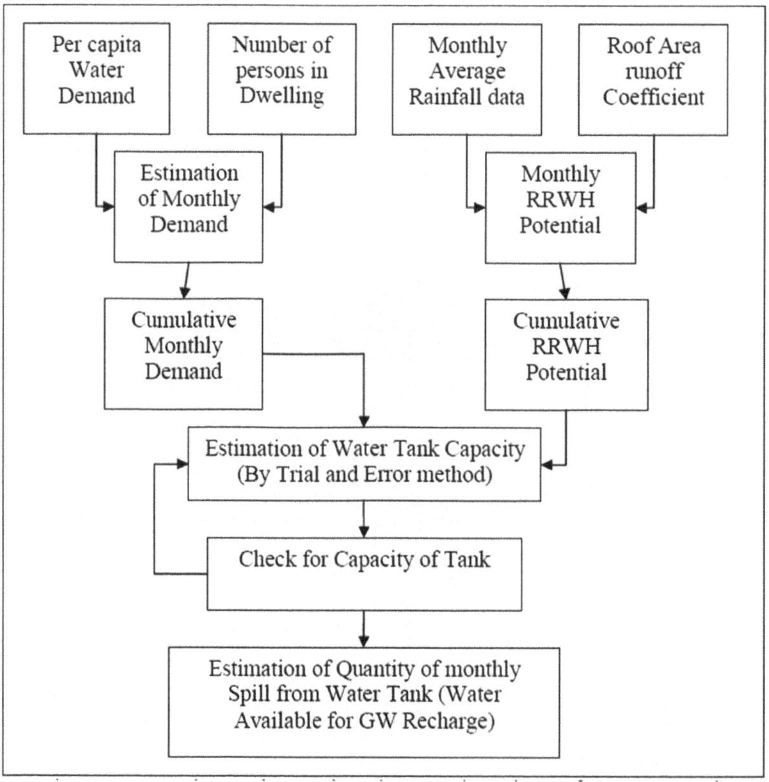

Fig. 27.2 Methodological flow chart (Dwivedi and Bhadauria 2009; Panhalkar 2011)

27.2.3 Roof-Top Rainwater Harvesting System Design and Planning

In a single household RRHW system, rainwater which falls from the roof-top of the building is conveyed to a storage tank suitably located on or under the ground, through a system of semi-circular channels of PVC or galvanized iron; so that it can be used during the period of water scarcity for domestic purpose. Usually this system is designed for drinking and cooking needs of an individual household level.

Slope of the roof tells how fast water runoff during a rain event will. Either it will shed runoff slowly or quickly depending on the pitch angle of a given particular roof. When angle of pitch increases, it causes water to move fast and reduces the risk of contamination to remain on the catchment surface area (Manyama 2006; Gupta and Khare 2016).

Calculation of roof catchment area is also referred to as the action of sizing a catchment area, whereby the "footprint" of the roof in an individual household is the base of calculating catchment size by finding the sum of built up area and its roof overhang. Therefore, the roof area as catchment will determine the probable amount of rainwater that can be harvested (Dwivedi and Bhadauria 2009; Pande and Telang 2014; Texas Water Development Board 2006). The catchment size ranges from 25 to 150 m^2, due to varying of the actual building surface area among family to family, house to house and their living standards dominated on the study area. Average household size is 4.0 based on 2012 census (National Bureau of Statistics 2013). This value is used to establish domestic water demand for a single household per day per month and per year.

27.2.4 Storage Capacity

The aspect of rainwater harvesting system design is very important to estimate minimum volume of storage for proper functioning of system (Gupta and Khare 2016; Pande and Telang 2014). It involves sizing of the rainwater tank required to store enough water to satisfy the appropriate household user's demand. The minimum required volume of storage tank is a function of many variables like catchment surface area, rainfall distribution and coefficient of runoff fixed in the rainwater supply with regard to demand pattern.

27.2.5 Potential of Roof-Top Rainwater Harvesting

The roof-top potential rainwater is the quantity of rainwater harvested from all the rainy days to a particular roof within a period of one year. In general, potential RRWH is normally known as the annual roof-top yield of which it is referred to as the mathematical relationship between mean annual rainfall and size of the roof by considering to the runoff loss factor of less than 20% (Hatibu and Mahoo 2000; Manyama 2006).

Quantification of RRWH potential depends upon the total amount of rainfall, type of roofing materials and roof area (Dwivedi and Bhadauria 2009; Lizárraga-Mendiola et al. 2015; Panhalkar 2011). It is simply the product of annual rainfall and roof area, and it can easily be calculated by adopting the following empirical relationship; $\{= R * A * \mathrm{Cr}\}$ (Dwivedi and Bhadauria 2009; Panhalkar 2011), whereby $S =$ rainwater harvesting potential (m^3), $R =$ mean annual rainfall (mm), $A =$ catchment area (m^2) and $\mathrm{Cr} =$ runoff coefficient.

27.2.6 Estimation of Water Tank Capacity and Groundwater Recharge

An estimation sample of rainwater tank capacity with respect to dominated rooftop area to fulfil a single household drinking and cooking water demands has been calculated, by adopting the method proposed by Dwivedi and Bhadauria (2009) and Panhalkar (2011), as it is shown in Fig. 27.2. The quantity of rainwater obtained that exceeds the actual tank capacity is known as the overflow and used for recharging aquifer (Of et al. 2015; Hatibu and Mahoo 2000).

27.3 Results and Discussion

27.3.1 Land Use

Land use practices receive considerable attention across the whole world including Arusha region in Tanzania. Water Development Division defines land use concerns, the benefits attained from land activities, its management and natural environmental modifications done by human being into built environment like commercial and residential land fields for recreation, passage, pastures, etc. Currently, about 67.7% of the study area is considered arable farmland, with 2.9% in bare land, 5.1% forests and woodland and 23.9% settlement areas (Fig. 27.1).

27.3.2 Rainfall Pattern

The meteorological data from five rainfall stations have been averaged for 40 years (1970–2009) to identify rainfall patterns. The results show that the region exhibits two main seasons of rainfall, with long rains commencing from January to May and

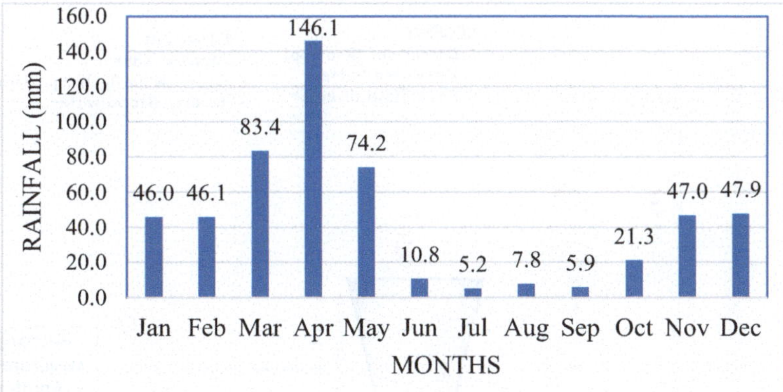

Fig. 27.3 Mean monthly rainfalls for five stations of the study area

the short rain from October to December. The amount of rainfall during the long rain months was observed to be higher than that of the short rain months for all the five stations. The wettest month for all the five stations was April followed by March and May for the long rain (*masika*) season (Fig. 27.3). There is very little rainfall during dry season commencing from June to October of which water becomes scarce. Mean rainfall amount during March, April and May has been found to be 83.4 mm, 146.1 mm and 74.2 mm, respectively, for long rain season as per Fig. 27.3. Similarly, for short rain season, mean rainfall amount during October, November and December has been observed to be 21.3 mm, 47 mm and 47.9 mm, respectively. The observations on Fig. 27.3 show that the study area experiences bimodal rainfall season, also revealed by other studies done. Hence, potential rainwater can be harvested during these two seasons.

27.3.3 Roof-Top Rainwater Harvesting System Design and Planning

Figure 27.4 recommends basic components of roof-top rainwater harvesting system which consists of a collection area (roof catchments), filter unit, cistern or storage facility and a conveyance system having gutters, down pipes and first flush diverter to prevent the entry of first monsoon shower in the storage tank which may contain roof contaminants.

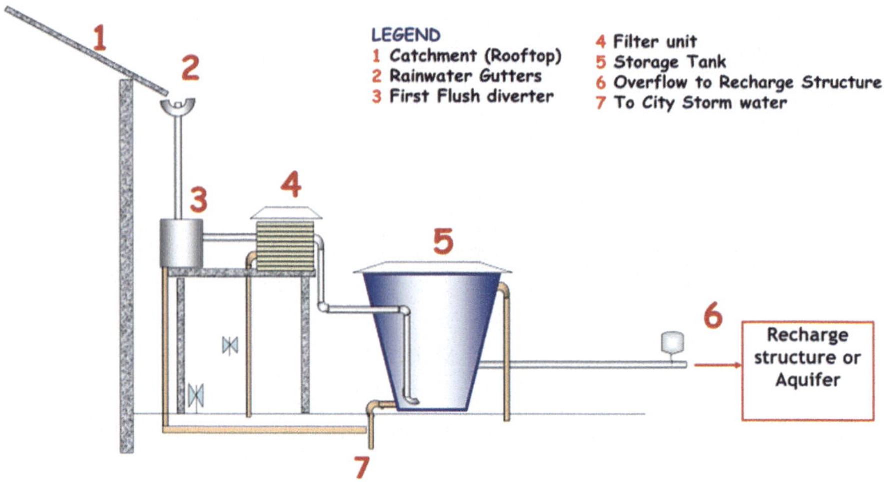

Fig. 27.4 Proposed schematic roof-top RWH system design flow diagram

27.3.4 Roofing Material

According to Tanzania Housing Board (THB) (NBS 2014; Takwimu 2010), 73.7% of houses were roofed with corrugated iron sheets, 17.4% of houses were roofed with dried grass/leaves, 7.8% of houses were roofed with mud and leaves, 0.5% of houses were roofed with tiles, 0.2% of houses were roofed with plastic material and same 0.2% of houses were roofed with asbestos, whereas concrete and box tent are used to roof 0.1% of houses each, as shown in the pie chart below. Roofing materials chart shown in Fig. 27.5 reveals that the most used type is corrugated iron sheets, and therefore, it is the most considered roof catchment material for rainwater harvesting

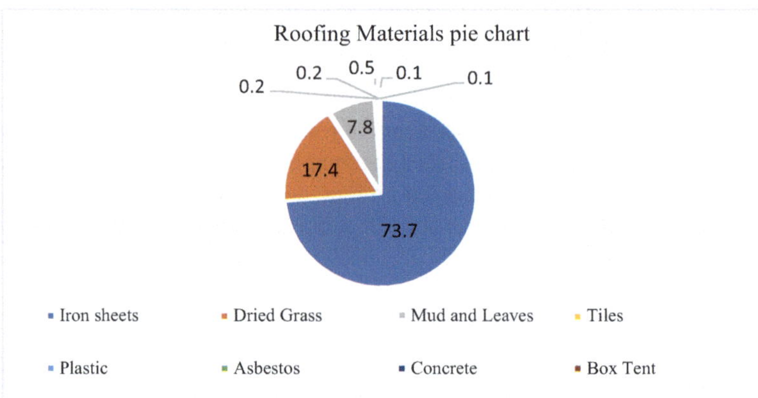

Fig. 27.5 Roofing material percentage distribution over the study area, Arusha region

in the study area. The literature also recommends a metal roof because they easily shed off contaminants within short period of time.

27.3.5 Estimation of Water Tank Capacity and Groundwater Recharge

An estimated sample of rainwater tank capacity for 50 m^2 roof-top area is shown in Table 27.1. Drinking and cooking water demand of four persons in a single family at 10 L per capita per day (Chandel and Sharma 2013), for each month, has been calculated. The demand at10 L per capita per day ranges from 1120 up to 1344 L per month as per Table 27.1. The cumulative demand of such family is also calculated in order to find out the total yearly water requirement of which it found to be 14,600 L. Similarly, a domestic water demand to fulfil a single family of four persons at 80 L per capita per day as per United Nations minimum requirement guide, for each month, has also been calculated. The demand has been found to be between 8920 and 9920 L per month. The overall cumulative domestic water requirement at 80 L per capita per day was found to be 116,800 L as total yearly water demand for a single family at 80 L per capita per day. The monthly roof-top rainwater harvesting potential of an individual house is estimated using mean monthly rainfall data of the study region by applying Gould and Nissen equation and presented on Fig. 27.6. The calculation on Table 27.1 shows monthly RRWH potential ranging from lowest 240 L for July to highest 7000 L for April month. The total RRWH potential is 24,960 and only 58.5 per is sufficient to satisfy a single household yearly water requirement at 10 L per capita per day. The graph of cumulative domestic water demand and cumulative potential RRWH for a complete cycle of months are shown in Fig. 27.7 with respect to Table 27.1, and it can be seen that potential RRWH is relatively high as compared to the cumulative domestic water demand.

End of each month quantity status of probable harvested rainwater available in the storage water tank is estimated from the difference between month-wise RRWH potential and water demands (Table 27.1). Monthly status of rainwater quantity available in the storage water tank was estimated to be 4000 L capacity for 50 m^2 roof areas as per Table 27.1, as shown in Fig. 27.6.

Dry season is between June to October, whereby rainfall is practically nothing in study area as per Table 27.1; hence, it is not enough to satisfy the total domestic water requirement throughout the year from potential RRWH of which water becomes scarce. Even during wet months, water demand cannot be met from roof having an area less than 75 m^2. Table 27.2 reveals that the gross water demand for wet months can be met from roof having area 75 m^2 and above.

Table 27.1 Estimating water tank capacity to fulfil household drinking and cooking water demand by DRWH and quantity of monthly water available for groundwater recharge from roof-top area 50 m^2

1	2	3	4	5	6	7	8	9	10	11	12	13	14	15
Months	**Days**	**Averaged Rainfall**	**Water demand (cooking & drinking) 10Lcpd @4soul**	**Cumulative water demand 10Lcpd @4soul**	**Overall water demand 80Lcpd**	**Cumulative Overall @80Lcpd demand**	**RWH potential**	**Cumulative RWH Potential**	**Monthly Potential Demand @10Lpcd**	**Surplus Column9 – Column 5**	**COMMENTS**	**End of Month Storage**	**DRWH Available for Ground Water Recharge**	**% of Total Demand Can Met from DRWH**
Jan	31	47	1240	1240	9920	9920	1880	1880	640	640	The drinking & cooking demand can be met from DRWH. The size of the storage tank = 3840~4000 Liters. The surplus water for GW Recharge = 10360 Litres.	640	0	18.95
Feb	28	35	1120	2360	8960	18880	1400	3280	280	920		280	0	15.63
Mar	31	81	1240	3600	9920	28800	3240	6520	2000	2920		2000	0	32.66
Apr	30	175	1200	4800	9600	38400	7000	13520	5800	8720		4000	4720	72.92
May	31	108	1240	6040	9920	48320	4320	17840	3080	11800		4000	3080	43.55
Jun	30	13	1200	7240	9600	57920	520	18360	-680	11120		3320	0	5.42
Jul	31	6	1240	8480	9920	67840	240	18600	-1000	10120		2320	0	2.42
Aug	31	8	1240	9720	9920	77760	320	18920	-920	9200		1400	0	3.23
Sep	30	7	1200	10920	9600	87360	280	19200	-920	8280		480	0	2.92
Oct	31	23	1240	12160	9920	97280	920	20120	-320	7960		160	0	9.27
Nov	30	60	1200	13360	9600	106880	2400	22520	1200	9160		4000	1200	25.00
Dec	31	61	1240	14600	9920	116800	2440	24960	1200	10360		4000	1200	24.60
TOT									**-3840**	**10360**			**10200**	

Note RRWH = Roof-top rainwater harvesting; DRWH = Domestic rainwater harvesting; RWH = Rainwater harvesting

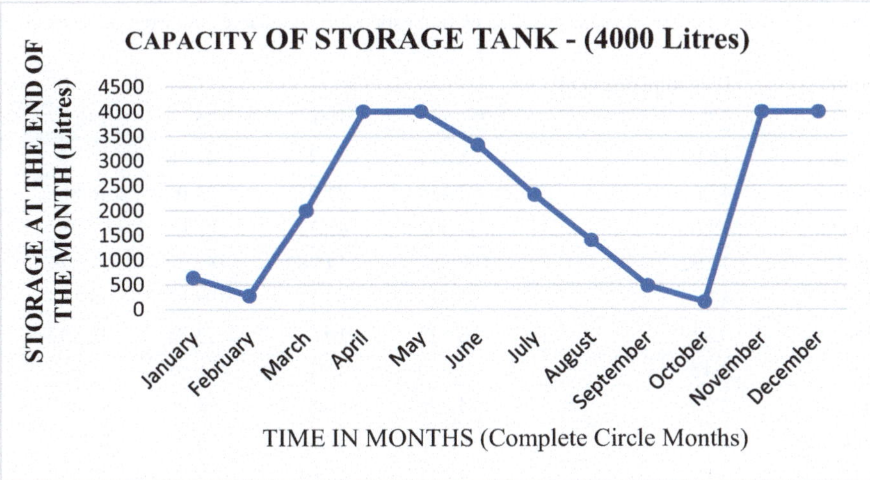

Fig. 27.6 Month-wise rainwater availability in storage tank throughout the year

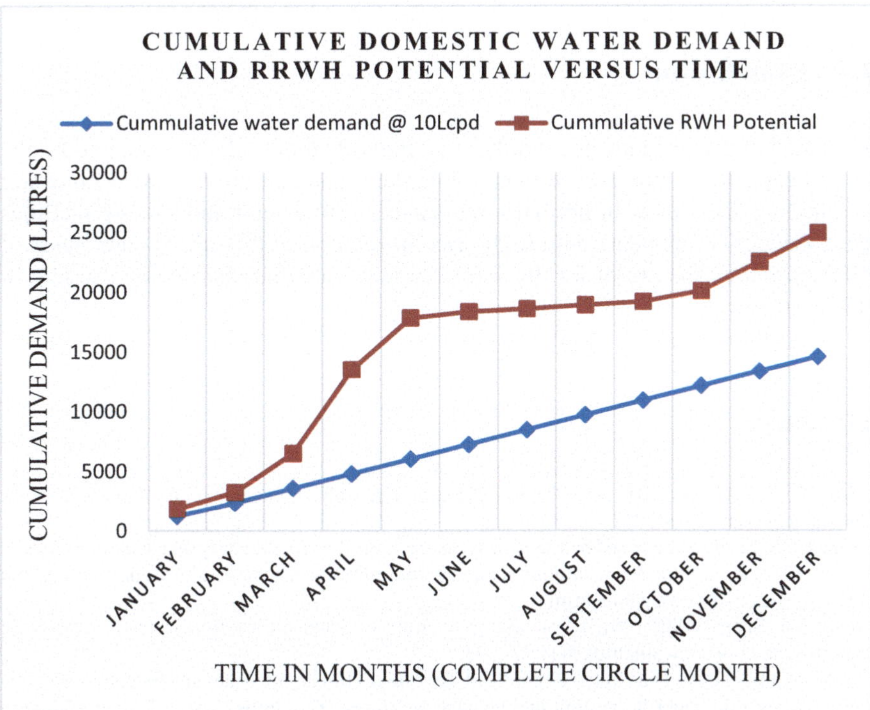

Fig. 27.7 Cumulative domestic water demand and cumulative potential RRWH for complete cycle months

Table 27.2 Water per cent demand fulfilled by domestic rainwater harvesting from different roof sizes

Month	Roof-top area in m^2				
	50	75	100	125	150
January	18.95	28.427	37.9	47.38	56.85
February	15.63	23.438	31.25	39.06	46.88
March	32.66	48.992	65.32	81.65	97.98
April	72.92	109.38	145.83	182.29	218.75
May	43.55	65.323	87.1	108.87	130.65
June	5.42	8.125	10.83	13.54	16.25
July	2.42	3.629	4.84	6.05	7.26
August	3.23	4.8387	6.45	8.06	9.68
September	2.92	4.375	5.83	7.29	8.75
October	9.27	13.911	18.55	23.19	27.82
November	25	37.5	50	62.5	75
December	24.6	36.895	49.19	61.49	73.79

27.4 Conclusion

The sustainability of rainwater supply in an individual household is assured by the size and capacity of the storage tank, while the quality of the harvested rainwater is assured by filter unit. In normal circumstances, filter unit and storage facilities are the critical components with high price in relation to the other components of RWHS, whereby the use of locally available materials diminishes the general cost of the systems.

References

AUWSA (2012) Arusha urban water supply and sewerage authority annual report and accounts for the financial year 2011/2012

Baguma D (2012) Rainwater and health in developing countries: a case study on Uganda. Retrieved 16 Jan 2018, from https://unu.edu/publications/articles/rainwater-and-health-in-developing-countries-a-case-study-on-uganda.html

Chandel RS, Sharma MR (2013) Potential and limits of domestic rooftop water harvesting in Hamirpur area of Shiwalik hills 5(1):97–102

David Manyama (2006) The United Republic of Tanzania, Ministry of Water and Irrigation (MoWI) guideline for rain water harvesting in Tanzania. Retrieved from https://www.academia.edu/24809528/

Deluxe ÌÌ, Mkomwa CN (2012) AICC Looking for a meeting venue …, (16)

Dwivedi AK, Bhadauria SS (2009) Domestic rooftop water harvesting—a case study. J Eng Appl Sci 4(6):31–38

Gupta S, Khare D (2016) Quality of roof-top harved rainwater stored in polythelene container. In DVR Vikas Garg, Prof. Dr. Vijay P. Singh (Ed.), Development of water resources in India. Springer International Publishing. Retrieved from www.springerprofessional.de/en/development-of-water-resources-in-india
Hatibu N, Mahoo HF (2000) Rainwater harvesting for natural resources management—a planning guide for Tanzania. Relma.
Lizárraga-Mendiola L, Vázquez-Rodríguez G, Blanco-Piñón A, Rangel-Martínez Y, González-Sandoval M (2015) Estimating the rainwater potential per household in an urban area: case study in Central Mexico. Water 7(9):4622–4637. https://doi.org/10.3390/w7094622
National Bureau of Statistics (2013) 2012 Population and housing census; Population Distribution by Administrative Areas. National Bureau of Statistics
NBS (2014) The 2002 population and housing census: basic demographic and socio-economic profile. National Bureau of Statistics, 44
NEPAD Water Centres N (2014) Tanzania; AUWSA needs 5 years to rework water system. Retrieved 20 July 2017, from https://nepadwatercoe.org/tag/arusha-city/
Of F, Tunnel S, Flood I, Flow T, In M, Lumpur K (2015) Int J Res Rev 2(May):256–268
Pande P, Telang S (2014) Calculation of rainwater harvesting potential by using mean annual rainfall. Surface Runoff Catchment Area 3(7):200–204
Panhalkar S (2011) Domestic rain water harvesting system: a model for rural development. Soc Sci Nature 2(3):861–867
Takwimu (2010) The United Republic of Tanzania. Gazette IV(0):1–24
Texas Water Development Board (2006) Rainwater harvesting potential and guidelines for Texas. Report to the 80th Legislature, (November), 45

Chapter 28
Rainwater Harvesting in Rural Communities: A Case Study of Ghana

Collins Andoh, Sakshi Gupta, and Deepak Khare

Abstract Water is a valuable resource to mankind and propels every nation's economy. The availability of a safe, reliable and impregnable source of water has been and continues being an extensive matter concerning to human populations. In Ghana, the challenges brought about by dry wells, as well as minerals in the groundwater resources make harnessing of rainwater for household use non-negotiable. Many researchers and scientists are deliberately looking into the suitability of harvested rainwater as the best, prudent and a reasonable substitute to conventional sources of water across the globe. Rainwater harvesting has proven beyond reasonable doubts as an assuring substitute for freshwater supply in the era of expanding shortage of water in Ghana as well as in different regions across the world. It is anticipated that if we adopt such rainwater harvesting systems, it would facilitate the augmentation of various water demands of the community households, control issues of flooding and erosion resulting in alleviation of poverty.

Keywords Rainwater harvesting · Rural communities · Alternative water source · Ghana

28.1 Introduction

The Sahel zone of sub-Saharan part of West Africa is undeniably among the less privileged regions in the world. This region which has been covered by Savannah grassland has one short rainy season (May to September), followed by a long period of dry weather normally called the dry season. Several people in rural communities are

C. Andoh (✉) · D. Khare
Department of Water Resources Development and Management, Indian Institute of Technology, Roorkee 247667, Uttarakhand, India
e-mail: ebo2000kwabena@gmail.com

S. Gupta
Department of Civil Engineering, Graphic Era Deemed to be University, Dehradun 248002, Uttarakhan, India

A. Pandey et al. (eds.), *Water Management and Water Governance*, Water Science and Technology Library 96,
https://doi.org/10.1007/978-3-030-58051-3_28

faced with chronic food insecurity coupled with inadequate access to water supplement and proper sanitation. Increasing demands of water for growing populations in the sub-Saharan Africa have resulted in insufficient pipe-borne water supply from main distribution outlets (Awuah et al. 2014). A study conducted in 2002 by New Partnership for Africa's Development concluded that about 50% of Africa's population would be hit by water scarcity by 2025 (Opare 2012). In order to have an environment contributing to overall economic development and encouraging progress of the concerned nation, the quest for an alternate yet affordable potable supply of water is belittling which can even reduce poverty. In this regard, the adoption of system pertaining to the harvesting of rainwater in progressing nations provides guarantee against shortage of water, especially in a country like Ghana.

Rainwater harvesting basically is the process of collecting the rainwater during a downpour and storing it in either the overhead tanks above the terrain or recharging the subterranean water to be used later. It can be done both in crowded cities and in accessible wider areas (Ali and Jain 2014). Collecting rainwater into buckets, tanks and pots has been the traditional practice of rainwater harvesting in Ghana. Out of all the techniques, the most widely accepted regime was collecting water from rooftops of buildings. The rainwater harvesting system was noted as much cheaper compared to ground catchment systems, owing to the reason that the rooftops which are to be used as the catchment requires no additional installation being already present in the areas and having certain rise they also safeguard against possibility of contamination.

It is known that during the colonial era in the country's history, missionaries and government residences have incorporated RWH in their design. Even prior to the inception of the water supply pipeline system in 1928 at Cape Coast, rainwater harvesting was universally practised. Traditional rainwater harvesting systems, which were then rudimentary declined and generally faded out of favour because communities have grown out to be more prosperous and the buildings were fabricated with the plumbing systems internally itself (Boakye and Nsiah 2016). The approach to addressing the increasing water demand by the non-governmental organizations and aid agencies has been on the sinking of boreholes in the rural communities. However, as the numbers of these boreholes increase, there are increasing fears with regard to the security of the groundwater resources especially with the observation wells drying up. Despite the numerous water supply projects by government and non-governmental organizations in communities, the situation remains dire.

28.2 Study Area

Agona Swedru simply referred to as Swedru is a town that is located in the south-eastern corner of the Agona West Municipality. It is the administrative, political and commercial capital of the municipality (Fig. 28.1). The town has a bimodal pattern of rainfall, with the maximum occurring in May/June and September/October. The annual rainfall of the study area ranges between 1000 and 1400 mm. The dry season

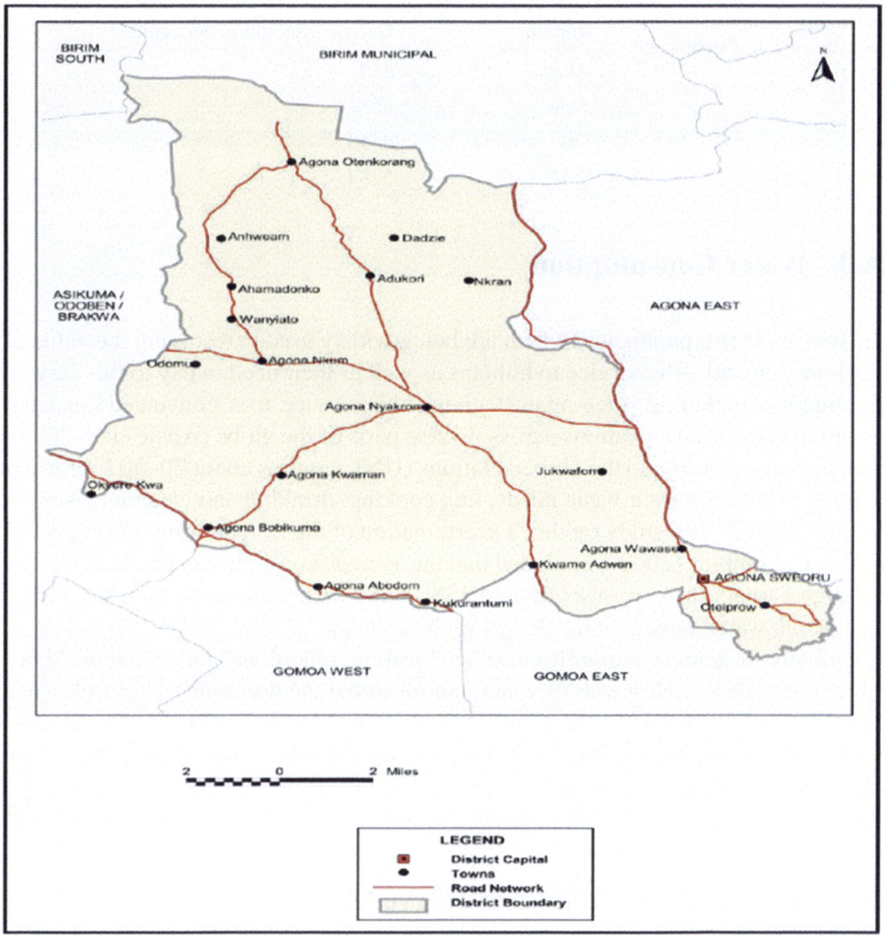

Fig. 28.1 Municipality context of Agona West (Ghana Statistical Service 2014)

felt within the months of December and March. The highest mean monthly temperature of 33.8 °C is observed between March and April. The lowest mean monthly temperature of 29.4 °C is, however, recorded in August (Osei 2013).

Agona Swedru is densely populated with a population of about 45,614. Agriculture is the major economic activity in Agona Swedru, and it engages vast proportion of its population. Cultivation includes tree/cash crops, food crops, vegetables and sugarcane. Fishing is done along the Akora River, but this activity is economically insignificant (Osei 2013).

Table 28.1 Domestic water usage by region (Department 2005)

Region	Daily consumption per capita
Africa	47 L
Asia	85 L
UK	334 L
USA	578 L

28.3 Water Consumption

The two important parameters which are being widely used in resolving the value of water are demand—the service to humans as well as their need to pay to this service and supply—including price against giving this service to a convinced capacity, condition and various locations across diverse parts of the globe (Abaje et al. 2009). Each person according to the United Nations (UN), requires about 20–50 L of water in a day to cater for their basic needs, i.e., cooking, drinking and cleaning (Apraku Gyampoh 2012). In a study on the "Determination of the water consumption pattern in Accra", Lamptey (2008) concluded that the average consumption per head for low and high earners in Accra was 66 and 138 l/h/d which were more than the values used by the water managers for design in the country.

Globally on comparing with the other locations, the domestic sector of Africa comes out to be a modest user of water, talking about the domestic sector of Europe it accounts to about double the consumption of the African levels which consumes about 47 L per capita per day. Table 28.1 explains domestic water usage by region (Department 2005).

28.4 The Need for Rainwater Harvesting

Water is known to be a restricted reserve and, hence, the needs for which are increasing rapidly. Populations in some cities, like Agona Swedru, are expanding swiftly resulting into the cause that water managers, i.e., Ghana Water Company Limited, must do everything possible to reserve the water supplies for future generations along with the current supplies (Moses 2011). This, therefore, renders rainwater harvesting a very important venture to undertake.

28.5 Significance of Rainwater Harvesting

With the new rainwater harvesting system framed and upgraded in context to the available water resources and conditions locally, the following are some of the perceived benefits that could be accrued (Kabo-bah et al. 2008).

28.5.1 Water Availability

This improved approach of harvesting rainwater would help in clarifying health problems related with water in poor communities, and it will forbid children and women in communities from walking long stretches in search of water whose quality status is unknown as well. Such water sources usually harbour disease-carrying organisms like Guinea worm.

28.5.2 Convenience

Implementing rainwater harvesting in homes gives people easy access to water. It is everyone's dream to have water supply near to their dwelling areas to have an ease in fetching and consuming. In some cases, it is also concerning the security of women as water available in close proximity will reduce the chances of any misconduct against them. Men also enjoy the benefits as they can give water to animals near their homes instead of travelling miles with its attendant.

28.5.3 Time is Saved

Women as well as young girls often spend long hours during the day time in search of water; hence, if the source of water is closer to the user, this situation will be avoided. The time freed can be best utilized in other useful activities like better care of kids, economic enhancement, more time for leisure and education. Time wasted in drawing water often gives room for girls of school-going age to skip school.

28.5.4 Injury Prevention

When girls frequently carry large amounts of water across long distances, it creates risk of spinal column pelvic injury coupled with other sufferings. Rainwater harvesting could mitigate these risks and associated suffering of young girls.

28.5.5 Money Saved

It is observed in few households of Ghana that they purchase water often at huge prices. Therefore, harvesting rainwater for usage also minimizes this practice, which directly benefits the households paying for water.

28.5.6 Improves Agriculture

It is evident that the quantities of water made available through water supply improvement schemes are used for such agricultural purposes, to which the community deem it beneficial. The new improved rainwater harvesting system could supplement existing sources of water supply to address the quest to irrigate and rear livestock.

28.6 Approach Adopted

Currently, rainwater harvesting is practised in other regions of the world with well-developed designs and standards. Figure 28.2 shows a system, which is highly adapted to the type of housing infrastructure typically found in both developed and developing countries. Though overall concepts and approaches to the practice of rainwater may be similar across the world, designs such as one shown in Fig. 28.2 may not be the most appropriate for communities in poverty-stricken areas such as Northern Ghana where there is a dire need for systems adapted to suit the local needs and challenges (Kabo-bah et al. 2008) (Fig. 28.3).

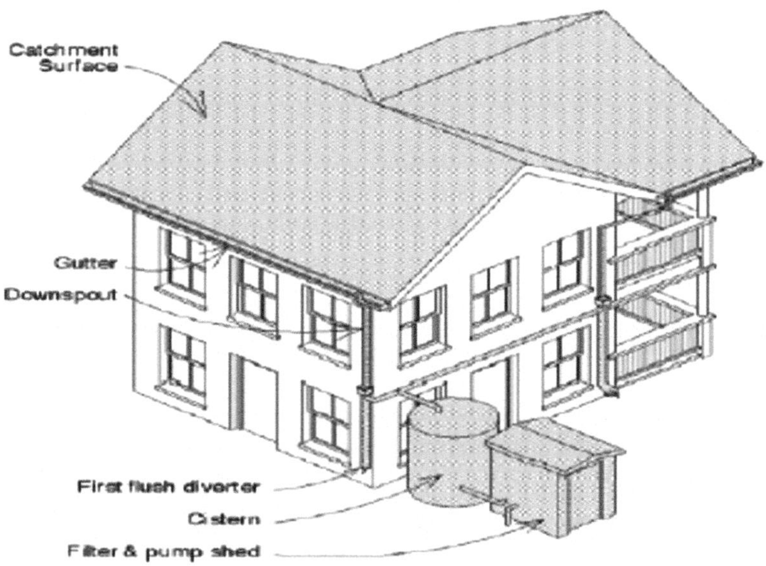

Fig. 28.2 Schematic rainwater harvesting installation system (Kabo-bah et al. 2008)

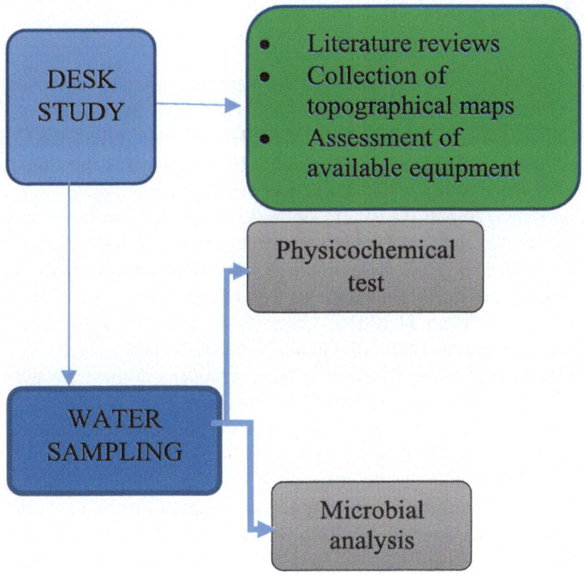

Fig. 28.3 Flowchart of design process

28.7 Design Process

The general description of all the facets of the design process is outside the scope of this paper.

28.8 Conclusion

The sub-Saharan region of West Africa where Ghana is located is among the poorest in the world with many rural people facing chronic food and water supply insecurity and sanitation (Kabo-bah et al. 2008). It is an undeniable fact that these perennial challenges across low-income West African nations may not go soon. Comments and reports from the numerous rainwater harvesting studies render it to be the most assuring interventions for freshwater supplies in this era of swift increase in water demand (Lundgren and Akerberg 2006).

The public is therefore entreated to cultivate the habit of harnessing rainwater because surface water sources have shrunken further with the rising threats of climate change. It is anticipated that the systems if adopted can facilitate the augmentation of basic and regular domestic water demands of poor households, control issues of flooding, erosion and alleviating poverty.

References

Abaje IB, Ati OF, Ishaya S (2009) Nature of potable water supply and demand in jema. A Local Govern Area of Kaduna State, Nigeria', Area, 1(1):16–21

Ali SR, Jain RK (2014) Rain Water Harvesting System for College of Engineering, Teerthanker Mahaveer University, Moradabad. Int J Innov Res Sci Eng Technol 3(8):15649–15657. https://doi.org/10.15680/IJIRSET.2014.0308077

Apraku Gyampoh B (2012) Geothinking: 2012, Goodnews on Groundwater in Africa: what next? Available at: https://benjigyampoh.blogspot.in/2012/. Accessed: 18 Apr 2018

Awuah E et al (2014) Assessment of rainwater harvesting as a supplement to domestic water supply: case study in Kotei-Ghana. Int Res J Public Environ Health 1(6):126–131

Boakye E, Nsiah JJ (2016) Quantifying rooftop rainwater harvest potential: case of Takoradi Polytechnic in Takoradi. Ghana 5(6):1004–1008

Department G (2005). research project report s domestic water consumption per capita : a case of study of selected households in Nairobi

Ghana Statistical Service (2014) 2010 population and housing census: district analytical report, Agona West Municipality, p 83

Kabo-bah A et al. (2008) Affordable rainwater harvesting systems : a collaborative research effort. In: 11th international conference on urban drainage, pp 1–10

Lundgren A, Akerberg H (2006) Rainwater harvesting in the peri-urban areas of Accra: status and prospects. Department of Geography and Resources Development, p 75

Lamptey F (2008) Determination of domestic water consumption pattern in Accra, MSc. Thesis (unpublished), KNUST, Kumasi, Ghana

Moses N (2011) Optimal production of potable water case study: Agona Swedru—Ghana Water Company Limited (Gwcl)

Opare S (2012) Rainwater harvesting: an option for sustainable rural water supply in Ghana. GeoJournal 77(5):695–705. https://doi.org/10.1007/s10708-011-9418-6

Osei C (2013) Evaluating post-flood disaster response strategies in Ashaiman and Agona Swedru (10226799), pp 9–38

Chapter 29
Dynamic Programming Integrated Differential Evolution Algorithm for Determining Optimal Policy of Reservoir

Bilal, Millie Pant, and Deepti Rani

Abstract An integrated approach by merging the classical dynamic programming and metaheuristic technique differential evolution is suggested for reservoir optimization problem for obtaining the optimum release policy. Two test cases are considered. In the first case, results are obtained for a single year data, while in the second case, results are obtained for ten years' data. Solutions found imply that the proposed integrated approach can be an alternative algorithm for reservoir operations.

Keywords Reservoir operation storage · Dynamic programming · Differential evolution · Release

29.1 Introduction and Literature Review

In past decades, optimization algorithms have become quite well known for solving complex optimization problems arising in various domains including reservoir operations. Dynamic programming (DP), a well-known classical optimization method, has been frequently used for solving reservoir operation models. A number of references emphasizing the application of DP include Bellman and Dreyfus (1962), Larson (1968a, b), Hall et al. (1969), Heidari et al. (1971), Chow et al. (1975), Labadie (1999), Kumar and Baliarsingh (2003), Srivastava and Awchi (2009), and Rani et al. (2016). However, it has also been observed that the computation time of DP is directly proportional to the complexity of the problem. That is the computational complexity

Bilal (✉) · M. Pant
Department of Applied Science and Engineering, Indian Institute of Technology, Roorkee 247667, India
e-mail: bilal25iitr@gmail.com

M. Pant
e-mail: millifpt@iitr.ac.in

D. Rani
National Institute of Hydrology, Roorkee, India
e-mail: deeptinatyan@yahoo.com

A. Pandey et al. (eds.), *Water Management and Water Governance*, Water Science and Technology Library 96,
https://doi.org/10.1007/978-3-030-58051-3_29

of DP increases with the increase in the size of the problem. Paradigm shift is more toward the generic algorithms like that of metaheuristics which are capable of solving a wide variety of problems with equal ease. The focus of this study is differential evolution (DE), a metaheuristic technique proposed by Storn and Price (1997). It has been effectively implemented for solving real-life optimization problems with varying complexities arising in different fields (Bilal et al. 2020a). In this paper, DE is applied for finding the ideal release strategy for the Mula reservoir of the state of Maharashtra.

Some references highlighting the application of DE in reservoir operation include Lakshminara and Subramanian (2006), Reddy and Nagesh Kumar (2007, 2008), Vasan and Raju (2007), Mandal and Chakraborty (2009), Qin et al. (2010), Suribabu (2010), Vasan and Simonovic (2010), Adeyemo and Otino (2010), Regulwaret al. (2010), Adeyemo (2011), Fallah-Mehdipour et al. (2012), Ahmad et al. (2014), Schardong and Simonovic (2015), Ochoa et al. (2016, 2017a, b), Bilal and Rani (2018), Bilal et al. (2020b).

A hybrid approach by integrating DP with DE is suggested in this paper to obtain the optimal release policy for the Mula River, Maharashtra, India. The present article is segregated into six sections. In Sect. 29.2, a brief description of the area of study is provided; in Sect. 29.3, methodology; in Sect. 29.4, the significance of integrating DP with DE is given; computations and results are given in Sect. 29.5; and finally, the study concludes with Sect. 29.6.

29.2 Region of Study: Mula Project

The present study considers Mula River, a tributary of Pravara River. The Mula project is an essential multipurpose project intended to provide irrigation water supply to the city of Ahmednagar. The map of the Mula River and project can be seen in Figs. 29.1 and 29.2 shows the average monthly inflow and demand.

Problem Formulation

This purpose of this study is to formulate an optimal and ideal release policy for reservoir operation while satisfying the assisting constraints.

The objective function is the minimized squared difference of target demand and optimal release policy, and the decision variables are the timely releases from the reservoir.

The associated constraints are bound constraints for release and storage along with the continuity equation, which needs to be satisfied for each time duration.

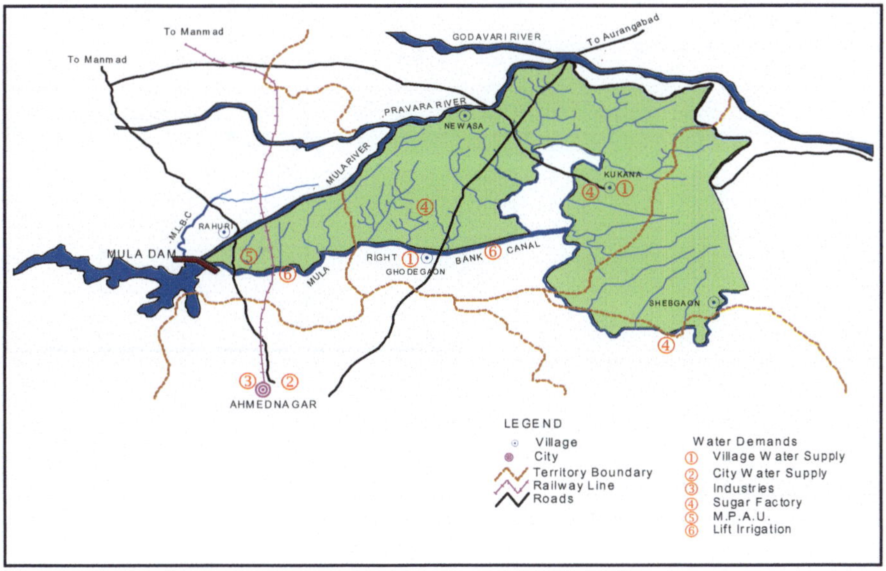

Fig. 29.1 Mula project

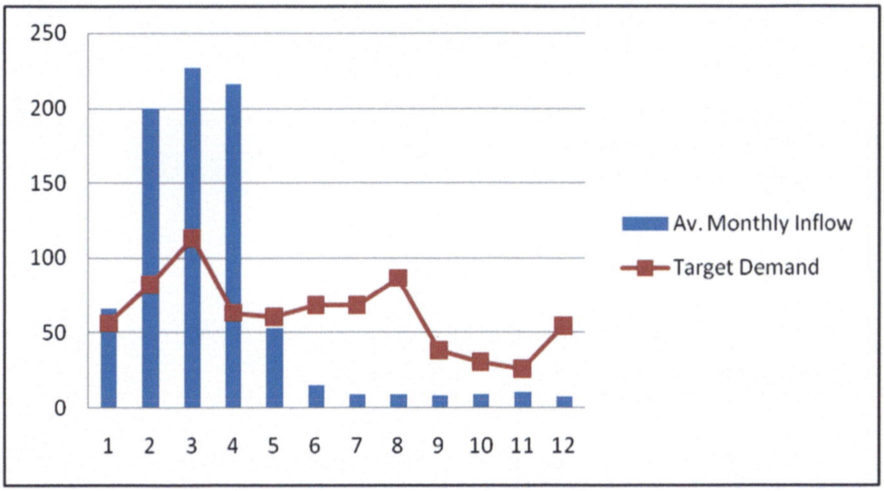

Fig. 29.2 Average monthly inflow and demand

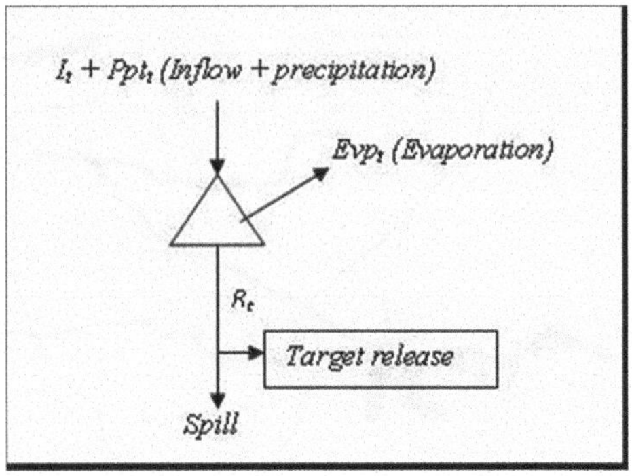

Fig. 29.3 Variables incorporated with a reservoir problem

A single reservoir system is provided in Fig. 29.3.

Mathematical Formulation
Generalized form of the problem can be shown as:

$$g_t(R_t) = \text{Min} \sum_{t=1}^{N} (Q_t - D_t)^2 \tag{29.1}$$

where:

$g_t(R_t)$ the function of release at time t
Q_t release for the time t and
D_t target demand for time t

Subject to constraints given below:

1. The continuity equation is represented as:

$$S_{t+1} = S_t + I_t - Q_t - Evp_t \, \forall t = 1, \ldots, N \tag{29.2}$$

where:

S_t, I_t and R_t symbolizes the storage, inflow and release, respectively

N denotes the time limit

Evp_t denotes evaporation that takes place during time t

2. The corresponding bound constraints are:

Limits on storage (S_t) and limits on release (Q_t):

$$S_{\min} \leq S_t \leq S_{\max} \quad \forall t = 1, \ldots N \tag{29.3}$$

$$Q_{\min} \leq Q_t \leq Q_{\max} \quad \forall t = 1, \ldots, N \tag{29.4}$$

$S_{\min}$, $S_{\max}$, $Q_{\min}$, and $Q_{\max}$ denote minimum and maximum storages and releases, respectively.

29.3 Methodology

In the beginning, DP Bellman (1957) is invoked to reduce the bounds of the variables, and in the second phase, DE is activated to obtain the optimum value to determine the optimal and ideal release policy for the concerned area of study, i.e., Mula River of Maharashtra.

29.3.1 DP Model for Reservoir Operation

The DP model is applied to approximate the ideal monthly releases using monthly inflow, to set an appropriate operation policy for the reservoir.

The assumptions are that the input to the reservoir is the current inflow, and the spill included in the reservoir released model is considered as the backward process with the following notations:

$f_r(S_r)$	minimum optimal return function from all the r stages to go provider the reservoir storage (state) to go is S_r in r stages
$g_r(s_r)$	return function at r stage to go
S_r	initial reservoir storage (state)
S_{r-1}	resulting reservoir storage (state)
I_r	inflow to the reservoir
EVr	evaporation lossed from the reservoir in r stages to go
Q_r	total reservoir release, which also includes the spill. It is the decision variable for the DP model in r stages to go
D_r and Sp_r	the target release and reservoir spill, respectively.

DP model is formulated as:

$$f_r(S_r) = \min_{Q_r}\left[g_r(S_r, Q_r) + f_{r-1}(S_{r-1})\right]\forall r \tag{29.5}$$

The return function is:

$$g_r(S_r, Q_r) = (Q_r - D_r)^2 \tag{29.6}$$

Subject to the constraints due to reservoir continuity (state transformation) given in Eq. (29.7) and constraint due to spill from the reservoir given in Eq. (29.8):

$$S_{r-1} = S_r + I_r - Q_r - EV_r, \quad \forall r \tag{29.7}$$

$$Sp_r = Q_r - D_r \text{ if } Q_r > D_r elseSp_r = 0 \quad \forall r \tag{29.8}$$

Besides these, the other constraints are due to the reservoir storage (state) limited to the live storage capacity (*Ya*) and non-negativity of the state and decision variables. These are given in Eqs. (29.9) and (29.10).

$$Q_r \geq 0 \quad \forall r \tag{29.9}$$

$$0 \leq S_{r-1} \leq Ya \quad \forall r \tag{29.10}$$

Beginning from the known initial storage, likely release decisions $\{Q_1, Q_1, \ldots, Q_N\}$ at each stage should be such that the final storage belongs to the set of discrete stages

$$\{S_{\min} = 0, \delta, 2\delta, 3\delta, \ldots, m\delta = S_{\max} = Ya\}$$

Smaller values of 'δ' will result in increasing the number of states if the reservoir is huge sized, thereby increasing the computational complexity for DP. Conversely, a rough increment size may result in reducing the precision of the optimal results, though the solution may be global optimal for the given increment.

Solutions obtained by DP model deliver a set of optimal releases $\{Q_1^*, Q_2^*, \ldots, Q_r^*, \ldots Q_N^*\}$ which provides an optimal policy for discrete increment 'δ' used for solving the DP model.

Differential Evolution

DE is a population-based iterative process for which the computational steps are defined below:

Initialization: This is the first step of DE, during which a uniformly distributed population set is generated as follows:

Let $S^G = \left\{X_j^G : j = 1, 2, \ldots, \text{NP}\right\}$ and NP denotes the population setfor generation G the population size, respectively. Here,X_j^G denotes a D-dimensional vector as $X_j^G = \left\{x_{1,j}^G, x_{2,j}^G, \ldots, x_{D,j}^G\right\}$.$X_j^G$ is generated as:

$$X_j^G = X_{\text{low}} + \left(X_{\text{upp}} - X_{\text{low}}\right) * \text{rand}(0, 1) \tag{29.11}$$

$X_{\text{low}}, X_{\text{upp}}$ indicate the lower and upper bounds, respectively, of the search space S^G and rand (0, 1) denote the uniformly generated a random number between 0 and 1.

(a) **Evolution: In this phase**, three operations, viz. mutation, crossover, and selection, are activated as follows:

Mutation: In this step, a mutant vector V_j^G is obtained for each target vector given as X_j^G as

$$V_j^G = X_{r1}^G + F * \left(X_{r2}^G - X_{r3}^G\right) \tag{29.12}$$

where F the scaling factor is generally varied between 0 and 1; $r1, r2, r3 \in \{1, 2, \ldots, \text{NP}\}$ are vectors, randomly selected such that they are mutually different from each other.

Crossover: crossover operation is performed after mutation. During this operation, a new vector is generated known as trial vector denoted as $U_j^G = \left\{u_{1,j}^G, u_{2,j}^G, \ldots, u_{D,j}^G\right\}$ is generated. The crossover operation happens between the target vector $X_j^G = \left\{x_{1,j}^G, x_{2,j}^G, \ldots, x_{D,j}^G\right\}$ and mutant vector $V_j^G = \left\{v_{1,j}^G, v_{2,j}^G, \ldots, v_{D,j}^G\right\}$. U_j^G is generated as

$$u_{i,j}^G = \begin{cases} v_{i,j}^G & \text{if } r \text{ and } j \leq Cr \\ x_{i,j}^G & \text{otherwise} \end{cases} \tag{29.13}$$

where $i \in \{1, 2, \ldots, D\}$ and Cr, the crossover probability $\in [0, 1]$.

Selection: This is done to select the target vector or trial vector as per their fitness values, and one with better fitness moves for further generations. The selection operation is defined as:

$$X_j^{G+1} = \begin{cases} U_j^G & \text{if } f\left(U_j^G\right) \leq f\left(X_j^G\right) \\ X_j^G & \text{otherwise} \end{cases} \tag{29.14}$$

The above three operations are repeated until a predefined termination criterion is obtained.

29.4 Significance of Integrating DP with DE

The hybridization of DP with DE is done mainly to contract the search space. The solutions obtained through DP are used as initial search ranges for DE. Increment used to discretize the storages sometimes results in overestimation or underestimation the solution. This particular characteristic of DP is exploited to narrow down the search space for implementation of DE.

The working scenario of DP-DE is as follows:
Step 1. Formulate an optimization model.
Step 2. Set the input parameters.
Step 3. Implement DP to incorporate over/under-estimated limits.
Step 4. Apply DE within limits obtained in Step 3 to get the final optimal solution.

29.5 Computations and Results

In the first case, we considered a small dataset for an average year only.

Implementation of DP resulted in a significant change on the release limits, leading to the narrowing down the search space for all the cases, shown in Table 29.1. Row 1 (Table 29.1) shows for month 1, and after applying the DP, lower limit has increased from 0 to 45.91. Now the search space is reduced from 0(lower limit)-56.45 (upper limit) to 45.91 (lower limit)-56.45 (upper limit). So, this change significantly reduces the solution search space making it easier for optimization algorithms like DE for determining the solution faster. Figures 29.4 and 29.5 show the search space before and after applying DP, respectively.

Table 29.2 shows how DE exploits the obtained reduced search space. The solution determined by DE before and after its integration with DP justifies the claim of this study. Figure 29.6 shows the combined release policies of the reservoir.

Case 2: Results for ten years of data
We further increased the data from one year to 10 years dataset (1972–73 to 1981–82). This dataset substantially increased the number of decision variables to 120, which

Table 29.1 Monthly lower and upper limits on release

Month (s)	Before applying DP		After applying DP	
	Lower limit	Upper limit	Lower limit	Upper limit
1	0	56.45	45.91	56.45
2	0	82.33	74.33	82.33
3	0	113.5	105.5	113.5
4	0	63.33	55.33	63.33
5	0	60.83	51.74	60.83
6	0	68.39	60.39	68.39
7	0	68.39	58.71	68.39
8	0	86.6	75.66	86.6
9	0	38.21	26.78	38.21
10	0	30.45	18.38	30.45
11	0	25.72	17.72	25.72
12	0	54.39	46.39	54.39

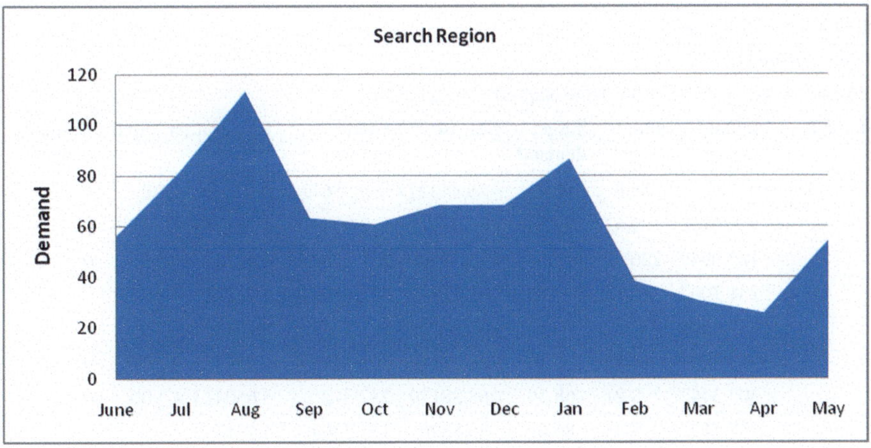

Fig. 29.4 Search space before applying DP

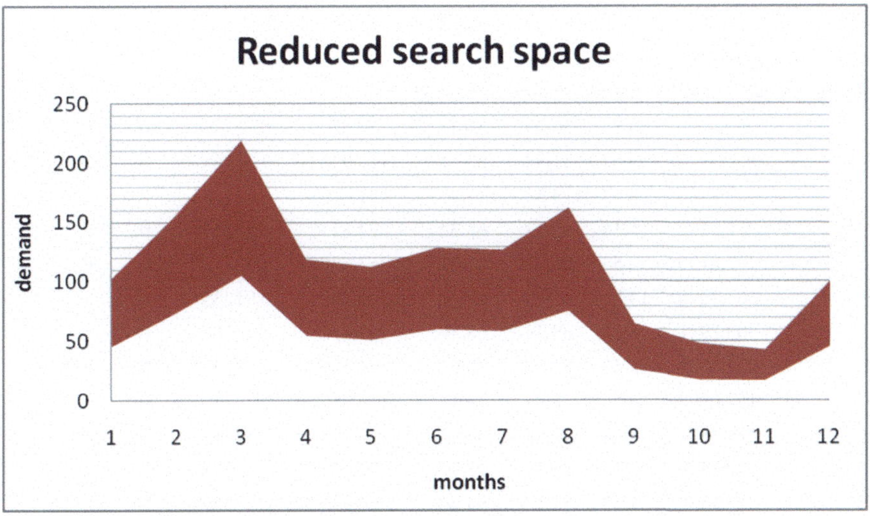

Fig. 29.5 Reduced search space after applying DP

made the problem more challenging though more practical as well. The optimal value attained by DP-DE for ten years is 358.01 in 1–2 s. Figure 29.7 shows the convergence of DP-DE.

Table 29.2 Results (gen = 500, pop = 20)

Month (s)	Initial storage	Inflow	Target demand	Release	Final storage	Evaporation	Spill	Irrigation deficit
Results using DE								
Objective value = 7.7017e-04, Time elapsed = 0.527234 s								
1	0	65.8	56.45	56.44322	9.356778	0	0	0
2	9.356778	199.67	82.33	82.32011	122.9949	3.711747	0	0
3	122.9949	226.52	113.5	113.4864	231.7062	4.322333	0	0
4	231.7062	216.18	63.33	63.3224	377.8509	6.71292	0	0
5	377.8509	52.5	60.83	60.8227	363.7737	5.754469	0	0
6	363.7737	14.87	68.39	68.38179	303.3442	6.917752	0	0
7	303.3442	8.21	68.39	68.38179	237.512	5.660423	0	0
8	237.512	8.43	86.6	86.5896	154.7306	4.621793	0	0
9	154.7306	7.5	38.21	38.20541	119.8609	4.164317	0	0
10	119.8609	8.48	30.45	30.44634	93.65207	4.242443	0	0
11	93.65207	9.7	25.72	25.71691	72.10274	5.532417	0	0
12	72.10274	7.32	54.39	54.38347	19.03401	6.005257	0	0
Results using DP-DE								
Objective value = 2.671e-32 Time elapsed = 0.245331 s								
1	0	65.8	56.45	56.43	9.35	0	0	0
2	9.35	199.67	82.33	82.31	122.9783	3.321682	0	0
3	122.9783	226.52	113.5	113.4	231.6762	4.712145	0	0
4	231.6762	216.18	63.33	63.32	377.8136	6.122543	0	0
5	377.8136	52.5	60.83	60.81	363.7295	5.644149	0	0
6	363.7295	14.87	68.39	68.37	303.2921	6.912357	0	0
7	303.2921	8.21	68.39	68.39	237.4521	5.662005	0	0
8	237.4521	8.43	86.6	86.6	154.6608	4.621324	0	0
9	154.6608	7.5	38.21	38.19	119.7871	4.163706	0	0
10	119.7871	8.48	30.45	30.43	93.57549	4.24598	0	0
11	93.57549	9.7	25.72	25.73	72.02437	5.532117	0	0
12	72.02437	7.32	54.39	54.40	18.95007	6.203648	0	0

29.6 Conclusion

In this article, integration of classical dynamic programming (DP) and differential evolution (DE) is done to provide an ideal and global release policy for the Mula reservoir. For the optimization algorithm, there is a risk of getting stuck in local optima in the search space. Integration of DP into DE contributes to reducing the search space resulting in the faster performance of DE. Two different cases investigated indicate that such integration helps in improving the performance of DE, more prominently for the larger test cases. Water resource problems are usually large-scaled in nature;

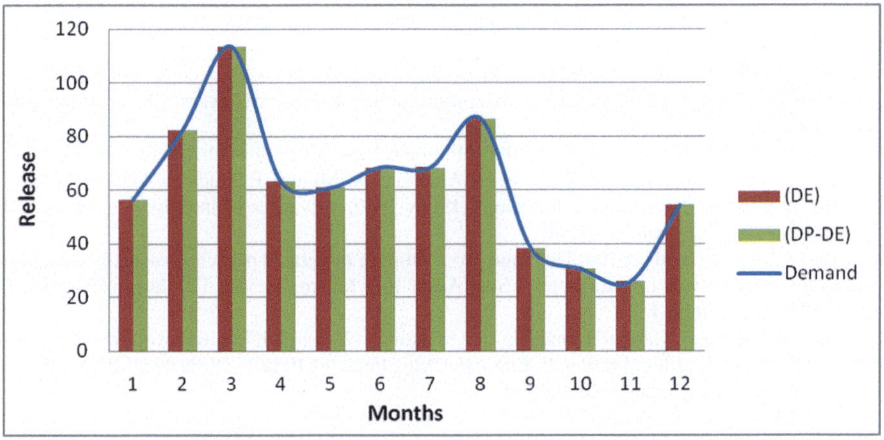

Fig. 29.6 Release policy for DE and DP-DE

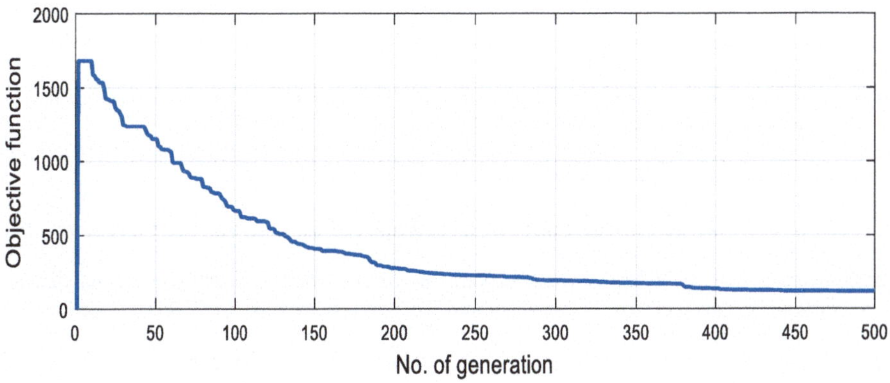

Fig. 29.7 Convergence graph as obtained by DP-DE

therefore, it would be interesting to observe the scalability of the proposed method, i.e., to identify the range of variables up to which this method is efficient and to identify the further enhancement needed for larger problems. Secondly, the methodology suggested here is generic in nature and integration of DP can be done with any of the metaheuristics to enhance its performance.

References

Adeyemo JA (2011) Reservoir operation using multi-objective evolutionary algorithms-a review. Asian J Sci Res 4(1):16–27

Adeyemo J, Otieno F (2010) Differential evolution algorithm for solving multi-objective crop planning model. Agric Water Manage 97(6):848–856

Ahmad A et al (2014) Reservoir optimization in water resources: a review. Water Resour Manage 28(11):3391–3405
Bellman R (1957) Dynamic programming. Princeton University Press, Princeton, New Jersey
Bellman R, Dreyfus S (1962) Applied dynamic programming. Princeton University Press, Princeton, New Jersey
Bilal, Pant M, Rani D (2018) Determining the optimum release policy through differential evolution: a case study of Mula irrigation project. In: Abraham A, Muhuri P, Muda A, Gandhi N (eds) Intelligent systems design and applications. ISDA 2017. Advances in intelligent systems and computing, vol 736. Springer, Cham
Bilal, Rani D, Pant M et al (2020a) Dynamic programming integrated particle swarm optimization algorithm for reservoir operation. Int J Syst Assur Eng Manag 11:515–529. https://doi.org/10.1007/s13198-020-00974-z
Bilal, Pant M, Zaheer H, Garcia-Hernandez L, Abraham A (2020b) Differential evolution: a review of more than two decades of research. Eng Appl Artif Intell 90:103479, ISSN 0952-1976. https://doi.org/10.1016/j.engappai.2020.103479
Chow VT, Maidment DR, Tauxe GW (1975) Computer time and memory requirements for DP and DDDP in water resource systems analysis. Water Resour Res 11(5):621–628
Fallah-Mehdipour E, Bozorg Haddad O, Mariño MA (2012) Extraction of multicrop planning rules in a reservoir system: application of evolutionary algorithms. J Irrigation Drainage Eng 139(6):490–498
Hall WA, Harboe RC, Yeh WW-G, Askew AJ (1969) Optimum firm power output from a two reservoir system by incremental dynamic programming. Contribution 130, Water Research Centre, University of California, LA
Heidari M, Chow VT, Kokotovi PV, Meredith DD (1971) Discrete differential dynamic programing approach to water resources systems optimization. Water Resour Res 7(2):273–282
Kumar DN, Baliarsingh F (2003) Folded dynamic programming for optimal operation of multireservoir system. Water Resour Manage 17:337–353
Labadie JW (1999) Generalized dynamic programming package: CSUDP version 3.3. Documentation and user guide, Department of Civil Engineering, Colorado State University, Ft. Collins
Lakshminarasimman L, Subramanian S (2006) Short-term scheduling of hydrothermal power system with cascaded reservoirs by using modified differential evolution. IEE Proc Gener Transm Distrib 153(6):693–700
Larson RE (1968a) State increment dynamic programming. Elsevier, New York
Larson RE (1968b) A dynamic programming successive approximations technique. In: Joint automatic control conference, University of Michigan
Mandal KK, Chakraborty N (2009) Short-term combined economic emission scheduling of hydrothermal power systems with cascaded reservoirs using differential evolution. Energy Convers Manage 50(1):97–104
Ochoa P, Castillo O, Soria J (2016) Fuzzy differential evolution method with dynamic parameter adaptation using type-2 fuzzy logic. In: 2016 IEEE 8th international conference on intelligent systems (IS). IEEE
Ochoa P, Castillo O, Soria J (2017a) A new approach for dynamic mutation parameter in the differential evolution algorithm using fuzzy logic. North American fuzzy information processing society annual conference. Springer, Cham
Ochoa P, Castillo O, Soria J (2017b) Differential evolution using fuzzy logic and a comparative study with other metaheuristics. In: Nature-inspired design of hybrid intelligent systems. Springer International Publishing, pp 257–268
Qin H et al (2010) Multi-objective cultured differential evolution for generating optimal trade-offs in reservoir flood control operation. Water Resour Manage 24(11):2611–2632
Rani D, Srivastava DK (2016) Optimal operation of Mula reservoir with combined use of dynamic programming and genetic algorithm. Sustain Water Resour Manage 2(1):1–12

Rani D, Srivastava DK, Gulati TR (2016) A set of linked optimization models for an inter-basin water transfer. Hydrol Sci J 61(2):371–392

Reddy MJ, Kumar DN (2007) Multiobjective differential evolution with application to reservoir system optimization. J Comput Civ Eng 21(2):136–146

Reddy MJ, Kumar DN (2008) Evolving strategies for crop planning and operation of irrigation reservoir system using multi-objective differential evolution. Irrigation Sci 26(2):177–190

Regulwar DG, Choudhari SA, Anand RP (2010) Differential evolution algorithm with application to optimal operation of multipurpose reservoir. J Water Resour Prot 2(06):560

Schardong A, Simonovic SP (2015) Coupled self-adaptive multiobjective differential evolution and network flow algorithm approach for optimal reservoir operation. J Water Resour Plann Manage 141(10)

Srivastava DK, Awchi TA (2009) Storage-yield evaluation and operation of Mula Reservoir, India. J Water Resour Plann Manage, ASCE 135(6):414–425

Storn R, Price K (1997) Differential evolution–a simple and efficient heuristic for global optimization over continuous spaces. J Global Optim 11(4):341–359

Suribabu CR (2010) Differential evolution algorithm for optimal design of water distribution networks. J Hydroinf 12(1):66–82

Vasan A, Raju KS (2007) Application of differential evolution for irrigation planning: an Indian case study. Water Resour Manage 21(8):1393

Vasan A, Simonovic SP (2010) Optimization of water distribution network design using differential evolution. J Water Resour Plann Manage 136(2):279–287

Chapter 30
Relationship of Catchment, Storage Capacity and Command Area for Rainwater Harvesting in the Farm Pond

R. S. Patode, M. B. Nagdeve, V. V. Gabhane, M. M. Ganvir, A. B. Turkhede, and G. Ravindra Chary

Abstract For construction of farm pond, the important factor is its location. Normally farmers dug out the farm ponds without considering technical aspects. At All India Coordinated Research Project for Dryland Agriculture, Dr. Panjabrao Deshmukh Krishi Vidyapeeth, Akola, demonstrations of site-specific farm ponds based on catchment area were planned and conducted. Based on runoff from the catchment area, the storage in the farm pond was assessed and the relationship was developed. Here, the results of the relationship of catchment, storage and command during the years 2014–2015 to 2016–2017 are presented. During 2016–2017, the runoff causing rainfall in the catchment area of 5 ha of farm pond was 301.3 mm which helps in accumulation of 2014.8 m^3 runoff in the farm pond. Therefore, the catchment–storage–command relationship for the season can be given as, from 5 ha catchment, 2014.8 m^3 water was stored in the farm pond which can irrigate (command) about 4.0 ha area. Moreover, the available water in the farm ponds was utilized for giving protective irrigations to different crops including vegetables. It was observed that due to utilization of stored farm pond water for protective irrigation, the yield of soybean during *Kharif*, chickpea during *Rabi* and vegetables during winter-summer had been increased.

Keywords Catchment · Farm pond · Protective irrigation · Runoff · Storage · Yield

R. S. Patode (✉) · M. B. Nagdeve · V. V. Gabhane · M. M. Ganvir · A. B. Turkhede
All India Coordinated Research Project for Dryland Agriculture, Dr. Panjabrao Deshmukh Krishi Vidyapeeth, Akola, Maharashtra 444104, India
e-mail: rspatode@gmail.com

G. R. Chary
All India Coordinated Research Project for Dryland Agriculture, ICAR-Central Research Institute for Dryland Agriculture, Santoshnagar, Hyderabad, Telangana 500059, India

A. Pandey et al. (eds.), *Water Management and Water Governance*, Water Science and Technology Library 96,
https://doi.org/10.1007/978-3-030-58051-3_30

30.1 Introduction

Available water resources all over the world are under pressure due to increased population therefore conservation as well as preservation of these resources should require to being undertaken immediately. Since ancient period water management is being practiced in the world, but at present it should be done on precedence basis. Several management options are available at the farm scale to increase rainfall use efficiency. Some of these are management of crop residues to improve infiltration and reduce sediment levels, construction of farm ponds for collection of excess rainfall flowing from the farm area, crop rotations and soil amendments (Freebairn et al. 1986). Several researchers have shown that on-farm runoff collection into dugout farm ponds and supplemental irrigation can increase and stabilize the crop production (Krishna et al. 1987). There is an abundant scope and opportunity for harvesting excess runoff in the rainfed region in different states of the country (Wani et al. 2003; Sharma et al. 2010).

All over the country the distribution of rainfall is not even under such circumstances the in-situ approach of conservation of rainfall should be adopted in which main emphasis is being given on more infiltration than the surface runoff. Once the soils get saturated then sub-surface as well as overland flow of rainwater started and this runoff should be stored with the help of reservoirs (Adhikari et al. 2009). Thus, rainfall is the most imperative input and based on average annual rainfall the surface runoff judgment is possible. Now a day's one or more region in our country is undergoing water scarcity. Sometimes intense rains occur but due to lack of proper conservation measures and lack of implementation the surface runoff water flows down to downstream side. Therefore, in such instances proper rainwater storage structures are required to be designed for collection of harvested runoff water. Thus, the entire rainfall runoff process to be channelized. When heavy rainfall occurs that time infiltration is less and rainwater started flowing from surface. If it is taken through drain lines to the storage structures then maximum collection of rainwater is possible (Jalal Uddin et al. 2017). If proper drain lines are not constructed then the surface water flows randomly and its collection at one place becomes difficult. At low lying areas the storage structure should be made by considering runoff volume from the catchment area. If we want to construct a farm pond then, by knowing the volume of water, we can calculate the dimensions of the farm pond. For different catchment areas different sizes of the farm ponds should be designed and accordingly technically proper implementation/construction should be done. Farmers are constructing the farm ponds without considering the technical facts therefore proper storage in the farm ponds does not occur. Thus, if any farmer desires to construct the farm pond then he should consider the advice of technical person. If proper advice is not taken then there may not be sizeable rainwater collection (Sthool et al. 2013). Since rainwater collection and its judicious use is the need of the day, everybody should adopt the technical things that are important for proper implementation and construction of farm ponds. The detailed features of the water contributing area, possible utilization of the stored water, suitable site for the pond and economics in terms of benefits have to be studied before making a farm pond. Catchment selection is one of the important aspects in designing the farm pond. The runoff from the catchment will depends on several factors viz. rainfall, topography, soil types and land use. The construction of runoff harvesting ponds involves consideration of yield of watershed, storage available at site, water requirements for different needs and groundwater conditions. The design considerations shall include relationship between watershed area and capacity of pond, choice and design of outlets, size of embankment and cost benefit ratio (Reddy et al. 2012). If farm pond of proper size and at proper location is constructed then the storage of water in the farm pond will definitely occur (Vora et al. 2008). Based on this approach the demonstrations of farm ponds of different sizes estimated from volume were undertaken and the harvested rainwater was reuse for different crops and vegetables as per need during kharif and rabi seasons (Nagdeve and Patode 2012, 2015; Patode et al. 2017), and results related to yield and cost economics are presented here.

30.2 Methodology

In order to deal with the problems of dryland farming appropriately, Indian Council of Agricultural Research (ICAR) started All India Coordinated Research Project for Dryland Agriculture (AICRPDA). Akola is one of the centres at Dr. Panjabrao Deshmukh Krishi Vidyapeeth, Akola, Maharashtra, India which is striving to solve location specific complexities through a cohesive, multi-pronged approach, emphasizing farmer's point of view.

In Vidharbha region of Maharashtra State about 90% of cultivable land is under rainfed farming. The region is renowned for large variations in monsoon rainfall and its uneven within-season distribution. Variation in the amount and timing of rainfall is a major challenge to crop management and the applicability of new technologies for yield improvements and competitiveness of rainfed crops. As such evidence is emerging globally that climate change is increasing rainfall variability and the frequency of extreme events. Decadal trend analysis of Akola location also showed consistent decrease in the annual (817, 832.2, 780.3 and 676.5 mm) and monsoonal (683.6, 682.3, 627.6 and 546.1 mm) rainfall across the past four decades (1971–2009), the decrease being more marked in the last decade. Similarly, monsoonal and winter minimum temperatures showed rise. The high risk for water-related yield loss makes farmers risk averse, influencing their other investment decisions, including labour, improved seed, and fertilizers. In the first place, technological investments that reduce water-related risks can build more resilience to face occurrence of droughts and dry spells under a changing climate. Hence, improving upon water conservation and thereby groundwater recharge, increasing water productivity, suitable crops and crop varieties will contribute to building resilience to climate change and allow for a more balanced development and sustainability of watershed in the region (Annual Report, AICRIPAM 2010).

30.2.1 Rainfall Analysis

Rainfall is one of the most important and critical hydrological input parameters for the design of rainfall harvesting structure. Its distribution varies both spatially and temporally in the semi-arid regions. The quantity of surface runoff depends mainly on the rainfall characteristics like intensity, frequency and duration of its occurrence. The high intense rainfall exceeding infiltration capacity of soil can produce more runoff than the event with low intensity for longer duration. Apart from the physical characteristics of the catchment area, the rainfall analysis is very critical for optimal economic design. The variability of rainfall in arid and semi-arid areas is considerable. An analysis of only 5 or 6 years of observations is inadequate as these 5 or 6 values may belong to a particularly dry or wet period and hence may not be representative for the long term rainfall pattern. The annual rainfall data available at Agro-meteorological observatory, Dr. Panjabrao Deshmukh Krishi Vidyapeeth, Akola was used.

30.2.2 Harvesting and Collection of Rainwater

The perennially flowing rivers were become dry and all over the region groundwater table was depleted. In a number of areas the depletion is concerning to about 30–50 m in the last 35–45 years (Sivanappan 2006). Floods and droughts are being occurred in the country at the same time in many states of our country. This is due to, no actual action was undertaken to safeguard, conserve and direct the rainwater resourcefully. Thus, there is necessity to take up the recharge process of the rainwater this can be achieved through water harvesting and conservation structures. In rainfed areas, some agronomic or other measures cannot conserve whole rainwater and a definite amount of runoff is hoped to occur and this runoff should be channelized to suitable storage structure (Taley et al. 2009). The runoff could be collected and harvested in tanks for a protective irrigation to dryland crops (Bangar et al. 2003).

30.2.3 Dimensions of the Farm Pond Based on Runoff Volume

Depending on the overland flow (surface runoff) from the micro-catchments, the storage capacity (volume) of the farm ponds was decided at AICRPDA, Dr. PDKV, Akola and construction of these farm ponds had been done.

30.2.3.1 Runoff Computation

The amount of runoff for the catchment was computed by using the following (Eq. 30.1) formula (Singh 1986).

$$R = \frac{kPA}{10} \tag{30.1}$$

where, R is Amount of runoff, cum, k is Expected per cent runoff of the total rainfall, P is Average annual rainfall in mm and A is Catchment area, ha.

Dimensions of the farm pond based on Runoff volume on the basis of the above equation are presented in Table 30.1.

30.3 Results and Discussion

30.3.1 Catchment, Storage Capacity and Command Area Relationship

During the year 2014, runoff events were recorded and presented in Table 30.2. The rainfall which causes overland flow in the catchment area of 4 ha of farm pond was 235.9 mm which helps in accumulation of 1776.96 m^3 runoff in the farm pond. The total runoff recorded was 44.41 mm, which was 18.82% of the runoff causing rainfall. From the stored pond water (1776.96 m^3) if protective irrigation of 5 cm

Table 30.1 Proportions of the storage structures (farm ponds)

Catchment area, A (ha)	Expected runoff in per cent of the total rainfall, k	Amount of runoff from the catchment area (cum)	Capacity of pond (cum)	Top dimensions (m × m)	Bottom dimensions (m × m)	Depth (m)	Side slopes
5.0	13	5154	2753	45 × 27	36 × 18	3.0	1.5:1

Table 30.2 Runoff during the season 2014 from catchment area of 4 ha in Farm pond

Date	Rainfall (mm)	Runoff	
		Accumulated in farm pond (m^3)	Recorded (mm)
12-06-2014	18.5	41.48	1.03
23-07-2014	136.4	1068.89	26.72
08-09-2014	81.0	666.59	16.66
Total	235.9	1776.96	44.41

Table 30.3 Runoff during the season 2015 from catchment area of 4 ha in Farm pond

Date	Rainfall (mm)	Runoff	
		Accumulated in farm pond (m^3)	Recorded (mm)
04-08-2015	194.0	1961.20	49.03
12-08-2015	28.0	120.0	3.00
15-09-2015	57.0	209.6	5.24
17-09-2015	78.5	288.4	7.21
Total	357.5	2579.2	64.48

depth is given then about 3.5 ha area can be irrigated. Therefore, the catchment–storage–command relationship for the season can be given as, from 4 ha catchment, 1776.96 m^3 water was stored in the farm pond which can irrigate (command) about 3.5 ha area.

During the year 2015, runoff events were recorded and specified in Table 30.3. The rainfall which was responsible for causing, the surface runoff in the catchment area of 4 ha of farm pond was 357.5 mm which helps in accumulation of 2579.2 m^3 runoff in the farm pond. The total runoff recorded was 64.48 mm, which was 18.04% of the runoff causing rainfall. From the stored pond water (2579.2 m^3) if protective irrigation of 5 cm depth is given then about 4.0 ha area can be irrigated. Therefore, the catchment–storage–command relationship for the season can be given as, from 4 ha catchment, 2579.2 m^3 water was stored in the farm pond which can irrigate (command) about 4.0 ha area.

During the year 2016, runoff events were recorded and given in Table 30.4. The rainfall (which contributes runoff) in the catchment area of 5 ha of farm pond was 301.3 mm which helps in accumulation of 2014.8 m^3 runoff in the farm pond. The total runoff recorded was 50.54 mm, which was 16.77% of the runoff causing rainfall. From the stored pond water (2014.8 m^3) if protective irrigation of 5 cm (depth) is given then about 4.0 ha area can be irrigated. Therefore, the catchment–storage

Table 30.4 Runoff during the season 2016 from catchment area of 5 ha in Farm Pond

Date	Rainfall (mm)	Runoff	
		Accumulated in farm pond (m^3)	Recorded (mm)
10-07-2016	91.0	691.6	17.30
11-07-2016	74.1	615.6	15.30
26-07-2016	38.2	228.0	5.70
27-07-2016	19.0	94.0	2.60
03-08-2016	21.0	108.0	2.70
17-09-2016	30.0	144.0	3.60
23-09-2016	28.0	133.6	3.34
Total	301.3	2014.8	50.54

capacity–command area relationship for the season can be given as, from 5 ha catchment, 2014.8 m^3 water was stored in the farm pond which can irrigate (command) about 4.0 ha area. The available water in the storage structures (farm ponds) was used for giving protective irrigation for soybean during *Kharif* and to chickpea during *rabi* season.

The catchment area, storage and command area relationship for the years 2013–2014 to 2016–2017 is depicted in Fig. 30.1. Based on these data, it can be estimated that from 1 ha catchment, on an average 544.52 m^3 water can be stored in the farm pond which can be utilized to irrigate (command) about 1.0 ha area with 5 cm (depth) of irrigation.

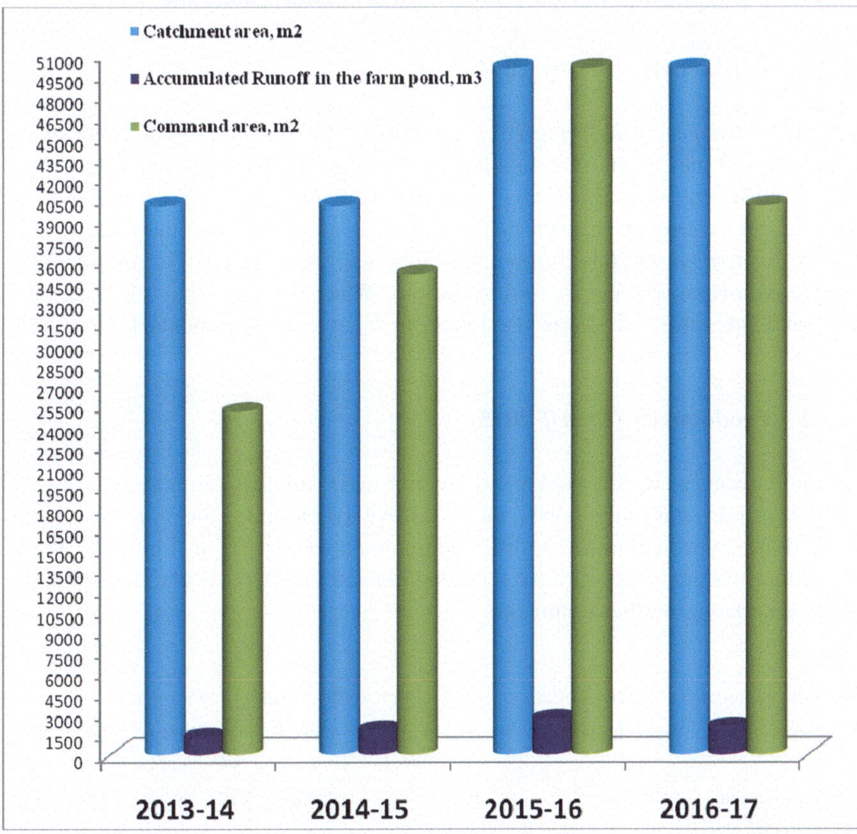

Fig. 30.1 Catchment area, storage in farm pond and command area relationship for the years 2013–2014 to 2016–2017

Table 30.5 Consequence of protective irrigation on yield of soybean and its economics

Treatments	Yield (kg ha^{-1})		Increase in yield (%)	RWUE (kg ha^{-1} mm^{-1})	Net returns (Rs ha^{-1})	B:C ratio
	With irrigation	No irrigation				
T_1 (Protective irrigation)	474	422	12.32	0.83	543	1.03
T_2 (Two protective irrigations)	1021	422	141.94	1.79	20,089	2.20

30.3.2 Productivity During Kharif *and* Rabi *Seasons*

30.3.2.1 Productivity (*Kharif* 2014)

All through *kharif* season during 2014, the rainfall received in crop growing stages was 570 mm. 2 major runoff events had been occurred. From the harvested water the required irrigation to the *Kharif* soybean crop was given. The obtained yield with different treatments is given in Table 30.5. It was found that the treatment T_2 (2 protective irrigations) have shown enhanced yield as compared to treatment, T_1 (1 protective irrigation) and T_3 (No irrigation). Water use efficiency (1.79 kg ha^{-1} mm^{-1}) and B:C ratio (2.20) was also higher in T_2 over the treatments T_1 and T_3.

30.3.2.2 Productivity (*Kharif* 2015)

Four runoff events were observed in 2015, which had resulted in rainwater harvesting. From the stored water in the ponds, the protective irrigation for the crops was given. The yields of soybean obtained in different treatments are given in Table 30.6. It was inferred that the treatment T_2 have recorded highest yield and B:C ratio (1055 and 1.68) as compared to other treatments.

Table 30.6 Outcome of different treatments on yield of soybean and its economics

Treatments	Yield (kg ha^{-1})	Percent increase in yield over T_3	RWUE	Net returns (Rs.ha^{-1})	B:C ratio
T_1-One protective irrigation (pod initiation)	910	13.46	1.48	12,355	1.49
T_2-Two protective irrigations (at pod initiation and pod filling)	1055	31.54	1.72	17,285	1.68
T_3-Without irrigation	802	-	1.30	8700	1.34

Table 30.7 Protective irrigation impacts on soybean yield and its economics

Treatments	Yield (kg ha^{-1})	% increase in yield over T_3	RWUE	Net returns ($Rs.ha^{-1}$)	B:C ratio
T_1-One protective irrigation (pod initiation)	1399	12.19	1.69	12,113	1.44
T_2-Two protective irrigations (at pod initiation and pod filling)	1848	48.19	2.24	24,217	1.87
T_3-Without irrigation	1247	–	1.51	8017	1.19

30.3.2.3 Productivity (*Kharif* 2016)

Overall, seven overland runoff events were occurred and therefore the rainwater harvesting occurred in the structures. From the harvested and stored water, the supplemental irrigations were given for *Kharif* crops. The yield of soybean with applied treatments is given in Table 30.7. The treatment T_2 has recorded highest yield and B:C ratio (1848 and 1.87) as compared to other applied treatments. The rainwater use efficiency (RWUE) was found to be highest (2.24) in the treatment T_2 as compared to other treatments.

30.3.2.4 Productivity (*Rabi* 2015–2016)

Storage was there in the farm pond. The sowing was done in the month of October. The germination in T_3 (without irrigation) was very poor owing to less residual moisture. However, the supplemental irrigations of 50 mm depth were given to chickpea and the yield of chickpea with protective irrigations is presented in Table 30.8. Highest yield and B:C ratio (730 kg ha^{-1} and 1.52) was noted in the two protective irrigation treatment of 50 mm depth each with sprinkler set (T_2).

Table 30.8 Yield of chickpea during 2015–2016

Treatments	Yield (kg ha^{-1})	% increase in yield over T_1	Net returns ($Rs.ha^{-1}$)	B:C ratio
T_1-One protective irrigation (after sowing)	510	–	9880	1.04
T_2-Two protective irrigations (after sowing and at flowering)	730	43.13	16,740	1.52

Table 30.9 Yield of chickpea during 2016–2017

Treatments	Yield (kg ha^{-1})	% increase in yield over T_1	Net returns (Rs.ha^{-1})	B:C ratio
T_1-One protective irrigation (after sowing)	1082	–	24,947	2.03
T_2-Two protective irrigations (after sowing and at flowering)	1538	42.14	45,103	2.81

30.3.2.5 Productivity (*Rabi* 2016–2017)

During the *Rabi* season 2016–2017, the germination in treatment T_3 (with no irrigation) was very poor due to less residual moisture in the soil. However, the supplemental irrigation of 50 mm depth was given to chickpea and the recorded yield of chickpea with protective irrigation is given in Table 30.9. The highest yield and B:C ratio (1538 kg ha^{-1} and 2.81) was recorded in the treatment two protective irrigations of 50 mm depth each with sprinkler set from stored pond water (T_2).

30.3.3 Use of Harvested Water for Vegetables

The stored farm pond water (2014–2015) was used for timely irrigation to various vegetables by using MIS, i.e. micro-irrigation systems. It was revealed that for the vegetables like brinjal, ridge gourd, cowpea, Indian round gourd (demshe), lady's finger, spinach and fenugreek (methi) the B:C ratio (Fig. 30.2) was in the range of 1.26 (Indian round gourd) to 1.94 (Ladies finger) and water use efficiency (Fig. 30.2) was in the range of 1.03–4.35 kg/m^3. The computed income from the vegetables crops comes to Rs. 78,092 ha^{-1} (Fig. 30.3).

From Table 30.10, it was found that in 2015–2016, the vegetable crops like Cluster Bean, Brinjal, Okra, Sponge Guard, Bitter Guard, Coriander, Carrot, Fenugreek, Spinach, Tinda and Radish the WUE (water use efficiency) was in the range of 1.05–4.50 kg/m^3. The received income from the vegetables plots during that season was Rs. 7868. Total income (computed) from these vegetables was Rs. 73395 ha^{-1}.

The micro-irrigation systems were used for application of stored farm pond water during the season 2016–2017 to different vegetables. From Table 30.11, it can be inferred that in the vegetable crops like Radish, Sponge Guard, Tinda, Spinach, Coriander, Fenugreek, Cluster Bean, Dolichus bean (wal), Brinjal Chilli and Cowpea the WUE was in the range of 2.50–5.60 kg/m^3. The income obtained from small plots during the season was Rs. 9375. Total computed income from vegetables crops was Rs. 101241 ha^{-1}.

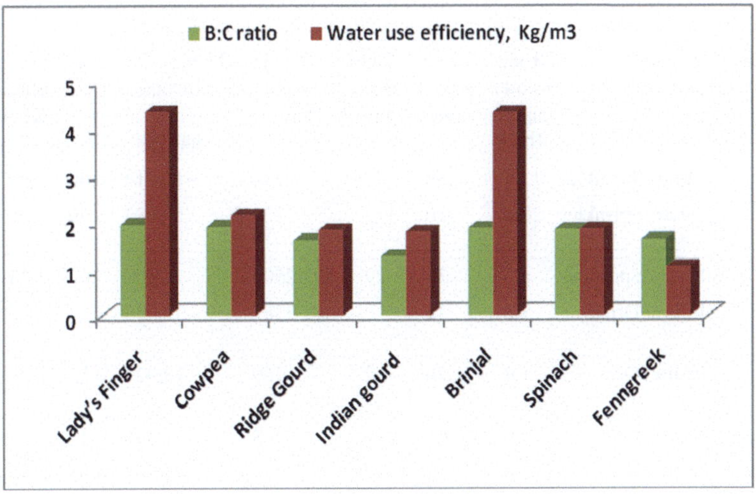

Fig. 30.2 B:C ratio and water use efficiency for different vegetables due to use of MIS during 2014–2015

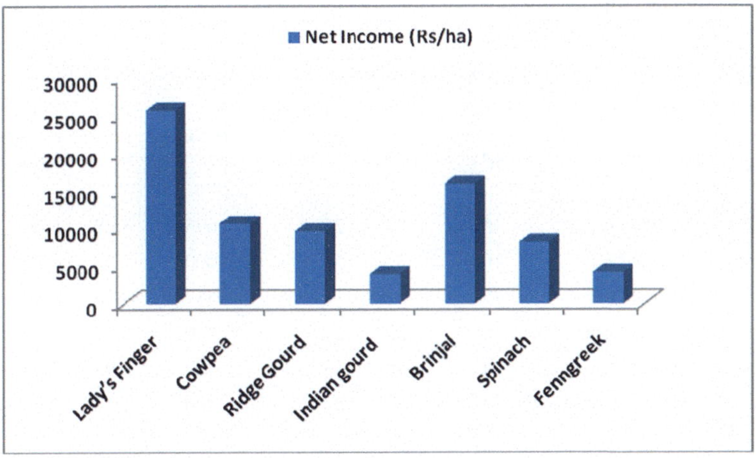

Fig. 30.3 Net income from different vegetables due to use of harvested water during 2014–2015

30.4 Limitations of Study

The rainfall runoff relationship depends mainly on the average annual rainfall for the particular location and if the rainfall is not up to that mark and if runoff events do not takes place then the rainwater harvesting and thereby the storage in the structures/reservoirs/farm ponds may not occur thus protective irrigation with the water from these structures will not be possible.

Table 30.10 Yield economics of different vegetables for the season 2015–2016

Vegetables	Irrigation system	Total water applied (m^3)	Plot Area (m^2)	Yield (kg $plot^{-1}$)	Computed Yield (kg ha^{-1})	Net Income (Rs ha^{-1})	B:C ratio	Water use efficiency, Kg/m^3
Okra	Micro-Sprinkler	12.0	280	20	714	−16,314	0.60	1.66
Cluster Bean	Micro-Sprinkler	28.0	410	112	2732	59,683	3.39	4.0
Brinjal	Micro-Sprinkler	4.0	36	9	2500	2500	1.05	2.25
Sponge guard	In-line Drip	7.0	42	10	2381	27,048	1.64	1.42
Bitter guard	In-line Drip	1.9	26	2	769	−18,923	0.55	1.05
Fenugreek	In-line Drip	7.0	65	15	2308	23,646	2.17	2.14
Spinach	In-line Drip	11.0	52	49	9423	165,962	6.38	4.45
Coriander	In-line Drip	6.0	45	8	1778	37,333	2.17	1.33
Carrot	In-line Drip	6.0	45	8	1778	1556	1.05	1.33
Radish	In-line Drip	10.0	45	45	10,000	246,000	5.59	4.50
Tinda	In-line Drip	2.0	26	3	1154	8846	1.25	1.50

Table 30.11 Use of MIS and yield economics of vegetables for the season 2016–2017

Vegetables	Irrigation system	Water applied (m^3)	Plot area (m^2)	Yield (kg $plot^{-1}$)	Net Income (Rs $plot^{-1}$)	B:C ratio	Water use efficiency, Kg/m^3
Coriander	In-line Drip	5	32	25	695	3.28	5.00
Fenugreek	In-line Drip	10	32	56	1000	3.50	5.60
Radish	In-line Drip	22	32	104	577	3.26	4.73
Spinach	In-line Drip	8	32	40	720	3.57	5.00
Cluster bean	In-line Drip	4	40	13	300	2.94	3.25
Dolichus bean (Wal)	Micro-Sprinkler	3	40	8.5	200	2.43	2.83
Sponge Guard (Dodke)	In-line Drip	4	60	15	390	2.86	3.75
Tinda (Dhemase)	In-line Drip	6	80	15	355	3.09	2.50
Brinjal	In-line Drip	10	80	43	770	3.52	4.30
Brinjal	Micro-Sprinkler	25	158	148	2760	3.94	5.92
Cowpea	Micro-Sprinkler	6	158	20.25	428	3.38	3.38
Chilli	Micro-Sprinkler	8	182	42	1180	3.36	5.25
Total		111	926	–	9375	–	–

30.5 Future Scope of Study

From this study the relationship between surface runoff storage and its reuse was established. Therefore farmers, extension workers, technical staff from State Agriculture Department will get the idea of calculation of surface runoff thereby fixing the exact dimensions of the farm ponds for a particular location so that assured rainwater will be collected in that particular farm pond and as per needs can be utilized for different purposes.

30.6 Conclusions

The catchment, storage capacity and command area relationship was developed at AICRP for Dryland Agriculture Centre, Dr. PDKV, Akola.

1. It was observed that from 1 ha catchment, on an average 544.52 m^3 water can be stored in the farm pond which can be utilized to irrigate (command) about 1.0 ha area with 5 cm (depth) of irrigation.
2. The collected rainwater in the storage structures/farm ponds was utilized for providing irrigation for protecting the crops under stress condition (one or two as per condition) during *Kharif* to the soybean crop and during rabi for chickpea and also to vegetables through micro-irrigation systems.
3. Overall, the treatment T_2 (two protective irrigations) for soybean as well as chickpea have performed better as compared to other treatments. Because during stress condition if crop gets water its growth as well as yield and economic returns will be more.
4. If the harvested farm pond water is used for protective irrigation to soybean crop during *kharif* season then on an average around 38.56% increase in yield was observed as compared to without irrigation treatment. Similarly, for chickpea the average increase in yield was 31.84% in protective irrigation treatments as compared to non-irrigated treatment.
5. Therefore, it can be concluded that while making the farm pond, the technical assistance must be taken by the farmers so that the farm pond technology will be feasible and beneficial to them. The farm pond if laid properly then it will be helpful for bringing the sustainability in Dryland agriculture.

Acknowledgements The authors are thankful for the financial support received by the ICAR—All India Coordinated Research Project for Dryland Agriculture, Central Research Institute for Dryland Agriculture, Santoshnagar, Hyderabad-500059, Government of India and Dr. Panjabrao Deshmukh Krishi Vidyapeeth, Akola, Government of Maharashtra. The authors are also thankful to the Chief Scientist, AICRPDA, Akola for permission towards undertaking the field experiments at AICRPDA, Dr. PDKV, Akola and for providing the required facilities.

References

Adhikari RN, Mishra PK, Muralidhar W (2009) Dugout farm pond—a potential source of water harvesting in deep black soils in Deccan plateau region. Rainwater harvesting and reuse through farm ponds. In: Proceedings of national workshop cum brain storming, CRIDA, Hyderabad, pp 100–108

Annual Report, AICRIPAM (2010) All India coordinated research project on agrometeorology under AICRPDA, annual report. Dr. Panjabrao Deshmukh Krishi Vidyapeeth, Akola, Maharashtra, India

Bangar AR, Deshpande AN, Sthool VA, Bhanavase DB (2003) Farm pond—a boon to agriculture. ZARS, Solapur (MPKV), pp 32–37

Freebairn DM, Wockner GH, Silburn DM (1986) Effect of catchment management on runoff, water quality, and yield potential from vertisols. Agric Water Manag 12(1):1–19

Jalal Uddin Md, Sravya V, Abdullah OB (2017) Rain water harvesting (farm pond). Int J Emerg Res Manag Technol 6(2):76–86

Krishna JH, Arkin GF, Martin JR (1987) Runoff impoundment for supplemental irrigation in Texas. Water Resour Bull 23(6):1057–1061

Nagdeve MB, Patode RS (2012) Protective irrigation through farm pond for enhancing crop productivity. Application technologies for harvested rainwater in farm ponds. In: Proceedings of national consultation meeting held at CRIDA, Hyderabad, pp 62–67

Nagdeve MB, Patode RS (2015) Farm pond for enhancing vegetable productivity in rainfed condition. National Seminar on Dryland Agriculture in Vidarbha: Priorities and Development Issues, Dr. PDKV, Akola, 36.

Patode RS, Nagdeve MB, Palaspagar NR, Chary GR (2017) Rainwater management through *In-Situ* soil and water conservation techniques and utilization of harvested water through farm pond. In: Proceedings of national conference on sustainable water and environmental management (SWEM-17) at JNTUH, Hyderabad, pp 168–172

Reddy KS, Kumar M, Rao KV, Maruthi V, Reddy BMK, Umesh B, Ganesh Babu R, Srinivasa Reddy K, Vijayalakshmi, Venkateswarlu B (2012) Farm ponds: a climate resilient technology for rainfed agriculture; Planning, design and construction. Central Research Institute for Dryland Agriculture, Santoshnagar, Saidabad, Hyderabad 500059, Andhra Pradesh, India, 60

Sharma BR, Rao KV, Vittal KPR, Ramakrishna YS, Amarasinghe U (2010) Estimating the potential of rainfed agriculture in India. Prospects for water productivity improvements. Agric Water Manag 97:23–30

Singh RP (1986) Farm ponds. CRIDA, Hyderabad Project Bulletin No. 6

Sivanappan RK (2006) Rainwater harvesting, conservation and management strategies for urban and rural sectors. National Seminar on Rainwater Harvesting and Water Management, Nagpur, pp 1–5

Sthool VA, Upadhye SK, Jadhav JD, Sanglikar RV, Rao VUM (2013) Farm pond—a boost for sustainability in Dryland under climate change situation. MPKV, Res. Pub. No. 80

Taley SM, Patode RS, Dikkar MG, Hedau VD (2009) Rainwater management in deep black soils under rainfed agro-ecosystem. Green Farm 2(12):816–820

Vora MD, Solanki HB, Bhoi KL (2008) Farm pond technology for enhancing crop productivity in Bhal area of Gujrat. J Agric Eng 45(1):40–46

Wani SP, Pathak P, Sreedevi TK, Singh HP, Singh P (2003) Efficient management of rainwater for increased productivity and groundwater recharge in Asia. In: Kijne et al. Water productivity in Agriculture. Limits and Opportunities for improvement. CABI publishing, Cambridge, USA

Chapter 31
Assessing Groundwater Recharge Potential Through Rainwater Harvesting in Urban Environment: A Case of Bhopal City

Mrunmayi Wadwekar and Rama Pandey

Abstract In this age of globalization, Indian cities are growing at an unprecedented rate and are using their natural resources without much thought to their regeneration and rejuvenation. The primary water requirement is now fulfilled by utilizing local and external resources. Bhopal city, in spite of having the ample water bodies, is facing a water crisis. Extensive groundwater extraction over the years has led to depletion and degradation of resource, and the water requirement is now fulfilled from Narmada River. This paper analyses the existing status of groundwater in the city and identifies the probable reasons for the depletion through a Cause/Effect relationship. The study explores suitable recharge locations through GIS based land suitability analysis and identifies techniques to improve the resource. Spatial planning interventions and policy measures are proposed to recharge and rejuvenate the groundwater resources through rainwater harvesting and improve the overall ecosystem health.

Keywords Groundwater recharge · Natural resources · Rainwater harvesting · Rejuvenation

31.1 Introduction

Groundwater plays a significant role in the development and growth of urban areas. Although the major user of groundwater is agriculture, urban areas are increasingly dependent on it due to discrepancy between the supply and demand of the municipal water.

M. Wadwekar (✉)
Architect and Environmental Planner, Bhopal, India
e-mail: ar.mrunmayi@gmail.com

R. Pandey
School of Planning and Architecture Bhopal, Bhopal, India
e-mail: ramapandey@spabhopal.ac.in

A. Pandey et al. (eds.), *Water Management and Water Governance*, Water Science and Technology Library 96,
https://doi.org/10.1007/978-3-030-58051-3_31

Groundwater while being an important source of potable water, also provides ecosystem services like providing base flows to streams, lakes and rivers during dry season. It is also a source of water for vegetation in dry seasons. It acts like a habitat for species and microorganisms that are otherwise part of the habitat. Groundwater while being a valuable resource is often neglected or ignored because of its hidden nature. Problems with the quality and quantity of the water are noticed only when they start affecting the human population depending on it.

Cases such as Plachimada water poisoning in Kerala by Coca Cola is a known example of severe over extraction and degradation of the resource. Similarly, the case of Love Canal in Niagara and Hinkley in California are landmark cases for groundwater pollution by private entities.

In this context, depletion and degradation of a resource such as this provides an important objective to improve and restore the environmental benefits available from the resource.

Bhopal city is facing water scarcity in the recent years and the groundwater is polluted as well as depleting. The city therefore has been selected for this study.

31.2 Study Area

Bhopal is the State capital of Madhya Pradesh and of former princely state of Bhopal. Bhopal is also known as "City of lakes" due to presence of numerous small and large waterbodies, the largest among them being "Bhojtal" or locally known as Upper lake/Bada talab.

The city is also infamously known as site of worst industrial disaster of Union Carbide in 1984. The incident and the subsequent deaths was a result of leaking of poisonous gas Methyl Isocyanate (MIC) from the storage tanks on the intervening night of 2–3 December 1984 (BMC 2017). This resulted in contamination of soil and groundwater in the surrounding areas which continues till today.

Efforts have been undertaken by various authorities to monitor and evaluate the contamination, but the pending legal case makes it difficult to implement corrective measures for the site and the surroundings.

31.3 Environmental Issues in Bhopal

Bhopal city has ample natural beauty with its undulating topography and many lakes. However, the urbanization trends and its subsequent destruction have left the city shorn of its well-known beauty. According to a study by researchers from IISc Bangalore, Bhopal lost its vegetative cover of 92% in 1977 to 21% in 2014; it predicts 11% of vegetative cover by 2018 and just 4% by 2030 if pre-emptive actions are not taken (Aithal et al. 2016).

There are also problems with the water quality of various lakes in Bhopal. According to a study by an independent newspaper in Bhopal, among the 31 registered lakes in and around Bhopal only 21 exist as of 2016 with 11 have been permanently lost (Team DB Post 2016).

31.4 Problems with Water Supply and Quality of Water

Upper lake (Bada talab) is the only lake within the city limits used for potable water supply. However, other lakes are in various stages of eutrophication as shown in Table 31.1 (LCA 2008; Pani et al. 2014). According to City Development Plan (CDP), Bhopal prepared under the JNNURM project, Rivers and streams within the city are increasingly being used as sewage disposal streams rather than for clean water. Groundwater charged by sewage streams is getting polluted with chemical and microbial contaminants. Surface water bodies are getting encroached and the water quality is also getting deteriorated.

The city thus faces various problems for water supply. The shortage of potable water is now being addressed by bringing water from Narmada River situated 100 km away. Despite the city having many water bodies, the dependency on a water source so far away is an irony.

Similarly, many parts of the city are dependent on tube wells, dug wells and hand pumps in absence of piped potable water. Most of these localities are on the outskirts of the city. According to one estimate, this accounts for 35–40% of total supply (CSE 2006).

31.5 Issues in Groundwater

Central Ground Water Board (CGWB) maintains 26 dug wells and four piezometers in Bhopal district for monitoring and assessment of the dynamic groundwater resources. The continuous assessment has suggested that Phanda block (Huzur tehsil) within which Bhopal city is situated is now in semi critical stage as shown in Table 31.3. The groundwater development has crossed almost 75% of available capacity. Among the four piezometers in the tehsil, the one situated at Ahmedpur near Barkatullah University has completely dried up in recent years (Jogdand 2017).

The fall in the groundwater levels is attributed to over exploitation of the resource as well as low infiltration (CSE 2006; CGWB 2013). Various researchers and activists have undertaken studies to identify issues with the water quality of surface and underground water sources of the city. The quality of the water has been found of poor quality at various locations with anthropogenic contaminants discovered in the samples (Trivedi and Kataria, 2012; Wanganeo et al.2007).

Mr. Subhash Pandey, a noted environmental activist in Bhopal has won a case filed in National Green Tribunal, Central Bench, Bhopal against Urban Administration

Table 31.1 Status of water bodies in Bhopal

Salient features of Bhopal Lakes				
S. No.	Name of water body	Water spread area (in ha)	Present use	Remark/status
1	Upper Lake	3100	Water supply, recreation and fisheries	Mesotrophic and part of the lake is Eutrophic
2	Lower Lake	129	Raw water supply and recreation	Advance stage of Eutrophic
3	Shahpura	39	Recreation and fisheries	Advance stage of Eutrophic
4	Motia Tank	1.89	Washing	Advance stage of Eutrophic
5	Siddiqui Hussain Tank	1	Abandoned	Bog Lake
6	Munshi Hussain Khan Tank	1.2	Fisheries	Eutrophic
7	Lendiya Talab	1.5	Recreation and fisheries	Advance stage of Eutrophic
8	Sarangpani	4.2	Recreation	Advance stage of Eutrophic
9	Laharpur Reservoir	350	Irrigation	Advance stage of Eutrophic
10	Hathaikheda Reservoir	113	Irrigation	Mesotrophic
11	Halali Reservoir	1625	Irrigation	Mesotrophic
12	Kerwa Reservoir	524	Irrigation	Mesotrophic
13	Kolar Reservoir	2850	Potable water supply and Irrigation	Mesotrophic
14	Char Imli Pond	1.2	Recreation	Eutrophic
15	Ayodhya Nagar Abandon stone quarry Ponds (4 Nos)	6	Recreation	Mesotrophic
16	Damkheda Village Pond	2.4	Potable water, Recreation	Mesotrophic

Source Lake Atlas of Madhya Pradesh (2008)

Table 31.2 Water quality study

GW quality sampling statistics						
S. No.	Parameter	KS1	KS2	KS3	KS4	BIS standards
1	pH	7.7	7.8	7.6	7.6	6.5–8.5
2	Dissolved oxygen (mg/L)	*6.4*	4.4	4.4	4	4–6
3	BOD (mg/L)	*12*	*20*	*4*	*12*	2
4	COD	40	64	16	44	

Source Annexure 11 of case filed in NGT Bhopal
Subhash Pandey versus Municipal Corporation Bhopal and 6 others
Case no. 34/2013(CZB) and M.A.No. 542/2014
Values in italics suggest that the parameters are above the suggested value as per BIS Standards

and Development Department (UADD) for failure to control the pollution caused by Shahpura Lake as demonstrated in Table 31.2. Shahpura Lake is an oxidation pond for waste water and has been found to pollute the surrounding wells and aquifer due to leaching of toxics and contaminants (Subhash Pandey vs. Municipal Corporation Bhopal and 6 others 2014).

31.6 Methodology

The study was conducted to identify the issues of groundwater in distinct parts of the city through literature and primary survey. In the first stage, a study area was selected within the city having issues of groundwater based on the following criteria. The city was divided into five zones of North, West, East, South and Central. Each of the zones were analysed for existing status of groundwater; Pollution statistics through literature; Current urban growth and Proposed growth and Existing status of water supply and dependency on groundwater.

In the second stage, the relationships between various factors that cause groundwater depletion were identified using a Cause/Effect relationship diagram as detailed in Sect. 31.8. Each of the cause were then analysed through Software tools such as ArcMAP and ERDAS and verifying them with primary survey.

In the third stage, an attempt has been made to explore potential recharge zones within the study area. An estimation of rainwater harvesting is carried out using water balance calculations and runoff estimation. This helps in demonstrating the potential of the study area for recharging groundwater.

31.7 Analysis of Study Area

A study area was identified from the city to assess the impacts of urbanization and identify the potential for groundwater recharge. The current and proposed growth direction of the city was identified based on a temporal study of the urban development. Also cases regarding pollution of groundwater were identified.

As depicted in Fig. 31.1, all the peri urban edges of the municipal area are unserved by piped water network and therefore depend on groundwater for their requirement. These areas have also seen rapid increase in population due to urbanization pressures. As depicted in Table 31.3, groundwater levels in the city are witnessing a decreasing trend in recent years. The locations mentioned in the study are scattered all across the city.

The current and proposed growth in the city is towards south direction. The presence of a water body like Bada talab limits the growth towards western part of the city. The south direction is the planned settlement preferred by most high income group localities. This area has therefore seen rapid development in recent years. Based on these factors, zone 13 comprising of four wards was identified as one undergoing most rapid change. A ward immediately close, Katara ward, from zone 19 showing similar characteristics was added to study area as shown in Fig. 31.2.

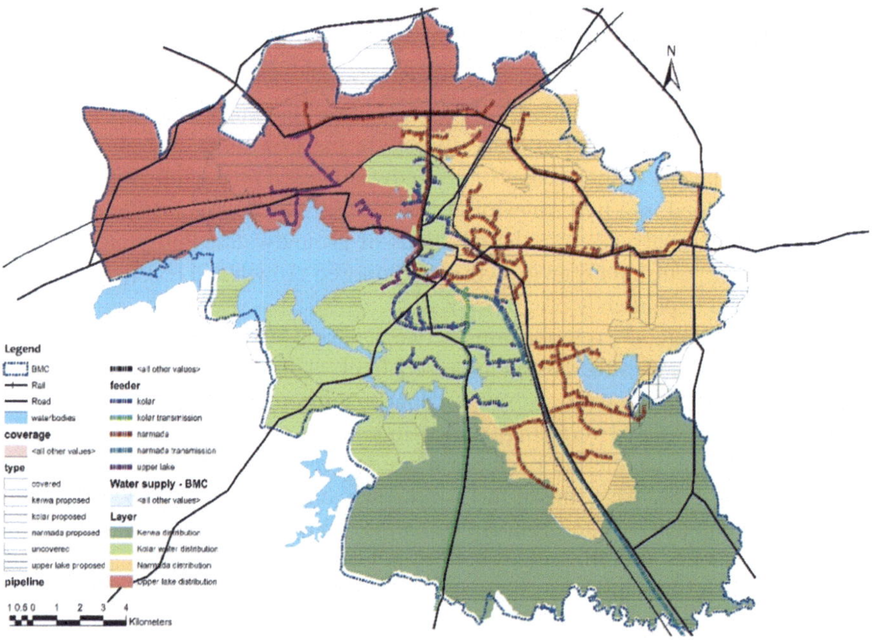

Fig. 31.1 Existing water supply network of Bhopal city. *Source* CDP—revised, Mehta and Mehta Associates, Indore (2006)

Table 31.3 Groundwater levels in Bhopal city

Sampling period	Ayodhyanagar				Bairagarh				Gandhinagar			
Year	Pre monsoon —Jan.	Pre monsoon —May.	Monsoon —August	Post monsoon —Nov.	Pre monsoon —Jan.	Pre monsoon —May	Monsoon —August	Post monsoon —Nov.	Pre monsoon —Jan.	Pre monsoon —May	Monsoon —August	Post monsoon —Nov.
1996												
1997												
1998												
1999												
2000												
2001												
2002												
2003												
2004				2		10.52		6.25		8.05		0.84
2005		6.15				10.52				2.38		
2006				5.8				7.37				2.24
2007						9.55		7.3		8		3.15
2008												
2009												
2010												
2011												
2012	2.35	3.04	0.95	1.74	7.92	9.95	4.68	6.02	5.16	7.74	0.41	2.6
2013	1.18	1.98	0.85	1.31	6.97	9.5	6.2	6.94	1.62	15.36	8.1	10.28
2014	0.46	8.4	4.82		0.61	1.63	2.3		8.9	2.21	0.9	
2015												

(continued)

Table 31.3 (continued)

Sampling period	Islamnagar				Misrod (S)				Shahpura			
Year	Pre monsoon —Jan.	Pre monsoon —May	Monsoon —August	Post monsoon —Nov.	Pre monsoon —Jan.	Pre monsoon —May	Monsoon —August	Post monsoon —Nov.	Pre monsoon —Jan.	Pre monsoon —May.	Monsoon —August	Post monsoon —Nov.
	11.78	15.9	9.65	11.1								
1996	13.4	16.5	7.8	11.4								
1997	10.7	15.5	10.07	9.5				4.5				
1998	12.35	15.88	10.63	10.81		8.73	7.93	4				
1999	12.39	16.47	11.6	12.1	5.53	8.83	4.7					
2000	14.1		7.75	12.27		15.21		10.96				
2001	13.8		9.56	13.5	12.38	18.58		12.05				
2002	15.51		6.72	4.27	14.47	17.38	12.81	9.78				
2003	12.8	18.4	9.65	11.6	11.22	11.26	11.87	9.26				
2004	13.3	18.4	10.84	8.86	9.56	13.12	10.74	8.91				
2005	13.8	16.36	8.4	12.89	9.38	16.62	10.41	6.21		2.34		
2006	13.8	18.4	8.5	13.38	5.7	9.42	8.88	8.99				1.75
2007	15.5	18.4	8	13.02	8.98		12.13	11.53		1.95		2
2008	16.4	18.4	13.44	9.9	13.48							
2009	12.8		9.3	12.9	10.98							
2010	12.4	18.4	9	10.25		18.98						
2011	11.5	18.4				12.78						
2012	18.4	16.9	9.1	12.06						1.46	0.6	0.93
2013	14.38	15.56	10.2	11.08					0.48	1.24	0.58	0.75
2014	11.95	16.8	10.5	12.8	6.41	8.81	4.08	4.88	5.1	7.65	0.33	
2015												

Source Generated from CWC, IndiaWRIS website

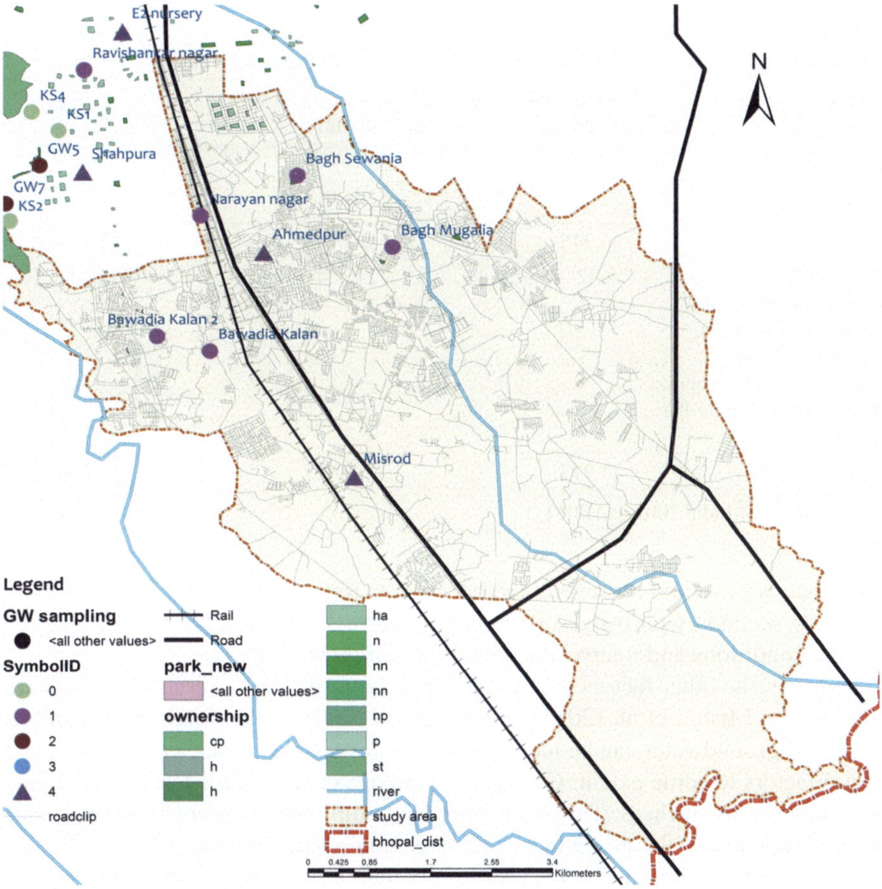

Fig. 31.2 Study area. *Source* Generated by Author

The major changes in the area have been of land use conversion. As identified from Google Earth images and through primary survey, it was noticed that majority of the agricultural fields are now sold and converted into high end residential townships. The area adjoining the main thoroughfare has transformed into a commercial zone with malls, shopping complexes and businesses dotting its edge.

The population for the study area has increased at a rate of 57% from 2001 to 2011 as depicted in Table 31.4 while that of Katara ward at 122%.

31.8 Identification of Drivers of Change

Groundwater is a resource that is an important part of hydrological cycle but is dependent on various factors for its effective quality and quantity. According to a study by Daiman and Gupta (2015), groundwater and its movement within the

Table 31.4 Population increase in study area

Village/ward	2001 (Census 2001)	2011 (Census 2011)	% increase
Misrod	Ward 53 + Ward 54 = 66,345	Ward 52 + Ward 53 = 98,027	
Jatkhedi			
Barkatullah			
Baghmugalia			
	66,345	98,027	47.75
Katara (Ward 85 as of 2016) comprising of Katara, Barrai, Bagli, Rapadiya, Maksi, Bhairopur, Chan Deepdi, Bangrasia and Samardha Kaliasot	9632	21,434	122.53
Total	75,977	119,461	57.23

Source Census of India 2001 and 2011

subsurface is governed by several factors such as topography, lithology, geological structures, secondary porosity, soil, drainage pattern, landforms, land use/land cover, climatic conditions and interrelationship between these factors.

However, the other factors that affect depletion of groundwater at local level are identified by Mishra et al. (2014) and WaterAid (2008). The factors that manifest in reduced groundwater can be identified as anthropogenic causes, urbanization and natural factors to some extent. Changes in land use mostly occur locally and create large impervious surfaces. This leads to reduced infiltration and increasing surface runoff which affects the groundwater storage and recharge potential.

A cause effect relationship diagram as depicted in Fig. 31.3 identifies the various causes that effect in reduced groundwater infiltration.

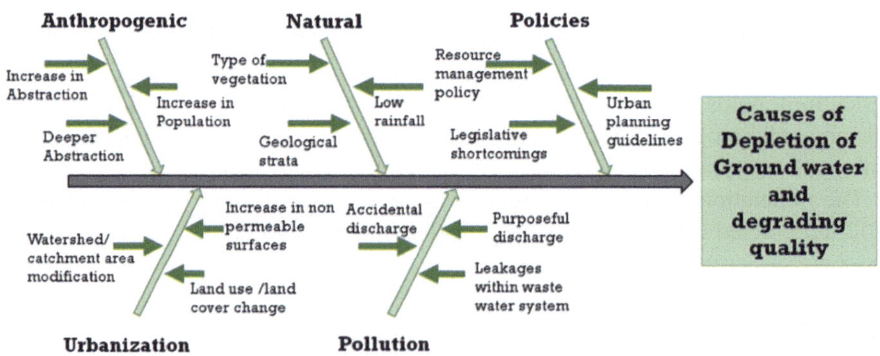

Fig. 31.3 Cause and effect relationship. *Source* Composed by Author

To study impacts of urbanization, remotely accessed data was used and interpreted using GIS software like ArcMap and ERDAS. LANDSAT images from various years were acquired through open source USGS and analysed. The acquired LANDSAT images were from 1972, 1992, 2000, 2011 and 2016.

The standard image processing techniques of extraction, layer stacking, geometric correction/geo-referencing and change detection were performed on the five Landsat TM images obtained for the month of October 1972, 1992, 2000, 2011 and 2016. The raster images were then interpreted to identify land use/land cover change. Built-up, barren surfaces, agriculture and others were the four categories that were identified.

The classified images were then analysed for changes in land cover wherein area calculations identified change in various land cover types. As demonstrated in Table 31.5 and which can be seen from Fig. 31.4, Agriculture reduced by 45% while built-up increased by almost 260% from 1972 to 2016.

Similarly, to assess the issues in the drainage network, a stream order map was generated using ArcMAP and a five-metre Digital Elevation Model raster image. The resultant stream order was overlapped with built-up area for all the assessment years identified previously during Land use/Land cover analysis as demonstrated in Fig. 31.5.

Table 31.5 Land cover change

Land use/land cover change						
	1972	1992	2000	2011	2016	% change
Agriculture	3524.03	3234.12	2968.42	2300.81	1906.86	−45.89
Built-up	594.86	678.84	1408.28	2205.86	2129.80	258.03
Barren	522.17	1198.76	661.44	456.84	1067.38	104.41
Other	1373.33	908.88	982.45	1060.76	911.81	−33.61

Source Computed by Author

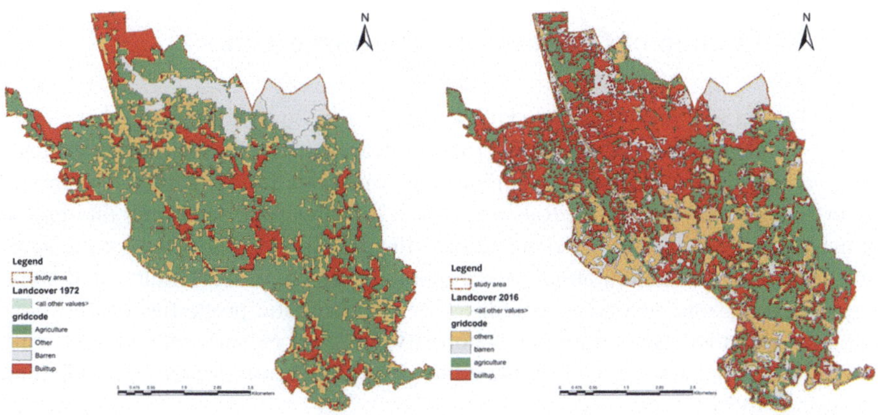

Fig. 31.4 LULC 1972 and 2016. *Source* Generated by Author

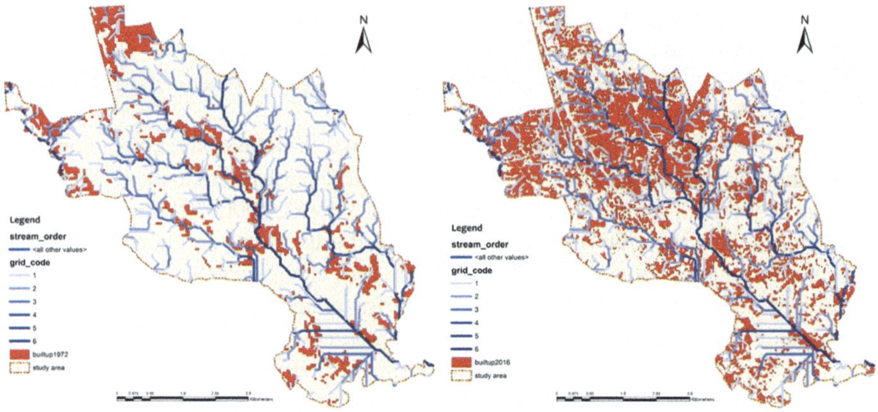

Fig. 31.5 Stream modification 1972 and 2016. *Source* Generated by Author

The analysis revealed that impervious surfaces have increased; directly leading to less surface area for groundwater infiltration. Also, several of the first and second order streams essential for carrying the storm flushes are blocked or modified due to built-up. The natural topography is disturbed thereby changing the flooding pattern and reducing infiltration.

The analysis as mentioned in Sect. 31.7 had revealed that the study area and almost all the peripheral areas of Bhopal city are dependent on groundwater as a resource. To quantify the increase in abstraction, a calculation was used to predict the absolute increase in groundwater extraction in the study area. Based on the increased population identified from Census data and assuming 135 L per capita demand as daily requirement, the total increase in groundwater abstraction in absolute terms was calculated as 2142.2 MCL/year from 2001 to 2011.

31.9 Estimation of Groundwater Recharge Zones

The study area was further assessed to estimate the amount of rainwater that can be harvested and utilized to recharge the groundwater resources. To identify the groundwater recharge potential zones, techniques and methods followed by (Daiman and Gupta 2015; Patil et al. 2014; Rais and Javed 2014) were adopted and interpreted for the study area. Figure 31.6 demonstrates the methodology adopted for the analysis. The base information about geomorphology, geology, soil textures and slope were collected from secondary sources. The characteristic properties of the study area were identified as medium to low porosity and good recharge potential in some pockets. These layers were then utilized to generate a groundwater potential zone map using ArcMAP software.

The resultant imagery created through ArcMAP is depicted in Fig. 31.7. It

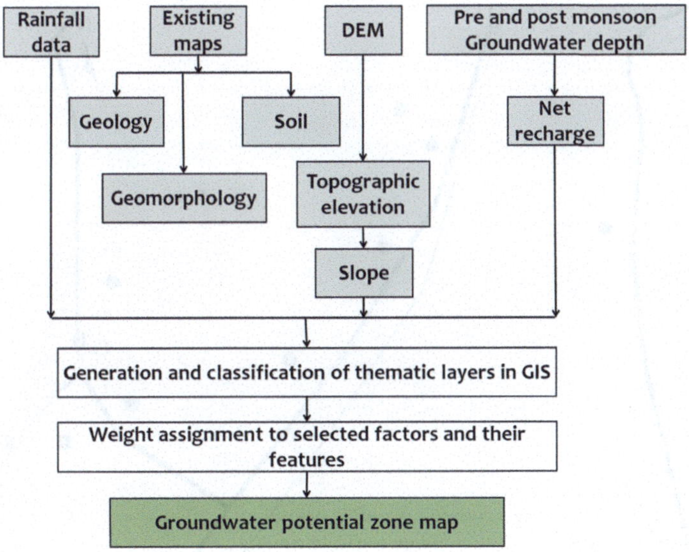

Fig. 31.6 Methodology adopted for GW recharge zone identification. *Source* Generated by Author

provided three types of recharge potential zones of good, moderate and poor infiltration. The map was overlapped with proposed master plan of 2031 to identify the techniques that can be applied to the study area.

Since, most of the existing as well as proposed built-up is of residential development, the recharge techniques are according to those structures. The recharge mechanisms suitable in this case are rooftop rainwater harvesting structures like rainwater trench, rainwater pits, recharge pits etc. Abandoned and dried wells and bore wells can also be utilised as recharge wells to direct water to the aquifer. The criteria of selection of recharge structures is as per by CGWB (2011).

31.10 Rainwater Harvesting Estimation

The study area has a deficit of 11 million L of water per year as identified from Water Balance calculations. The factors considered include various existing land uses and required water for the uses. The calculations are detailed in Table 31.6.

Bhumi Vikas Adhiniyam, Government of Madhya Pradesh specifies that every built structure with a roof area of 150 m^2 should be used for roof top rainwater harvesting. The study area has particularly high built-up and some Government institutes covering vast tracts of land.

An approximate estimate can be identified by using a sample roof top size of 150 m^2. As identified in Table 31.7, a single building with roof top area of 150 m^2 can save water by 109,650 L every year at 75% rainfall probability.

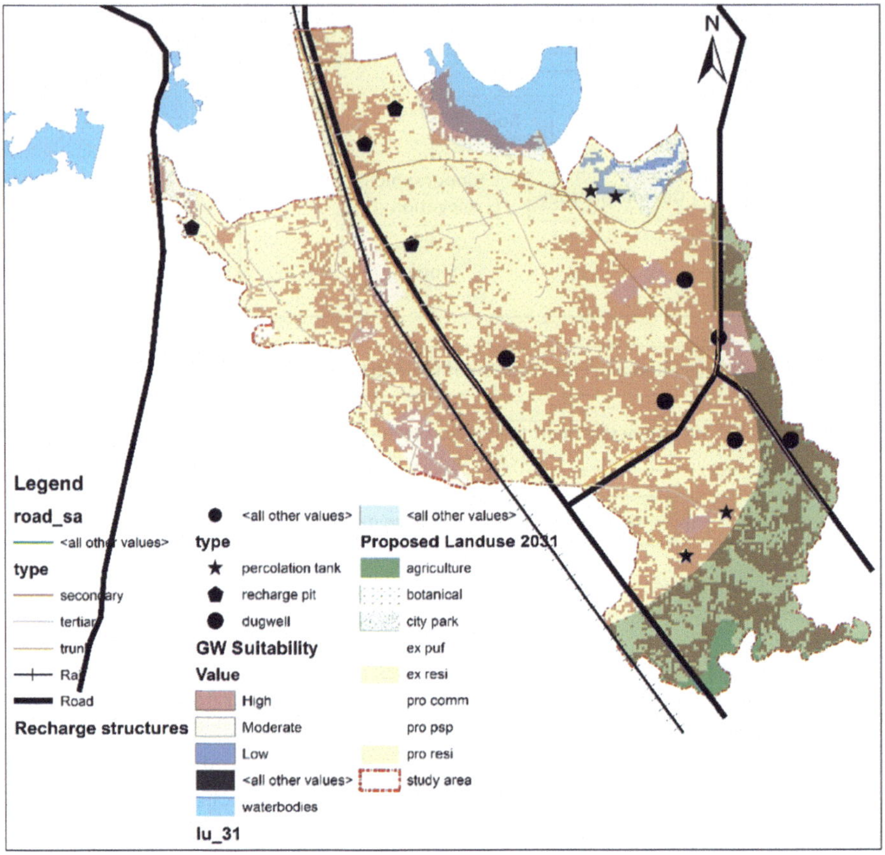

Fig. 31.7 Groundwater recharge potential map. *Source* Generated by Author

For assessing the total potential of study area, the building footprint and ownership was identified. There are some Government institutes in study area which can be utilized for Rainwater harvesting. These institutes cover approximately 192 ha out of the 6000 ha of study area.

Using the rainfall at 75% dependability and runoff coefficient for tiled surface and open land surface, the total rainwater harvesting potential for all the Government land is as mentioned in Table 31.8.

Comparing the harvested water potential to the deficit, it is observed that, only saving the water from Government institutes will reduce the deficit by almost 90% for the study area. Using the same principle, if Rooftop rainwater harvesting is implemented at 75% efficiency, the study area will yield an excess of approximately 74 million L of fresh water every year that can be utilized to recharge the groundwater sources through identified means.

Table 31.6 Water balance calculations

Water requirement—2011			
Total available water			
Total precipitation	Rainfall × study area	76,201.78	KL/year
Total groundwater draft	No. of wells × yield	12,053.25	KL/year
Total municipal supply	Domestic + commercial + educational	54,40,544.73	KL/year
Total		5,528,799.76	KL/year
Total requirement			
Total municipal requirement	Domestic + Commercial + Educational	66,48,933.08	KL/year
Agriculture	Major crop—Wheat and Soyabean	24,700	KL/year
Total requirement		66,73,633.08	KL/year
Water balance		-11,44,833.32	KL/year

Source Computed by Author

Table 31.7 Rainwater estimation potential

Rainwater harvesting recharge estimation	
Avg. Annual Rainfall	1146.7 mm
Rainfall at 75% dependability	860.02 mm
Roof area	150 m^2
Vol. of water over roof area	129,000 L
Runoff Coeff. (0.85)	109,650 L

Source Computed by Author

Table 31.8 Recharge through Government institutes

Recharge through Govt. Institutes	
Total plot area of Govt. land	192 ha
Volume of water at 75% rainfall and 0.6 runoff coefficient	990,786 K L
Volume of water from rooftops	40,683 K L
Total volume of water	10,31,470 K L

Source Computed by Author

31.11 Conclusion

Urbanization creates lasting impacts on the environment and the natural cycles. Human intervention alters the natural systems irreparably, however, also has the potential to manage and restore the damage to some extent. In contrast to the traditional top-down approach, measures such as rooftop rainwater harvesting can help manage this problem at a smaller and manageable scale. Also, as in case of Bhopal, the legislative mechanism is already available, however what lacks is the implementation and how it can help reduce the issue of water scarcity and improve the restoration needs to be done. Existing Bhumi Vikas Adhiniyam of Madhya Pradesh do not include potential recharge means such as road side culverts, permeable paving areas, public open spaces as probable sources. Including them in the existing policy can help develop a holistic mechanism. Also, on a broader scale, the Groundwater development in the country is currently mostly unregulated. Measures to monitor and limit extractions and policies to avoid overuse and misuse can help maintain the resource for the longer duration while also enabling its natural rejuvenation at a beneficial pace. The Ministry of Water Resources and Ganga Rejuvenation has brought out a model bill in 2016 to monitor and regulate use of Groundwater in the states. The implementation of this bill is currently miniscule and needs to be implemented on an immediate basis.

References

Aithal B, Ramachandra TV, Shivamurthy V (2016) Agent based modelling urban dynamics of Bhopal, India. J Settl Spat Plan 7:1–14. https://dx.medra.org/10.19188/01JSSP012016

Bhopal Municipal Corporation (BMC) (2017) About Bhopal. https://www.bhopalmunicipal.com/city-information/about-bhopal.html. Accessed on 15.04.2017

Central Ground Water Board (CGWB) (2011) Case studies on Rain water harvesting and Artificial recharge. Available at India Water Portal https://www.indiawaterportal.org/data/case-studies-rainwater-harvesting-and-artificial-recharge-2011. Accessed on 23.03.2017

Central Ground Water Board (CGWB) (2013) District information booklet—Bhopal. Available at https://cgwb.gov.in/District_Profile/MP_districtprofile.html. Accessed on 12.02.2017

Centre for Science and Environment (CSE) (2009) Contamination of soil and water inside and outside the Union Carbide India Limited, Bhopal. Available at https://cdn.cseindia.org/attachments/0.63821200_1499406577_Bhopal-Report-Final.pdf

Centre for Science and Environment (CSE) (2006) The waste water portrait. Centre for Science and Environment. Available at https://cdn.cseindia.org/userfiles/portrait.pdf

Daiman A, Gupta N (2015) Identification of groundwater recharge zones and rainwater harvesting sites using geoinformatics: a case of Karwan watershed, Sagar, Madhya Pradesh. J Geomatics 198–202

Jogdand M (2017) Improving ecosystem health through planning for groundwater and vegetation—a case of Bhopal City. Thesis (unpublished), School of Planning and Architecture Bhopal

Lake Conservation Authority of Madhya Pradesh (LCA) (2008) Lake Atlas of Madhya Pradesh. Published by Lake Conservation Authority, Bhopal

Mehta and Mehta Associates (2006) Bhopal City development plan under JNNURM. Available via academia.edu https://www.academia.edu/4365091/Bhopal_CDP_Final. Accessed on 26.10.2016

Mishra N, Kumar D, Gupta KK, Shukla R (2014) Impact of Landuse change on Groundwater—a review. Adv Water Resour Protect 2:28–41

National Green Tribunal (NGT), Central Zone Bench, Bhopal (2013) M.A.No. 542/2014 and Original Application No. 34/2013 (CZ) Subhash Pandey vs. Municipal Corporation Bhopal and 6 others. Available at https://www.indiaenvironmentportal.org.in/files/Shahpura%20Lake%20NGT%2018Sep2014.pdf. Accessed on 9.12.2016

Pani S, Dubey A, Khan MR (2014) Decadal variation in microflora and fauna in 10 water bodies of Bhopal, Madhya Pradesh. Curr World Environ 9(1). https://doi.org/10.12944/CWE.9.1.20

Patil SG, Mohite NM (2014) Identification of groundwater recharge potential zones for a watershed using remote sensing and GIS. Int J Geomatics Geosci 4(3):485–498

Rais S, Javed A (2014) Identification of artificial recharge sites in Manchi basin, Rajasthan using remote sensing and GIS techniques. J Geogr Inf Syst 6(2):162–175. https://doi.org/10.4236/jgis.2014.62017

Sharma S, Bharat A, Das VM (2013) Statistical change detection in water cycle over two decades and assessment of impact of urbanization on surface and sub surface flows. Open J Mod Hydrol 3:165–171. https://doi.org/10.4236/ojmh.2013.34020

Team DB Post (2016) The city of vanishing lakes. Bhopal edition: DB Post published on 20 June 2016, pp 1–2

Trivedi S, Kataria HC (2012) Physio-chemical studies of water quality of Shahpura lake, Bhopal with special reference to pollution of groundwater of its fringe areas. Curr World Environ 7(1):139–144. https://doi.org/10.12944/CWE.7.1.21

Wanganeo A, Naik AA, Sheikh IA (2007) Variation in some chemical parameters of underground water resources of Bhopal. Curr World Environ 2(1):85–88. http://www.cwejournal.org/vol2no1/variation-in-some-chemical-parameters-of-underground-water-resources-of-bhopal/

Wateraid (2008) Urbanisation and water. British Geological Survey. Available at https://washmatters.wateraid.org/publications/urbanisation-and-water-2006. Accessed on 24.12.201

Chapter 32
Advancement Plans for Revitalization and Development of Ankobra River Basin in Ghana

Benjamin Lawortey, Thanga Raj Chelliah, and S. K. Shukla

Abstract Illegal small-scale mining infamous in Ghana, as 'Galamsey', has been uncontrolled for a considerable length of time. Even though artisanal and small-scale gold mining (ASGM) offers financial advantages and business opportunities to about 4.5% of the Ghanaian people, its effect on human well-being has prompted a high frequency of health hazards, deforestation, pollution and loss of biodiversity. This study identifies the current challenges faced by the Ankobra River Basin and its environment in the Tarkwa Nsuaem Municipality and suggests measures for advanced plans for revitalization and development of the Ankobra River Basin to be improved significantly. Analysis of satellite images on land use/land cover of the same geographic territory reflects changes for the past 30-years. From 1986 to 2016, multi-temporal images of the area classified indicate that the Ankobra River Basin is largely covered by dark forests and recent decline at 2.37%. Out of the 100 distributed questionnaires to the experts and stakeholders in the field, 80 completed questionnaires were analyzed. The results show that the main indicators of 0.344 are very poor and need urgent attention to revitalize and develop the river basin. Buffer zones positioned close to the source of surface water pollution are more likely to succeed in controlling water quality. Hence, support assessments for watershed land use change, pollutant discharge, management practices on water quality, encouraging environmentally friendly mining practices with automation and harnessing green electricity from the perennial stream and biogenic forest produce to phase-out thermal power will be the criterion to develop and enhance revitalization and development of the Ankobra River Basin. Thus, Rivers Revitalization Advancement Plan has been formulated.

Keywords Ankobra river basin · Ghana · Illegal small-scale gold mining · Pollution · Land use land cover · Reclamation · Revitalization and development

---

B. Lawortey (✉) · T. R. Chelliah · S. K. Shukla
Department of Water Resources Development and Management, Indian Institute of Technology Roorkee, Roorkee 247667, Uttarakhand, India
e-mail: benjaminlawortey@gmail.com

A. Pandey et al. (eds.), *Water Management and Water Governance*, Water Science and Technology Library 96,
https://doi.org/10.1007/978-3-030-58051-3_32

32.1 Introduction

Other than the normal flow of a river with seasonal variability, anthropogenic effects influence lithology, geography and superficial water quality including land use (Attua et al. 2014; Alvani et al. 2011; Singh 1989; Hu et al. 2005; Li et al. 2017). For an ever-increasing population with more stress on natural resources and growth (Macdonald et al. 2015; Owusu and Waylen 2009) ('Environmental Impacts of the Akosombo Dam and Effects of Climate Change on the Lake Levels' 2001; Gyau-Boakye and Tumbulto 2000; Li et al. 2017), today Africa needs its water more than any other continent.

Ghanaian rivers offer eco-friendly and other economic benefits like fish harvesting, irrigation and drinking water source. The perennial river system of Ghana is a living ecosystem, an essential component of the natural hydrological cycle and its continuous clean flow has been the character of its health. In the aftermath exploitation of basin, for various development purposes, combined with disintegration in the quality of the water, causing demand rise as rainfall continues to decline (Macdonald et al. 2015), streams experiencing tension or pressure.

Globally, artisanal small-scale gold mining (ASGM) is becoming significantly industrialized, generally run unlicensed, has created jobs for at least 15 million of the rural population directly and about 100 million in rural communities across more than 70 countries, reaching beyond Africa Union (Macdonald et al. 2015). Numerous ASGM operations emerge close to streams and waterways for simple access to alluvial metals (gold dust found in soil sediments), in addition to disposing of water utilized to stream as a part of preparing and as an acceptable need for mine processing waters. Even though ASGM adds to rural economies, it regularly inflicts ecological, security and social threats because of the simple mining techniques used (Macdonald et al. 2015).

The approaching danger of Ghana's water assets running dry in the years to come gets scarier day by day. Progressively, water treatment plants are being closed down. This is due to the contamination of the rivers by illicit gold miners' discharges in the stream and damaging the environment.

Illegal 'Galamsey' has been uncontrolled for a considerable length of time in Ghana.

Even though ASGM offers financial advantages and business opportunities to about 4.5% of the Ghanaian people, its effect on human wellbeing has prompted a high frequency of health hazards, deforestation, pollution and loss of biodiversity.

It is understood that self-supporting stream structures give basic characteristic and social products and ventures to human life (Janeiro 2011; Palmer et al. 2005). Waterway rejuvenation in urban situations has its chief focus on the rebuilding of the sidelong availability with the riverbanks and its tributaries, the reclamation of the waterway common stream system, just as the extension of the degree of a chance of the stream. Urban watercourse rebuilding is a test for administrators, scientists, specialists and residents (Janeiro 2011; Palmer et al. 2005).

32.2 Expert Opinion

The mineral quarrying impacts within the Ankobra Basin is given in Fig. 32.1. Owusu et al. (2016) reported that water shortage is turning into a noteworthy worry for individuals around the globe and there is need to secure the current water stock and discover ways or intents to give safe water to people world over in sufficient amounts, remembering necessities of who and what is to come. Water formed the bases of life, connected to loads of direct or indirect services, for example, the well-being of human including the socioeconomic, welfare and monetary improvement of a group or nation. Exploring Ghana's water assets management is fundamental.

Obiri et al. reported that the economy of Ghana is profiting from gold mining with negative eco-friendly and socioeconomic effects on the host groups correlated to gold mining have subjugated these economic escalations. This calls for money-saving advantages investigation of mining before experts allow new mining leases.

Odum portrayed indicate that riparian zones are fundamental assets looking specifically at water and land. The requirement for riparian zone dictates the choice of species. The recovery processes not requiring human contributions can be an alternative to river rehabilitation.

Fig. 32.1 Mineral quarrying impacts within the Ankobra Basin

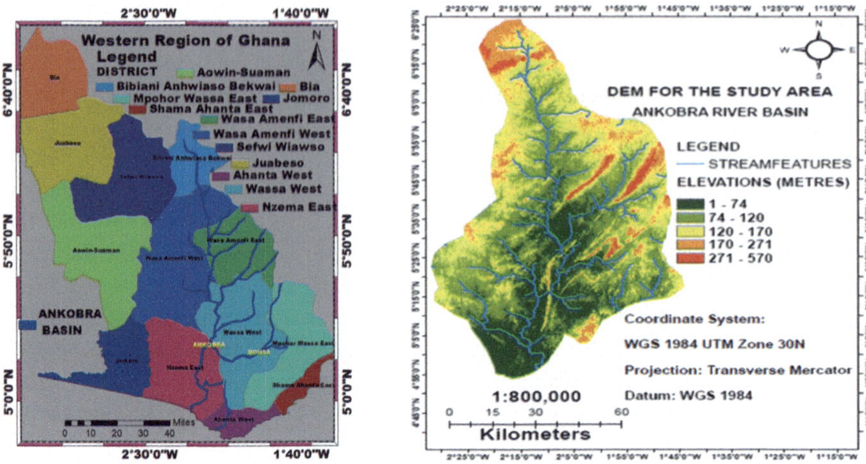

Fig. 32.2 Western region of Ghana, districts in the region with Ankobra River Basin, DEM (Ankobra River Basin)

32.3 Materials and Methods

32.3.1 Description of Study Area

Ankobra River Basin is classified on longitude 1° 50′ W and 2° 30′ W and latitude 4° 50′ N and 6° 30′ N (Fig. 32.2). The basin is constrained toward the little shoreline front Butte Basin. The basin has a place in the physiographic zone called the forest plateau.

Numerous times of escalated disintegration have diminished the region fairly, consistently showing up level largely at a height of 260–300 m. Overwhelming precipitation ensuing timberland averts disintegration and clarifies the articulated analyzed level.

Ankobra River is a bit of the Western River System and ranges a region of around 8460 km^2. The stream takes its source from the inclinations north of Basindare (close Bibiani) and streams fittingly 260 km largely due south before it enters the Gulf of Guinea at Asanta a few kilometers west of Axim (Ampomah 2009).

32.3.2 Methods of Data Collection

To accomplish this phase for comprehensive Ankobra River Basin advancement revitalization plan, the following activities were achieved through different approaches.

Within this activity, the documentation was made from numerous sources of information:

1. Scientific literature—books, articles and other scientific publications
2. Official website of Ghana Water Commission on Ankobra River Basin and international sites.
3. Transcribed news coverage from television, radio interviews, newspaper publications and websites of the municipality
4. Between the periods of 1986–2016, three satellite images were downloaded.
5. Questionnaires of standards to stakeholders and experts on main and sub-indicators based on an average weight for revitalization and development of the Ankobra River Basin.

32.3.3 Data Acquisition

From the United States Geological Survey (USGS) and the Global Land Cover Facility (GLCF), three satellite images cover the era of 1986–2016 attained.

Land use and land cover (LULC) data were obtained by image classification of a Landsat 8 OLI/TIRS under collection 1 Level 1 and 7 ETM+ C1 Level-1 taken on December 29 2015–January 5, 2016, and January 6–15, 2002, respectively and downloaded free from the USGS website. Also from Landsat 1–5 multi-spectral scanner, images are taken from January 18–27, 1986, under Global Land Cover Facility (GLCF).

32.3.4 LULC Analysis (Landsat Enhanced Thematic Mapper (Etm+) Imagery of 2017

Study of satellite images to recognize land use/land cover changes relies upon the assumption that the recorded electromagnetic radiation, which is the reason for requesting land covers, is altered as the land use/land cover of the equivalent geographic region changes (Aduah et al. 2015; Lu and Weng 2007; Nori and Elsiddig 2008).

Post-classification comparison, spectral–temporal, multi-data classification combined analysis and unaided change identification were the consisting classification methods (Aduah et al. 2015; Kumi-Boateng et al. 2012; Singh, 1989).

32.3.5 Pre-processing Image

The geometric contortions coming about because of sensor and stage blunders, just as the turn of the Earth corresponding to the sensor, are revised by image registration

(Butt et al. 2015). Geometric adjustment changes over the satellite image geometry to genuine facilitate and projection frameworks (Aduah et al. 2015; Kumi-Boateng et al. 2012), to empower the correlation of at least two images, the extraction of data for use in a Geographic Information System (GIS) and to overlay images with maps from various sources (Aduah et al. 2015). The goal of radiometric remedy in change identification is to guarantee that accessible images are looked at based on comparable radiometric properties (Aduah et al. 2015; Lu and Weng 2007). Image pre-handling in this examination comprised of just geometric revision. All the images had been geo-referenced to the UTM WGS84 Zone 30 North projection by the information providers, yet they didn't coordinate with the geo-referenced topographic guide of the investigation region (Nori and Elsiddig 2008).

32.3.6 Image Classification with Its Accuracy Assessment

Image characterization was directed by producing spectral signatures, utilizing preparing samples made for each satellite image. The preparation samples were made by haphazardly choosing 70% of the sample class information for each satellite image (Lu and Weng 2007; Eastman 2001). Five topical classes were chosen to represent the land cover of the Ankobra Basin, utilizing the USGS's territory cover characterization plot for Landsat information (Kumi-Boateng et al. 2012). The classes utilized are water, evergreen timberlands, settlements, bushes/ranches and mining territories (Table 32.1). After creating the spectral signatures, the separability of the thematic classes was checked using the Jeffries–Matusita's matrices (Aduah et al. 2015; Butt et al. 2015). In multi-dimensional perceptions, numerous characterization calculations (unsupervised) require the choice of ideal groups in which the classes are generally unmistakable. The Jeffries–Matusita (JM) separate is generally utilized as a distinguishableness standard for ideal band choice and assessment of characterization comes about lastly, and the maximum likelihood classification (MLC) calculation was utilized to all the images using the generated spectral signatures.

Table 32.1 Land cover image classifications

Land cover	Description
Dark forest	It includes Africa mahogany, walnut, bamboos, afrormosia, etc., and species such as duiker and bongo baboons
Built-up	It comprises playgrounds, industrial and residential structures, road infrastructures, etc.
Agriculture/farmland	Tree crops, cereals, vegetables, cash crop, exposed land, etc.
Mining area/bare land	It includes all areas of land where mining activities are practiced
Water body	It includes pools, wetland or more rarely puddles, rivers and streams

32.3.7 Change Detection

The land cover maps for 1986, 2002 and 2016 created after image characterization were renamed and consolidated to shape various bi-worldly land cover change maps for the periods 1986–1996, 1996–2006 and 2006–2016 utilizing Erdas and GIS spatial investigation.

At long last, the general deforestation proportions were resolved, by blending perennial and auxiliary woodland classes in every period and figuring the extent of changes every year, between the three time frames, utilizing the Food and Agricultural Organization's (FAO) meaning of timberlands (FAO 1976).

The change identification in this analysis depends on the hypothesis that land cover classes for the years 1986 and 2016 continued as before (Figs. 32.3, 32.4 and

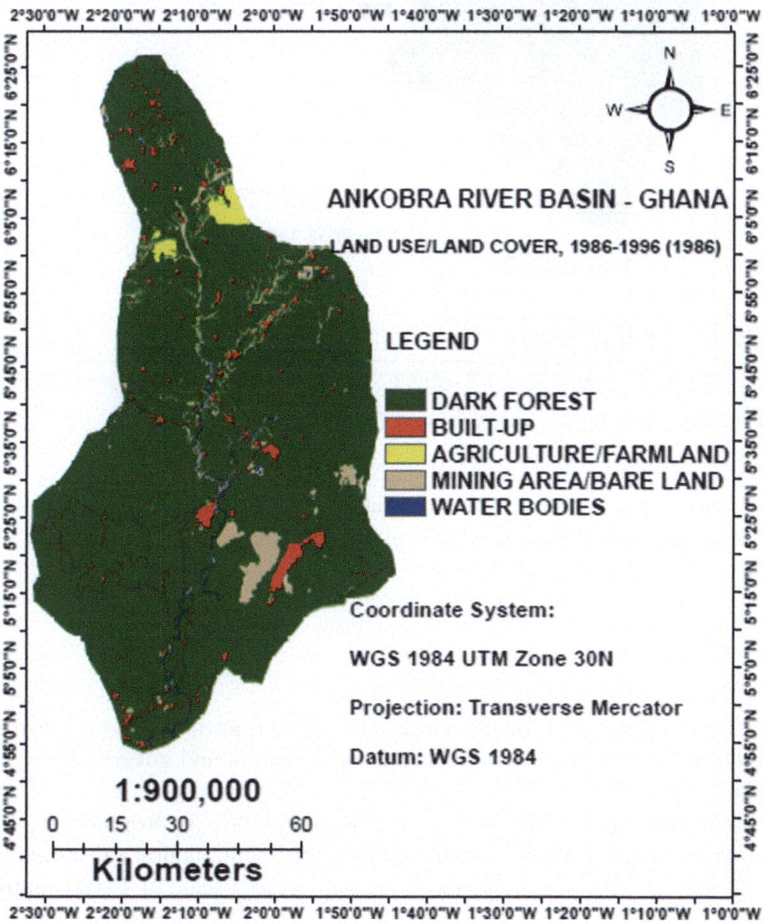

Fig. 32.3 Land use/land cover class for the year 1986–1996 (1986)

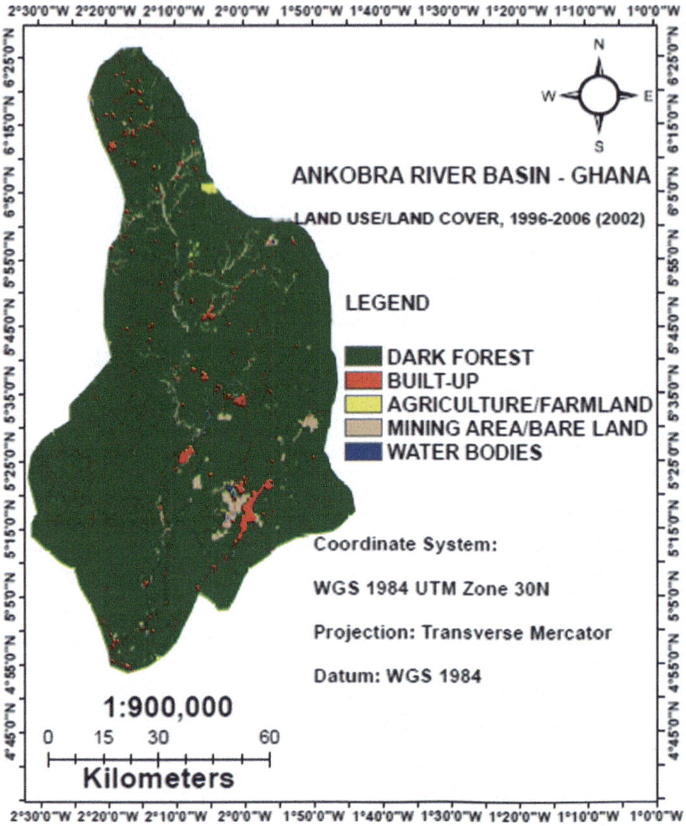

Fig. 32.4 Land use/land cover class for the year 1996–2006 (2002)

32.5). Land use change matrix is given in Tables 32.2, 32.3 and 32.4. The proportion of land cover between 1986 and 2016 is given in Table 32.5.

32.3.8 Questionnaires

For authenticating the data for the study, a designed questionnaire was distributed randomly to the selected non-governmental organization and government agencies for 100 individuals.

Main indicators such as mining, navigation, agriculture, hydropower, biological degradation, urbanization and infrastructure development, natural causes and water quality, as well as other sub-indicators were subject to a scale of 1–10 (minimum–maximum).

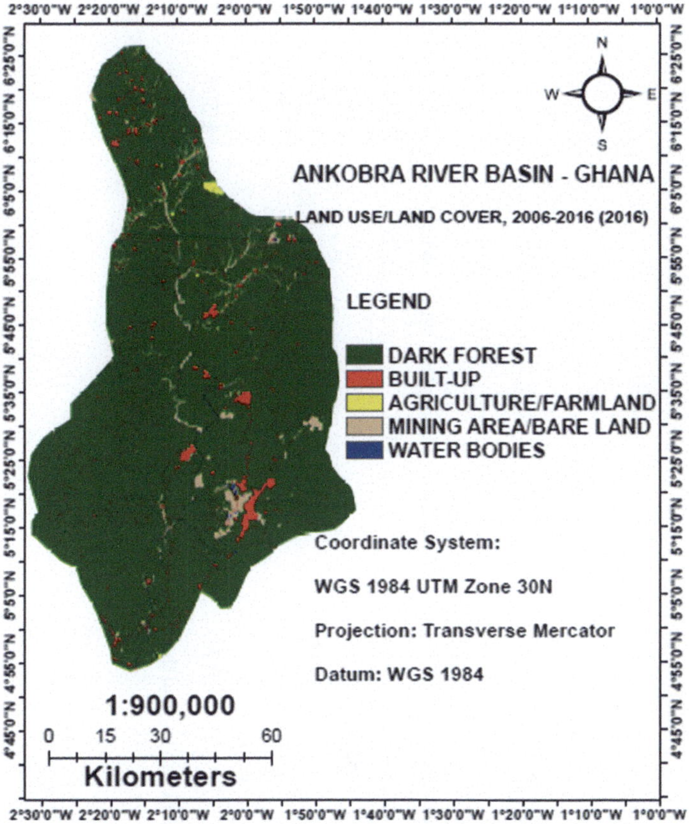

Fig. 32.5 Land use/land cover class for the year 2006–2016 (2016)

32.3.9 Riparian Buffer Zones

Area of drainage flanking a stream is called the riparian zone and is of precarious importance to the function, as well as the protection and management of a river (Vyas et al. 2012). Buffer zones are set at different distances of 500 and 600 m on both sides of the river basin to improve the efficiency of particulate trapping and provide surface litter to facilitate the assimilation of dissolved nutrients and toxic materials.

Table 32.2 Land use change matrix 1986–2006 (km^2) **from 1986, *to 2006

Land use/land cover change matrices							
	From 1986						
	Land cover	Dark forest	Built-up	Agriculture/farmland	Mining/bare land	Water body	Total** km^2
In 2006*	Dark forest	**7781**	6.4	0	11.62	1.75	7800.8
	Built-up	60.84	**185.3**	0	1.62	0.13	247.82
	Agriculture/farmland	6.97	0	**9.78**	0	0	16.75
	Mining/bare land	79.49	1.37	0	**184.6**	0.21	265.68
	Water body	3.34	0.13	0	0.17	**15.91**	19.55
	Total** km^2	7931.6	193.15	9.78	199	18	8350.6

Bold represent the land use for the same land cover from the year under review and the present year understudied

Table 32.3 Land use change matrix 2006–2016 (km^2) **from 2006, *to 2016

Land use/land cover change matrices							
	From 2006						
	Land cover	Dark forest	Built-up	Agriculture/farmland	Mining/bare land	Water body	Total** km^2
In 2016*	Dark forest	**7647.17**	39.94	1.91	42.74	1.66	7733.41
	Built-up	71.10	**204.93**	0.01	4.98	0.08	281.10
	Agriculture/farmland	7.72	0.03	**14.83**	0.00	0.00	22.78
	Mining/bare land	68.84	2.59	0.00	**216.18**	1.31	288.92
	Water body	5.72	0.34	0.00	1.78	**16.51**	24.35
	Total** km^2	7800.75	247.82	16.75	265.68	19.55	8350.56

Bold represent the land use for the same land cover from the year under review and the present year understudied

Table 32.4 Land use change matrix 1986–2016 (km^2) **from 1986, *to 2016

Land use/land cover change matrices							
	From 1986						
	Land cover	Dark forest	Built-up	Agriculture/farmland	Mining/bare land	Water body	Total** km^2
In 2016*	Dark forest	**7707.62**	7.58	0.04	16.52	1.65	7733.41
	Built-up	93.72	**184.63**	0.00	2.68	0.07	281.10
	Agriculture/farmland	13.04	0.00	**9.73**	0.00	0.00	22.78
	Mining/bare land	109.87	0.63	0.00	**178.12**	0.30	288.92
	Water body	7.38	0.31	0.00	0.67	**15.99**	24.35
	Total** km^2	7931.63	193.15	9.77	198.00	18.00	8350.56

Bold represent the land use for the same land cover from the year under review and the present year understudied

Table 32.5 Proportion of land cover between 1986 and 2016

Proportion of land cover between 1986 and 2016							
	1986		2006		2016		
Land cover	Area km^2	%	Area km^2	%	Area km^2	%	% increased
Dark forest	7931.63	94.98	7800.75	93.42	7733.4	92.61	**−2.37**
Built-up	193.15	2.31	247.82	2.97	281.1	3.37	**1.07**
Agriculture/farmland	9.78	0.12	16.75	0.20	22.78	0.27	**0.16**
Mining/bare land	198	2.37	265.68	3.18	288.92	3.46	**1.09**
Water bodies	18	0.22	19.55	0.23	24.35	0.29	**0.08**
Total	8350.6	100	8350.6	100	8350.6	100	

Bold represent the annual deforestation rate from 1986–2016 or the decreased and increased in percentage of land cover from 1986–2016

32.4 Results

32.4.1 Change Detection

32.4.2 Questionnaires

Among the respondents were researchers, engineers, water quality experts, laboratory managers, environmentalist, hydrologist, directors, hydrogeologist, commissioners and policy analysts.

Out of the 100 distributed questionnaires, a total of 80 completed questionnaires were returned and then analyzed statistically.

This procedure is applied to items having a scale of minimum to maximum (1–10) where the respondents were instructed to provide their personal opinion or perception on the scale. Thus, major indicators (mining, navigation, agriculture, hydropower, biological degradation, urbanization and infrastructure development, natural causes and water quality) and sub-individual items were given equal weight by

- Finding the weight of each main indicators
- Finding the average weight for each main indicators.

Thus, indices of main indicator (IMI) = indices of sub-indicator under main indicator (ISI) × weight average of main indicator (WAI).

For sub-indicators—attributes.

- Percentages for each individual sub-indicator were calculated, i.e., (weight/total) * 100
- Average percentages of each sub-indicators.

Thus, total index = sum of average indices of main indicators.

Table 32.6 Weight index of main indicators for revitalization and development of Ankobra River Basin

Main indicators	Average indices	Percentage indices	Total index
Mining (Illegal)	0.040	4.0	**0.344**
Navigation	0.042	4.2	
Agriculture	0.036	3.6	
Hydropower	0.053	5.3	
Biological degradation	0.038	3.8	
Urbanization and infrastructure development	0.054	5.4	
Natural causes	0.043	4.3	
Water quality	0.038	3.8	

Bold denote the sum of the average indices

The results show that the main indicator (0.344) is very poor (Table 32.6) and needs urgent attention to revitalized and develop the Ankobra River Basin.

32.4.3 Riparian Buffer Zones (RBZs)

Although riparian zones exist along the Ankobra River, it is limited to the basin and with limited distance. Survey of RBZ will be done in two separate distance of 500 and 600 m with all survey parameters evaluated accordingly Figs. 32.6 and 32.7.

32.4.4 Hydropower Potential

As tabularized below, the 90% reliable yearly flow for the 20-year flow sequences for the period of 1996–2016 are given in Table 32.7. The flow duration curve is presented in Fig. 32.8.

The yearly discharge data have been designed in descending order for the year 1996–2016. The rate likelihood of a flow size being equivalent or exceeded been assessed by the Weibull's formula. The plotting position of any discharge 'Q' and 'n' numbers of data used is calculated as follows:

$$P = (m/(n + 1)) * 100\%$$

where m = order number of flow

n = absolute number of data and

P = rate likelihood of flow extent being equivalent or exceeded.

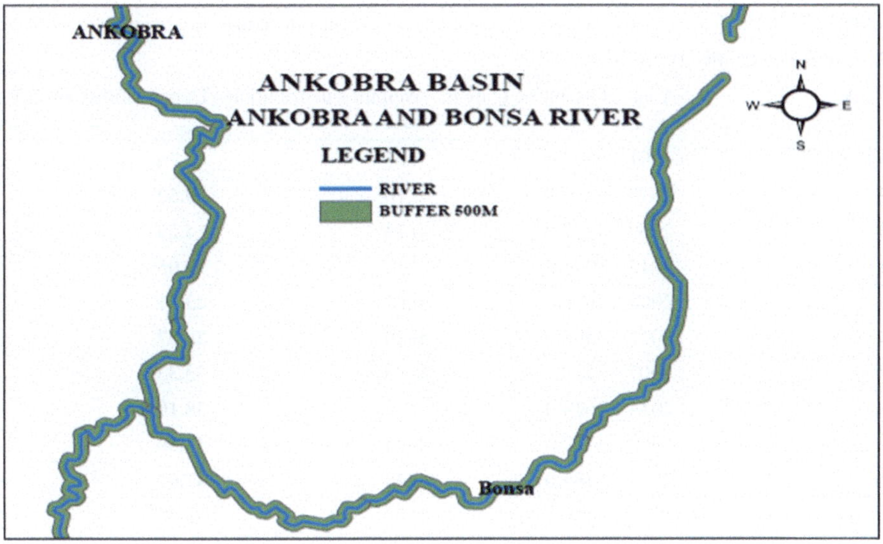

Fig. 32.6 Riparian buffer zoning for 500 m

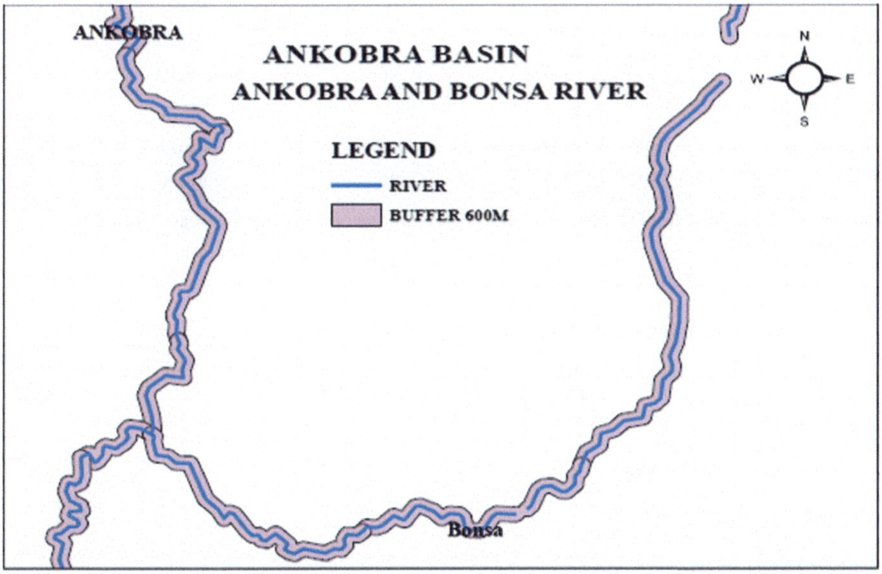

Fig. 32.7 Riparian buffer zoning for 600 m

Table 32.7 Water Research Commission—Ghana report, annual dependability from 1996 to 2016

Annual dependability 1996–2016				
Rank	Year	Discharge Q m^3/s	Weibull's distribution $m/(n+1)$	Dependability (in %)
1	2001	5744.0	0.05	**4.76**
2	2010	5378.0	0.10	**9.52**
3	2009	4827.0	0.14	**14.29**
4	2013	3996.0	0.19	**19.05**
5	2008	3896.0	0.24	**23.81**
6	2011	3885.0	0.29	**28.57**
7	2007	3722.0	0.33	**33.33**
8	2014	3644.0	0.38	**38.10**
9	1999	3619.0	0.43	**42.86**
10	2012	3465.0	0.48	**47.62**
11	1998	3230.0	0.52	**52.38**
12	2006	2990.0	0.57	**57.14**
13	2003	2699.0	0.62	**61.90**
14	2004	2450.0	0.67	**66.67**
15	2015	2301.0	0.71	**71.43**
16	2000	2219.0	0.76	**76.19**
17	1997	2045.0	0.81	**80.95**
18	1996	1978.0	0.86	**85.71**
19	2002	1823.0	0.90	**90.48**

Bold denote the annual rainfall dependability in percentage concerning the years from 1986–2016 in ascending order

The power produced is assessed as:

$$P = Q * H * g * \eta/1000 \quad (32.1)$$

where P—Power in 'MW'

Q—Discharge in 'm^3/s'

H—Net Head in 'm'

g—Acceleration due to gravity. '9.81 m/s^2'

η—Combined efficiency '87.36%' (turbine and generator etc.).

The estimated power potential and discharges with an average elevation of 38 and 421 m are presented in Tables 32.8 and 32.9, respectively.

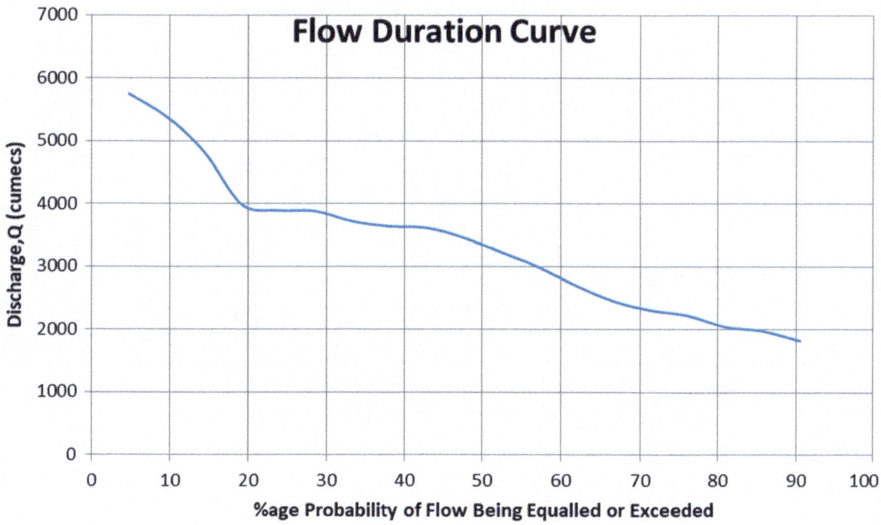

Fig. 32.8 Flow duration curve

32.5 Discussion and Conclusion

32.5.1 Discussion

32.5.1.1 Drivers of Changes Associated with Land Cover

Presently, affirming to the early study by Kusimi, the dark forest has reduced drastically to 2.3% between 1986 and 2016 due to the deforestation in the Ankobra River Basin, while mining areas increased 1.09% and settlements 1.05%.

The outcomes further show that, even though the deforestation rate has been expanding with time, over half of the land cover in the Ankobra River Basin persisted unaltered somewhere in the range of 1986 and 2016. The largest class in the dark forest (93.42% between 1986 and 2006, 92.61% between 2006 and 2016), while agriculture and farmland were the most divided classes. The recorded land changes demonstrate that a few zones, which were initially secured by dark forests or agriculture/farmland, have been changed over to other types, and the reforestation during the three periods is insignificant, contrasted with the general deforestation.

Anthropogenic drivers, viz. global and local influences on the Ankobra River Basin, are the main causes of land use/land cover changes. Urbanization, agriculture, international trade systems and politics are the global influences whereas population growth, immigration, urbanization, government policy and economic development are also local influences (Aduah et al. 2015). Human encroachment aided the further deforestation of the Ankobra River Basin, horticultural, lumber logging and expanded superficial mining since 1986, when the utilization of the fundamental

Table 32.8 Power potential and discharges with an average elevation of 38 m

Months	2011			2012			2013		
	Q m^3/s	Net head m	Power P (MW)	Q m^3/s	Net head m	Power P (MW)	Q m^3/s	Net head m	Power P (MW)
Jan.	50	38	16	38	38	12	41	38	13
Feb.	130	38	42	246	38	80	57	38	18
Mar.	203	38	66	522	38	169	133	38	43
Apr.	834	38	270	694	38	225	562	38	182
May	904	38	293	750	38	243	1114	38	361
Jun.	117	38	38	112	38	36	262	38	85
Jul.	395	38	128	197	38	64	90	38	29
Aug.	586	38	190	455	38	147	744	38	241
Sept.	453	38	147	246	38	80	796	38	258
Oct.	91	38	29	90	38	29	74	38	24
Nov.	68	38	22	66	38	21	81	38	26
Dec.	54	38	17	49	38	16	42	38	14
	3885			3465			3996		

Table 32.9 Power potential and discharges with an average elevation of 421 m

Months	2011			2012			2013		
	Q m³/s	Net head m	Power P (MW)	Q m³/s	Net head m	Power P (MW)	Q m³/s	Net head m	Power P (MW)
Jan.	50	421	180	38	421	137	41	421	148
Feb.	130	421	468	246	421	887	57	421	205
Mar.	203	421	732	522	421	1881	133	421	479
Apr.	834	421	3005	694	421	2501	562	421	2025
May	904	421	3258	750	421	2703	1114	421	4015
Jun.	117	421	422	112	421	404	262	421	944
Jul.	395	421	1423	197	421	710	90	421	324
Aug.	586	421	2112	455	421	1640	744	421	2681
Sept.	453	421	1632	246	421	887	796	421	2869
Oct.	91	421	328	90	421	324	74	421	267
Nov.	68	421	245	66	421	238	81	421	292
Dec.	54	421	195	49	421	177	42	421	151
	3885			3465			3996		

change program of the World Bank, changed Ghana's economy and besides, extended outside mining, timber and direct enthusiasm for the cultivating organizations.

Along these lines, both adjoining and worldwide drivers cause anthropogenic land cover changes, while the effects are to a great extent felt at the nearby scale.

Buffer and greening; creating of buffer zones were carried out across a width of 500 and 600 m area of riparian buffer zones of the selected reach of the basin. Some areas were subjugated by illegal mining activities and agricultural practice, but fewer areas were wrapped with forests and trees. It was found that in some areas people were converting areas protecting the basin into mining zones.

Thus, the successive application of the riparian buffer zones will ensure revitalization water quality through enhanced flood storage, to ease back stream speeds to empower reintroduction of vegetation; improved water quality, through local scale stormwater treatment at river conjunctions, and limited 'treatment porches' at storm deplete outfalls; improved free inside the channel through patios and inclines, little pocket parks and ponded zones; and a reestablished riparian biological community. These rules for the stormwater administration and the reasonable building will propel the City's 'green plan.'

Keeping up or setting up vegetation in, or on the banks of, waterways impressively improve stream scenes.

- The roots of tree increase the resistance quality of the waterway tier and secure it in contradiction of disintegration.
- Undergrowth along these lines balances out floodplains and channels and lessens disintegration.

For hydropower install capacity, studies specify that installed capacity of 12–243 MW for a net head of 38 m and 148–4015 MW for a net head of 421 m, comprising of 3 producing units of 81 MW and 1338 MW each respectively, would be the most ideal alternative among others.

Due to the accessibility of average discharge on the Ankobra River Basin with the main tributary of Bonsa River, the basin is more reliable for development of hydropower project. The 90% derivative of the annual dependability is from the power potential study.

Thus, a design flood of 1823 m^3/s of the basin is recommended based on the annual dependability of the basin. The anticipated hydropower project can be a runoff-the-river scheme making safer from large inundation.

Water quality: Occasional changes in stream and temperature and reacting natural movement cause water quality to vary between rivers because of climate and the major land acknowledgement as the year progressed. The impact of human activity on water quality and quantity changes the drainage outlines to rivers and the corresponding chemical elements and sediments the water may be carrying.

Rivers used as sewage or liquid waste dumps and nourish stormwater depletes that have depleted messy urban zones into them. Ailment causing specialists, oxygen-requesting squanders, water-dissolvable inorganic chemicals, inorganic plant supplements, natural chemicals, dregs or deferred issue, water-solvent radioactive isotopes and warm toxins are for the most part because of contamination in the stream basin.

Diminishing plenitudes and biodiversity of the basin influence the aquatic biological communities' capacity to work proficiently. Unless all these are appropriate, arranged restoration might be unsuccessful.

Improve water quality to recondition impaired waters, associations with Environmental Protection Agency (EPA), the agency's capability to validate results in watershed management, the ability to use adaptive supervision in land management plan implementation and the Improving National Environmental Policy Act scrutiny's and compliance are collective drivers desired as an active tool for the agency to achieve a standardized National Best Management Practice (BMP).

Automation will improve waterway renewal and advancement through; making Mining Safer (Automation gives mines more superior control over their fabrication processes and, subsequently, permits them to deliver an advanced quality completed item.), Safer Mining through Better Oversight (Automating mining forms permits the mine condition overall to be all the more firmly observed. Environmentally Friendly Mining (the superior degree of control over the mining practice offered by automation allows mines to assess their environmental impact more accurately. Restricting waste delivered by the mining procedure and lessening outflows caused by the unneeded task of hardware is only a couple of the manners by which mechanized innovation can help mines to wind up greener). Moreover, Fewer Hazards for Workers (Automation also ensures that tasks are completed correctly and consistently every time. Accordingly, the 'human mistake' factor caused by the erroneous task of a machine or a slip by in consideration can be disposed of.).

32.6 Conclusion

From 1986 to 2016, multi-temporal images of the area classified indicate that the Ankobra River Basin is largely covered by dark forests and recent decline at 2.37%. The analysis additionally demonstrates the possible drivers that incorporate universal exchange, neighborhood populace development, horticulture escalation and urbanization.

Small-scale miners termed Galamsey using mercury poisons the stream for their short-term economic advantages. The small-scale miners may be influenced through customized training and educational programs accompanied by technical and financial assistance to abandon the practice.

Support zones situated near the source of surface water contamination are bound to prevail concerning controlling water quality.

Three main activities or measures formed the basis for a successful rehabilitation effort. To moderate the impacts of hard designing practices, these were utilized because they were earth delicate river conservation activities, soil bioengineering or delicate biotechnical engineering practices.

In terms of hydropower, it is a more secure alternative; hydroelectric power can be depended upon availability of green energy, and it can be re-utilized, it is a dependable, highly established technology, supportive to the grid and remote inaccessible area, less expensive, no fuel cost, averts floods, greater employment opportunities and as tourist attraction sites. Hence, the basin is recommended depending on the annual dependability with the design flood of 1823 m^3/s.

Automation can enhance river revitalization and development through decreased process variability, higher quality product, lower operating costs and better monitoring.

Subsequently, River Revitalization Plan run-downs an open framework program that will encourage the reclamation, redevelopment and revitalization the area. Revitalized rivers will enhance water quality, give green space to accelerate health progress in poor and socially excluded groups, enhance surge assurance and diminish our reliance on imported water, restore a functional ecosystem, enhance river identity, foster civic pride, provision of hydropower, and improve the quality of life.

A revitalized stream with support to the population for its livelihood is the future to hold the promise of human beings living along the riverbanks dependent on each other peacefully in nature. The proposed schemes brought in practice will deliver in that direction.

A holistic approach to ensure suitable management packages beneficial for biodiversity and reducing flood risk, erosion, habitat loss and damage to any structural integrity etc., as well as improving access where necessary and enhancing the aesthetics of the area by programs tailored to requirements is to be adopted.

References

Aduah MS, Warburton ML, Jewitt G (2015) Analysis of land cover changes in the bonsa catchment, Ankobra Basin, Ghana. Appl Ecol Environ Res 13(4):935–955. https://doi.org/10.15666/aeer/1304_935955

Alvani J, Boustani F, Tabiee O, Hashemi M (2011) The effects of human activity in Yasuj area on the health of stream city. World Acad Sci, Eng Technol 5(2):341–345

Ampomah B (2009) Water resources commission, Ghana Ankobra River Basin—Integrated Water Resources Management Plan, pp 9–57

Attua EM, Ayamga J, Pabi O (2014) 'Relating land use and land cover to surface water quality in the Densu River basin, Ghana. Int J River Basin Manag. https://doi.org/10.1080/15715124.2014.880711

Butt A, Shabbir R, Ahmad SS, Aziz N (2015) Land use change mapping and analysis using Remote Sensing and GIS: a case study of Simply watershed, Islamabad, Pakistan. Egypt J Remote Sens Space Sci Authority for Rem Sens Space Sci 18(2):251–259. https://doi.org/10.1016/j.ejrs.2015.07.003

Eastman JR (2001) Introduction to remote sensing and image processing. idrisi32: Guide to GIS and Image Proces 1:17–34. https://doi.org/10.1111/j.1540-5915.1994.tb01870.x

Environmental Impacts of the Akosombo Dam and Effects of Climate Change on the Lake Levels (2001) pp 17–29

FAO (1976) A framework for land evaluation, FAO soils Bulletin n. 32, M-51
Gyau-Boakye P, Tumbulto JW (2000) The Volta Lake and declining rainfall and streamflows in the Volta River Basin. Environ Dev Sustain 2:1–10. https://doi.org/10.1023/A:1010020328225
Hu Q, Willson GD, Chen X et al (2005) Effects of climate and landcover change on stream discharge in the Ozark Highlands, USA. Environ Model Assess 10:9–19
Janeiro RD (2011) River revitalization in the context of urban stormwater management: the Case of Acari River Basin, Brazil. In: 12th international conference on urban drainage, pp 11–16
Kumi-Boateng B, Mireku-Gyimah D, Duker AA (2012) A spatio-temporal based estimation of vegetation changes in the tarkwa mining area of Ghana. Res J Environ Earth Sci 4(3):215–229. Available at: https://dspace.knust.edu.gh/bitstream/123456789/7339/1/Kumi-Boateng%2CB.pdf
Li H, Wang C, Zhong C, Zhang Z, Liu Q (2017) Mapping typical urban LULC from Landsat imagery without training samples or self-defined parameters. Remote Sens 9(7):1–23. https://doi.org/10.3390/rs9070700
Lu D, Weng Q (2007) A survey of image classification methods and techniques for improving classification performance. Int J Remote Sens 28(5):823–870. https://doi.org/10.1080/01431160600746456
Macdonald K, Lund M, Blanchette M (2015) Impacts of artisanal small-scale gold mining on water quality of a tropical river (Surow River Ghana). In: 10th international conference on acid rock drainage (ICARD) and IMWA 2015 annual conference
Nori W, Elsiddig EN (2008) Detection of land cover changes using multi-temporal satellite imagery. Remote Sens Spat Inf Sci 2004–2009. Available at: https://www.isprs.org/proceedings/XXXVII/congress/7_pdf/5_WG-VII-5/36.pdf
Owusu K, Waylen P (2009) Trends in spatio-temporal variability in annual rainfall in Ghana (1951–2000). Weather 64(5):115–120. https://doi.org/10.1002/wea.255
Owusu PA, Asumadu-Sarkodie S, Ameyo P (2016) A review of Ghana's water resource management and the future prospect. Cogent Eng 3(1):1164275. http://doi.org/10.1080/23311916.2016.1164275
Palmer MA, Bernhardt ES, Sudduth E (2005) Standards for ecologically successful river restoration. J Appl Ecol 42(2):208–217
Singh A (1989) Review article: digital change detection techniques using remotely-sensed data. Int J Remote Sens 10(6):989–1003
Vyas V, Kumar A, Wani SG, Parashar V (2012) 'Status of Riparian buffer zone and floodplain areas of River Narmada. Int J Environ Sci 3(1):659–674

Chapter 33
Groundwater Governance and Interplay of Policies in India

Akshi Bajaj, S. P. Singh, and Diptimayee Nayak

Abstract India is the leading consumer of groundwater resources in the world. Moreover, groundwater resource accounts for 85% of rural water supply and 84% of the net irrigated area in India. The groundwater level in the north-western states of India is depleting rapidly in such a manner that they are labeled as 'over-exploited' by the Central Groundwater Board (CGWB), India. Hence, the regulation of such a common-pool resource is imperative to arrest its unsustainable free-riding uses. This paper analyzes the trends and present status of groundwater usage in India, the structure of groundwater governance including policies at the central level and three north-western states of India—Punjab, Haryana, and UP. It also examines the interplay of policies of different departments influencing the groundwater resources. We follow qualitative and descriptive analysis by reviewing official documents like government acts, regulatory frameworks, reports, policies and model bills of the center and respective state governments. The study finds that the policies and bills are comprehensive enough to layout a firm action plan on the ground. Although the policies of different concerned departments are well-framed, there is a trade-off among them apparently due to lack of integrated approach. Further, due to the status quo of the groundwater management bills in these states for the past decade, the guidelines framed by the center have no role to play so far. The study suggests that these bills and governance mechanisms must be in tune to adhere to the present groundwater scenario and establishing proper coordination between the center and states for managing this fragile common-pool resource.

Keywords Groundwater · Water governance · Sustainable management

A. Bajaj (✉) · S. P. Singh · D. Nayak
Department of Humanities and Social Sciences, Indian Institute of Technology Roorkee, Roorkee 247667, Uttarakhand, India
e-mail: abajaj@hs.iitr.ac.in

S. P. Singh
e-mail: s.singh@hs.iitr.ac.in

D. Nayak
e-mail: dipti.nayak@hs.iitr.ac.in

A. Pandey et al. (eds.), *Water Management and Water Governance*, Water Science and Technology Library 96,
https://doi.org/10.1007/978-3-030-58051-3_33

JEL Classification Q25 · Q28

33.1 Introduction

Hardin (1968) states that '*Therein is the tragedy. Each man is locked into a system that compels him to increase his herd without limit – in a world that is limited. Ruin is the destination toward which all men rush, each pursuing his own best interest in a society that believes in the freedom of the commons. Freedom in a common brings ruin to all*' (p. 1244). The seminal article on common property resources and collective actions, i.e., 'Tragedy of Commons' by Garett Hardin explains the essence of the ruin of all the natural resources. Our ecological system has largely failed to operate in its natural market mechanism because it is finite but the human needs are infinite. We have reached the stage where even renewable resources have also started showing signs of depletion. The carrying capacity of our natural ecosystem is on the verge of its potential stage. The question of sustainable use of natural resources though raised in the late nineteenth century but the governance structure or appropriate institutional framework that can ensure sustainability has not been defined per se. Therefore, groundwater governance is one of the key mechanisms to control groundwater depletion. In this context, groundwater governance can be defined as '*the process by which groundwater is managed through the application of responsibility, participation, information availability, transparency, custom, and rule of law. It is the art of coordinating administrative actions and decision making between and among different jurisdictional levels–one of which may be global*' (Varady et al. 2013).

India is one of the largest consumers of groundwater. Today, India's groundwater use for agriculture is the largest in the world with an estimated almost 30 million groundwater structures in use. The governance of groundwater resources has become a matter of concern in India for the past two decades. The groundwater alone accounts for 84% of the net irrigated area, 85% of rural water supply and 45% of urban water supply (World Bank 2010). The ubiquitous and reliable nature of groundwater makes it a crucial resource to sustain the growth of the agricultural economy. Despite being replenishable in nature, the annual average groundwater level has been depleting rapidly. There has been a dramatic increase in the groundwater extraction units in India because of several social, economic and political reasons (Kulkarni et al. 2015). The shallow tube wells increased from 4.8 million in 1987–88 to 9.1 million in 2006–07 but decreased to 5.9 million in 2013–14 and on the other hand, the number of deep tube wells increased from mere 0.1 million in 1987–88 to 2.6 million in 2013–14 (Government of India 1993, 2017a, b). These figures show a huge demand for groundwater resources which makes the resource unsustainable.

One of the most important causes of unsustainable use of groundwater is lack of effective water governance (Water Governance Facility 2013; World Bank 2010; Prasad 2008; Shah and van Koppen 2006). The model bills and some policies proposed at the national level could not prove useful at ground level as "water" as a subject is enlisted under the state jurisdiction. Moreover, after 73rd constitutional amendment the panchayati raj institutes (PRIs) were given the authority for micro irrigation, water management and watershed development. The larger picture shows governance failure as root cause of the problem as a majority of states with high number of over exploited administrative assessment units are yet to pass any formal law. Thus, the delegation of power to Panchayati Raj institutes (PRIs) is a far-off dream as of now. Secondly, the focus is still on the management of groundwater as a separate unit. Cooperation and coordination across departments and functional units are often lacking. Consequently, the policies framed 'in silos' lead to competing and often conflicting actions (Varady et al. 2013). These days the research is largely on what kind of institutions or governance structure can ensure the sustainability of natural resources. Though the answer to this question is imperative, before that, it is important to know what exactly our present system is and what the loopholes are, so that these aspects may be imbibed to frame a governance model based on equity, efficiency, and sustainability.

In this context, this study primarily focuses on the present groundwater governance at the Centre, the three north-western states and interrelations between some policies related to groundwater governance. The study is based on secondary sources, specifically the official sites of the respective state government ministries. It deals with the governance issues of groundwater resources in India as this common-pool resource is crucial to insure the farmers from variability in crop yield, to make agriculture more resilient to climate change, to ensure safer drinking water, to feed the drying rivers, and to provide a reliable source of water for industries. The sustainable use of groundwater resources is important to ensure the food and water security of India.

The study is outlined in the following four sections. First, it presents a brief picture of the trends in groundwater usage and its present status in India. Second, it highlights the structure of groundwater governance at the central level and the three northern states, which are known as the food grain baskets of India, viz., Punjab, Haryana, and Uttar Pradesh. The third section discusses the interplay of policies and programs of different departments influencing groundwater resources. The fourth section briefs the concluding remarks and suggestions for formulating a holistic approach for groundwater sustainability.

33.2 Trends and Present Condition of Groundwater Usage in India

The groundwater exploitation in India is rising at an increasing rate since the 1960s. These trends in India can be studied by several statistics. One, an increase in the share of tube wells in comparison to other sources of irrigation (Fig. 33.1). Second, the increase in the percentage of over-exploited and critical assessment units by the Central Ground Water Board (CGWB) (Table 33.1). Third, the percentage of wells

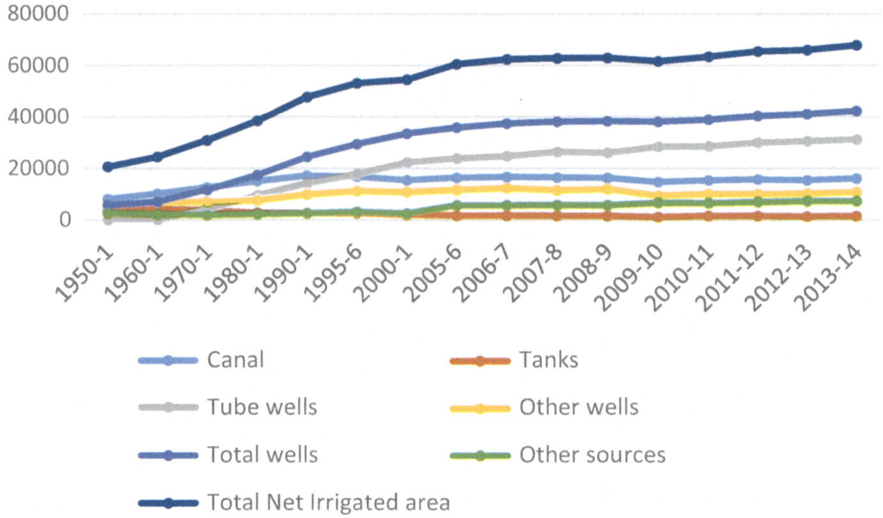

Fig. 33.1 Share of different sources of irrigation in India ('000 ha). *Sources* Authors' own representation: data compiled from Agricultural statistics at glance 2015, Government of India (2015); data.gov.in.

assessed that have shown a fall in the present water level in comparison to the decadal average water level (Table 33.2). These trends highlight the intensity of the problem that demands urgent attention to implement an efficient, equitable and sustainable governance system.

There is a vast literature that explains the major causes of such trends (Dhawan 1993, 1995; Pant 2004; Banerji et al. 2006; Singh and Singh 2006; Shah et al. 2008; Srivastava et al. 2009; Pandey 2014; Varady et al. 2013; Kulkarni et al. 2015). We have reviewed the existing literature to explore *raison d'etre* for the over-exploitation of groundwater, which has been summarized as follows:

First, a formal governance system is foremost for the efficient utilization of any resource. At the policy level, there is no void of research studies and expert reports with special focus on the situation: the awareness is high that large parts of the country are already severely stressed with a bleak future under a business as usual paradigm. Yet there is insufficient action beyond the reports and policies (Water Governance Facility 2013).

Second, the green revolution was a package deal that could have not turned successful without heavy irrigation facilities. It was promoted only in those areas which could afford intense irrigation from groundwater. The flat power tariff for pump sets militates against any desire for achieving the economy in groundwater use because the marginal cost of pump set operation is practically zero (Dhawan 1993).

Third, the free boring scheme by Uttar Pradesh government in 1984–85 which assisted with cash amounting Rs. 3000 and subsidy of up to 50% to small and

Table 33.1 Comparative status of level of groundwater development

Level of groundwater development	Explanation	% of district in 1995	% of district in 2004	% of district in 2009	% of district in 2011	% of district in 2013	% of district in 2017
0–70% (Safe)	Areas which have groundwater potential for development	92	73	72	71	69	64
70–90% (semi-critical)	Areas where cautious groundwater development is recommended	4	9	10	10	10	14
90–100% (Critical)	Areas which need intensive monitoring and evaluation for groundwater development	1	4	4	4	4	5
>100% (Over-exploited)	Areas where future groundwater development is linked with water conservation measures	3	14	14	15	16	17

Source *Compiled from* Dynamic Groundwater Resources of India, 2017, Government of India (2017a, b)

marginal farmers for diesel and electric pump sets promoted the installation of water extraction mechanisms by a large number of farmers in the state (Pant 2005).

Fourth, in the 1980s the groundwater markets emerged as a reliable source of irrigation for small and marginal farmers. The poor farmers also resorted to these markets even if they provided costlier services (Mukherji 2004; Pant 2004). The groundwater markets also altered the cropping pattern.

Fifth, the farmers lacked knowledge about the optimal utilization of groundwater available from these markets. In this scenario, the lack of any defined policy or law worsened the matter (Srivastava et al. 2009). Sixth, with the increase in population dependent on agriculture from 0.4 ha/person in 1900 to less than 0.1 ha/person in 2000, the means of irrigation was required to intensify and diversify the land use. In this condition, small mechanical pumps and boring rings provided a technological breakthrough (Shah 2014).

The literature explicitly speaks about government policies, lack of clear guidelines and legal frameworks as the major reasons for farmer's rising inclination toward

Table 33.2 State wise fluctuation from premonsoon 2018 to decadal mean [Premonsoon (2008–17)]

Rank	State	No. of wells surveyed	No. of wells (fall in water level)	No. of Wells (fall in water level)%
1	Punjab	216	181	84
2	Uttar Pradesh	563	466	83
3	Himachal Pradesh	86	65	76
4	Haryana	273	205	75
5	Uttarakhand	28	20	71
6	Tamil Nadu	528	318	60
7	Andhra Pradesh	714	423	59
8	Chhattisgarh	458	271	59
9	Madhya Pradesh	1330	782	59
10	Meghalaya	22	13	59
11	Bihar	619	350	57
12	Gujarat	756	401	53
13	Maharashtra	1632	857	53
14	Rajasthan	929	454	49
15	Telangana	568	268	47
16	Kerala	1431	654	46
17	Karnataka	1343	542	40
18	Odisha	1254	485	39
19	Arunachal Pradesh	8	3	38
20	Assam	154	56	36
21	Jharkhand	255	87	34
22	West Bengal	614	205	33
23	Goa	70	20	29
24	Tripura	25	4	16

Source *Compiled from* Ground Water Scenario in India Premonsoon, 2018, Government of India (2018)

higher groundwater pumping. The groundwater is a more reliable and efficient source for irrigation than the surface water but farmers believed it to be an infinite source which leads to its inefficient use. The present condition puts reflection upon *Jevons Paradox, which* refers to a form of induced demand wherein efficiency improvements in the use of a resource causes increased consumption of the resource rather than a decrease in its use. Therefore, governance is the prior most dimension that can economize the present status of groundwater and arrest this paradoxical scenario.

33.3 Governance Structure of the Government of India

33.3.1 Legislative Framework

The constitution of India confers the jurisdiction of 'water' on the state government but it does not explicitly state anything about the 'groundwater.' Therefore, groundwater is taken into account under the generic term 'water.' With the phenomenal growth of groundwater irrigation units, the central government stepped in to propose a broad framework for groundwater governance by the states as it is highly correlated to the prosperity of the agricultural economy. The legislative framework remained untouched for a long time. The Indian Easement Act of 1882 was the first step in the British period to bring groundwater usage under any defined law. However, the question of sustainability did not arise at that time leading to no interference by any authority on the groundwater rights of an individual or society as a whole.

As severe scarcity and unsustainability in groundwater use have been experienced and reported, there is an urgent need to evaluate the developments in the legal frameworks by the central government, which would help the policy-makers to find the effectiveness and limitations of the existing measures in addressing the groundwater issues. The important legal frameworks by the central government related to this issue have been identified for further evaluation. They are Indian Easement Act 1882, Water (prevention and control of pollution) Act 1974, The Environment (Protection) Act (EPA) 1986, Model Bill for the Conservation, and Protection, Regulation and Management of Groundwater, 2016. They are analyzed to highlight the pros and cons of these groundwater-related documents.

33.3.1.1 Indian Easement Act 1882

This Act is considered to be a very vital legal framework to address the issue of groundwater rights. It states, 'The right of every owner of land to collect and dispose within his own limits of all water under the land which does not pass in a defined channel and all water on its surface which does not pass in a defined channel' (Indian Easement Act 1882, as cited in Prasad 2008, p. 27). This act referred to groundwater as private property. The ownership of land meant the complete right to exploit the groundwater beneath it (Prasad 2008). The term 'defined channel' was ignored by the then authority concerned since the sustainability of this resource was out of the question at that time. The Easement Act does not permit landowners ownership of groundwater if it is passing in a defined channel. This underscores the fact that a proactive step to managing any resource is still a far-off dream.

33.3.1.2 Water (Prevention and Control of Pollution) Act 1974

This act mandates the formation of Centre and State Pollution Control Boards. This authority led to the beginning of monitoring the agents polluting the natural resources including water. The act provided the action plan for regulating, preventing and prohibiting the pollution of resources, particularly the water resource, but did not mention about the groundwater explicitly. The term 'well' stated in the act is not defined, leading to a major focus on surface water only which has been defined as 'stream.' Moreover, the pollution sources of groundwater have also been omitted. It is important to notify in the act that the over-extraction of groundwater is also a source of pollution. The amendment of this act in 1978 also did not mention this issue. It was affordable to ignore these aspects at that time but the prevailing condition demands urgent amendments in this act.

33.3.1.3 The Environment (Protection) Act (EPA), 1986

Under this act, two important authorities were constituted: Water Quality Assessment Authority (WQAA) and Central Ground Water Authority (CGWA). WQAA is authorized to issue direction and to take measures for investigations and research; establish or recognize environmental laboratories and institutes; collect and disseminate information and prepare manuals, codes or guides relating to the prevention, control, and abatement of water pollution (WQAA, GoI). The institution of CGWA was the first major step to bring regulation, protection, and conservation of groundwater under a strong legal framework. The authority was assigned with administrative and quasi-judicial functions as well to regulate the groundwater resource. It is entitled to organize activities like registration of wells, the grant of permission for the replacement of the existing or the construction of new wells, organizing rooftop rainwater harvesting without causing any inconvenience to the people, regulate indiscriminate boring and withdrawal of groundwater in the country and to issue necessary regulatory directions to preserve and protect the groundwater (CGWA, GoI). However, overlapping functions and lack of coordination with the state authorities have left these authorities more or less redundant to arrest the problem at the grass-root level.

33.3.1.4 Model Bill for the Conservation, Protection, Regulation, and Management of Groundwater, 2016

The need for restricting excessive exploitation of groundwater was realized by the Centre as early as about 40 years ago. But the Centre could not do much since the regulation of groundwater was supposed to be a 'State subject' (Prasad 2008). The first model bill was introduced in 1970 which directed the state governments to regulate the groundwater abstraction in the notified areas, as declared by the groundwater authority, except for the drinking water purpose. The provision to constitute a groundwater authority was made for the first time in this bill itself. This bill framed broad guidelines to bring groundwater resources under the legal framework but the major bottleneck was that it was only advisory in nature. The model bill of 1970 has been revised multiple times in these years—1992, 1996, 2005, 2012, 2016—to incorporate stringent measures as per the demand of the consistently deteriorating condition of the groundwater. Now, the question is not only of quantitative depletion but also of the qualitative degradation of groundwater.

Further, the revised bills included the exempted categories also. The model bill of 1992 covered all the abstraction units for regulation in the notified areas but exempted the small and marginal farmers provided they use groundwater for personal activities, not for commercial purposes. In the subsequent major revisions of 1996 and 2005, the marginal and small farmers were also included keeping only manually operated wells out of groundwater authority's purview and all wells would require registration even in the non-notified areas respectively. The model bill 2005 also emphasized upon rainwater harvesting, mass awareness and training of the concerned stakeholders. But these bills did not incorporate the provisions for decentralization of authority, integrated use of water, ensuring behavioral changes among users and maintaining a strong database about groundwater resources.

The model bill of 2012 clearly stated groundwater as a common-pool resource and mandated the concerned authority to ensure that the use of groundwater by any person on their land does not deprive other persons of their right to groundwater for life, in case these persons are dependent for their right to groundwater for life on the same aquifer. The recent model bill of 2016 is comprehensive and progressive in nature. It included the aspects to ensure the principle of non-discrimination and equity, subsidiarity and decentralization, protection, precaution and prior assessment and integrated approach. The hierarchy of structure of groundwater governance proposed by the model bill 2016 has been presented in Fig. 33.2.

As provided in the previous model bill, this bill also sets the priority for groundwater allocation and accepts the verdict of the Supreme Court to ensure water as a fundamental right under the right to life (Article 21) of the constitution of India. The first priority and charge on groundwater shall be meeting the right to water for life, followed by allocation for achieving food security, supporting sustenance agriculture, sustainable livelihoods, and ecosystem needs, subjected to local circumstances (except the right to life). The bill also states the social audits, grievance redressal mechanism, and groundwater protection zones. Although the bill has tried to ensure equity, efficiency, and sustainability of groundwater usage, still some of the limitations are observed by analyzing the bills. These are stated in the following lines.

There is no defined system as to how the authorities will ensure the equitable distribution of groundwater on the basis of priority. The provision is ideal but not practical in nature; there is provision for penalizing the offenders but the bill did not take into account the state governments which have not yet initiated any legislation for groundwater conservation despite the critical condition of their groundwater resources; the groundwater abstraction units can be registered but there is no mechanism to monitor the volume extracted by each user; the social audits are also not feasible as of now because of lack of detailed data about the resource. Moreover,

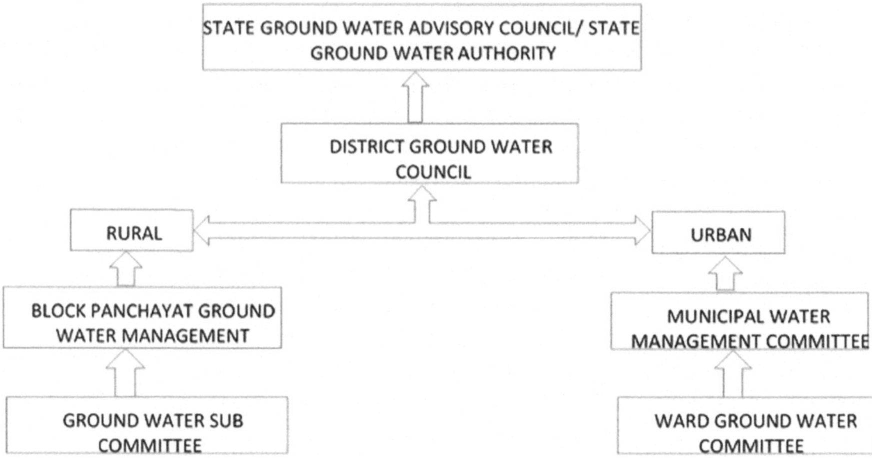

Fig. 33.2 Hierarchy of Structure of Ground Water Governance Proposed by the Model Bill 2016. *Source* Author's own representation: compiled from the Model Bill 2016, Government of India (2016a)

there is no universal mechanism to ensure efficient groundwater governance. The decentralized governance system cannot assure the sustainability of groundwater. A number of factors like hydrogeological patterns, community needs, land usage patterns, political issues, and others need to be analyzed before adopting any defined governance system.

33.4 Governance Structure in the Three Northern States

33.4.1 Why Three Northern States?

It has become important to gauge the laws and policies for groundwater governance of the three northern states, namely, Punjab, Haryana, and Uttar Pradesh. These states have served as a food grain basket of India but unnoticed growth in private tube wells to reap maximum benefit out of the existing system has bought them on the verge of rural economic crisis. A majority of blocks in the north-western region are categorized as over-exploited by the CGWB. Some of the highlights are described here.

One, the stage of groundwater development is very high in the states of Delhi, Haryana, Punjab, and Rajasthan, where it is more than 100%, which implies that in these states the annual groundwater consumption is more than annual groundwater recharge and for Uttar Pradesh, it is more than 70%. Second, the number of over-exploited assessment units, where the stage of development is more than 90%, is 79, 61 and 11% in Punjab, Haryana and UP respectively (Government of India 2019).

Third, Punjab has the highest number of deep tube wells (depth of bore more than 70 m) and UP ranks first in the number of shallow tube wells (depth of bore up to 70 m), according to the 5th Census of Minor Irrigation Schemes. In this context, the water governance system in these states must be critically analyzed to fix their groundwater extraction issues. Hence, in the following section, we have given an overview of the present groundwater governance system in these states.

33.4.2 Punjab

In Punjab, water resource management is broadly assigned to two departments—Department of Water Supply and Sanitation and the Department of Irrigation.

The Department of Water Supply and Sanitation: The department is assigned to provide water supply and sewerage connections in rural areas of Punjab, to provide safe drinking water and sanitation facilities to rural habitations on a sustainable basis, to ensure permanent drinking water security in rural areas, to promote conjunctive use of ground and surface water and rainwater harvesting to achieve sustainability, to involve the rural community in planning and execution of rural water supply schemes, and to make the community capable of operation and maintenance of schemes on their own. However, there is no mention of the action plans to execute these targets. Also, the official websites do not provide the dataset or the specifications of any program/scheme/project for the same.

The Department of Irrigation or Water Resource Department: This department has constituted a Water Resources Organization which is primarily entrusted with research and development activities relating to ground and surface water. Various activities being undertaken by this organization are groundwater monitoring, collection of groundwater data, preparations of various maps to depict groundwater level fluctuations, groundwater investigations, dynamic groundwater estimation, collection and storage of rainfall data and surface water data, up-gradation of hydro-meteorological and surface water observation sites. In addition to this, this organization also deals with framing of policy matters like groundwater legislation, state water policy and other matters relating to water resources referred by the state government from time to time. All the activities related to groundwater are being carried through Water Resources and Environment Directorate (Department of Irrigation, Government of Punjab).

Some projects were carried out under this department. First, Groundwater Investigation and Integrated Utilization of Water Resources in Punjab state started in 1971. The various works carried out under this project were: Installation of exploratory bores to determine subs-oil lithology, thickness, and nature of aquifer and its hydraulic characteristics in the entire state. Second, Punjab was one of 13 states of India where the World Bank Aided Hydrology Project Phase-II (2006–2012) was undertaken. This Project is to improve institutional and organizational arrangements, technical capabilities and physical facilities available for measurement, validation, collation, analysis, transfer, and dissemination of hydrological, hydro-meteorological and water quality data and in the development of comprehensive, easily accessed and user-friendly databases covering all important aspects of the hydrological cycle for long-term planning and development of water resources.

Punjab Water Resources Management and Development Corporation Limited (PWRMDC) formed in 1970 under this department is mandated to provide irrigation facilities in the backward Kandi region of Punjab by the installation of deep tube wells.

In addition to these projects, the Department of Power with the support of World Bank launched ‘Pani Bachao Pesa Kamao’ scheme on pilot basis in June 2018 in 6 agriculture feeders. Under this scheme, farmers are provided with a fixed electricity allocation through metered connection for a particular irrigation period. In case the electricity consumption is lesser than the fixed allocation, a pre-determined amount per unit of power saved is directly transferred to the farmer’s bank account. Further, no extra amount is charged in case of higher consumption. The scheme aims to incentivize farmers for saving both electricity and groundwater. However, how this scheme impact the groundwater levels in long run still need to be examined for implementing it at large scale.

The legislative picture for groundwater management is quite bleak. The Punjab Ground Water (Control and Regulation) Bill, 1998 was prepared on the basis of Model Bill but failed to become an Act because the political lobby thought of it to be harsh on the farmers. The attempt of Punjab Preservation of Sub Soil Water Act 2009 (Government of Punjab 2009) to delay rice plantation from May 10 to June 10 was a progressive step to check the excessive exploitation of groundwater for sowing and transplanting paddy. Many studies have shown positive savings on groundwater level and energy consumption because of this policy. The fall in the water table can be checked by about 30 cm, which is about 65% of the long-term rate of fall, by delaying the transplanting beyond 10th June, as envisaged in the Act. The total savings in electricity due to the Act come to 276 million kWh and the saving in electricity subsidy is of Rs. 108 crore (Singh 2009).

Despite having dedicated organization for groundwater assessment and management since the 1970s, the condition of groundwater resources has deteriorated gravely to the stage that Punjab ranks first in the number of over-exploited blocks as per the report of CGWB. It is to be noted that the concerned departments did not have any legislative backing. The management cannot be ensured without defined regulation, especially in the case of common-pool resources like groundwater. Till date, no state-level groundwater authority has been formed on the model bill guidelines.

33.4.3 Haryana

There are no legal mechanisms, dedicated authority, or department for groundwater management in Haryana but there are some schemes and programs for the same which are spread across three departments—Department of Agriculture, Department of Irrigation, and Department of Rural Development.

Department of Agriculture: Three schemes are running under this department related to groundwater management. The first scheme is for providing assistance on the adoption of water-saving technologies. It mainly focuses on artificial recharge structures, monitoring by piezo metric tubes, promoting micro-irrigation technology and awareness among farmers for judicious use of groundwater. The second scheme is Integrated Watershed Development and Management Project. It is a continuous scheme that targets the enhancement of agricultural productivity and production in a sustainable manner. The action plan also includes building water harvesting structures, percolation tanks, check dams, drop structures, etc. The third one is Scheme for Technology Mission on Sugarcane which targets to promote production and better productivity of sugarcane as farmers are replacing sugarcane by paddy cultivation. The reasons cited for this trend are assured returns and better marketing, short duration and less labor-intensive nature of paddy. It is to be noted that this scheme is completely opposite to the above two schemes as sugarcane is a water-intensive crop. The concerned authority should promote diversification of the farmers' portfolio for less water demanding crops.

Department of Irrigation: This department has introduced the Haryana Pond Development, Protection and Conservation, Pond Water and STP Treated Affluent Utilization Authority Bill 2017. One of the clauses of the bill states the objective to arrest the encroachment of ponds for other activities that leads to an acute shortage of water and rapid depletion of groundwater. Undoubtedly, the restoration of the pond's ecosystem is positively correlated to a better groundwater ecosystem.

Department of Rural Development: This department implements the Pradhan Mantri Krishi Sinchai Yojana—Watershed Development Component (PMKSY-WDC). Under this program, a sub-scheme for water conservation and water harvesting in over-exploited blocks (as per the notification of Central Ground Water Board) is being implemented.

33.4.4 Uttar Pradesh

Though late, UP has initiated strong steps to arrest the groundwater depletion problem. After the failure of enacting UP Groundwater Conservation, Protection and Development (Management, Control, and Regulation) Bill 2010, the government has passed Uttar Pradesh Ground Water (Management and Regulation) Act, 2019 as per the model bill guidelines. The Ground Water Department (GWD) has also initiated the State Ground Water Conservation Mission in 2017. Some of the important activities undertaken by this department are groundwater mapping (GIS-based), promoting rooftop rainwater harvesting on government buildings, groundwater mass awareness, and publicity, etc.

Uttar Pradesh has also been part of various projects sponsored by the World Bank. All these projects are also implemented by the groundwater department. First, UP water sector restructuring project phase-2: the project action plan includes up-gradation of the existing monitoring system, development of comprehensive groundwater management plan for the state, development of aquifer management and conjunctive use plan for selected blocks in Fatehpur district, adopting basin approach and capacity building/institutional development (GWD, UP). The second is the National Hydrology Project, whose major objective is to organize a strong water resources database and enhance institutional capacity. The third is National Ground Water Management Improvement, which particularly focuses on stressed blocks of Bundelkhand and western UP. The main activity is to sustain groundwater by promoting community participation and self-regulation.

UP Ground Water Policy 2013 (Government of Uttar Pradesh 2013) has been traced with a comprehensive approach. Its objective includes all the major components: aquifer mapping, conjunctive use, recharge programs, identification of polluted groundwater zones, effective legal structure, research, and training, etc. The policy also mentions about the action plans. Some of the plans highlighted in the policy are planned management in urban and rural areas, saturate each micro watershed of Bundelkhand-Vindhyans, groundwater data bank and information system, panchayat to panchayat in rural and school to school in urban- training, publicity extension, and capacity building.

The policy and programs being implemented in UP have high potential to manage the resource before reaching to governance crisis stage but it is to be seen that how effective legal mechanism is imposed since the groundwater regulation has been enacted now.

33.5 Policy Interplay

A number of studies have shown empirically the impact of some of the government policies that created a trade-off in the rural economy (Dubash 2007; Singh 2009; World Bank 2010). The flat tariff system reduced the burden of electricity tariff on farmers but in the long run, increased the economic cost by pumping from deeper tube wells on one hand and the other hand severely impacted the groundwater level. Not only groundwater, but the surface water can also perish, specifically the seasonal water structures, in long run in the areas where groundwater is depleting rapidly as is evident from changing regimes of seasonal river systems in the north-western belt. Thus, the first policy that created trade-off was the electricity policy in rural areas. The reversal in this system is possible only through strong political will and providing farmers a progressive incentive so that it may not lay the unnecessary burden on their economic well-being. The *Gujarat Jyotigram model* has become an ideal alternative for this problem (Shah et al. 2008).

Second, the unregulated groundwater markets: the poor or non-functioning of public tube wells has led to more inclination of small or marginal farmers toward groundwater markets which are mainly operated by large landowners. These local groundwater markets provide a reliable source of water for farmers (Banerji et al. 2006; Shah and Chowdhury 2017; Singh and Singh 2006; Srivastava et al. 2009) but it is to be noted that the water sellers pump extra units for economic gain in return of ecological loss. On the other hand, the buyers tend to flood their fields with water for the whole duration of water supply irrespective of the crop requirement and soil moisture content, this practice has led to excessive soil salinity. Despite abundant sources of knowledge about the growing role of these unorganized local water markets in the rural economy since two–three decades, the governance of these markets has largely been ignored.

Third, National Water Policy 2012 (Government of India 2012) states as one of the guidelines, 'It needs to be ensured that industrial effluents, local cesspools, residues of fertilizers and chemicals, etc., do not reach the groundwater.' The solid waste management rules 2016 (Government of India 2016b) states about the proper site selection and pollution prevention measures for landfill sites but both the policy measures do not provide a strong legal mechanism that will be followed in case of any failure in implementation. Moreover, there are other unaddressed issues, such as: Which authority will be accountable in case of pollution by the landfill sites? How it can be ensured that chemicals and fertilizers do not penetrate the soil and pollute groundwater?

Fourth, National Water Policy 2012 mentions water pricing as one of the measures for economic use of water resources. It is important to note that groundwater accounts for 45% of total urban water supply which is further expected to rise in the near future with a change in lifestyles. The water tariffs in urban areas are nominal and non-progressive in nature which hardly puts any burden of marginal cost on the users. Before metering the rural water abstraction units, the urban water pricing policy needs a revision that is in congruence with the national water policy.

Fifth, The National Rehabilitation and Resettlement Policy, 2007 (Government of India 2007) quotes 'as far as possible, projects may be set up on wasteland, degraded land or un-irrigated land. Acquisition of agricultural land for non-agricultural use in the project may be kept to the minimum; multi-cropped land may be avoided to the extent possible for such purposes, and acquisition of irrigated land, if unavoidable, may be kept to the minimum.' In complete contrast to this provision, a majority of real estate projects mushrooming in the western UP are on the agricultural land. The complete hydrogeological cycle gets affected by these structures.

Sixth, another important policy factor is the subsidy on fertilizers. The higher fertilizer use demands higher irrigation. Although the crux of such a crisis lies in the problem of lack of awareness among farmers about the optimum quantity of inputs required, the policies need to be framed keeping in view all the socio-economic factors of a particular region.

Seventh, solar energy is being promoted as a captive source of energy in rural areas. On one side, the renewable source of energy provides a sustainable alternative but on the other side, the same measure gives a free hand to the farmers to use an unregulated energy source for groundwater abstraction. This interplay needs to be considered before any policy formulation. It can be analyzed from a number of studies that short-term welfare measures can prove disastrous on the same class of beneficiaries in the long run.

33.6 Conclusions and Suggestions

The present study shows that the governance structure proposed by the central government has included all possible principles for providing sustainable groundwater governance. But the responsibilities of different authorities are overlapping in nature. It must be mentioned that everyone's responsibility is no one's responsibility. Also, it is important to introduce required amendments in the legal measures so that the inaction on part of states may be addressed. An inter-departmental council can be formed at both the Centre and the state level to discuss policies of different departments affecting the water resources.

The groundwater governance in Punjab, Haryana and Uttar Pradesh is in the evolution stage. Punjab and Haryana did not enact any law for groundwater regulation so far. Also, there is no dedicated groundwater authority in these states though it was proposed in the model bill way back in 1970. In comparison to the other two states, the groundwater governance structure of UP is far more defined as per the secondary sources. It is important to note that a majority of programs being implemented in these states focus on rainwater harvesting, which is recharging of groundwater. However, this 'bucket approach' cannot yield sustainable results for groundwater management (Srinivasan and Lele 2017). The groundwater is not an isolated unit, rather an interrelated one. The major focus is on groundwater as a resource unit not on the whole resource system. There is a need to revise the programs in light of integrated use of ground and surface water as proposed by the National Water Policy. Therefore, the optimum use of groundwater is not solely confined to the demand and supply management of groundwater but also takes into account other resources management.

The following are highlights of different policies:

- It is not feasible to manage the paradoxical condition unless we have a reliable and rigorous database about the water resources that can be shared among different departments.
- It is more of a lack of awareness problem among farmers than the interplay of policies. A large section of farmers assumes that more water and fertilizer mean more production. There is a dearth of scientific and economic knowledge among farmers about the groundwater resources. It has become imperative to deploy an agricultural scientist in every village to guide farmers in this regard.
- Several policies of different departments address the issue of water resource quantity and quality maintenance like real estate act mandates the concerned builder to have the Environment Impact Assessment (EIA) certificate before beginning the project. Under this EIA, there is provision for confirming the groundwater availability and demand for the concerned project. However, the problem is largely with the poor implementation system rather than the policy framework.

There are no defined universal guidelines for governance of common-pool resources like groundwater that can ensure sustainability but it has been empirically proved in many studies that a governance structure that ensures maximum community participation is successful in the long run. It is suggested to have a group of trained personnel or specialized agency to introduce farmers with the dynamics of groundwater. The sustainability quotient in governance can be addressed by the endogenous factors like capacity building, community participation and awareness first, only then can technology play its role to ensure equity, efficiency, and sustainability in any governance system.

References

Banerji A, Meenakshi JV, Khanna G (2006) Groundwater irrigation in North India: institutions and markets. SANDEE Working Paper No. 19-06

Dhawan BD (1993) Ground water depletion in Punjab. Econ Polit Wkly 2397–2401

Dhawan BD (1995) Magnitude of groundwater exploitation. Econ Polit Wkly 30(14):769–775

Dubash NK (2007) The electricity-groundwater conundrum: case for a political solution to a political problem. Econ Polit Wkly 45–55

Government of India (1993) Report on census of minor irrigation schemes 1986–87. Ministry of Jal Shakti, Department of Water Resources, River Development and Ganga Rejuvenation, Minor Irrigation (Statistics) Wing, New Delhi

Government of India (2007) The National rehabilitation and resettlement policy 2007. Ministry of Rural Development, Department of Land Resources, New Delhi

Government of India (2012) National water policy 2012. Ministry of Jal Shakti, Department of Water Resources, River Development and Ganga Rejuvenation

Government of India (2015) Agricultural statistics at a glance 2015. Ministry of Agriculture and Farmers Welfare, Department of Agriculture, Cooperation and Farmers Welfare Directorate of Economics and Statistics

Government of India (2016a) Model bill for conservation, protection, regulation and management of groundwater, 2016. Ministry of Jal Shakti, Department of Water Resources, River Development and Ganga Rejuvenation

Government of India (2016b) Solid waste management rules 2016. Ministry of Environment, Forest and Climate Change (MoEFCC), New Delhi

Government of India (2017a) Report of 5th census of minor irrigation schemes. Ministry of Jal Shakti, Department of Water Resources, River Development and Ganga Rejuvenation, Minor Irrigation (Statistics) Wing, New Delhi

Government of India (2017b) Dynamic ground water resources of India, (as on March 2013). Central Ground Water Board, Ministry of Jal Shakti, Department of Water Resources, River Development & Ganga Rejuvenation, Faridabad

Government of India (2018) Ground water scenario in India, Premonsoon, 2018. Central Ground Water Board, Ministry of Jal Shakti, Department of Water Resources, River Development and Ganga Rejuvenation

Government of India (2019) Dynamic ground water resources of India, 2017. Central Ground Water Board, Ministry of Jal Shakti, Department of Water Resources, River Development and Ganga Rejuvenation, Faridabad

Government of Punjab (2009) The Punjab preservation of sub soil water preservation Act 2009. Department of Legal and Legislative Affairs, Punjab

Government of Uttar Pradesh (2013) UP ground water policy 2013. Ground water department, Uttar Pradesh

Hardin G (1968) The tragedy of commons. Science 162(3859):1243–1248
Kulkarni H, Shah M, Vijay Shankar PS (2015) Shaping the contours of groundwater governance in India. J Hydrol Reg Stud 4:172–192
Mukherji A (2004) Groundwater markets in Ganga-Meghna-Brahmaputra basin: theory and evidence. Econ Polit Wkly 3514–3520
Pandey R (2014) Groundwater irrigation in Punjab: some issues and way forward. National Institute of Public Finance and Policy (NIPFP), New Delhi
Pant N (2004) Trends in groundwater irrigation in eastern and western UP. Econ Polit Wkly 3463–3468
Pant N (2005) Control of and access to groundwater in UP. Econ Polit Wkly 40(26):2672–2680
Prasad K (2008) Institutional framework for regulating the use of groundwater in India. Final report submitted to Ministry of Water Resources, Government of India, Institute for Resource Management and Economic Development, New Delhi
Shah T (2014) Groundwater governance and irrigated agriculture. Global Water Partnership, p 69
Shah T, Chowdhury SD (2017) Farm power policies and groundwater markets contrasting Gujarat with West Bengal (1990–2015). Econ Polit Wkly
Shah T, van Koppen B (2006) Is India ripe for integrated water resources management? Fitting water policy to national development context. Econ Polit Wkly 3413–3421
Shah T, Bhatt S, Shah RK, Talati J (2008) Groundwater governance through electricity supply management: assessing an innovative intervention in Gujarat, western India. Agric Water Manag 95(11):1233–1242
Singh K (2009) Act to save groundwater in Punjab: its impact on water table, electricity subsidy and environment. Agric Econ Res Rev 22(conf):365–386
Singh DR, Singh RP (2006) Structure, determinants and efficiency of groundwater markets in western Uttar Pradesh. Agric Econ Res Rev 19(347–2016–16764):129–144
Srinivasan V, Lele S (2017) From groundwater regulation to integrated water management. Econ Polit Wkly 52(31):107
Srivastava SK, Kumar R, Singh RP (2009) Extent of groundwater extraction and irrigation efficiency on farms under different water-market regimes in Central Uttar Pradesh. Agric Econ Res Rev 22(June):87–97
Varady RG, van Weert F, Megdal SB, Gerlak A, Iskandar CA, House-Peters L, McGovern ED (2013) Thematic Paper No 5: groundwater policy and governance. In: Groundwater governance: a global framework for country action GEF ID 3726
Water Governance Facility (2013) Groundwater governance in India: stumbling blocks for law and compliance. WGF Report No. 3, SIWI, Stockholm, 30
World Bank (2010) Deep wells and prudence: towards pragmatic action for addressing Groundwater Overexploitation in India. The International Bank for Reconstruction and Development/The World Bank, p 120

Chapter 34
Developing Values and Ethics in Leadership for Effective Water Governance

Nanditesh Nilay

When there was neither kingdom nor king, there was neither governance nor governor, the people were protected by Dharma.

Abstract Ethical water literacy is the call of Mother Earth. The society has to be literate in understanding the value of water. Being educated has always been a responsibility which expects an enlightening answer from the generations globally. The aim of education has been to develop an individual's ethics and values towards nature and people across. The moment society creates an educated mindset; simultaneously it knits behaviour with the fabric of values and ethics. However, the problems like global warming, pollution and water crisis all are interrelated and urgently demand an ethical and value-based approach in water governance. This paper qualitatively explores those aspects of human values and ethics in leadership for effective water governance.

Keywords Leadership · Ethics · Water governance · Water management

34.1 Introduction

Water is more than a wonderful liquid. It is life for those who are walking on earth. The water cycle being taught in schools just tries to sensitize children on the importance of water. However, the world water day signifies the value of water for all of us and inspires us to treat water with the spirit of cleanness and thus makes it accessible for everybody globally. 2018 is being celebrated as a year which will inspire us to treat the nature qualitatively rather than quantitatively in handling water challenges globally. There will be so many options we may learn from nature to make water

N. Nilay (✉)
Doctoral Fellow -Gandhian Studies (ICSSR), New Delhi, India
e-mail: nanditeshnilay@gmail.com

Author, Speaker & Founder (Training Requirement & New Concept), Greater Noida, India

A. Pandey et al. (eds.), *Water Management and Water Governance*, Water Science and Technology Library 96,
https://doi.org/10.1007/978-3-030-58051-3_34

purest but we cannot forget that between nature and water there is a powerful presence of human nature too.

We all have witnessed the Kedarnath flood, rapid drought and water pollution in India, and such wrath of nature has been a sad story of degrading human nature towards nature and water per se. Though unenviable Capetown water crisis has broken our luxurious slumber a bit yet as per the BBC survey, Bangalore will also face the same fate if people will not come out from their materialistic approach towards life. It has been written in the Bodhisatwa that the first person get drowned in the flood is the sleeping person. Therefore, it is highly desired to develop ethical and value-based leadership for effective water governance.

As per UN projects projections, global demand for freshwater will surpass supply by 40% in 2030. The new research has been more alarming as it confirms that 1.8 billion people will suffer severe water scarcity for at least half a year. Only 2.53% of world's water is freshwater and India with its population of 1.3 billion shares only 4% of the global freshwater reserves. Thus, only a small percentage of freshwater is available for drinking water and food production, which, too, is diminishing over time with spiralling growth of population. In 1947, the per capita availability of freshwater in India was 5000 m^3/year, and in 2000 it was 2000 m^3/year and is likely to fall below 1000 m^3/year by 2050 reaching the threshold value for water scarcity. Climate change, global warming and seasonal rainfall changes are further exacerbating this critical situation with changes in hydrologic cycles.

In the above background, it is urgent to treat water crisis more intelligently and empathetically. Human beings have become more a consumer than a citizen and not a happy being under the lap of nature. Nobody can deny that today between man and nature, entropy reigns supreme. The grey is becoming more selfish and dominant with the least concern for green. Water is being enslaved by rich in the urban lifestyle, and poor has to receive the contaminated and unfiltered. Its presence and absence are not observed in the connection with nature rather than just with the giant purchasing power of rising middle and upper middle class. One is treating water as a market product as there is a rising feudal urban class in India. I and mine have taken human being distance from nature and so from the sensitivity of treating water as a thirst of human being across. No one is concerned for roots of trees and heart of soils. Prof Avdhesh Pratap, an eminent global expert in water law, believes that water literacy is the urgent call for our nation and one does not need to look towards the west for all solutions. India has been the leader from years and years in water resource management. Even great water management expert, Prof. Arjen Hoekstara, believes that among all environmental problems water scarcity is at the top.

There is no way one can ignore this crisis. The present paper will explore and look into various aspects of human values and ethics in building leadership for effective water governance.

34.2 Water Governance

Water governance requires ethical and value-based leadership capacities in decision-making (Scholz and Stiftel 2005; Pahl-Wostl et al. 2008; Kante 2011). It is the responsibility of leadership and government to improve water management facilities and ensure that water is available to everybody (Roberts 2003). Though lot of international institutions come forward for developing water management resources, it is the responsibility of national government and leadership to provide those facilities (Durant et al. 2004). The global world needs comparative assessments across countries with time (Dinar and Saleth 2005) and a way forward for developing capacities in ethical and value-based leadership. It will help all countries to go for value-driven leadership and thus enhance quality in water governance. The modern social and environmental discourse is putting a lot of emphasis on such aspects. Though incidents and charges of corruption even in water management have no ends, yet one cannot deny that character and conduct of leadership of various institutions are being judged and marked by values nowadays.

34.3 Values in Leadership for Water Governance

Values are what we value for. The journey from the child stage to maturity stage evolves a person with basics and desires. Values belong to the fundamental zone. It is the founding stone for becoming a human being. Leaders like Mahatma Gandhi, Sardar Patel, Jawaharlal Nehru worn the west at the time of their youth but their heart was not synthetic. It was full of compassion and concern for starving millions in India. In the imperfect society, their leadership was perfect with values. Therefore, it is highly needed to build values in leadership for effective water governance.

Leadership determines the personality of a person as well as it equips a person to analyse the situation well and with larger perspective (Drucker 1954). In the water governance, one needs those leadership capacities in employees of an organization to create the desired future where everybody lives conductively (Toubiana and Yair 2012). Governance needs strategic directions, and this is being done by a leader only. Leader's role is to analyse and examine the situation and develop appropriate capacities which will safeguard the future of people and society across (Deshpande 2012). In the field of water governance, the focus is too much in present and lot of times mediocrity in analysing the future crisis leads the environment in a dead land. In the world of climate change and global warming, water literacy can only reach to maximum if governance is moving with ethical and value-based capacities in the leadership initiatives. In the midst of competing in turbulent and unpredictable environments, water governance requires those capacity building initiatives in leadership which must focus on ethics and values.

It is really a worry sign for India when human values have to suffer due to hand-picked cronies who influence the process of executive and snatch away the due benefit from the last person. The list of cities is increasing where water crisis will become national emergency. The lack of values in governance has been a big reason in developing water crisis across. On the other hand, Folke and others treat water as a necessity of the well-being of human beings and environment. Leadership has to take into consideration the value aspects in personality and governance. These contrasts illustrate the issues of value judgement that becomes prominent once we try to understand and assess water dynamics impact on human and socio-ecological systems across. Further, value dimensions introduce the proper meaning and understanding of water in different societies and culture too (Blatter and Ingram 2001).

We are comfortable in using the word 'value' in the context of money or profile value but the moment the term Human values appear, generally, it is treated something good for discussions but not worthy to be practised. However; when we are exposed to water unavailability or organizations/nations are cursed by nature, then we talk endlessly about values. The reason is values are best realized when they are snatched or violated. Values have always served as a benchmark for leadership. It has been easy to rule the country or run a company but it has been a great challenge to lead with values and ensure that water is for everybody. Value is the protocol of behaviour that enhances the trust, confidence and commitment among the members of a community. When we go through the pages of history, we find the process of transformation begins with the foundation of values.

34.4 Ethics in Leadership for Water Governance

Ethics has been one of the most discussed concepts of human civilization. Scholars across the world have come to their own understanding of ethics. Still, scholars have not agreed on any one definition of Ethics (Ayee 1998).

The word "ethics" is derived from the Greek word Ethikos, meaning custom, conduct and character that are valued. It creates a distinction between wrong behaviour and right behaviour. Good or bad human conduct has been the benchmark of ethics as it provides the fundamental principles of right conduct (Chapman 1993). These principles can be exercised through law, code of conduct or merely an individual's interpretation of good or bad conduct. This constitutes the standards that people as well as institutions strive to follow for a solid foundation. In other words, ethics can be considered as a set of values which we develop throughout in our life influenced by our upbringing, parents and teachers, etc., who shape and guide our world (Adeleke 2015).

The world of ethics is all about piercing right as well as wrong actions. It leads us towards choosing appropriate actions as per social standards. When we go through the debate and the argument among Greek philosophers, we realize that ethics is being received and explained as a guiding force in taking right action in a given situation (Freakley and Burgh 2000).

Ethical leadership is highly desired in water governance. Cost–benefit ethics ensures that leader will confer benefit towards society. However, the term benefit brings different type of connotation, and it is possible that leader can justify it by unethical behaviour (Toubiana and Yair 2012). We have already been exposed to such situations where in the name of development water is drying fast from the surface of the earth. In the twentieth century, the focus of ethical leadership was more utilitarian and it reached to all corners of world (Feldman 1995). The utilitarian focus on consequences was closer to cost–benefit ethics and that depended on what produces the greatest good. It also helped the process of water governance due to its objective approach and rationality (Blatter and Ingram 2001). In all integrated water resource management systems, the focus is cost–benefit and utilitarian only. The ultimate objective of leadership is to maximize economic and social welfare with hurting the environment (Agarwal 2003). Therefore, from normative perspective, such type of approach in ethical leadership will assist the process of water governance. We must see the water governance not only from the aspect of utility but what values we put across. Therefore it is important to balance the mean as well as the end for an effective ethical leadership.

Ethical leadership also has to bring aspect of prudence in governance which needs more careful approach towards decisions (Toubiana and Yair 2012). This is because the distribution of water and that with equanimity is highly desired in governance. This is an ethical issue that water has to be for everybody. It should not be used as private property, and therefore ethics in leadership is needed urgently (Schorr 2005). Even the issue of corruption in pricing of water and using prudence in its distribution brings the case of values and ethics in water governance (Robertson 2007).

Ethics in leadership has to incorporate Confucian ethics for better water governance. This model of ethics does not only impose responsibility on leaders but also on of parents and employers. This model endorses for equality of obligation and vice versa (Baker and Comer 2011). Ethics and value-based leadership recognizes the role of emotion in governance and decisions in water governance influences people across and here Confucian ethics become more important as it includes individuals per se.

34.5 Conclusion

In this paper, we discussed two approaches for effective water governance—values and ethics in leadership. Though few discussions are overlapping yet highly effective in the present context as there are few works where water governance has been observed and explored from this point of view. Through this paper, we can conclude that water governance depends on ethics and values in leadership. For a country like India, water education is a matter of primary concern for the Government of India, and it is the need of the hour for the leadership to sensitize people on value and ethics. However, there has been no express provision to deliver water education in the Constitution of India except Article 51-A(g) in part IV-A containing fundamental duties which was inserted by the Constitution (42nd amendment) Act, 1976 with effect from 3-1-1977, envisages that “it shall be the duty of every citizen of India to protect and improve the natural environment including forest, lakes, rivers and wildlife and to have compassion for living creatures".

This education has to incorporate human values and ethics among people across India, and governance has to build such value-driven capacities. There is rising attitude of apathy among urban middle class towards water. It is only being used and exploited as it is our birthright. Nature as well as democracies has always been considerate towards their citizens, but are we? We need values literacy along with water literacy. One has to shape world not only flat in products and feelings but overt for conducive living and harmonious existence. We must save boundaries but cannot ignore water. Water cycle begins with water and ends up with water only but every time it serves nature as well as human beings. This aspect of water has to be realized but not with the consumer character but as an enlightened soul. If not, then there will be missed calls from nature which will destroy all of us. Let us respect and realize the Mother Nature by purifying our human nature.

References

Adeleke A (2015) Work ethics, values, attitudes and performance in The Nigerian public service: issues, challenges and the way forward. J Pub Adm Gov. ISSN 2161-7104

Agrawal A (2003) Sustainable governance of common-pool resources: context, methods and politics. Ann Rev Anthropol 32:243–262. https://dx.doi.org/10.1146/annurev.anthro.32.061002.093112

Ayee J (1998) Ethics in the public service. A paper delivered at the Second Pan—African conference of the Ministers of Civil Service, Rabat, Morocco

Baker SD, Comer DR (2011) Business ethics everywhere: an experiential exercise to develop student's ability to identify and respond to ethical issues in business. J Manag Educ 36:95–125. https://doi.org/10.1177/1052562911408071

Blatter J, Ingram H (eds) (2001) Reflections on water: new approaches to transboundary conflict and cooperation. MIT Press, Cambridge, Massachusetts, USA

Chapman RA (ed) (1993) Ethics in public service. Edinburgh University Press

Deshpande AR (2012) Workplace spirituality, organizational learning capabilities and mass customization: a integrated framework. Int J Bus Manag 7(5):3–5. https://doi.org/10.5539/ijbm.v7n5p3

Dinar A, Saleth M (2005) Can water institutions be cured? A water institutions health index. Water Supply 5(6):17–40

Drucker PF (1954) The practice of management. Harper, New York, NY

Durant RF, Fiorino DJ, O'Leary R (eds) (2004) Environmental governance reconsidered: challenges, choices and opportunities. MIT Press, Cambridge, MA

Feldman D (1995) Water resources management: in search of an environmental ethic. John Hopkins University Press, Baltimore, Maryland, USA

Freakley and Burgh (2000) Engaging with Ethics: ethical enquiry for teachers. Social Science Press, Australia

Kante B (2011) Shaping an international governance system for environmental sustainability. In: United Nations, first preparatory meeting of the world congress on justice. Governance and Law for Environmental Sustainability. 12–13 Oct 2011. Kuala Lumpur, Malaysia

Pahl-Wostl C, Gupta J, Petry D (2008) Governance and the global water system: a theoretical exploration. Glob Gov 8(4):419–435

Roberts T (2003) Water sector governance in Africa, Vol 1. Theory and practice. Water Partnership Program

Robertson M (2007) Discovering price in all the wrong places: the work of commodity definition and price under neoliberal environmental policy. Antipode 39:500–526. https://doi.org/10.1111/j.1467-8330.2007.00537.x

Scholz J, Stiftel B (eds) (2005) Adaptive governance and water conflict: new institutions for collaborative planning. Resources for the Future Press, Washington, DC

Schorr D (2005) Appropriation as agrarianism: distributive justice in the creation of property rights. Ecol Law Q 32:3–71

Toubiana M, Yair G (2012) The solution of meaning in Peter Drucker's oeuvre. J Manag Hist 18:178–199. https://doi.org/10.1108/17511341211120684

Chapter 35
Let's Discuss Water Leadership: Enabling Adaptive Governance for Evolving Waterworlds

David M. N. Gosselin

> "The global water crisis is a crisis of governance"
> OECD (2011)

Abstract Among the 400 papers and presentations delivered at the International Conference on Sustainable Technologies for Intelligent Water Management (STIWM) in Roorkee, India, in February 2018, over 40 addressed management issues, four had governance in their title, but only one had leadership in it. Although many experts and panels called for "leadership" action, it was couched within a management paradigm. This paper suggests that the concepts of water governance are largely based on management principles that stem from normalized water decision-making and administrative circumstances, not leader-based principles that focus on individuals who engage the interpersonal and steward the linkages they create within complex, ambiguous, uncertain, and unpredictable water, environmental, and decision-making circumstances. This paper will argue that the concept of leadership, as an interpersonal dynamic aimed at engaging personal potential, is lacking in the governance debate, and because leadership is a people-centric activity, basic concepts of what constitutes water leadership need to be popularized and debated. Further, it will argue that to fully grasp what is meant by water leadership requires a renewed view of hydropolitics and water governance through the holistic lens of simultaneously appreciating water's materiality (hard power) and non-materiality (soft power). Several cases, taken from the author's lengthy military career, and fieldwork in researching Namibian meaning of sustainable water, will be used.

Keywords Water · Leadership · Governance · Management · IWRM

D. M. N. Gosselin (✉)
Strategy Prime, 20715 Timberlane Drive, Omaha, NE 68022, USA
e-mail: david.mn.gosselin@gmail.com

WARI Fellow IIT Roorkee, Roorkee, India

A. Pandey et al. (eds.), *Water Management and Water Governance*,
Water Science and Technology Library 96,
https://doi.org/10.1007/978-3-030-58051-3_35

35.1 Introduction and Scope

It is widely accepted that water governance schemes that embrace the 1992 Dublin Water Principles (Solanes and Gonzalez-Villarreal 1999) and Integrated Water Resources Management (IWRM), advocating for broad-based participation across political levels and economic sectors, will facilitate the achievement of sustainable water. Conceptually this is a powerful image, an integrated network-of-networks, polycentric formal and informal institutions, all subscribing to the common goal of sustained economic development in balance with social and environmental needs. A priori, the presumed goal of water governance is twofold: first, to achieve sustainable water, that amount of water that meets the qualitative and quantitative needs of today without compromising the needs of tomorrow (in perpetuity); and two, to achieve water security, defined simply as the safe and assured access to water in quantities sufficient to address human needs. But the governance literature diverges on what constitutes its goals, and rarely discusses leadership as the active ingredient that enables "broad-based participation," while insinuating that it is achieved through administrative means.

Woodhouse and Muller (2017) point out that this view of water governance is not beyond criticism. They argue that it emphasizes a water scarcity narrative and encourages commoditization; it minimizes government responsibilities and neglects environmental issues; and it underappreciates the legislative and capacity gaps at the community level. Of note, the term leadership appears twice in their review; each time it referred to a structural institution, such as China's "central leadership" (Woodhouse and Muller 2017), and a call for a centralized "think-tank and leadership" (p. 12) to solve global cosmopolitan water problems.

In 2015, the Organization for Economic Cooperation and Development (OECD) Ministerial Council adopted twelve water governance principles and highlighted the necessity to close seven governance gaps: policy, accountability, funding, capacity, information, administrative, and objective (Akhmouch and Correia 2016). Noticeably, absent is no direct mention of the terms leader and leadership in the entire article, nor of a leadership gap, although it could be assumed. Admittedly a small sample, but both review articles are noteworthy as they are recent and recount the salient governance debates of the past three decades while representing the views of 30 countries and over 100 delegates.

Among the 400 papers and presentations delivered at the International Conference on Sustainable Technologies for Intelligent Water Management (STIWM) in Roorkee, India, over 40 addressed aspects of management, four had governance in their title, and one had leadership in its title. Notwithstanding debates about leadership issues writ large, concerns emerged along two themes: Calls for leadership appear substantially different than calls for management; and there appears to be a communication gap between water experts and key leaders and decision-makers. A priori, leadership, defined as the art of influencing others to do what you want them to do, is lacking from the governance debate. And because leadership is about engaging others, if "broad-based participation" is important to water governance, then basic concepts of what constitutes leadership need to be popularized and debated.

My main proposition is that because water is a total social event (Orlove and Caton 2010) in that it touches all aspects of society and civilization, water leadership is considerably different than leadership writ large. The purpose of this paper is to engage water-related disciplines in a debate that focuses on water as a social dynamic, subject to political whims and power struggles, with a view to inspire a leadership culture that enhances multi-disciplinary professional collaboration.

This paper will make three broad arguments:

- First, that current concepts of water governance reflect management principles and assume normalized water, decision-making and administrative circumstances, whereas contemporary circumstances are not normalized, especially in developing countries. Hence, lacking are leader-based principles that focus on engaging the interpersonal within ambiguous, uncertain, and unpredictable water, environmental, and decision-making circumstances.
- Second, it will argue that this limitation is largely due to a bias toward appreciating water's physicality (H20) over its non-materiality. Therefore, to fully grasp what is meant by water leadership requires a renewed view of hydropolitics and water governance through the lens of appreciating water's simultaneous materiality (hard power) and non-materiality (soft power).
- And third, examples from the author's lengthy military career, fieldwork in Namibia, and other secondary sources will be used to make a case that water leadership can free untapped organizational potential to adapt to changing, complex, circumstances, such as climate change and increased populations.

Finally, four intractable topics will be broached during the discussion section as a means to stimulate further debate about institutionalizing a water leadership culture: water as a human right; the need to establish multi-disciplinary communication strategies; the need to create multi-disciplinary training and development facilities; and the need to create spare human capacity in water disciplines.

35.2 Water Governance Without due Consideration for Water

Research into water governance has increased in the past decades,[1] yielding little more than common themes between definitions. Many institutions and scholars have become increasingly prescriptive in naming the actors, the functions they should perform, the goals they should accomplish, and the role that leadership (as an entity) and wide spread stakeholders should embrace, and especially, the degree that governments' role is central or not. Three definitions are provided to highlight their focus

[1] Woodhouse and Muller (2017) provide an excellent review of the research strands and critical debates in water governance since the 1980s, which I will not attempt to duplicate.

on management principles and normalized circumstances, and that the differences between them are marginal, subtle, and slightly misleading; the underlining and italics are mine to identify key themes:

- The Global Water Partnership (GWP) defines water governance as "the range of political, social, economic and administrative *systems* that are in place to *develop and manage* water resources, and the delivery of water services, at different levels of society" (Akhmouch and Correia 2016).
- The United Nations Development Program's (UNDP) definition "encompasses the political, economic and social *processes and institutions* by which governments, civil society, and the private sector make *decisions* about how best to use, *develop and manage* water resources" (UNDP 2004).
- The Organization for Economic Cooperation and Development (OECD) states that it "formally refers to the range of political, institutional and administrative rules, practices and processes (formal and informal) through which *decisions* are taken and implemented, stakeholders can articulate their interests and have their concerns considered, and *decision-makers are held accountable* for *water management*"(Akhmouch and Correia 2016).

At face value, each definition adequately introduces the magnitude, complexities and mostly, the interdependencies of water-related issues across sectors (economic, social, environmental, and political). Arguably, each adheres to Ostrom's (1990, 2005) first principle of establishing common property regimes with "clearly defined boundaries in which the identity of the group and the boundaries of the shared resource are clearly delineated" (Ostrom 1990, p. 90). But these definitions prompt further queries:

- Does the GWP definition intimate a global society with a common approach and meaning to water?
- Does this intimate a global water governance scheme, independent of state sovereignty or augmenting them?
- Does it recognize informal institutions that exist at local levels, or that emerge ad hoc during a particular crisis or event?
- Does the UNDP definition preclude the importance of planning, decentralization, and sustaining relationships as it focuses on the decision-making processes?
- And does the OECD's definition, the most recent addition to the Pantheon, deliberately add the decision-makers' accountability clause as a descriptor or a prescriptor of what constitutes water governance?
- Was this lacking from previous incarnations?

These queries highlight that there is no consensus on what constitutes governance and how it fits into broader hydropolitics. Regardless how these queries are addressed, which is not the purpose of this paper, three themes emerge from the above definitions. First, governance structures and dynamics are viewed as socially interacting systems that manage outcomes; second, these systems focus on making decisions; and third, decision-makers are responsible and accountable for the outcomes of their decisions.

Together this intimates that governance is the management of the decision-making processes that influence the way water is managed. But, these definitions seem to assume a common purpose and understanding of what constitutes water. So, what insight do they proffer on how to govern and "best use, develop, and manage" a resource that has different intersubjective meanings? This is more than a philosophical question; it is also a very practical one.

Increasingly, researchers are querying the very nature of water (Linton 2010) and asking themselves what are the "missing links" in water governance (Pahl-Wostl et al. 2013; Woodhouse and Muller 2017). Some are querying how water problems and policy trajectories defy and test geo-political boundaries (Ferreyra et al. 2008; Pahl-Wostl et al. 2013) and institutions (Schnegg and Bollig 2016), suggesting that problem-sheds and policy-sheds (social issues) are more appropriate than river-sheds and basins as management units (geo-hydrological issues). This challenges traditional research into how and what constitutes successful governance and its emphasis on managing it (water) and its services (Akhmouch and Correia 2016; Ostrom 2005; Woodhouse and Muller 2017). Similar to Woodhouse and Muller (2017), I assume that the context within which humans and societies interact with water will frame the relationship, and consequently, will influence the way they organize to manage it.

35.3 The Limits of Management and Genesis of Leadership

Ironically, the OECD declares that "there is no global water crisis, but a crisis in governance" (OECD 2011), and as illustrated above, they define governance in broad management terms. A priori, management intimates known inputs and outcomes to known processes, yet rainfall in many areas of the globe is deemed scarce and unpredictable. Even though global scarcity may be a debunked myth (Wolf 1998), the unpredictability and high variability dimension of rainfall may be sufficient to justify that leadership that entails leading teams and stakeholders through ambiguity and complexity needs to be considered an important attribute of water governance. Further, if water is indeed in a crisis, then this acknowledgement alone reflects non-normalcy; a crisis calls for something different than normalized management.

In their process analysis of global water governance, Pahl-Wostl et al. (2013) highlight several "normalized" functions (policy framing, conflict resolution, rule-making, resource mobilization, and knowledge-generation) as required in a functioning governance scheme, but insist that legitimacy, leadership, representativeness, and comprehensiveness are the performance drivers, the measures-of-success, that crystallize these functions. Regarding leadership and stewardship they quote:

"Complex governance processes are characterized by self-organization and emergence. However, self-organization without leadership may fail to produce tangible outcomes. This may be formal leadership of a governmental body, but it may also be emergent leadership that develops from an actor's influential role in a network….

We also pay attention to the key role of stewardship, defined as facilitation and the forging of inclusive linkages as processes unfold" (Pahl-Wostl et al. 2013).

The emergent leadership they refer to demands greater debate because it challenges preconceived notions of positional authority by highlighting personal authority that ability for an individual to stand out as a guide to others around them. This accentuates their point on legitimacy, for which they claim "refers to the validity and broadly based acceptance of the authority of an actor or event, making it possible for those actors and events to play an influential role in the overall process" (Ibid., p. 5). They cite several case studies to validate their claims, and therein lies the challenge of empirical evidence regarding leadership: It is oftentimes difficult to operationalize and generalize. Further, ego plays a large part in impeding leadership development and the creation of an inclusive leadership culture.

To overcome ego, modern management predicated on the drive for efficiency has adopted two broad organizational design principles: one, for the sake of continuity of service: one staff can be replaced with another with equal qualifications; and two, for the sake of organizational resilience: different degrees of authority (power to carry out tasks), responsibility (duty to fulfill tasks), and accountability (liable to others for expenditure of capital and outcome) are assigned to individuals across organizational schemes (hierarchical, concentric, polycentric, or loosely networked, etc.).

The significance of highlighting these attributes of organizational design is that the OECD governance definition calling for increased accountability intimates that capital expenditure decisions are the most important; lending credence to the criticism that modern governance has endorsed a scarcity and economic narrative to the detriment of other facets of water, namely human capacity, social and environmental (Woodhouse and Muller 2017).

Gallup, a global performance-management company, conducted a study based on World Poll data that represented 120 countries and 95% of the world's population, concluded that approximately 60% of employees worldwide are deemed "not engaged" in their work (Harter and Agrawal 2011). Assuming that this applies to the water sector writ large, this is an incredible source of untapped potential who are already getting paid.

Shiva (2016) provides examples of where Indian lumber-cutting and bottling companies have followed established processes and the letter of the law, but still inflict damage on the environment and communities. In Namibia, the public sector is the country's largest employer, yet many claim it's unaffordable and not delivering what is required (Ngutjinazo 2018). The point is that most institutions are not immune to the phenomena of relying on outdated processes and management functions, which result in untapped potential as they fail to adapt to changing needs. On the one hand claiming to be efficiency driven, but objectively, being operationally ineffective. To build on the OECD claim, there is no governance crisis, there is a leadership crisis.

I submit that individual and localized leadership is the understudied and misrepresented ingredient that coalesces the complex and interdependent management functions recognized by researchers (Harter and Agrawal 2011; Ostrom 2009; Pahl-Wostl et al. 2013); the necessary agent at the center of the decision-making principles recognized as critical (Woodhouse and Muller 2017); and, the key to unlocking existing organizational potential and wholly appreciate the hidden meaning of water.

35.4 What is Water?

Linton (2010) addresses this question in his book, *What is water? The History of a Modern Abstraction*, and concluded that water has a defining and unifying feature in many cultures and takes on varying significance over time; sometimes based in myth, sometimes in hard facts. He dubbed this as the hydro-social cycle, a society's continued relationship with water over time. My native Canada is the world's third largest freshwater steward and framed by three oceans. Its identity is closely related to the many meanings that water has taken since its early pioneer days, far beyond its mere functionality.

A priori, most Canadians share a strong emotional attachment to their waters: fresh and saltwater ocean. Poets, songwriters and artists comment on its abundance and beauty, its cleanliness and its accessibility, and most citizens will react negatively to accounts of industrial water pollution, acid rain, or large bottling companies absconding entire river flows. Economically, Quebec's wealth (one of the provinces in Canada) is largely derived from its hydropower enterprise, which also dominates the national narrative, and its history and lore is heavily steeped in the stories of the *coureur-de-bois* (fur traders) who adopted native ways and ventured deep into the unknown territories during the early years of our nation (1600–1750); traveling in traditional birch bark canoes via the countless uncharted rivers and streams to trap the European-coveted beaver pelts. These narratives convey water as the facilitator, antagonist, and protagonist, of a romantic, courageous, and adventuresome lifestyle; core features of contemporary Canadian identity. But they also convey the hydropolitics of British colonial power and the role that water played in its execution. To drive the point home, it is said that Native Inuit, the ethnic groups of the Canadian north, refer to snow in multiple ways, each capturing its functionality (wet, dry, soft, etc.) and its mythical character. Our relationship with water has a past, present, and imagined future: this gives us order!

To the native-Americans who lived in the Americas before the Europeans started their migration circa early sixteenth century, the water and rivers represented a gift from their Almighty and a spiritual connectedness to nature; many cultures express water's origin in their respective creation story. In India, this degree of connectedness with water is displayed every day with spiritual cleansing rituals taking place at key points along the Ganga River; while simultaneously, the Ganga River and its canals are fundamental to the agricultural success in Uttarakhand, one of India's northern states.

It should be no surprise that state identities share functional and constructivist perspectives of water. Water is vital to every aspect of human and organic life for which there is no substitute, making this simple compound unique; a uniqueness that plays out politically, economically, environmentally, and socially. However, as argued in *Water and Post-Conflict Peacebuilding,* the failure to consider the social and constructivist aspects of water has led to the failure of many hydro-related projects during post-conflict peacebuilding (Weinthal et al. 2014). The authors relate cases where project staff failed to consider the degree that religious customs were important in North Afghanistan regarding gender roles, or the extent that a negative stigma attached to victims of sexual and gender-based violence had on choosing the location of a community water pump in the Democratic Republic of the Congo (DRC).

Meissner (2016a) in *Hydropolitics, Interest Groups and Governance: The Case of the Proposed Epupa Dam* relates the challenges that the Ovahimba faced, a traditional ethnic group who inhabit both sides of the transboundary Kunene River between Angola and Namibia, when both governments planned a hydroelectric dam at the Epupa Falls. A subtle message that emerges from his book is that the influences that arise from dedicated interest groups, often created ad hoc for a particular event, disappear altogether or get morphed into other non-profit organizations' messages and approaches after an accord has been reached, leaving local communities to deal with the aftermath of infusing new ideas and new forms of protest into their local water narrative.

In my research into the meaning of sustainable water in Namibia, the Namibian government has retained high degrees of legitimacy among its population, even though it has not rectified many of the pre-1990 independence apartheid-era inequities and much of its water infrastructure development has stalled. Agriculture uses over 70% of its scarce water and contributes only 6.6% of the GDP (CIA Factbook, Accessed: 4 May 2018). However, it also supports over 70% of the population and a third of its workforce, so redirecting water to other sectors is largely a non-starter politically. To typify this complexity, in 1998, tourism used 1.5% of Namibia's water and resulted in N$ 113/m3 of water, whereas irrigation used 44% and produced a mere N$ 0.20/m3 (MAWF and DRFN 1998).

From this trend, Namibians now talk about water in two ways: It is discussed directly in a water-centric economic model (e.g. for agriculture, industry), and indirectly in a land-centric economic model that promotes community forests, conservancies and tourism (Gosselin 2018a). These pit two different water governance approaches: one state-centric, the other community and traditional authority based. Anecdotally, the success of the former relies heavily on technological breakthroughs, the latter on leadership engagement: Both are necessary to rectify past Apartheid inequities, but the latter is cheaper and contributes more to GDP. As a note, Namibia has democratically elected three presidents since 1990, all from the South West African People's Organization (SWAPO) political party with a resounding 70% vote (Melber 2015). Each of the above examples reflect a positivist and constructivist view of water: it's hard and soft power.

35.5 A Framework to Appreciate Water's Social and Environmental Complex Interdependencies

To fully understand the Namibian tale and what the other examples typify, it is necessary to adjust how we frame water's significance to societies. Orlove and Caton (2010), both anthropologists suggest that research into water should adopt a holistic approach that recognizes waterworlds, the complete social connections of water, to appreciate its role in all aspects of society. My proposition, stated earlier, is that water simultaneously assumes a material and non-material form; this is experienced individually, then collectively. Consequently, water governance and water leadership, being a social construct, should embrace all perspectives: subjectivity, objectivity, intersubjectivity, and interobjectivity, as expressed in Fig. 35.1.

A priori, I adopt many of these terms from Harari's (2015) book, *Sapiens: A Brief History of Humankind*, because of their straightforwardness: "An objective phenomenon exists independently of human consciousness and human beliefs, for example, radioactivity, gravity, and a river. Subjectivity is something that exists depending on the consciousness and beliefs of a single individual, which are derived from its ontology and epistemology; changing or vanishing if that particular individual changes their beliefs. The intersubjective is something that exists within the communication network linking the subjective consciousness of many individuals; it underpins cultures and institutions. If a single individual changes his or her beliefs, or even dies, it is of little importance" (Ibid. p. 117). I use the term interobjective to represent the dynamics between objects. A simple example may be the dynamic between an apple falling and the earth, which has little meaning except the one we attribute to it; water flowing down river has little significance, except if we assign meaning to the ecosystem's requirements, food production, riparian rights, aesthetics, or power generation.

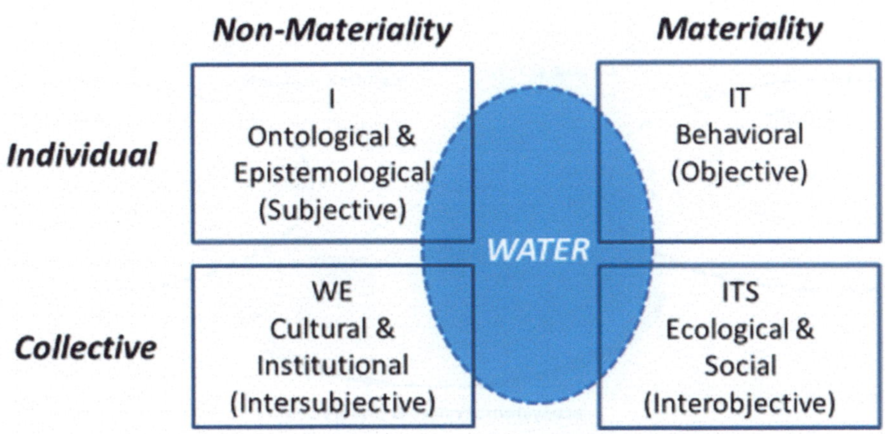

Fig. 35.1 Water's dual nature: individual and collective perspectives of water

As Harari (2015) stated, many of history's most important drivers are intersubjective, i.e., they provide an imagined order to an otherwise chaotic reality. Law, money, gods, nations, and power hold sway over people and societies because they provide a sense of coordinated reality in which decisions and influence can be effected: "in order to dismantle an intersubjective belief, we need to imagine something more powerful; there is no way out of the imagined order" (Ibid., p. 117).

As noted by the political scientist Alexander Wendt (1992), "meanings, which lie at the basis of interests and identities, are created through the process of signaling, interpreting, and responding between actors—it is through reciprocal interaction that we create and instantiate the relatively enduring social structures in terms of which we define our identities and interests." This becomes the realm of water leadership, governance, and hydropolitics: influencing each of these perspectives within teams, across communities, and between societies.

For completeness though, arguably the environment has its own dynamic and language that must be considered. Figure 35.2 adapted from Liehr et al. (2017) and Ostrom (2009) depicts the interaction space between society and the environment which embraces the above holistic perspective of water, and recognizes that the interplay between environment and society behaves like a complex adaptive system (CAS), defined as "a complex, nonlinear, interactive system which has the ability to adapt to a changing environment" (Pahl-Wostl et al. 2013). Suffice to illustrate that when a new technology is introduced (an object) into a social-ecological system (SES), for example, an irrigation scheme, it has both an interobjective reality (how the tool is being integrated functionally within the broader system) and an intersubjective reality (what its insertion means to humans collectively).

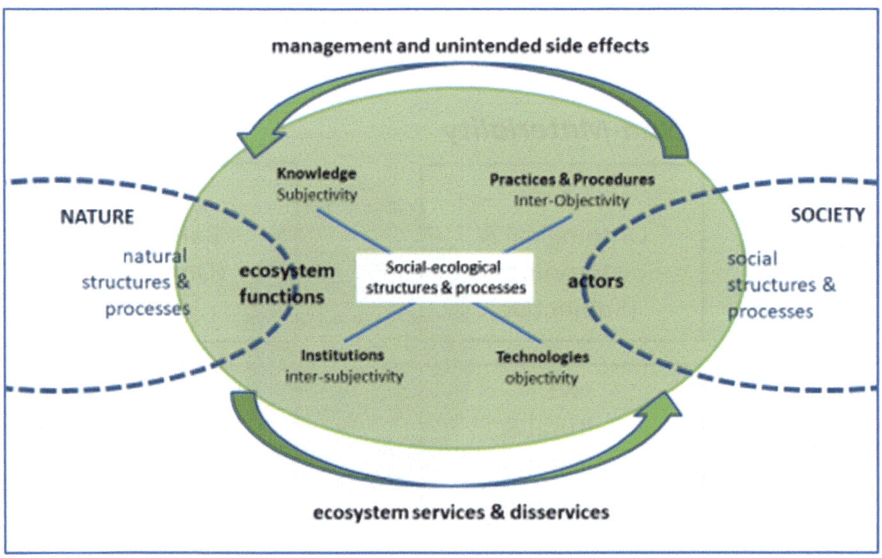

Fig. 35.2 Social–ecological systems framework (adapted from Liehr (2017) and Ostrom (2009))

For argument's sake, any technological insertion (or modification) holds little meaning for the environment even though it may alter it significantly, and nature's response may inadvertently change the society that inserted the technology in the first place, even though it has no intent or malice, just the drive to seek ecological balance (between organic and inorganic matter). These unintended consequences are illustrated in Fig. 35.2 by the feedback arrows between both sub-systems.

35.6 A Case for Leadership

Deeming interpersonal relationships as communication networks across interpersonal and community space, these practical networks become replete with intersubjective meaning, significance, debate, compromise, transformation, influence, etc. Combining this depiction with Harari's assertion that intersubjective meaning and imagined order is highly resilient to individual changes and that aspects of the interobjective hold sway over these networks, the question is how are these human networks, especially the interface between the political and professional ones, are influenced to exact social unity of effort and action? A priori, leadership as the art of influencing others to do what we want them to do, intimates that water leadership is the art of influencing individual and communities' shared water perspectives, especially the intersubjective meaning (which is the most difficult to change).

The governance definitions presented earlier insist that "decision-makers are held accountable for water management", but do not specify how this should play out, especially, how and which decisions are more important than others when circumstances are not normalized; such as the Epupa Dam example above. In fairness, many countries do develop principles to guide their decision-making processes. Namibia has enshrined these principles in its 1990 Constitution. In 1998, water specialists from pre-Independence South West Africa (cum Namibia) created *Namibia's Water: A Decision-Maker's Guide* to assist the nascent nation (MAWF and DRFN 1998). The 2007 Environmental Management Act (EMA) has embraced twelve principles to help rationalize social and environmental imperatives within sustainable development decisions. But realistically, all stakeholders within a river basin do not share the same burden of justifying their decisions and outcomes to all constituents; some are held accountable to their local communities, while others to the global one.

In contrast to the management-based governance explications, where authority, responsibility, and accountability are diffused across multiple people and structural layers, a leader-based governance structure establishes a sense of direction to a group of individuals engaged in a common endeavor; they influence individual drives to unify their efforts into unified action. And most notably, authority, responsibility, and accountability will often reside in a single person: the leader.

In short, leaders often inspire the creation of a new community-of-interest or practice and (help) establish a common purpose. They are key agents of change and transformation whose personal reputations are their key assets and their acceptance of risking that reputation on a given enterprise are some key distinguishing features. While managers administer an organization's efforts toward a common purpose, leaders yield the power to change that purpose.

35.7 Leadership and Power

A priori, leadership as the art of influencing others to do what we want them to do, reflects Dahl's (1957) classical definition of power as the "ability of state A to get state B to do something it doesn't want to do." Many classical political scientists assert that "most definitions of politics involve power" (Baldwin 2013). Moore (2008) insists that politics involves the decision-making processes and debates that result in "who gets what, when, and how", while power is "how people get what they want". Dahl's classical definition of power is nearly verbatim the leadership definition above with two remarkable differences: First, the agents of such power are states; and second, he highlights that the precondition of doing something is not wanting to do it. Dahl (1957), however, does elaborate on the important idea that power is a relationship, and relationships are not unilateral mandates, e.g. the other, state B, can always respond asymmetrically to A's demands by balancing other forms of power: cyber, resistance, or trade.

Note the use of the term art in the leadership definition above, in contrast to Dahl's term ability. Art intimates creativity, and in the interhuman domain, building trust and a relationship with the "others" that a leader is attempting to influence often requires artful tact. Whereas Dahl's "ability" is borrowed from Morgenthau who defined power somewhat ambiguously as "anything that establishes and maintains control of man over man" (Morgenthau 1948). Alternatively, (Mearsheimer 2001) defines power very simply as the "specific assets or material resources that are available to a state."

These realist views of power formed much of the earlier views of hydropolitics, which was defined as "the systematic study of interstate conflict and cooperation over transboundary water resources" (Elhance 1997). The study of shared rivers and the relative interests and strengths of the riparians were used to study whether water scarcity between states caused conflict, especially if the upper riparian happens to be both a water hegemon with a stronger military than the lower riparian. But Wolf (1998) found only one case in 4500 years of a true water war. Nevertheless, realists do emphasize the materiality of water as a central element to hard power and politics.

But beyond water's materiality (hard power) lies its non-materiality (soft power). Wolfers (1962) defines power as the ability to influence others "by threat or infliction of deprivations, or through promises or grants of benefits," while Nye (2002) defines it as the "ability to effect the outcomes you want, and if necessary, the ability to change the behavior of others". But Nye goes on to talk about soft power as an indirect way to obtain what you want because others value what you value. By setting the agenda you can "get others to want what you want" (Ibid. p. 552) by coopting people rather than coercing them. Nye's definition of soft power provides some clarity as to how leaders can influence water's intersubjective meaning: by building bridges of meaning from point A to point B.

These definitions provide some insight into the nature of power and why it is so contentious. For one, it can be understood simultaneously as "power as a resource," i.e. a state seeking to increase its power, and "power as a relation," i.e. the way a state uses its power in relation to another state to get what it wants. Secondly, power can also intimate a structural power, which "refers to the unintentional power or power with respect to the creation and/or control of structures" (Baldwin 2013) because the power held by one entity may be viewed as a threat or advantage by another; as in the case of an upstream riparian who also controls a major water dam, or a governance structure that advantages one stakeholder over another. Finally, as a dynamic as entities and polities attempt to balance their relative differences in power.

35.8 Hydropolitics Shapes Water Governance; Water Leadership Brings it to Life

When combined, these definitions and descriptions of leadership, power, and politics are insightful for enriching the concepts of water, hydropolitics, water governance, and water leadership as they combine elements of hard and soft power. According to Mearsheimer's (2001) earlier definition of power, water can be viewed as an element of national power (a means) available to accomplish national interests, and when combined with Morgenthau's (1948) ideas of power, it can be viewed as an instrument (a way) of accomplishing a national interest, and alternatively as the interest (end) itself. Thus water as either a means, a way, or an end, is the focal point of hydropolitics, and in its ideal state entails dynamics across the entire water enterprise: education centers, water production and distribution installation, political ministries and agencies, private stakeholders and farmers, etc. Hydropolitics then shapes a governance structure and IWRM approach that befits the myriad hydropolitical interests, and water diplomacy can be viewed as extending hydropolitics (Pangare 2013).

Specifically building on Dahl's (1957) definition of power, hydropolitics seeks to balance power (hard and soft) among polities with a view to "enhance political stability, regional security, economic prosperity, and environmental sustainability" (Kehl 2011). Kehl concluded that the use of soft power and economic power were more effective than hard power for long term river and regional stability. But to illustrate how soft power relates to hard power and political verve, recently in the Hindustan Times, 16 April 2018, General Bajwa, Pakistan's Army Chief, claimed that the only way to peace between India and Pakistan was through "comprehensive and meaningful dialogue." Yet closer examination of the article reveals that the obstacles that need to be overcome are related to soft power: that the overture to dialogue is not seen as weakness, that sovereignty and dignity be respected, and that the use of soft power is backed by a willingness to use hard power (Ahmad 2018).

Therefore, hydropolitics is fashioned by the political environment and is the overarching concept that shapes the water sector and its interdependencies across societies. It is understood as the interactions between stakeholders—states, non-state actors and individuals within the national and international domains—to influence the authoritative allocation and use of water resources to achieve their interests, goals, and objectives through the exercise of hard and soft power. Consequently, due to the exigency of public accountability, governance is predominantly a top-down, all-inclusive management and stakeholder scheme, while IWRM is a bottom-up, resource-based and process-driven management approach to water, often set up within geo-hydrological boundaries, such as river basins (Ferreyra 2008).

In light of reinforcing a difference between management and leadership functions, water leadership is at the forefront of responsibly using and developing national power and establishing a water governance scheme that provides a unified vision of its purpose and clarifies the bounds of its functions and responsibilities for the betterment of the national social–ecological system (of systems). Ultimately, it builds trust, legitimacy, and imagination that emerges within a network of relationships, but more importantly, it embraces the possibility that water means something different across cultures, societies, and people (appreciating the intersubjective meanings). The essays in *Hydro Diplomacy: Sharing Water Across Borders*, conceptualized and written by the IUCN Water Program in Bangkok and edited by Ganesh Pangare (2013), bring many of these ideas to light. It stems to reason that, if a particular society has a particular relationship with water, then their approach to managing it and the problems they deem requiring decisions will be different. Many of the concepts presented by Meissner (2016b) in *Paradigms and theories in water governance*, surmises that the ontology with which one views water's relationship and reality will affect the way they derive solutions to perceived problems.

35.9 Discussion: Leadership and the Messiness of Being Human

The purpose of this paper is to occasion debate within and across water communities-of-interest. Because leadership begins and ends with introspection, four issues were selected for further discussion based on their contentiousness and intractability which require leadership to solve: water as a human right; communication bridges across communities-of-interest and professional bodies; incorporating an intra- and inter-disciplinary leader-based training and development philosophy across (and beyond) the water sector; and engaging existing teams and technologies to their full potential.

Water as a Human Right. The SES framework presented at Fig. 35.2 has been used to map the progress of inserting technologies and implementing policy changes (Liehr et al. 2017). In countries, such as Namibia, the natural environment is the limiting constraint to all development (Brown 1992). Some constraints are flexible, while others may seem intransigent; one such constraint deemed intransigent by many is the 2010 UN General Assembly declaration affirming that "the right to safe and clean drinking water and sanitation as a human right that is essential for the full enjoyment of life and all human rights" (UN 2010). However, the pragmatic side of water governance, typified by interpersonal relationships, leadership priorities, and available resources, highlights the difference between supporting the UN Resolution, believing in its tenets, and implementing it, therein lies the messy realm of hydropolitics and water leadership.

Soyapi (2017) investigated how four South African countries, Botswana, Zambia, Zimbabwe, and South Africa, either embraced this right in their constitution or read it into existing laws, claiming that the former held more political sway. In those that didn't, Botswana and Zambia, there have been several legal cases where the right to live with dignity has been sufficient to occasion lasting change to water distribution schemes and redress water quality. Arguably, establishing a human right to water is an attempt to regulate and normalize water distribution and water security (quality, quantity, and consistency) across the globe. However, without full awareness of the impact of their decisions across the SES (Liehr et al. 2017), active leadership will fail to engage limited local capacity, balance scarce resources with development aspirations, and occasion transformation in an iterative, transparent manner to minimize the unintended consequences to nature and society.

Multi-Disciplinary and Interdisciplinary Communications. Oftentimes when discussing hydro-social issues with academic and water professionals, they intimate that the missing link in exacting water efficiencies is to educate the farmers and population at large. In discussing governance with farmers, they cite politicians' and academics' lack of knowledge, empathy, and trust; and when discussing these matters with junior engineers and water experts, they claim that seniors make the decisions and that they must bide their time. And finally, when discussing with political leaders, they bemoan complexity. Yet Falkenmark (1990) insists that oversimplifying complex dynamics between the environment and society is at the core of causing "water stress" between actors. Together these claims embody Einstein's quote that "we can't expect different results by using the same thinking that created the problem in the first place".

His warning and these claims merely highlight that all communities have inherent biases and established water intersubjective meanings; the leadership challenge is to break through them and find bridges of meaning across communities. Two ideas are submitted (Gosselin 2018d). First, leaders within single branches of expertise, e.g. irrigation engineers, water treatment technicians, farmers, sanitation service managers, academics, etc. should develop the means to question their own professional myths and biases: those constructs that limit their scope, yet ironically, provide them immeasurable expertise and prestige.

Second, leaders should institutionalize interdisciplinary knowledge, i.e. the new sets of knowledge that emerge when multiple disciplines engage in the resolution of a common problem, akin to advocating for a Unified Theory of Hydro-Social Knowledge. I wholly recognize the amplitude of what is being advocated when expertise is deliberately parsed into partial theories and knowledge. Unifying existing expertise into new intelligence requires effort and imagination. As a minimum, leaders (oftentimes seniors in their field) must become aware of their biases, guard against their resistance to innovation, and devote portions of their budgets toward the future. Leadership starts and ends with personal introspection; this is not easy, but nor is it idealistic and frivolous.

Leader-Based Training and Development. Creating multi-disciplinary individuals and teams requires active pursuit of opportunities to extend current knowledge-bases to allow for emergent bodies of knowledge and expertise to be captured and further enhanced in a methodic way. This may appear as a call for normalization and management, which in a sense everything is. But with a special caveat: Improvement and progress depend on understanding and appreciating the fringe of disciplines, their overlaps and synergies over time (Gosselin 2018b). Although a priori, this can be accomplished within campuses that share water-related discipline training and leader-type education, such as various Master's in Business Administration (MBA) programs co-located with university and college-level system engineering schools. But other solutions may be more novel. Raucher et al. (2018) found that many water providers throughout the USA were well postured to become the vanguard leaders in addressing broader climate change challenges because they already enjoyed a high degree of legitimacy and trust with the citizenry, and hence, were trusted with the responsibility of their future welfare. This example highlights that water is a total social event and that water leadership seeks to extend and cross-pollinate across professional boundaries from one area to others.

Engaging New Technologies and Uncovering Spare Capacity. During my research in Namibia on sustainable water, the Government's decentralization policies established local Water Point User Associations and Committees responsible for collecting user fees and maintaining water infrastructure, such as pumps and well integrity. As well intended as these initiatives are, they accentuated the fact that many local communities have neither the capacity to manage local water resources, integrate new technologies, or the administrative tail required to maintain them. These stories are not uncommon: subordinate managers and leaders claim that there is no spare capacity to operate, let alone transform. Leaders must uncover how to reverse this plight.

In an earlier professional capacity, I led the transformation of a multi-disciplinary cyber and information management training and education facility that engaged six primary trade fields, from apprentice to master technicians; junior and middle management and leader tracts; and specialized senior manager operations and maintenance short duration courses. The operating budget was approximately $80 million per year, with over 300 teaching cadre and a student throughput exceeding 2000–3500 per year. During the early stages of discovery, i.e. pre-formal strategy development phase, most first level managers insisted that there was no money or spare staff to effect change. This is a common cry from managers worldwide when tasked to transform, and it is not without its merits, except in light of the earlier Gallup finding that found that approximately 60% of employees are not really engaged in their work (Harter and Agrawal 2011). Without recounting the intricacies of that entire transformational activity, two points are relevant here.

First, there is always spare capacity if the leader and leadership team is willing to reflect on the organization's biases, resistance to change, and irrelevant activities. Second, communication and engagement across disciplines is essential. To the first comment, the relevant point here was how to gain legitimacy and trust from staff that new tasks won't be added to old ones. One of the most unsettling decisions is to drop irrelevant tasks that bear no fruit within a future vision, while understanding that for many, those (perceived) outdated behaviors may constitute a large part of their identity and comfort zone, and that they may be needed again in the future.

Hazlitt (1946) in his famed book *Economics in One Lesson* recounts how during the late 1880s English workers in the cloth weaving and needle manufacturing industry were threatened by automation; contemporary examples include automation and redistribution of jobs in the coal mining and energy sector, or automobile manufacturing sector. The significance of these examples, whether it's a low-tech change, such as the introduction of cost recovery measures or rain harvesting technologies in a village, or introducing a high-tech water distribution and equalization plant in a modern-day smart city, is that all of these changes affect people and livelihoods. The effective implementation of policies, laws, and projects will depend on how well these changes are embraced by those they are intended to serve: This is a leadership responsibility, and open communication and building trust are paramount.

Summary. These four topics are exemplary of the breadth and depth of leadership issues that take place during normalized times but are accentuated during times of change, crisis, and transformation. Water as a human right is used as a flashpoint to illustrate that almost all water-related projects will highlight inequities between groups and that dependent on the dominant culture and existing power centers, their outcome and implementation may be subject to lengthy resistance, such as litigation. Water diplomacy and leadership are attempts to mitigate these perils; even though sometimes litigation may be necessary to highlight a grave social injustice. Communication, uncovering spare capacity, and continual development topics were raised to illustrate that professional leaders seek ways to realign scarce resources to effect greater change that benefits larger swaths of society, and that given that all circumstances change, even the definition and understanding of what constitutes sustainable water will change; especially in light of climate change, population growth, and expected increased demand (Gosselin 2018c).

35.10 Concluding Thoughts: Water, Power, Hydropolitics, Governance and Water Leadership

This paper was meant as a discussion primer centered on how many of the water sector's tenets are based on normalized circumstances, management principles, and water's materiality. This was found to be insufficient to address the complex, adaptive, and unpredictable circumstances facing many polities. Instead, a leader-based approach that embraced water's dual nature as a material (hard power) and non-material (soft power) was advocated to redefine and explore how water leadership and governance can resolve many of the intractable challenges facing various polities.

These dynamics were framed within a hydropolitical sphere that is characterized by an adaptive and dynamic balance of hard and soft power among actors. The first contention by many is that considerable managers are called upon to fill leadership roles and functions. This is not being disputed, as there is a need to unleash untapped potential within all institutions. However, if the intent is to transform the water sector to ensure that the narrative of a global water crisis doesn't become a reality of a global water crisis, then it is worthwhile to debate how leaders can be identified and developed to provide the impetus for change and take advantage of the vast wealth of water governance knowledge, such as synthesized in the twelve principles espoused by the OECD (Akhmouch and Correia 2016), and bring them to their fruition.

By way of concluding and consolidating the insights from this paper, a proposed definition of what constitutes water leadership is:

> The art of developing those personal traits and relationships that influence a group's inter-subjective waterworld and assists the group in moving forward from a momentary impasse. It is a calling to stand out from the masses with a personal dynamic that becomes evident within a group that allows the group to discover a solution and way forward, either through persuasion, collaboration, coercion, cooption, or inspiration.

Acknowledgements I would like to thank Dr. A. Pandey and Dr. M. L. Kansal of the Water Resources Development and Management Department, IIT Roorkee, for being patient sounding boards for many of the ideas expressed herein; to Dr. Ross Miller, Dr. Alice Kang, and Dr. Patrice McMahon of the Political Science Department, at the University of Nebraska in Lincoln (U.S.), for reviewing elements of this manuscript and providing encouragement and suggestions for relating water governance to broader political themes; and to Adam Gosselin (Kiewit) and Jacque Merritt (Gallup) who patiently provided timely feedback, suggestions, and encouragement.

References

Ahmad I (2018, April 16). Gen. Bajwa: Peace with India only through dialogue. Hindustan times, p 14

Akhmouch A, Correia FN (2016) The 12 OECD principles on water governance—when science meets policy. Utilities Policy 43:14–20. https://doi.org/10.1016/j.jup.2016.06.004

Baldwin D (2013) Power and international relations. In: Carlsnaes W, Risse T, Simmons B (eds) Handbook of inernational relations. Los Angeles: SAGE

Brown CJ (1992) Namibia's green plan. Ministry of Wildlife, Conservation and Tourism, Windhoek

Dahl RA (1957) The concept of power. Behav Sci 2(3):201–215. https://doi.org/10.1002/bs.3830020303

Elhance AP (1997) Conflict and cooperation over water in the Aral Sea basin. Stud Confl Terrorism 20(2):207–218. https://doi.org/10.1080/10576109708436034

Falkenmark M (1990) Global water issues confronting humanity. J Peace Res 27(2):177–190. https://doi.org/10.1287/mksc.l090.0490

Ferreyra C, de Loë RC, Kreutzwiser RD (2008) Imagined communities, contested watersheds: challenges to integrated water resources management in agricultural areas. J Rural Stud 24(3):304–321. https://doi.org/10.1016/j.jrurstud.2007.11.001
Gosselin DMN (2018a) A quest for sustainable water: a study to understand namibia's water governance. Ann Arbor: MI: ProQuest L.L.C
Gosselin DMN (2018b) Working paper: water leadership: training and developing for success (Water Leadership and Governance No. 5). Roorkee, India
Gosselin, DMN (2018c) Working paper: water leadership and a water profession (Water Leadership and Governance No. 2). Roorkee, India
Gosselin DMN (2018d) Working paper: water leadership and building bridges across communities of influence (Water Leadership and Governance No. 3). Roorkee, India
Harari YN (2015) Sapiens: a brief history of humankind, 1st edn. Harper, New York
Harter JK, Agrawal S (2011) A worldwide study of employee engagement and its relationship to wellbeing, health, and giving. Omaha
Hazlitt H (1946) Economics in one lesson (2008 Ed.). Auburn: AL: Ludwig von Mises Institute
Kehl JR (2011) Hydropolitical complexes and assymetrical power: Conflict, cooperation, and governance of international river systems. J World-Syst Res 18(1):218–235. https://doi.org/10.1017/CBO9781107415324.004
Liehr S, Röhrig J, Mehring M, Kluge T (2017) How the social-ecological systems concept can guide transdisciplinary research and implementation: addressing water challenges in central northern Namibia. Sustainability (Switzerland) 9(7):1–19. https://doi.org/10.3390/su9071109
Linton J (2010) What is water? The history of a modern abstraction. UBC Press, Vancouver
MAWF and DRFN (1998) Namibia's water: a decision makers' guide. In: Heyns P, Montgomery S, Pallett J, Seely MK (eds) DWA, MAWRD, and DRFN
Mearsheimer JJ (2001) The tragedy of great power politics. WWW and Norton Company, New York
Meissner R (2016a) Hydropolitics, interest groups and governance—the case of the proposed epupa dam. Springer International Publishing
Meissner R (2016b) Paradigms and theories in water governance: the case of South Africa's national water resource strategy, second edition. Water SA, 42(1):1–10. https://doi.org/10.4314/wsa.v42i1.01
Melber H (2015) From Nujoma to Geingob : 25 years of presidential democracy, 18:49–65
Moore B (2008) The critical spirit and comparative historical analysis. In: Munck GL, Snyder R (eds) Passion, craft, and method in comparative politics. JHU Press, pp 86–112
Morgenthau H (1948) Politics among nations. Alfred Kopf, New York
Ngutjinazo O (April 3, 2018) Govt advised to freeze salaries. The Namibian. Retrieved from https://www.namibian.com.na/65938/read/Govt-advised-to-freeze-wages-jobs
Nye JS (2002) Limits of American power. Polit Sci Q 117(4) (Winter, 2002–2003):545–559
OECD (2011) Water governance in OECD countries: a multi-level approach. https://doi.org/10.1787/22245081
Orlove B, Caton SC (2010) Water sustainability: anthropological approaches and prospects. Ann Rev Anthropol 39(1):401–415. https://doi.org/10.1146/annurev.anthro.012809.105045
Ostrom E (1990) Governing the commons: the evolution of institutions for collective action. In: Alt JE, North DC (eds) Cambridge: UK: Cambridge University Press. https://doi.org/10.1017/CBO9781316423936
Ostrom E (2005) Understanding institutional diversity. Public choice vol 132. Princeton and Oxford: Princeton University Press. https://doi.org/10.1007/s11127-007-9157-x
Ostrom E (2009) A general framework for analyzing sustainability of social-ecological systems. Science 325(July):419–422. https://doi.org/10.1126/science.1172133
Pahl-Wostl C, Conca K, Kramer A, Maestu J, Schmidt F (2013) Missing links in global water governance: a processes-oriented. Ecolo Soc 18(2):33. https://doi.org/10.5751/ES-05554-180233
Pangare G (ed) (2013) Hydro diplomacy—sharing water across borders. Academic Foundation, New Delhi, India

Raucher K, Raucher R, Ozekin K, Wegner K (2018) The opportunities and needs of water utility professionals as community climate–water leaders. Weather, Climate, Soc 10(1):51–58. https://doi.org/10.1175/WCAS-D-17-0061.1

Schnegg M, Bollig M (2016) Institutions put to the test: community-based water management in Namibia during a drought. J Arid Environ 124:62–71. https://doi.org/10.1016/j.jaridenv.2015.07.009

Shiva V (2016) Water wars: privatization, pollution, and profit (Second). North Atlantic Books, Berkeley

Solanes M, Gonzalez-Villarreal F (1999) The dublin principles for water as reflected in a comparative assessment of institutional and legal arrangements for integrated water. TAC Background Papers No. 3

Soyapi C (2017) Water security and the right to water in Southern Africa: an overview. PER: Potchefstroomse Elektroniese Regsblad 20(1):1–26. https://doi.org/10.17159/1727-3781/2017/v20n0a1650

UN (2010) General assembly resolution 64/292. New York

UNDP (2004) Water governance for poverty reduction: key issues and the UNDP Response to Millenium Development Goals, 93

Weinthal E, Troell J, Nakayama M, Swain A (2014) Water and post-conflict peacebuilding. Hydrol Sci J 61(7):1313–1322. https://doi.org/10.1080/02508060.2011.561772

Wendt A (1992) Anarchy is what states make of it: the social construction of power politics. Int Organ 46(2):391–425. https://doi.org/10.1017/S0020818300027764

Wolf A (1998) Conflict and cooperation along international waterways. Water Policy 1(2):251–265. https://doi.org/10.1016/S1366-7017(98)00019-1

Wolfers A (1962) Discord and collaboration: essays on international politics. John Hopkins Univ Press, Baltimore

Woodhouse P, Muller M (2017) Water governance—an historical perspective on current debates. *World Development*, *xx*(In Press), 17. https://doi.org/10.1016/j.worlddev.2016.11.014

The manufacturer's authorised representative in the EU is Springer Nature Customer Service Centre GmbH, Europaplatz 3, 69115 Heidelberg, Germany. If you have any concerns regarding our products, please contact ProductSafety@springernature.com

Printed and bound by CPI Group (UK) Ltd, Croydon, CR0 4YY

30/09/2026

02225462-0018